Lost Knowledge

Technology and Change in History

Series Editors

Adam Lucas
Steven A. Walton

VOLUME 16

The titles published in this series are listed at *brill.com/tch*

"The Road to Singularity" by Abayomi Louard-Moore

Lost Knowledge

The Concept of Vanished Technologies and Other Human Histories

By

Benjamin B. Olshin

BRILL

LEIDEN | BOSTON

Cover illustration: "A Pause to Consider the Arrogance of Machines", copper plate etching, by James Skvarch, 2005. © James Skvarch

Library of Congress Cataloging-in-Publication Data

Names: Olshin, Benjamin B., author.
Title: Lost knowledge : the concept of vanished technologies and other human histories / by Benjamin B. Olshin.
Description: Leiden ; Boston : Brill, [2019] | Series: Technology and change in history, ISSN 1385-920X ; volume 16 | Includes bibliographical references and index.
Identifiers: LCCN 2018060003 (print) | LCCN 2019000440 (ebook) | ISBN 9789004352728 (Ebook) | ISBN 9789004352711 (hardback : alk. paper)
Subjects: LCSH: Technology and civilization—History—To 1500. | Technology and civilization—History—To 1500—Sources. | Technology—History—To 1500. | Communication of technical information—History—To 1500. | Knowledge, Theory of—History—To 1500. | Civilization, Ancient.
Classification: LCC CB478 (ebook) | LCC CB478 .O47 2019 (print) | DDC 303.48/3—dc23
LC record available at https://lccn.loc.gov/2018060003

Typeface for the Latin, Greek, and Cyrillic scripts: "Brill". See and download: brill.com/brill-typeface.

ISSN 1385-920X
ISBN 978-90-04-35271-1 (hardback)
ISBN 978-90-04-35272-8 (e-book)

Contents

Preface

I became interested in the subject of "lost knowledge" due to several experiences in my life. First, there was my stumbling across cataclysmic theories of history — really, *conjectures* rather than theories — when I heard of Immanuel Velikovsky in my early teens. Even then, such ideas struck me as science fiction, because the conjectures were quite elaborate while the evidence quite thin. At the same time, I respected the fact that there were people out there who were not necessarily academics, or who were experts in fields other than history, but who nonetheless wrestled with the question of the fate of civilizations and human knowledge.

Many years later, in graduate school, I came across similar ideas of lost, advanced ancient knowledge as I carried out research for my Ph.D. thesis in the history of cartography. In the course of my research, I came across several conjectures about such knowledge: first, there was Charles Hapgood's belief in an ancient mapping of Antarctica, outlined in his 1966 book, *Maps of the Ancient Sea Kings: Evidence of Advanced Civilization in the Ice Age*. Hapgood was no fool: he had studied history, and he had taught college courses in that field and other subjects. His interest in maps was quite genuine, but his speculations, I found, wandered too far afield from the actual evidence. Similarly, I stumbled upon the researchers collectively known as the "Escuela Argentina de Protocartografía": Paul Gallez, Enrique de Gandía, and Dick Edgar Ibarra Grasso. These writers argued that South America was mapped in its entirety long before the arrival of Spanish and Portuguese explorers. Enrique de Gandía was a highly respected historian, and his 1942 book entitled *Primitivos Navegantes Vascos* was the first to articulate the idea that the peninsula of Asia found in Ptolemy's ancient maps actually was a very early cartographic rendering of South America. Gallez, in turn, was kind enough to carry on an extensive correspondence with me in the early 1990s about these ideas, and I found them intriguing — but as he himself noted, they were conjectural.[1] In the end, the evidence for an idea so outside the well-established understanding of Ptolemy's maps did not strike me as sufficient to change any paradigms.

Later, after the start of my academic career, I came upon other works which spoke of advanced feats in a much more remote past, such as the arguments by René Adolphe Schwaller de Lubicz for different interpretations of ancient history. Schwaller de Lubicz, although possessing some scientific training, spent

1 See the discussion in William A.R. Richardson, "South America on Maps before Columbus? Martellus's 'Dragon's Tail' Peninsula", *Imago Mundi* 55 (2003): 25–37.

much of his time studying the occult, which led to an interest in the temples of Egypt. He spent over a decade in that country, and was the founder of alternative ideas about its early history and culture. Schwaller de Lubicz's interpretations, expressed in his 1958 book, *Le Roi de la Théocratie Pharaonique*, influenced more recent writers such as John Anthony West. West argued — without strong evidence — that the Sphinx is many thousands of years older than the date claimed by traditional archaeology. Then there were the works of Graham Hancock, which made related arguments, but on a much broader scale. His 1995 book, *Fingerprints of the Gods: The Evidence of Earth's Lost Civilization*, speculated that early cultures such as those of ancient Egypt and Central America descended from a highly advanced civilization that had existed thousands of years before them.

My impression was then — and continues to be now — that such interpretations are fantasy, although at the same time having the capacity to compel readers to think about history in ways that transcend academic conventions. More particularly, over many years of examining these ideas, I came to understand several key points: First of all, looking at alternative models of history is important, since our current model is just that — a model. In addition, the pitfall of most of these modern writers on "alternative history", such as Hapgood and Hancock, was that they drew too many conclusions from too little evidence. Finally — at the same time, in fact — there was a great deal of interesting evidence floating around that was being ignored, misrepresented, or not thoroughly analyzed in academic works. I realized that a different approach was needed. The debates go around and around because the same evidence, such as the famous Sphinx, can be examined and interpreted in so many different ways.

This book works with a different kind of evidence — early texts — to see how writers from Plato to Chaucer wrestled with the nature of technology and technological knowledge that might have been found then lost across the vast span of human existence. Instead of going back and forth on the same piece of evidence — that peculiar Sphinx — this book presents interesting evidence in an organized, systematic way, something that many of the speculative writers, and even some academic historians, fail to do. This book, indeed, is more about assembling evidence and providing some analysis than making any kind of far-reaching claims.

Another vitally important factor that shaped the initial conception of this book was a particular realization concerning the nature of knowledge transmission. Having grown up in the contemporary West, my model of knowledge transmission was always a formal one — I, like millions of others in the modern world, had gained much of my knowledge through schooling, didactic instruction, textbooks, and so on. Even in my youth, however, I began to realize that a great deal of knowledge is transmitted from generation to generation through other, non-formal means — apprenticeships, tutors, or just simple observation and participation.[2] As a teenager I had learned, for example, to fix bicycles by working at a shop, not through the study of any book on the subject. When I went as an undergraduate to Sri Lanka as part of a project, I was exposed for the first time to whole sectors of a society that were not "literate" in the modern sense of the word — but that managed to have perfectly well-organized lives, with ample knowledge of the environment around them. They had gained that knowledge informally — through experience and active learning — but also quite soundly.

Many years later, in Ghana, I was exposed to the same phenomenon, and put together a clearer picture of how in these societies knowledge about everything from agriculture to astronomy was transmitted over the centuries without any of the modern "apparatus" to do so. Not long after graduate school, I heard the writer William Irwin Thompson speak about the same topic, and became intrigued by his suggestion that non-literate peoples use the vehicles of myth and folktale to preserve and transmit their culture's essential knowledge about the natural world and how to survive in it.

Finally, all this was brought home by a very personal experience: having a child. Again, being a product of modern, North American culture, even before our child was born, my wife and I were inundated with gifts of "how to" books on child-rearing. So much reading to do! But as a historian I realized, of course, that for millennia there had been no such books, and yet men and women across many cultures had managed to raise children successfully. It became clear to me that the ancient methods of knowledge transmission must have been rather sophisticated, because the amount of knowledge just to have a child and raise that child is quite significant. Yet for all of human history, there were individuals and societies who had such knowledge, and well-established non-literate means of passing it on from generation to generation.

A great deal of knowledge concerning various crafts and technologies was also passed down through the centuries without a literate setting. One writer, in a work on the complex geometrical designs in early Islamic architecture, states:

---

2 A seminal work on this subject is David Turnbull, *Masons, Tricksters and Cartographers: Comparative Studies in the Sociology of Scientific and Indigenous Knowledge* (London: Routledge, 2003). For recent specialized research on this subject see, for example, Jamshid J. Tehrani and Felix Riede, "Toward an Archaeology of Pedagogy: Learning, Teaching and the Generation of Material Culture Traditions", *World Archaeology* 40.3 (2008): 316–331.

> [I]t seems that Western critics have seriously underestimated the ability of native craftsmen to retain large amounts of empirical knowledge on pattern designs and construction in the absence of any understanding of the theoretical background which a professional mathematician might bring to bear on these problems.[3]

This concept of a much more traditional, "organic" mode of knowledge transmission is important to the historian because it means that one must look at myths and other kinds of stories and narratives in a new way, since those were the bearers of knowledge for far longer than our present methods of formal pedagogy, textbooks, and instruction manuals.

Finally, what is really needed is a look at the question of "lost knowledge" from a broader perspective — what is it that we are really talking about here? Are we talking about the possibility of a completely new model of history? Perhaps one approach is to think about how human knowledge really works over the vast reaches of time. What one finds is that it can be a rather complex matter: knowledge — even rather sophisticated technological knowledge — often was discovered, then lost, and then re-discovered.

But at the same time, there was technological knowledge that was not really "discovered" at all, but — rather oddly — speculated upon by all kinds of early cultures, from the ancient Greeks to the remote Ainu people. Delving into this history means becoming immersed in the histories, literature, and folklore of these cultures. The result is this book: a journey into "lost knowledge" as it manifests itself over the vast reaches of time, the millennia that saw human beings both build technologically sophisticated societies and ponder their rise and fall.

---

3 See p. 182 of A.J. Lee, "Islamic Star Patterns", *Muqarnas: An Annual on Islamic Art and Architecture* 4 (1987): 182–197; also note John Noble Wilford, "In Medieval Architecture, Signs of Advanced Math", *The New York Times* (27 February 2007), F2, comprising a brief summary of Peter J. Lu and Paul J. Steinhardt, "Decagonal and Quasi-Crystalline Tilings in Medieval Islamic Architecture", *Science* 315.5815 (23 February 2007): 1106–1110.

Acknowledgements

This book would not have become a reality without the encouragement of a number of individuals. These include especially my wife Charlotte Lin, as well as my friend Nuno Baptista. Nuno insisted that I write this book, no matter what — even if he would be the sole reader. I also wish to thank another friend, Alex Schein, for the extended conversations that he and I have had on and off for many years about indigenous knowledge, the transmission of knowledge, and related subjects. A series of conversations with Don Bohn and Stephen Milton were also very helpful in shaping my thoughts about the idea of knowledge and its fate over the millennia.

Mikhail Sergeev connected me with the staff at Brill, and I am grateful for his collegial attitude toward research and writing. I wish to extend a special thanks to Steve Walton and Adam Lucas, for their careful and detailed comments on the manuscript, which were vital in my refining the material to communicate more clearly the central theme in this book. I also wish to thank Eric van Broekhuizen, Bram Oudenampsen, Wendel Scholma, Michiel Thijssen, and Rosanna Woensdregt at Brill. In addition, I would like to thank the anonymous manuscript reviewers, whose comments and edits were essential in making this a better book.

Finally, I would like to express my appreciation for those who assisted me in obtaining and working with some of the source material needed in my research. These include Lee Ihnhee, Victor Mair, Thomas Radice, Nathan Sivin, and Jidong Yang, all at the University of Pennsylvania; Robert P. Goldman at the University of California at Berkeley; Seong-Rae Park of the Hankuk University of Foreign Studies, Seoul, Korea; Brad Inwood at Yale University; Alexander Jones at the Institute for the Study of the Ancient World at New York University; Dane T. Daniel at Wright State University, Gábor Kósa at the East Asian Institute at Eötvös Loránd University; Cheng Fangyi at Tsinghua University; Mary Louise Castaldi and Sara J. MacDonald at the University of the Arts Libraries; Xin Lu, an artist in Philadelphia; Jim Ascencio, a translator in Taipei; and Chris G. Whipple at Ramboll Environ, Inc.

Figures

CHAPTER 1

Speculations and Fantasies

1 Lost Knowledge, Technology, and the Patterns of History

What *is* "lost knowledge"? As human civilizations arise and develop, they accumulate knowledge. That knowledge has many forms, from the pragmatic to the theoretical. In most cultures, there is a significant body of what one might call "organic" or indigenous knowledge: an understanding of plants and natural medicines, for example, or the ability to track an animal, or the methods for building a structure out of earth, wood, stone, or other material. That knowledge is accumulated in various informal ways over centuries and even millennia, while in the rapid progression to modernity it can be become lost in just a generation or two.

This book, however, focuses on a specific kind of knowledge: technological knowledge, and ideas about machines and devices in particular. Human beings seem to have a natural affinity for technology in this regard: even the most remote of our ancestors *made* things — arrowheads, flint tools, and so on. We are not just *Homo sapiens* but also *Homo faber*: we create and use technology, and in turn it defines us as human beings. Without doubt, other animals use tools, but the extent to which technology shapes human cultures is unique.

What is interesting is that the idea itself of humans as technological beings is quite ancient. We tend to think of technology and the human use of technology as a modern phenomenon, but our engagement with machines and devices — and speculations about them — goes back millennia. Early writers in a wide range of genres discussed technological themes. But this means that ideas about technology were often contextualized and transmitted in a way that was far less specialized than our contemporary modes of the Internet, textbooks, and subject-specific technical instruction. In early sources, even serious considerations of technology are found embedded — and sometimes placed almost in passing — in historical chronicles, fantastical tales, and other types of written and oral sources.

Early writers also frequently pondered the specific idea that there was knowledge of technologies that had been lost in the remote past. Writers from Plato to the Daoist scholar Ge Hong (葛洪) ruminated on lost civilizations and lost technological capabilities. In a marked contrast to our present view of technological development as linear, these early thinkers often articulated a cyclical model of knowledge, and attributed to their ancient predecessors

 | DOI:10.1163/9789004352728_002

technological abilities that we might consider as modern and sophisticated as our own, such as mechanized flight.

Investigating these traditions concerning "lost knowledge", however, is a messy business. As noted earlier, the process has been beset with ill-formed hypotheses, idle speculations, unfounded conjectures, and outright absurdities. The road through studies of the concept of "lost knowledge" is strewn with everything from Immanuel Velikovsky's cataclysms to hidden — and likely imagined — numerological patterns in the design of the pyramids of Egypt. But the fundamental narrative is defined clearly: there was a "golden age", a technologically and scientifically advanced civilization in the most remote of times, a civilization that rose and then fell and disappeared. The basic requirement of this model is also clear: the existence of human beings on the planet is long enough for this process to have happened. That is, the time of *Homo sapiens* on this planet is sufficiently long to have allowed a great civilization, one perhaps even as technologically advanced as ours, to develop ... and then to collapse, and to disappear so completely that virtually no trace survived.

From the outset, the situation concerning the evidence to support a model is obvious — there are virtually no archaeological finds of any kind that indicate the presence of such an advanced civilization. Archaeological digs have revealed no traces of prehistoric skyscrapers, or power plants, or airplanes, or anything else of comparable technological sophistication.[1] Of course, there are early civilizations that certainly did leave behind grand architectural structures: these may have affected the viewpoint of some early writers, who were

1 For a look at applying traditional archaeological methods to check such extreme claims of lost civilizations, see Alice Beck Kehoe, *Controversies in Archaeology* (Walnut Creek, CA: Left Coast Press, 2008). A popular critique on the idea of lost civilizations can be found in the magazine *Lost Civilizations Uncovered* by the freelance writer Jason Colavito. Another work that examines such ideas is Ronald H. Fritze's *Invented Knowledge: False History, Fake Science and Pseudo-religions* (London: Reaktion Books, 2009). Fringe claims concerning lost civilizations continue unabated: one such claim is that Arkaim, an archaeological site in the Ural Mountains, was the seat of a highly advanced culture millennia ago. On this idea, see, for example, p. 46 of Victor A. Shnirelman, "Russian Response: Archaeology, Russian Nationalism, and the 'Arctic Homeland'", in Philip L. Kohl, Mara Kozelsky, and Nachman Ben-Yehuda, eds., *Selective Remembrances: Archaeology in the Construction, Commemoration, and Consecration of National Pasts* (Chicago: University of Chicago Press, 2007), 31–70, and idem, "Archaeology and Ethnic Politics: The Discovery of Arkaim", *Museum International* 50.2 (April–June 1998): 33–39. More serious discussions concerning the use of scientific evidence to investigate the existence of lost civilizations in the remote past are presented in Adam Frank and Gavin A. Schmidt, "The Silurian Hypothesis: Would It Be Possible to Detect an Industrial Civilization in the Geological Record?", *International Journal of Astrobiology* (2018): 1–9, https://doi.org/10.1017/S1473550418000095.

inspired to believe that something significant in terms of technological development had happened in the past, and then became lost.

Indeed, the viewpoint of early writers themselves — writers whose ideas are presented at length in this book — was that their historical predecessors had knowledge that later disappeared, specifically technological knowledge. The belief, then, was perhaps not only that there had been advanced *architectural* technology in the distant past, but *other* technology, as well, from medical to aeronautical. Furthermore, these early writers were not naïve: they understood that they were writing in the context of the vast historical timeline of human existence, and they ably posed fundamental questions about how civilizations both transmit and lose technological knowledge, the reliability of sources concerning the past, and the consequences of technology.

But if the archaeological evidence is non-existent, then why the belief in the existence of a long-vanished, technologically advanced civilization? Certainly, the postulation of a past "golden age" by a writer on contemporary society would allow a benchmark or standard by which they could critique the present age and its decline. Yet there are also more elaborate stories of such golden ages and the lost knowledge associated with them. Past civilizations were not just a rhetorical tool — but rather, more profoundly, an element in a belief system concerning a different kind of history: a cyclical one in which mankind's achievements once before had flourished and then disappeared. Technology is a key component of that belief system, as advanced technology can be used to define an "advanced civilization". This definition held even for ancient writers, who believed that the degree of sophistication and development of a society could be measured by its technological achievements. As noted later in this book, these same writers were also aware that a sophisticated technology could be accompanied by or even incur unsophisticated and corrupt morals. As discussed in a subsequent chapter, when a group of adventurers in *The Arabian Nights* find the technologically astonishing "City of Brass" with its automata, they also find all the inhabitants dead.

The present study examines the belief system of a wide range of writers and storytellers, by showing how they wrote about apparently advanced technologies that do not quite "fit" in our traditional way of looking at history. Again, the idea that the human history is linear, going from the primitive to the advanced, is only a model, and a relatively recent one. Of course, in the broad sense, this teleological model of history and human development is certainly the one supported by archaeological evidence: again, no one has found the ruins of ancient aircraft in the Gobi desert. But it is also clear that on another level even ancient peoples were aware of the vagaries of civilization — the Greeks pondered the construction of the pyramids, and asked why the Egyptians of their time no longer built such structures. Technical abilities and technological knowledge *were* lost over time, and there were eras when human beings built vast cities followed by eras when those structures became empty and engulfed by jungle or desert. From a certain perspective, then, archaeological evidence indeed *does* tell us that the linear model of the history of civilization is a simplification, and that there have been periods of great human organization, technological expertise, and industriousness, followed by periods of decline or cataclysmic collapse.

Moreover, the textual record is intriguing, with accounts as fanciful as Plato's Atlantis, but also as sober as the strange South Asian descriptions of flying machines. All attribute a special knowledge of technology to the past. If nothing else, the sociological and psychological aspects of the persistence of this alternative model — one where an advanced civilization rose and fell, or where advanced knowledge was developed and then lost — should be addressed. Again, one can also safely argue that in a certain sense archaeological evidence does, in fact, present the possibility of cyclical rises and falls of civilizations. Great ruins everywhere from Central America to the Middle East show that great civilizations — with advanced architecture, for example, and distinct technologies in building, metalworking, and so on — rose, fell, and then were succeeded by much less sophisticated ones.

The goal of this book, then, is to look at a range of examples — primarily textual, since that material has been less examined than the archaeological evidence — of a cyclical model of technological history across the millennia. What did writers such as Plato actually say about the rise and fall of civilizations? What does evidence from folktales and myths tell us about early views of technology? What kinds of alternative narratives about technical knowledge have been suggested by various thinkers through history? This book uses technology as a kind of index for these historical narratives. Writers across the centuries — including Chinese chroniclers, Arab storytellers, and even the famed Chaucer — did so, using technology as a way of talking about knowledge that different cultures had possessed, knowledge that then had become obscured over time.

For these writers, technologies that were almost magical, such as the power of flight, were attributes of an advanced civilization. The problem of course, is that when writers explored these realms, they talked about various machines and devices in a way that is often fragmentary — in one story, a flying machine is said to be launched by pressing a "pin", for example, but no other details are given. Moreover, as noted earlier, the writers' attitudes towards this idea of technologies from previous, advanced civilizations are ambiguous: there is both an expression of wonder at the potential power of such magical machines and a feeling of trepidation at their potential corrupting influence.

In this book, the reader can explore a selection of possible histories and experience the range of ways that human beings have conceptualized their own existence as technological creatures, and the rise and fall of their societies across the millennia. In a sense, this book is a "collectanea", but purposefully so: many relevant primary sources on the idea of lost technologies are gathered together for the first time here, with texts ranging from ancient folklore to the work of Francis Bacon. The goal is to provide the reader with an exposure to the varied and intriguing way early sources looked at technology and the patterns of history.

2 The Nature of Ancient Knowledge

We accept almost unquestioningly that civilization moves inexorably forward — yet this is a profoundly mistaken idea. Our definition of civilization, from at least the late nineteenth century to the present day, has been based largely on science and technology. We live in an age where scientific discovery and invention progress rapidly. "Faster," "more efficient," and "more powerful" are everyday phrases. And so we assume that this march will continue, onward and upward, into the future.

Yet looking back at human history, the evidence indicates that our present age is an unusual one. In fact, human beings lived for most of their existence in periods of persistent stasis. For millennia, they were hunter-gatherers. After the development of agriculture, humans raised crops, tended animals, and lived directly off the land for thousands of years. In many places on the planet, people still live very simple, day-to-day existences in this same manner, far removed from any sense of an imperative to advance technologically.

In addition to this stasis, there is ample evidence for a cyclical pattern in the development of civilizations. The classic example is ancient Roman culture, which possessed a sophisticated bureaucratic administration, elaborate residential and monumental architecture, paved roads, heated baths and indoor plumbing, and other technologies and organizational systems. Such developments were not repeated at any significant scale until many centuries after the breakup of the Western Roman Empire. Civilizations rise and fall.[2] There

2 This idea was made popular through Edward Gibbon's famed *The History of the Decline and Fall of the Roman Empire*, 6 vols. (London: Strahan & Cadell, 1776–1789).

have been "regressions" in a number of societies: where China was far ahead of Europe in the Middle Ages, for example, in terms of urban planning and invention, it "fell behind" as countries such as Italy experienced the Renaissance, a literal "rebirth" of learning and discovery, manifested in part through great technological projects, such as Brunelleschi's famous dome in Florence.

This naturally leads one to wonder about other such cycles in the more remote past. Were there other cultures of which we have no records, cultures that had sophisticated civilizations, as well as administrative, scientific, and technological systems that disappeared or fell into decline? And if so, how would one know? A number of ancient texts touch upon this theme. Tales from the ancient Greeks, the Chinese, and other cultures recount lost cities, strange inventions from the distant past, and the fundamental idea that in remote antiquity highly advanced civilizations had risen and fallen.

Whether such highly advanced civilizations actually existed is less important than the fact that the writers of these early texts believed that this was the case. Their beliefs are also evidence of a particular and quite interesting epistemology. Examining this complex idea involves questions of knowledge, the transmission of that knowledge, and the role of myth and folklore. Even ancient texts talk about how knowledge had been lost over time. That loss might happen when a society collapses, or when the fragile chain for the transmission of knowledge is broken. When discussing the transmission and loss of ancient knowledge, there are several aspects to consider:

- Sophisticated mechanical devices are invented, the knowledge is then completely lost, and the device is re-invented centuries later. The classic example of this is the Antikythera Mechanism, an ancient Greek astronomical device with advanced gearing systems.[3]
- Cultures develop technical processes that are refined over generations, but then become completely lost over time, or are preserved only by a small elite. Examples include processes as varied as Japanese sword-making techniques and Polynesian methods for sailing across vast stretches of the Pacific Ocean using "stick maps."
- Early technical or scientific knowledge is encoded in a myth, folklore, epic, or ritual for the purposes of preservation and transmission, but then the myth, folklore, epic, or ritual becomes "separated" from its content. The

3 Certain Greek devices were not completely lost, in the sense that the Muslim world preserved some ancient engineering knowledge, for example in the ninth-century *Kitab al-Hiyal* (*"Book of Ingenious Devices"*). For a recent, careful study of the Antikythera Mechanism, see Alexander Jones, *A Portable Cosmos: Revealing the Antikythera Mechanism, Scientific Wonder of the Ancient World* (Oxford: Oxford University Press, 2017).

mode of transmission survives, but the content is lost and the myth, folklore, or epic comes to be seen as pure fiction, and the ritual continues to be engaged in, but empty of meaning. Sometimes this can happen even to an archaeological site: for example, the ritual surrounding Stonehenge has become lost, and the meaning of the structure itself is now unclear.

- Records of ancient technical knowledge and mechanical invention become corrupted over time, so that the original purpose, use, and description of the invention being discussed become obscure. Similarly, as the description becomes obscure, the technology comes to be read as fantastic or imaginary. One can see an example of this in the automata that protect the "City of Brass" in *The Arabian Nights* tale, examined later in this book.

All of these areas merit further study. This book focuses on the idea of lost knowledge of technologies in historical, philosophical, mythical, and folkloric texts.

As noted earlier, there is another approach to the question of lost knowledge: through the examination of archaeological evidence. There are clearly inventions from the past that are remarkable in their sophistication. The classic example is the Antikythera Mechanism, mentioned above, with its sophisticated gearing dating from the first century BCE. It is certainly true that there exist general early references to astronomical devices, but as Alexander Jones notes, "such devices were never very common, and that most ancient references don't reflect personal experience of them."[4] In his book on the Antikythera Mechanism, Jones argues that it was at the same time both an object that indeed fit its historical and cultural context, and an object that was the extreme of what that historical and cultural context could design and craft.

Jones is careful in showing that the Antikythera Mechanism certainly "can be set in a context of Greco-Roman complex mechanical inventions, but the specific parallels for its gearwork ... are limited to a few technical descriptions of much simpler devices and to vague, nontechnical allusions to planetaria."[5] He also shows that the history of mechanical devices relating to astronomy continued for some time after the period of the Antikythera Mechanism, at least up to around 200 CE.[6]

The Antikythera Mechanism shows advances in several areas: machining, gearing, astronomy, and mathematical calculation. There remain few records or accounts to help us understand the mechanism in context, although Cicero mentions a perhaps similar machine that "produces ... behavior for the Sun

4 Jones, private communication, 24 May 2018.
5 Jones, *A Portable Cosmos*, 234.
6 Ibid., 240–242.

and Moon and five wandering stars [i.e., planets]".[7] However, Jones notes that, "*how* the motions of the planets were shown, and whether there was any representation of their spatial relations relative to each other and to the Earth, we are not told."[8]

> The device itself demonstrates that complicated models were within the possibilities of ancient technicians and that computing complexity and the precision of measures were not so far from the theoretical and practical knowledge of some skilled technicians who were alive during the Hellenistic age.[9]

Moreover, there are allusions to gearing in ancient texts in Western sources.[10] But the lack of technical descriptions more broadly in early textual sources adds to the dissonance between how we conceive of the ancient world and its abilities, and what it actually was capable of producing. As Jones notes:

> Concealing the mechanism meant giving emphasis to the effect rather than to the means by which it was achieved, and it also protected the exclusivity both of the possessor and the makers of an original device from imitators.[11]

When one combines this purposeful occlusion with the fact that writers who talked about technological subjects did not necessarily possess the knowledge to describe them accurately, it is little wonder that the texts — such as those presented in this book — include accounts that are lacking in detail. Other archaeological evidence of ancient technological abilities includes Stonehenge and similar early, advanced structures that reflect an ingenuity and that we may perceive — incorrectly — as incommensurate with the technological capacity normally attributed to early societies. The Antikythera Mechanism is at least an example where there is some contextual discussion that survives, such as the mention by Cicero of an astronomical device, and the more general

7 Jones, 183; also see his discussion of Cicero on pp. 239–240. In addition, see the comments on p. 60 of Giovanni Di Pasquale, "Vitruvius's Image of the Universe: Architecture and Mechanics", in Paolo Sanvito, ed., *Vitruvianism: Origins and Transformations* (Berlin: Walter de Gruyter, 2016), 49–64, and the discussion in James Evans, *The History and Practice of Ancient Astronomy* (New York: Oxford University Press, 1998), 82.

8 Jones, 183.

9 Di Pasquale, 61.

10 Jones, 200–201.

11 Jones, 202.

works from the period concerning astronomy.[12] In this book, however, many of the textual accounts treat objects — such as flying vehicles — that do not have any archaeological evidence to match them.

Since the archaeological evidence has been examined in cases such as the Antikythera Mechanism, Stonehenge, and other ancient devices and structures, again this book focuses on the peculiar textual accounts of advanced technological knowledge. In looking at the preservation and transmission of knowledge, it is critical to examine carefully the appearance of "modern" ideas and concepts in early texts. Once more, although it may appear that knowledge and scientific discovery follow a linear progression through history, in fact, ideas have been formulated, lost, found again, or re-formulated from scratch many times over the millennia. Sometimes the fact that a piece of knowledge was possessed by earlier societies escapes us because we do not understand the form in which it appeared. As already mentioned, the result may be a passage in a text that appears to be a myth or a piece of fiction, but actually may be the embodiment or poetic "encoding" of scientific or technical knowledge.

In some texts, however, the language is much clearer than we might suppose, and the very "modern" ideas they contain are apparent to the careful reader. For example, there were early writers who observed natural phenomena, and did their best to describe and explain them. Plutarch, in his *Life of Lysander*, touches upon the concept of rotational motion at several points in the text. In one passage, there is a mixture of early, speculative cosmological concepts, such as "aether", with careful, empirical observation of phenomena:

> There were some who declared that the Dioscuri [i.e., Castor and Pollux] appeared as twin stars on either side of Lysander's ship just as he was sailing out of the harbour against the enemy, and shone out over the rudder-sweeps. And some say also that the falling of the stone was a portent of this disaster; according to common belief, a stone of vast size had fallen from heaven at Aegospotami, and it is shown to this day by the dwellers in the Chersonese, who hold it in reverence. Anaxagoras is said to have predicted that if the heavenly bodies should be loosened by some slip or shake, one of them might be torn away, and might plunge and fall down to earth; and he said that none of the stars was in its original position; for being of stone, and heavy, their shining light is caused by friction with the revolving aether, and they are forced along in fixed orbits by the whirling

12 Research continues into the precise structure and function of the Antikythera Mechanism; see especially Jones, *A Portable Cosmos*, cited earlier, and M. Allen, et al., *Almagest: Special Issue — The Inscriptions of the Antikythera Mechanism* 7.1 (May 2016).

> impulse which gave them circular motion, and this was what prevented them from falling to our earth in the first place.[13]

There is a clear parallel here to our present understanding that objects stay in orbit and do not fall to Earth because they are moving at a sufficient velocity.[14] Of course, some of the ideas presented here in Plutarch's description have no equivalents in our modern physics: the "whirling aether", and the idea that there is an "impulse" which can give an object circular motion. But if we read further in the same passage, we see a critical examination of the evidence:

> But there is a more plausible opinion than this, and its advocates hold that shooting stars are not a flow or emanation of aetherial fire ... nor are they an ignition and blazing up of a quantity of lower air which has made its escape into the upper regions; but they are plunging and falling heavenly bodies, carried out of their course by some relaxation in the tension of circular motion, and falling, not upon the inhabited region of the earth, but for the most part outside of it and into the great sea; and this is the reason why they are not noticed.[15]

Here, Plutarch is approaching the idea that once an object loses sufficient velocity, it will fall. One does not see here any sort of ancient concept of gravitational force — that is, the idea that bodies fall due to some kind of force between those bodies and the Earth. But the passage refers to a "relaxation in the tension of circular motion", which can bee seen as an understanding, perhaps, of centrifugal or centripetal forces. Moreover, the passage describes the concept that bodies out in space can fall in just the same manner as those bodies near to the Earth; indeed, Plutarch goes on to discuss a rock that is torn from a mountaintop by a storm that then plunges to the ground.

---

13 Plutarch, *Lysander* 12.1–4 in *Plutarch's Lives*, 11 vols., trans. Bernadotte Perrin (London: W. Heinemann, 1914–1926), 4.261–263.

14 See Thomas J. McLaughlin, "Aristotelian Mover-Causality and the Principle of Inertia", *International Philosophical Quarterly* 38.2 (1998): 137–151; this article presents an argument for the recognition of Aristotle's "mover-causality principle" as connected to the principle of inertia presented in Newton's first law. Also note James F. O'Brien, "Some Medieval Anticipations of Inertia," *New Scholasticism* 44 (Summer 1970): 345–371. For the clear influence of Plutarch on Newton, see the discussion in Lucio Russo, *The Forgotten Revolution: How Science was Born in 300 BC and Why It Had to Be Reborn* (Berlin: Springer, 2004), 369 et ff.

15 Plutarch, *Lysander* 12.1–4, trans. Perrin, 4.263.

What is also interesting to note is that Plutarch presents a number of interpretations of a given physical phenomenon — that is, falling bodies — and the fact that Plutarch, as he does in his other writings, treats scientific themes in what are otherwise historical works.[16] Plutarch discusses the ideas of Anaxagoras in another work, *Concerning the Face which Appears in the Orb of the Moon*. He attributes to Anaxagoras a number of concepts that are quite sophisticated and modern, such as the fact that the light of the Moon is actually reflected light from the Sun.[17]

Later in this book, some lesser-known technical discussions are presented. Throughout, the point is that there *was* ancient knowledge, some of it quite sophisticated, but often it was not in a context or format that we would recognize. There is little doubt that various indigenous cultures possessed knowledge of astronomy, but stored it and transmitted it in forms rather unfamiliar to us.[18] Even when knowledge about technology appears in textual form in an early source, the text itself often is primarily a historical chronicle, a philosophical discussion, or a folktale. In the examples taken from Plutarch, as we have seen, thoughtful conjectures concerning astronomical phenomena are found in a historical and philosophical text. As we will see, other descriptions are found in similarly non-scientific and non-technical works.

3 A New Approach

In abandoning the idle speculations of pseudo-science and pseudo-history, we must be careful not to abandon the question of lost technology entirely. In fact,

16 For an interesting example of the preservation of scientific ideas in early non-scientific works, see the discussion in Wesley M. Stevens, "Science at the Edge of the Earth", in José Saldaña, ed., *Science and Cultural Diversity: Filling a Gap in the History of Science* (Mexico City: Sociedad Latinoamericana de Historia de las Ciencias y la Tecnología, 2001), 51–58. For a look at another discussion of the possibility of science in an apparently non-scientific culture, see the paper, comments, and reply in Alvar Ellegård, "Stone Age Science in Britain?", *Current Anthropology* 22.2 (April 1981): 99–125.

17 See the discussion in Plutarch, "Concerning the Face which Appears in the Orb of the Moon", 929b et ff., in *Moralia*, 15 vols., trans. Harold F. Cherniss, et al. (Cambridge, MA: Harvard University Press, 1927–1969), 12.100 et ff.

18 See, for example, Michelle Gantevoort, Duane W. Hamacher, and Savannah Lishick, "Reconstructing the Star Knowledge of Aboriginal Tasmanians", *Journal of Astronomical History and Heritage* 19.3 (2016): 327–347; also note Duane W. Hamacher and Ray P. Norris, "Meteors in Australian Aboriginal Dreamings", *WGN — Journal of the International Meteor Organization* 38.3 (June 2010): 87–98.

this book addresses the question directly, and does so through passages in extant texts. Academic historians might be tempted to categorize all the passages that this book examines as myth — and myth, in that usage, means pure invention. But to do this is to commit a disservice to the material. To be dismissive of these writings is as damaging to the material as is the construction by Graham Hancock and others of rickety, pseudo-scientific theories.

The material itself is waiting; there are ancient texts with curious contents, presenting us with unanswered questions — and we are obliged at least to consider those questions:

1. Why was the idea of pre-existing, technologically advanced knowledge so pervasive?
2. What are the ancient texts actually telling us? Do the authors of those texts make any differentiation between fiction, myth, and historical record in discussing such knowledge?
3. How was information concerning technological knowledge recorded and transmitted in ancient times? Is there a way that such information could have become lost over time, or even purposely hidden?

This last question is intriguing; obviously, in ancient times, there were no technical journals, but technological information — on agricultural practices, meteorology, astronomy, engineering, medicine, and many other subjects — was being transmitted from generation to generation. How? Certainly, there were ancient technical texts, such as Galen's work on medicine, and Vitruvius' work on architecture and engineering. But the majority of technical information was not passed down in those relatively "modern" formats.

Other authors have examined this question of information transmission over the millennia. Our study will focus on the other two points made above: the idea that earlier civilizations might have had some kind of knowledge that we might consider advanced or even markedly anachronistic, and the differentiation between fiction, myth, and historical records in discussing technical knowledge. But the question concerning the transmission of knowledge — our third point above — can provide us some key insights into the relationship between fiction, myth, and knowledge. Furthermore, the study of this transmission process helps us realize that information about technology can co-exist with other kinds of information in a given narrative structure. As a book on this subject, *When They Severed Earth from Sky: How the Human Mind Shapes Myth*, points out:

> We moderns tend to assume that the events and details in a story are all of equal "value" in the words — that the narrative contains either *all* "truth" or *no* "truth". And if we don't happen to believe that volcanoes

> are to be explained as a mountain that a spirit is using as his home ... we throw the whole thing out as nonsense.[19]

In other words, our modern reductive tendencies compel us to view stories in only one way, leading us to attempt to differentiate rigorously between fact and fiction. In doing this, we forget that fiction can convey fact, a function still served by various kinds of fictional works. Somehow, however, when we look at the past, we view all texts and orally transmitted accounts as having to be categorically fact *or* fiction. Again, we tend to forget that ancient peoples had important information — facts — that they needed to transmit; they were not interested simply in telling tall tales. As *When They Severed Earth from Sky* also notes

> [I]n shifting from hunting/gathering to an agricultural life, people had to work out a solar calendar, and since they were nonliterate, they had to encode this information orally into rememberable "myths".[20]

This "encoding" is important, because it implies that there is "decoding" that can be done, to find some of this ancient knowledge.

We also need to look at this word "myth", since in contemporary speech we tend to conflate myth with fiction: "The word *myth*, originally Greek *mythos* 'a traditional explanatory saying', has now become synonymous with 'untruth'":[21]

> [W]e have argued that 'myths' were not intended as fiction in our sense, but as carriers of important information about real events and observations. We have also given much evidence that contrary to prevailing belief today, this information can sometimes survive intact for thousands of years.[22]

This concept — of technical information surviving in a wide range of forms, including myth — is one that we shall return to later in this book.

---

19 Elizabeth Wayland Barber and Paul T. Barber, *When They Severed Earth from Sky: How the Human Mind Shapes Myth* (Princeton: Princeton University Press, 2004), 27.

20 Ibid., 179.

21 Ibid., 149.

22 Ibid., 244.

4 The Sources

In looking at ancient texts, historians must work with the material that is available. Speculation is only that — an assemblage of ideas of what "could have been known" or "might have been known". In engaging in such speculation, the extant material itself is often ignored or misread — willfully or not. This is the critical fault of the pseudo-historians, because it is the historical material itself that holds all the clues. Indeed, the ancient texts often tell us more than is apparent on a first reading, especially if the reader considers the material's subtlety.

This book suggests the need for a new method of historical analysis for certain ancient texts, particularly texts that contain anomalous and anachronistic information concerning technological or scientific concepts. The study here focuses on written texts — the archaeological evidence such as the Antikythera Mechanism has earned separate studies — interpreted with a different approach. That approach is a reading of the text from the perspective of the original author. In doing this, we ask why the authors chose the material that they did, and why they wrote with the particular style, focus, and emphasis that they did.

This strategy helps us begin to sort out what these writers really *meant*. Were they just writing fanciful tales? Or were they writing in metaphorical language about historical facts? In our first chapter here, on flying vehicles, we will see this distinction rather clearly: there are tales about flying beings and so on, but these are quite different in style and tone from texts of the same period describing *mechanical* devices used for flying.

This question of how to look at texts is part of a more thoughtful way of considering historical evidence, a way that one study calls the "camera angle", citing the historian of science Mott T. Greene:

> Mott Greene … in his perceptive book *Natural Knowledge in Preclassical Antiquity*, notices the effects of Camera Angle — for example in his essay on Egyptian fractions. He points out that whereas *our* mathematical efforts move toward finding the most *general* solution to the use of numbers, the Egyptians had "no concept of numbers in abstraction from things numbered".[23]

What this means is that for the Egyptians, mathematics was directly tied to the real world, and to pragmatic tasks such as measuring and cutting stone.

23 Ibid., 58.

Therefore, the Egyptians felt no need to develop a more refined or abstracted system:

> To understand why the Egyptians never moved beyond their seemingly cumbersome system of "unit fractions" ... we need to see what problems they were looking at. We need to see the world from their camera angle.[24]

In terms of the analysis in this book, the "camera angle" is the cultural context in which a particular society develops its view towards technology and knowledge about technology. The cultural context determines how a culture adopts technology and to what degree, and how it frames technical knowledge and its transmission. The "camera angle" in this study also includes the perspective of the particular ancient author who is writing about technology — how the writer frames the question of mechanical devices, be they special mirrors, flying vehicles, or something else.

Citing Greene again, the discussion in *When They Severed Earth from Sky: How the Human Mind Shapes Myth* reaches an important conclusion:

> In fact, after tackling several more problems that have baffled scholars because they had inappropriate camera angles ... Greene concludes, "there is a wisdom in assuming that the people whose works one studies [such as myths and other literature] are fully capable of describing their world", and that one is obliged to make that assumption at the very least "until one knows enough about that world to assay what they said about it". In other words, the onus is on *us* to find the appropriate position from which to assess their reports, rather than simply dismissing the reports as empty — like a deaf man dismissing the phenomenon of sound as an old wives' tale.[25]

This means that early texts must be parsed, in the sense of a careful reading that sets aside modern preconceptions.

In this parsing of the texts, the present study also explores the interplay of mythical language and historical description. The point that texts should be examined carefully, and understood within their cultural context, is an important one. This is a relatively new strategy, as one scholar has noted:

---

24 Ibid., 58.

25 Ibid., 58–59.

> [R]ecently, historians have approached texts and disciplines in terms of the particular historical cultures in which they developed and have also investigated how disciplinary categories themselves have shifted over time. Central to this latter approach is the notion that a text — or any other historical artifact for that matter — should be understood within its own culture in terms of the categories that belong to historical actors. The meaning of a text ... will be misunderstood without careful, detailed attention to the fabric of meaning that exists within the culture to which that text initially belonged.[26]

Similarly, another writer notes:

> There has been a marked tendency in the history of science ... not only to reject this "whiggish" approach to the history of science (judging the past in terms of present-day values), but also to emphasise the importance of the historical context for properly understanding the thought of past thinkers.[27]

In paying attention to cultural and historical context — or at the very least noting that ancient cultural contexts were quite different from ours — we can come to realize that a text that seems fanciful or mythical in the contemporary sense actually might bear some kind of veridical historical or technical knowledge.

This book examines a selection of early texts concerning the idea of "lost knowledge". These texts provide the reader with an intriguing range of topics: ancient accounts of flying vehicles, mirrors with peculiar technical abilities, a land lost beneath the seas, and a ring with a special power. In each case, the texts themselves are at the center of analysis. What do the texts actually *say*? What was the authors' purpose in writing the texts? How do the authors themselves view history and the question of knowledge?

This last question is another key element to our study. One is sometimes tempted to think of early writers simply as "objects" in history, and view those authors as unaware of their own presence in the vast realm of what has been called "deep time". But the discussion in this book shows that the writers

---

26 Pamela O. Long, *Openness, Secrecy, Authorship: Technical Arts and the Culture of Knowledge from Antiquity to the Renaissance* (Baltimore: Johns Hopkins University Press, 2001), 3.

27 John Forrester and John Henry, eds., *Jean Fernel's* On the Hidden Causes of Things*: Forms, Souls, and Occult Diseases in Renaissance Medicine* (Leiden: Brill, 2005), 16, n. 45; also see the comments on p. 64 in this same work.

indeed thought about history, the rise and fall of civilizations, the authority and reliability of the historical sources from which they were drawing their information, and where they as authors stood in the grand sweep of it all. Early writers appreciated the complexity of how knowledge behaves through time. One commentator notes the following in talking about the idea of scientific progress as articulated in early sources:

> The epistemology of progress differed from the common conception of history of knowledge in ancient times. Ancient literature offers many passages that used to be interpreted as a description of the idea of general progress of civilisation and knowledge. Likewise, one can find numerous quotations rejecting the progress of society and knowledge, proclaiming decline or the alternating cycles of prosperity and decline. The issue related to the existence of the idea of progress in ancient times must therefore be treated with some care.[28]

There was no one view of the past held by early writers, and no assumption that all human knowledge was moving towards some final, fixed goal. Texts from the past reflect a keen understanding of the vagaries of human knowledge, along with an understanding that technological capacities might be lost over time as a culture declined, and that sophistication in terms of technological progress did not always equate with a culture's moral growth and development.

Moreover, the view *from* the past of the even *deeper* past was complex:

> It must also be noted that the belief in progress was limited to the area of knowledge about nature and technology and it did not transcend to any other areas. The recognition of progress achieved both in the knowledge of nature and in technological development was sometimes even associated with the idea of decadence.[29]

It is important to understand this early view, since it is rather different from our current view. Most discussions of "progress" in contemporary society are framed in terms of technology; there is little discussion of moral or ethical progress outside of philosophy or religion. Moreover, technological development is infrequently connected to the idea of decadence, even though one

---

28 See p. 381 of Daniel Špelda, "Veritas filia temporis: The Origins of the Idea of Scientific Progress", *Annals of Science* 73.4 (October 2016): 375–391.

29 Ibid., 381–382.

may cite some exceptions.[30] Most critically, there is the common belief that our current technological growth will continue unabated, with even serious climactic and other environmental limitations being seen as surmountable by some future technological innovation.

This is in contrast to the ancient Greek view, for example, wherein "knowledge is expanded and grows, but ... its development is (cyclically) interrupted."[31] Even Aristotle believed that, "every craft and philosophy has been invented repeatedly, and has been destroyed again."[32] Again, this differs markedly from our current view of technological development and knowledge as almost exclusively linear and cumulative. One modern writer who captured this contrast precisely was the famous economist Robert L. Heilbroner:

> The future over most of mankind's experience has not been that benign and congenial environment which we unthinkingly anticipate it to be, but has been formidable and overwhelming, unalterable and often unanswerable ... It is difficult for us to imagine the state of mind that such an exposure to history might have given rise to. But no sooner do we attempt to capture this state of mind than an unexpected fact attracts our notice. It is that over by far the greater part of man's existence the importance of the future assumed much smaller proportions than it does in our day. Indeed, until a few centuries ago in the West, and until relatively recent times in the East, it was the past and not the future which was the dominant orientation to historic time.[33]

The future remains the focus of the present, in both social and technological terms. Furthermore, despite all the current problems in Western society, there is an underlying belief that it has built up sufficient dynamism to progress in terms of societal development and that technological capacity will continue to grow. This belief includes the assumption that in a broad sense, then, we *know more* than civilizations of the past. As Heilbroner notes, this attitude is an anomaly:

---

30 See, for example, Sherry Turkle, *Alone Together: Why We Expect More from Technology and Less from Each Other* (New York: Basic Books, 2011), and idem, *Reclaiming Conversation: The Power of Talk in Digital Age* (New York: Penguin Press, 2015). Also note Tom Wessels, *The Myth of Progress: Toward a Sustainable Future* (Hanover, NH: University Press of New England, 2013).

31 Špelda, 382.

32 Ibid., 382.

33 Robert L. Heilbroner, *The Future as History* (New York: Harper & Row, 1960), 18.

> Modern man, who typically sets his life goals in what is to come, displays an attitude quite the converse of earlier days. Ancient Egypt, Greece, Rome, the vast Asiatic civilizations, even the Renaissance, did not look ahead for the ideals and inspirations of their existence, but sought them in their origins, in their ancient glories, their fabled heroes, their pristine virtues real of fancied.[34]

Heilbroner wrote these words in 1959, but the contrast between modern and ancient perspectives that he describes still holds true. The scholar George Steiner notes that in a historical sense, prior to the First World War, the West indeed might have looked back on a "century of progress, of liberal dispensations, and reasonable hope."[35] Contemporary society still focuses on the potential for societal progress and technological advancement. Despite Steiner's subsequent clear warning about our continuing construction of a "philosophical edifice of hope ... of Cartesian rationality", in the West at least the view remains the same: history is linear and we are moving "forward". Given that, the idea of a cyclical pattern wherein a past civilization might have been more sophisticated than our own strikes us as absurd.

5 Technology in the Remote Past: the Case of Frederick Soddy

Of course, there have been many speculations on the theme of highly advanced ancient knowledge, leading to an abundant number of works that border, unintentionally, on fiction. As noted earlier, authors such as Charles Hapgood, for example, put forward conjectures as to lost ancient civilizations that were capable of mapping the entire globe. While his work was motivated by a genuine interest in the principles of early mapmaking, Hapgood's speculations far exceeded the available evidence.[36] Other writers, such as Graham Hancock, have weaved vast and complex tales about the past through a series of tenuous connections and elaborate speculations.[37] In some cases, the ancient material

34 Ibid., 18–19.

35 George Steiner, *Grammars of Creation* (New Haven: Yale University Press, 2001), 4.

36 Hapgood's speculations can be found in Charles H. Hapgood, *Maps of the Ancient Sea Kings: Evidence of Advanced Civilization in the Ice Age* (Philadelphia: Chilton Books, 1966). A critical examination of Hapgood's interpretations of cartographical evidence can be found in Gregory C. McIntosh, *The Piri Reis Map of 1513* (Athens, GA: University of Georgia Press, 2000).

37 See, for example, Graham Hancock, *Heaven's Mirror: Quest for the Lost Civilization* (New York: Crown Publishers, 1998). Hancock also came out with other books on this

itself — be it textual, cartographical, or archaeological — actually *does* cry out for new interpretations or at least further investigation, but writers such as Hancock use the material as no more than a foundation for pure conjecture or fantasy.[38]

These kinds of fantastic visions into the remote past of human civilizations — and the idea of lost technological knowledge — have a long and peculiar history, engaging not only writers on the fringe, but also sober scientists. Radium, despite its distinction of being radioactively luminescent, is not a substance that one readily connects with mysticism. But through a peculiar series of connections between science, speculation, and spiritual beliefs, one can trace a direct line from a scholarly work on the characteristics of radium by the Oxford professor Frederick Soddy — *The Interpretation of Radium and the Structure of the Atom* — all the way to the Theosophical doctrines of Blavatsky and ideas on lost knowledge.

Helena Petrovna Blavatsky was born Helena von Hahn in what is today Ukraine, and later was known often simply as "Madame Blavatsky". Details of her early life after her departure from Russia are uncertain, although there are various accounts of her traveling everywhere from Canada to Tibet. In 1873, she came to New York, and it was there that her spiritual activities began in earnest. During her subsequent career, she made many claims concerning psychic abilities, including telepathy and clairvoyance. Not long after her move to New York, she put together the first of several books, *Isis Unveiled*.

In 1875, Blavatsky and two colleagues founded the Theosophical Society. The term "theosophy" is derived from the ancient Greek *theos* ("god") and *sophia* ("wisdom"). As a single word, "theosophy" can be understood as "divine wisdom", and throughout history, there have been a number of different groups — from the Manichaeans to the Cathars — who have held related

---

same theme of lost civilizations; note his book *Underworld: The Mysterious Origins of Civilization* (New York: Crown Publishers, 2002), and, with Robert Bauval, *The Message of the Sphinx: A Quest for the Hidden Legacy of Mankind* (New York: Crown Publishers, 1996). Another book positing the existence of a highly advanced civilization over 8,000 years ago is Walter Cruttenden, *Lost Star of Myth and Time* (Pittsburgh: St. Lynn's Press, 2006); also see Philip Coppens, *The Lost Civilization Enigma: A New Inquiry Into the Existence of Ancient Cities, Cultures, and Peoples Who Pre-Date Recorded History* (Pompton Plains, NJ: New Page Books, 2013).

38 There have also been sociological and literary studies as to the origins of these conjectures and fantasies; see, for example, the essays in Alain-Michel Boyer, ed., *Modernités: Mondes Perdus* (Talence: Presses universitaires de Bordeaux, 1991). In addition, note the critique from the perspective of archaeology in John R. Cole, "Cult Archaeology and Unscientific Method and Theory", in Michael B. Schiffer, ed., *Advances in Archaeological Method and Theory*, 11 vols. (New York: Academic Press, 1978–1987), 3.1–3.33.

beliefs and practices under this broad definition. Such beliefs include the importance of mystical experiences, and the concept that there is a deeper reality accessible by meditation, revelation, or other methods. Most relevant to the discussion here is the emphasis on esoteric knowledge: modern Theosophists in particular believe that there are secret "inner teachings", and believers have engaged in attempts to decipher meaning that is apparently concealed in various sacred texts.

Not surprisingly, then, Blavatsky was deeply interested in the origins and meaning of various world religions, the beginnings of human civilization, and the connection between esoteric knowledge and contemporary science and technology. She believed in Atlantis as an historical fact, and held that part of the human race had its origins there. Towards the end of her life, she traveled to India, and spent her final years in England. Blavatsky left behind a number of books, and the Theosophical Society — and various offshoots — still exists today.

In fact, Theosophy's rise in a period of intense scientific and technological development — the late nineteenth century — is not as contradictory as it might seem. As one historian of science notes:

> One of a number of systems of belief that came to prominence in this period as alternatives to organized religion and scientific rationalism, theosophy drew on ideas from Eastern philosophy, mysticism and ancient occult traditions dating back to Pythagoras. Its blend of esoteric wisdom and spiritual philosophy ... appealed to Victorian audiences disenchanted by the materialism of much modern science ... In particular, theosophy's emphasis on esoteric wisdom gave it a strong appeal to intellectuals. They saw in it a way of exploring and expressing hidden realities in an increasingly materialistic world without moral or spiritual values.[39]

Such a goal — the exploration and expression of some more fundamental or transcendent reality — is not without merit. But in our current period in history, we are wary of any overlap of science and non-scientific thinking. For us, any knowledge, including esoteric knowledge, must have roots in evidence of some kind.

For Frederick Soddy (1877–1956), perhaps, the evidence was simply that ideas and symbols found in earlier cultures seemed to suggest ancient knowledge of a technological nature. Soddy graduated from Oxford in 1898 with a

39 See p. 33 of Jeff Hughes, "Occultism and the Atom: The Curious Story of Isotopes", *Physics World* 16.9 (September 2003): 31–35.

degree in chemistry, and went on to work with Ernest Rutherford on studies of radioactivity, and investigated in particular radium, a radioactive element. Soddy continued these studies in London with the Scottish chemist William Ramsay, and he later went on both to teach physical chemistry at the University of Glasgow and carry out research concerning radioactive decay and isotopes. In 1921, he received the Nobel Prize in chemistry.[40] Despite this honor, and the fact that in 1919 he had become a professor of chemistry at Oxford, he gradually moved away from science, and began involvement in economic, social, and political activities.[41]

At first glance, Soddy's book, *The Interpretation of Radium and the Structure of the Atom*, appears to be simply a collection of lectures on radioactivity and the nature of matter.[42] But in a chapter peculiarly entitled "Radioactivity and the Evolution of the World", the reader encounters a brief section with the intriguing heading, "Radioactivity and Mythology":

> The world probably being of much greater antiquity than physical science has thought to be possible, it is interesting and harmless to speculate whether man has shared with the world its more remote history.
>
> In this connection it is curious how strangely some of the old myths and legends about matter and man appear in the light of the recent knowledge. Consider, for example, the ancient mystic symbol of matter, known as Ouroboros — "the tail devourer" — which was a serpent, coiled into a circle with the head devouring the tail, and bearing the

40 Concerning Soddy's career as a scientist, see George B. Kauffman, ed., *Frederick Soddy (1877–1956): Early Pioneer in Radiochemistry* (Dordrecht: D. Reidel, 1986), as well as Thaddeus J. Trenn, *The Self-Splitting Atom: The History of the Rutherford-Soddy Collaboration* (London: Taylor and Francis, 1977). Soddy himself wrote a number of books on science, including *Matter and Energy, Radio-Activity: An Elementary Treatise*, and *The Chemistry of the Radio-Elements*.

41 For an excellent biography that discusses Soddy's life as both scientist and citizen, see Linda Merricks, *The World Made New: Frederick Soddy, Science, Politics, and Environment* (Oxford: Oxford University Press, 1996). Also see Mansel Davies, "Frederick Soddy: The Scientist as Prophet", *Annals of Science* 49.4 (July 1992): 351–367, and Herman E. Daly, "The Economic Thought of Frederick Soddy", *History of Political Economy* 12.4 (1980): 469–488. Soddy wrote several works on economic theory and its relation to politics that articulated his own perspectives, including *Wealth, Virtual Wealth, and Debt: The Solution of the Economic Paradox* and *Money versus Man: A Statement of the World Problem from the Standpoint of the New Economics*.

42 Frederick Soddy, *The Interpretation of Radium and the Structure of the Atom*, 4th ed. (New York: G.P. Putnam's Sons, 1920). Despite its somewhat staid title and dry contents, this book was a key influence for H.G. Wells, as he based his 1914 novel *The World Set Free* on this work by Soddy; see Merricks, 67–68.

> central motto, "The whole is one." This symbolises evolution; moreover, it is evolution of matter — the very latest aspect of evolution ... The idea which arises in one's mind as the most attractive and consistent explanation of the universe in the light of present knowledge is, perhaps, that matter is breaking down and its energy being evolved and degraded in one part of a cycle of evolution, and in another part, still unknown to us, the matter is being again built up with the utilisation of the waste energy. If one wished to symbolise such an idea, in what better way could it be done than by the ancient tail-devouring serpent?[43]

It seems that Soddy is arguing for advanced ancient technical knowledge. Reading on, in fact, we find that Soddy develops his argument further and with more specificity:

> Some of the beliefs and legends which have come down to us from antiquity are so universal and deep-rooted that we are accustomed to consider them almost as old as the race itself. One is tempted to inquire how far the unsuspected aptness of some of these beliefs and sayings to the point of view so recently disclosed is the result of mere chance or coincidence, and how far it may be evidence of a wholly unknown and unsuspected ancient civilization of which all other relic has disappeared.[44]

Soddy makes the suggestion, then, of an advanced civilization in the distance past. He muses further:

> It is curious to reflect, for example, upon the remarkable legend of the philosophers stone, one of the oldest and most universal beliefs, the origin of which, however far back we penetrate into the records of the past, we do not probably trace to its real source. The philosopher's stone was accredited the power not only of transmuting the metals, but of acting *as the elixir of life*. Now, whatever the origin of this apparently meaningless jumble of ideas may have been, it is really a perfect and but very slightly allegorical expression of the actual present views we hold to-day. It does not require much effort of the imagination to see in energy the life of the physical universe, and the key to the primary fountains of the physical life of the universe to-day is know to be transmutation. Is, then, this old

43 Soddy, 181.

44 Ibid., 181–182.

> association of the power of transmutation with the elixir of life merely a coincidence?[45]

Soddy concludes:

> I prefer to believe that it may be an echo from one of many previous epochs in the unrecorded history of the world, of an age of men which have trod before the road we are treading to-day, in a past possibly so remote that even the very atoms of its civilisation literally have had time to disintegrate. Let us give the imagination a moment's further free scope in this direction, however, before closing. What if this point of view that has now suggested itself is true, and we may trust ourselves to the slender foundation afforded by the traditions and superstitions which have been handed down to us from a prehistoric time? Can we not read into them some justification for the belief that some former forgotten race of men attained not only to the knowledge we have so recently won, but also to the power that is not yet ours?[46]

A "forgotten race of men"? This is an odd digression, no doubt, in a book on physics. Indeed, in other respects, *The Interpretation of Radium and the Structure of the Atom* is a methodical, if popular, treatment of physical science. So, why would Soddy include this discussion on vanished civilizations? Indeed, the passages presenting this idea — that a "former forgotten race of men" possessed great technological power — persisted in the first through the fourth edition of this work.[47]

In Theosophist writings, there are some very similar ideas. Blavatsky's *The Secret Doctrine: The Synthesis of Science, Religion, and Philosophy* outlines a complex picture of lost civilizations, including lengthy discussions of Atlantis and Lemuria.[48] In *Isis Unveiled*, Blavatsky writes, "The oldest esoteric traditions …

45 Ibid., 182.

46 Ibid., 182.

47 The first edition of *The Interpretation of Radium and the Structure of the Atom* came out in 1909, and the fourth edition, revised and enlarged, came out in 1920; in the 1909 edition, these speculations can be found on pp. 239 et ff. In a 1993 biography, Constance Reid discusses the influence of Soddy's book on the mathematician and science fiction writer E.T. Bell; however, Reid states incorrectly that these peculiar passages by Soddy, which were an important influence for one of Bell's novels, were "dropped from subsequent editions" of Soddy's book. See Constance Reid, *The Search for E.T. Bell: Also Known as John Taine* (Cambridge: Cambridge University Press, 1993), 179.

48 Helena Petrovna Blavatsky, *The Secret Doctrine: The Synthesis of Science, Religion, and Philosophy*, vol. 2, *Anthropogenesis* (London: The Theosophical Publishing Company, 1888).

teach that, before the mystic Adam, many races of human beings lived and died out, each giving place in its turn to another."[49] The book goes on to discuss how stories among the Quiché people tell of an advanced race of people who could "reason and speak", and mentions similar accounts among the Kalmyks and tribes in Siberia. Elsewhere in the same book, Blavatsky speaks of the "oldest traditions which claim an incredible antiquity for our race".[50]

In a discussion of ancient sources including Aristotle and Plato, Blavatsky presents a clear articulation of her ideas about lost civilizations and lost knowledge. She comments that early philosophers

> divided the interminable periods of human existence on this planet into cycles, during each of which mankind gradually reached the culminating point of highest civilization and gradually relapsed into abject barbarism.[51]

Another Theosophist, Elizabeth Winter Preston, wrote a work entitled *The Earth and Its Cycles*, which built upon the ideas in *The Secret Doctrine*; Preston's book expanded this concept of cyclical patterns in the narrative of humanity.[52] Presenting a catastrophic history of human civilizations, the author outlined vast reaches of Earth history, up to and including "glimpses of Aryan civilisations after the flood (10,000 BC)".[53]

The Theosophists offered further perspectives on ancient advanced civilizations. In *The Secret Doctrine*, one of the sources cited is a work known as the *Book of Dzyan* or *Stanzas of Dzyan*; this was supposedly an ancient Tibetan work, one that Blavatsky claimed to have seen firsthand.[54] It recounts, in very mysterious language, knowledge from the remote past, but goes up to and includes a treatment of Atlantis and Lemuria. The text was praised by one Theosophist this way:

---

49 Idem, *Isis Unveiled: A Master-Key to the Mysteries of Ancient and Modern Science and Theology*, vol. 1, *Science*, 6th ed. (New York: J.W. Bouton, 1892), 2.

50 Ibid., 3.

51 Ibid., 5.

52 Elizabeth Winter Preston, *The Earth and Its Cycles: A Comparison Between the Results of Modern Geological and Archaeological Research and the Statements of the Secret Doctrine* (London: Theosophical Publishing House, 1931).

53 Ibid., 118.

54 For the text of *The Stanzas of Dzyan*, see Blavatksy, *The Voice of the Silence and Other Chosen Fragments from the Book of Golden Precepts* (Sioux Falls, SD: NuVision, 2007), 53 et ff.

> The Stanzas set forth a cosmogenesis and anthropogenesis which, in their sweep and detail, leave far behind any existing record of such things from the past; they cannot be explained as the clever piecing together of the disconnected archaic fragments still preserved in sacred books and classical authors; they have an individuality of their own, and yet they bear the hall-mark of an antiquity and the warrant of an economy which the Western world thinks to have long passed away. Further, they are set in an atmosphere of commentary apparently translated or paraphrased from Far Eastern tongues, producing a general impression of genuineness that is difficult for a scholar who has sufficiently overcome his initial prejudices to study them, to withstand.[55]

While the exact nature of Blavatsky's actual source material is not entirely clear, there seems indeed to be little evidence, of course, for surviving manuscripts dating back to the time of Atlantis — or before.[56]

An initial examination reveals no explicit connection between Soddy's ruminations and Theosophical speculations about lost advanced civilizations. In fact, one modern commentator, Mark S. Morrisson, notes that although

> Soddy was clearly not a card-carrying Theosophist or in any way a follower of Madame Blavatsky, the Atlantis vogue of the late nineteenth century and especially the occultists' vision of Atlantis — both as the source of a secret wisdom and as a civilization that had destroyed itself through its own knowledge — are unmistakably present in Soddy's thinking.[57]

He also notes that Soddy was not "involved in occult groups".[58] However, the same author points out that, "Soddy formulated theories that must have had roots in occult lore of his day — about ancient races, for instance."[59] Morrisson also notes that

---

55 G.R.S. Mead, *Concerning H.P. Blavatsky* (Kila, MT: Kessinger, 1997), 16–17.

56 Concerning Blavatsky's sources, see Nicholas Goodrick-Clark, ed., *Helena Blavatsky* (Berkeley: North Atlantic Books, 2004), 75–76. Also see the preface in Tim Maroney, *The Book of Dzyan* (Oakland, CA: Chaosium Books, 2000), and Wouter J. Hanegraaff, *New Age Religion and Western Culture: Esotericism in the Mirror of Secular Thought* (Albany: State University of New York Press, 1998), 453. Another discussion of *The Book of Dzyan* is found in L. Sprague De Camp, *Lost Continents: The Atlantis Theme in History, Science, and Literature* (New York: Dover, 1970), 55–58.

57 Mark S. Morrisson, *Modern Alchemy: Occultism and the Emergence of Modern Atomic Theory* (Oxford: Oxford University Press, 2007), 162.

58 Ibid., 98.

59 Ibid., 109.

> it is difficult to assess Soddy's exposure to occult writings during any period of his life. He destroyed most of his personal correspondence that did not relate directly to science and economics, and he did not keep a diary or leave documents that might suggest what kinds of nonscientific reading he did or even much of what he did in his spare time. His scientific outlook and the surviving historical records of the occult revival make it all but impossible to believe that Soddy was ever involved in a secret Hermetic society like the Golden Dawn or is successors, or even in something like the Theosophical Society.[60]

So, what was the connection? Why are Soddy's ideas so much like those of the Theosophists?

First of all, while Soddy may not have been a Theosophist, his writings — notably, *The Interpretation of Radium* — were not unknown to them. In fact, his book is cited directly in a 1923 issue of *The Theosophical Quarterly*. There, both Soddy's discussion of the Ouroboros and more particularly his statements about "some former forgotten race of men" having achieved "not only to the knowledge we have so recently won, but also to the power that is not yet ours" are quoted directly.[61] The author of the article, Charles Johnston, takes Soddy's text and compares it to a passage in a book entitled *The Occult World*:

> Ages before the Royal Society found itself becoming a reality upon the plan of the "Prophetic Scheme," an innate longing for the hidden, a passionate love for, and the study of, Nature, had led men in every generation to try and fathom her secrets.[62]

Even to the casual reader, this reference does indeed seem to parallel Soddy's mentions of a lost race, but there is more. In fact, *The Occult World* is a very peculiar book, and its author, Alfred Percy Sinnett, was a British journalist and devotee of the occult. He was an important figure in the first inception of the Theosophical Society, and had met Blavatsky in India in 1880.

Sinnett, writing several decades before Soddy, cites an even more obscure account of a "forgotten race of men", stating

60 Ibid., 160.

61 See pp. 23–24 of Charles Johnston, "Theosophy", *The Theosophical Quarterly* 21.1 (July 1923): 21–24.

62 Ibid., 24, quoting A.P. Sinnett, *The Occult World*, 4th ed. (London: Trübner, 1884), 67. Johnston was a member of the Indian Civil Service, a scholar of Sanskrit, and a founder of the Dublin Lodge of the Theosophical Society; interestingly, the poet and dramatist William Butler Yeats also was a member.

> The *Vril* of the *Coming Race* was the common property of races now extinct. And as the very existence of those gigantic ancestors of ours is now questioned — though in the Himavats … we have a cave full of skeletons of these giants — and their huge frames, when found are invariably regarded as isolated freaks of Nature — so the *vril*, or *akas* as we call it, is looked upon as an impossibility — a myth.[63]

Some clarity is brought to this strange passage when we understand that it is actually referring to an 1871 science fiction novel by the English writer Edward Bulwer-Lytton entitled *The Coming Race* (or *Vril: The Coming Race*).[64] The novel portrays a kind of "master race" living underground, and possessing a form of energy called "vril". These people are descendants of an antediluvian civilization, and possess advanced technology.[65] Although the book is fiction, it cites scientists of the time, such as Michael Faraday and Louis Agassiz. Blavatsky ended up believing in the reality of "vril", and argued as such in her books *Isis Unveiled* and *The Secret Doctrine*. In the latter, she states:

> The name vril may be a fiction; the Force itself is a fact doubted little in India as the existence itself of their Rishis, since it is mentioned in all the secret works. It is this vibratory Force, which, when aimed at an army from an *Agni Rath* fixed on a flying vessel, a balloon, according to the instructions found in *Ashtar Vidya*, reduced to ashes 100,000 men and elephants, as easily as it would a dead rat. It is allegorised in the *Vishnu Purana*, in the Ramayana and other works.[66]

Here, Blavatsky is taking some passages in early Indian texts such as the *Rāmāyaṇa*, and interpreting them as evidence of advanced forms of warfare among these ancient peoples.[67]

---

63 Sinnett, 67.

64 Edward Bulwer-Lytton, *The Coming Race* (Edinburgh and London, W. Blackwood and Sons, 1871).

65 A similar concept of an advanced, underground civilization is found in the peculiar stories and other work of the American writer Richard Sharpe Shaver.

66 Blavatsky, *The Secret Doctrine: The Synthesis of Science, Religion, and Philosophy*, vol. 1, *Cosmogenesis*, 2nd ed. (London: The Theosophical Publishing Company, 1888), 563.

67 Such a strange reading of the classical Sanskrit epic of India, *The Rāmāyaṇa of Vālmīki*, with claims that India in ancient times had advanced aeronautical and military technology, is not unusual — see Dileep Kumar Kanjilal, *Vimana in Ancient India* (Calcutta: Sanskrit Pustak Bhandar, 1985) and Rishi Bharadvaja, *Vymaanika-Shaastra Aeronautics* (Mysore: Coronation Press, 1973). The latter book has a particularly peculiar premise and bizarre claims concerning these early stories of aerial vehicles. More modest claims for

While Soddy may not have been a Theosophist, the Theosophists, working from a long conceptual tradition of lost civilizations, brought his ideas along. In fact, we read in another Theosophical work:

> "The Knowledge" and "The Power" that Prof. Soddy suggests as having been possessed by "some forgotten race of men," *may never have passed from the Human Race* — and the day appears to be approaching when Mme. Blavatsky will be universally recognised as having been one of its custodians.[68]

In fact, not only Soddy, but another scientist who became engaged in occult pursuits, Sir Oliver Joseph Lodge, was also cited by the Theosophists in support of their idea of lost knowledge:

> It may appear strange, but is nevertheless true, that a number of statements regarding Man and nature made some years ago in the classic literature of Theosophy and Ancient Wisdom are now, year after year, being corroborated by science ... But if it is true that a large number of recent scientific discoveries have been anticipated in so many directions by the Ancient Wisdom, which Theosophy embodies; or, as Sir Oliver Lodge has put it, that modern science is rediscovering some of the truths of ancient science; or, again, in the words of Professor Soddy, that we are treading today the road which the ancients trod in the unrecorded history of the world, then there must be another method of investigation of which the Ancient Wisdom was the result, and it would be pertinent to inquire what that method is and who the persons are who use it.[69]

Lodge was not a surprising choice for the Theosophists. Although an esteemed physicist, he was interested in telepathy and other psychic phenomena. After

---

advanced technology in ancient India are made in Prashant Prabhaka Holay, *Machines in Samskrita Literature* (New Delhi: Samskrita Bharati, 2000). For a critique of these interpretations of ancient history in this region, see Sumathi Ramaswamy, *The Lost Land of Lemuria: Fabulous Geographies, Catastrophic Histories* (Berkeley: University of California Press, 2004).

68 See p. 60 of W.A. Dunn, "The Ancient and Modern Sciences of the Radioactive Transformations of Nature: A Vindication of the Teachings of H.P. Blavatsky — Part III)", *The Theosophical Path* 14.1 (January 1918): 53–61.

69 D.D. Kanga, *Where Theosophy and Science Meet: A Stimulus to Modern Thought — Part 1* (Adyar, Madras: Adyar Library Association, 1938), 1.

losing his son Raymond in World War I, Lodge turned to writing books on life after death and spiritualism.

Soddy himself, too, was not isolated from occult ideas:

> However secular his worldview was … Soddy's writings betray the influence of some major strands of the occult revival … some of his work contains speculations that locate him directly in a current of late-nineteenth-century occult thinking that posited a lost Atlantean civilization as the origin of much alchemical and other occult knowledge. Soddy incorporated elements of the occult narrative of Atlantis as a source of a secret tradition of advanced magical or scientific knowledge and the notion that the Atlanteans had brought about their own destruction through this knowledge.[70]

Both the Theosophists and Soddy used Atlantis to support their ideas about lost civilizations and the cyclical nature of human history. They took Atlantis as a historical reality, and took Plato's account as a solid basis for believing not only in a civilization that had disappeared in the vague recesses of the past, but as evidence for its advanced technical knowledge. Such a belief in Atlantis was held by other quite scientifically minded people in this period; in 1923, the pioneering radio inventor, Reginald A. Fessenden, published the first part of what would come to be a three-part book entitled *The Deluged Civilization of the Caucasus Isthmus*.[71]

Soddy was also connected to Theosophy in other ways — through his contemporaries in the scientific world. In 1921, he won the Nobel Prize in chemistry for his research into radioactive substances and his studies of isotopes; the very next year, Francis W. Aston won the Nobel Prize, also in chemistry. In 1919, Aston had invented the mass spectroscope; he had also determined that certain isotopes were not radioactive. Soddy knew Aston, and, in fact, cites a personal communication from him in his book *Science and Life*.[72] In that same book, Soddy again articulates his idea of lost civilizations at length:

> So long as the world was supposed to be six days older than man, and man a creature of the last 6000 years, the idea that we were "the first

70 Morrisson, 160–161.

71 Reginald A. Fessenden, *The Deluged Civilization of the Caucasus Isthmus* (Boston: T.J. Russell, 1923); for a discussion of this work and Fessenden's beliefs, see Helen M. Fessenden, *Fessenden: Builder of Tomorrows* (New York: Coward-McCann, 1940), 286–296.

72 Soddy, *Science and Life* (New York: E.P. Dutton, 1920), 142, n. 1.

> that ever burst" into the silent sea of science was pardonable enough. Possibly we were not. Just as no one would feel qualified to write a history of this country from materials gleaned from the newspapers of the present century, so no one ought to be so bold as to attempt to write a history of the human race from such written records as now exist, the most ancient of which go back to a time when the race was quite inappreciably younger than it is to-day. Neither is there any very valid ground for the belief that the startling advance civilisation had made in the past hundred or so years is in any way the climax or natural culmination of the slow and by no means even continuous progress previously. It seems rather a sudden forward leap apparently unconnected with and certainly not culminating necessarily out of the periodic ebb and flow of human fortune of which history tells. It is the work of a mere handful of men. The mass probably are little more scientific to-day than they were two thousand years ago, and this being the case, the advance does not appear to be the inauguration of the millennium, nor, indeed, of any other prolonged period of stable regime. Nothing but the most sublime egoism, the unconscious constitutional disability of the natural man to conceive of a universe not revolving around himself, can make it appear improbable that what occurred so suddenly and mysteriously in the past few centuries of recorded history may not have occurred before, not once but perhaps many times during the vastly longer period of which no record has yet been interpreted. It is only right to consider the possibility that the command exercised over Nature in the twentieth century may have been attained, possibly exceeded, previously.[73]

Aston, we know without doubt, was familiar with Theosophy. In fact, in a 1913 paper announcing a new scientific discovery, he cited a book entitled *Occult Chemistry: A Series of Clairvoyant Observations on the Chemical Elements* — a book written by two important British Theosophists, Annie Besant and Charles Leadbeater.[74] Soddy took Aston's research and integrated it into his own work on isotopes, so it is clear that he was aware of Aston's additional Theosophical inspirations.[75]

In fact, Soddy himself reaffirms the Theosophical concept of various rises and falls of civilizations through time, but without mentioning the

73 Ibid., 86–87.

74 See Hughes, 32–33.

75 Ibid., 34.

Theosophists by name. However, he alludes to the alchemical powers of a lost race, a Theosophical theme:[76]

> Science has reconstructed the story of the past as one of a continuous Ascent of Man to the present-day level of his powers. In face of the circumstantial evidence existing of this steady upward progress of the race, the traditional view of the Fall of Man from a higher former state has come to be more and more difficult to understand. From our new standpoint the two points of view are by no means so irreconcilable as they appeared. A race which could transmute matter would have little need to earn its bread by the sweat of its brow … The legend of the Fall of Man, possibly, may be all that has survived of such a time before, for some unknown reason, the whole world was plunged back again under the undisputed sway of Nature, to begin once more its upward toilsome journey through the ages.[77]

Although, as just pointed out, Soddy does not allude to the Theosophists directly, it is worth noting that Blavatsky cites a very similar passage in *The Secret Doctrine*. She highlights an excerpt from Louis Jacolliot's 1874 *Histoire des Vierges: Les Peuples et les Continents Disparus*:

> Whatever there may be in these traditions and whatever may have been the place where a civilization more ancient than that of Rome, of Greece, of Egypt, and of India, was developed, it is certain that this civilization did exist, and it is highly important for science to recover its traces, however feeble and fugitive they may be.[78]

The Theosophists used science to push forward their ideas, and Soddy and others, in turn, as scientists felt free to speculate on lost civilizations in their ruminations on the nature and transmission of advanced knowledge.

But an even closer relation between Soddy and the Theosophists is found in discussions surrounding an apocryphal work known as the *Book of Enoch*. The work is considered pseudepigraphical, since it is falsely attributed to Enoch, a character found in Genesis. There survive fragments in Aramaic from the Dead Sea Scrolls, as well as some passages in Greek sources, and "an Ethiopic

76 See Morrisson, 70.

77 Soddy, *The Interpretation of Radium*, 182–183.

78 Blavatsky, *The Secret Doctrine: The Synthesis of Science, Religion, and Philosophy*, vol. 2, *Anthropogenesis*, 223.

version", which is the most complete.[79] Dating the work is complicated; as one study notes,

> what the Book presents to the reader is a bizarre variety of often disparate and overlapping traditions, containing units of narrative and discourse which could be as early as the second century BC (the date assigned to the oldest portion of the Aramaic), side by side with redactional supplements etc. which are centuries later — each and all of them traceable to "authors" of different periods and persuasions.[80]

Moreover, there are actually two books attributed to Enoch. Various Jewish legends had arisen about this figure, and he was described has having special knowledge. A collection of stories comprises the book known as the *Ethiopic Enoch*, including a text about the Flood. The so-called *Slavonic Book of Enoch* or *Book of the Secrets of Enoch* recounts Enoch's travels in heaven.[81]

Works concerning Enoch became popular with practitioners of the occult such as Aleister Crowley, and, of course, Blavatsky. Blavatsky discusses Enoch in *The Secret Doctrine*.[82] In particular, she notes that Josephus talks about Enoch and says that this Old Testament figure "concealed his precious Rolls or Books under the pillars of Mercury or Seth".[83] Blavatsky outlines a further connection between Enoch and lost knowledge, and the "Great Deluge" that destroyed Atlantis.[84]

---

79 Matthew Black, *The Book of Enoch Or I Enoch: A New English Edition* (Leiden: Brill, 1985), 1.

80 Ibid., 8; also see the discussion of redactions in Kelley Coblentz Bautch, *A Study of the Geography of 1 Enoch 17–19: "No One Has Seen What I Have Seen"* (Leiden: Brill, 2003), 17 et ff.

81 See Michael A. Knibb, *The Ethiopic Book of Enoch: A New Edition in the Light of the Aramaic Dead Sea Fragments*, 2 vols. (Oxford: Clarendon, 1978), and Ephraim Isaac, "New Light upon the Book of Enoch from Newly-Found Ethiopic Mss", *Journal of the American Oriental Society* 103.2 (April–June 1983): 399–411. Also note Nathaniel Schmidt, "The Two Recensions of Slavonic Enoch", *Journal of the American Oriental Society*, 41 (1921): 307–312, and idem, "Traces of Early Acquaintance in Europe with the Book of Enoch", *Journal of the American Oriental Society* 42 (1922): 44–52.

82 Blavatsky, *The Secret Doctrine: The Synthesis of Science, Religion, and Philosophy*, vol. 2, *Anthropogenesis*, 557 et ff. A more modern — but also highly speculative — book again draws from the Enoch narrative; see Christopher Knight and Robert Lomas, *Uriel's Machine: The Ancient Origins of Science* (London: Arrow, 2000).

83 Blavatsky, *The Secret Doctrine: The Synthesis of Science, Religion, and Philosophy*, vol. 2, *Anthropogenesis*, 558.

84 Ibid., 562.

Perhaps Soddy was not a Theosophist, but he also drew from the *Book of Enoch*, and in a very similar way.[85] In his book *The Story of Atomic Energy*, Soddy looks at the *Book of Enoch* in a discussion of an ancient destruction of a human civilization, and "discusses the Book of Enoch to ascribe the destruction of mankind to ... forbidden alchemical knowledge".[86] Soddy's book notes that

> there is an ancient legend, before, and as the reason for, the Flood, the art of ennobling metals [i.e., alchemy] was brought down to earth by demons. Traces of this legend are to be found ... in a greatly expanded form ... in the Apocrypha, Book of Enoch, VI–IX.[87]

Despite such biblical references, Soddy always had science in mind. In fact, Soddy's interest in the apparent fall of past civilizations stemmed in part from his concern about his very own field of study, that is, atomic science. In a very prescient manner, Soddy had stated — as far back as 1915 — that, "atomic energy might be used destructively".[88] He also held a belief that there had been, in fact, "a former atomic age".[89]

Blavatsky left a legacy of mysterious writings and occult beliefs, and there are still various "schools" of Theosophy today.[90] But Soddy has slipped into obscurity — unjustly so. While he articulated some of the same peculiar ideas about lost civilizations, vanished peoples, and hidden knowledge, Soddy did so as part of an attempt at a larger understanding of human beings and their relationship to science and technology. As one commentator has put it:

---

85 See Morrisson, 162.

86 Ibid., 167.

87 Soddy, *The Story of Atomic Energy* (London: Nova Atlantis, 1949), 2. A very similar statement is found in Ernst von Meyer, *A History of Chemistry from Earliest Times to the Present Day*, trans. George McGowan (London: Macmillan, 1906), 26.

88 See p. 179 of Richard E. Sclove, "Alchemy to Atomic War: Frederick Soddy's 'Technology Assessment' of Atomic Energy, 1900–1915", *Science, Technology, & Human Values* 14.2 (Spring 1989): 163–194.

89 Ibid., 175.

90 Connections between Theosophy and science also still exist, with one physicist, Stephen Phillips, arguing for links between modern subatomic theories and the "findings" in Annie Besant and Charles W. Leadbeater, *Occult Chemistry: Clairvoyant Observations On the Chemical Elements* (London: The Theosophical Publishing Company, 1919): see Hughes, 35, as well as Morrisson, 92–95. The use of Soddy's text to support a belief in lost, advanced civilizations also continues: see Colin Wilson, *Atlantis and the Kingdom of the Neanderthals: 100,000 Years of Lost History* (Rochester, VT: Bear & Co., 2006), 292.

> [Soddy] used nonscientific knowledge to help establish a broader context within which his penetrating intuition and logic into the social implications of science could emerge … Soddy ably fulfilled the suggestion of psychologists and philosophers that we strive to balance dispassionate intellectual inquiry with periods of intense creativity and emotional engagement.[91]

From a modern perspective, any connection between science and the occult seems peculiar. But Soddy was interested in some of these marginal ideas because they gave scientists, and the wider public, some sense of the profundity of scientific and technological knowledge and the consequences of that knowledge.

In fact, Theosophy's rise in a period of intense scientific development — the late nineteenth century — is not as contradictory as it might seem. As cited earlier, one historian of science points out that Theosophy's "blend of esoteric wisdom and spiritual philosophy" was appealing because it gave people a way of "exploring and expressing hidden realities in an increasingly materialistic world without moral or spiritual values".[92] The goal of "exploring and expressing hidden realities in an increasingly materialistic world" is not without merit. But in our current period in history, we are wary of any attempt to bring the language of science into the realms of non-scientific thinking.

6 Speculations and Methods

The "grandfather" of the more recent speculations about vanished civilizations was Immanuel Velikovsky. In his 1950 book, *Worlds in Collision*, he introduced the idea of catastrophes that had occurred in the remote past. In addition, he argued that particular catastrophes had occurred in the course of human history, so dreadful that they had destroyed existing civilizations.[93] Velikovsky also sought to explain parallel myths of widespread or global catastrophes by seeking actual historical events as origins for these tales. Velikovsky was highly educated, and worked as a psychiatrist. He was well-read, and his speculations on catastrophe and history were, in part, an outgrowth of his extensive readings

91 Sclove, 181.

92 Hughes, 33.

93 Immanuel Velikovsky, *Worlds in Collision* (Garden City, NY: Doubleday, 1950); for a thoughtful analysis of both Velikovsky's ideas and their — mostly hostile — reception by the scientific community, see Michael D. Gordin, *The Pseudoscience Wars: Immanuel Velikovsky and the Birth of the Modern Fringe* (Chicago: University of Chicago Press, 2012).

of Freud's works on religion and the history of myth. However, Velikovsky's understanding of cosmology — critical to his arguments about catastrophic destruction of civilization — was inaccurate, and subject to extensive criticism. Later, the author Zecharia Sitchin carried on with Velikovsky's assumption that ancient myths could be read at face value as historical accounts of early events — and particularly, catastrophic events.[94]

All the works of such writers can be categorized as conjecture and speculation — and in some cases, as science fiction and fantasy. The only service, perhaps, that they have provided is to get readers interested in history, and to prod a few historians to examine the evidence of ancient texts again. If historians, and especially historians of science and technology, fail to include in their studies some investigation of these accounts of lost technologies and lost civilizations — non-academic as the subject may seem — then these speculative 'researchers' will come rushing in. The evidence for this is clear, since in the past two decades, there has been a flurry of poorly-researched books (new and republished) espousing the idea of lost advanced civilizations: the re-issue of Rene Noorbergen's *Secrets of the Lost Races: New Discoveries of Advanced Technology in Ancient Civilizations*, Graham Hancock's *Underworld: The Mysterious Origins of Civilization*, David Hatcher Childress' *Technology of the Gods*, Jonathon Gray's *Dead Men's Secrets: Tantalising Hints of a Lost Super Race*, Christopher Knight and Alan Butler's *Civilization One: The World is Not as You Thought It Was*, and so on.[95] The historian has a role in applying proper research techniques to these matters, and the result can be both good scholarship and good reading. Back in 1969, the historian Robin W. Winks presented such a goal in his book *The Historian as Detective: Essays on Evidence*, "What I have tried to explore in this anthology is the relationship between the colorful world of fictional intrigue … and the good gray world of professional scholarship, and to show how much of the excitement, the joy, and even the color properly belongs to the latter."[96]

---

94 See, for example, Zecharia Sitchin, *The 12th Planet* (New York: Avon, 1976), and idem, *The Wars of Gods and Men* (New York: Avon, 1985); Sitchin has written a number of other books on his concept of an advanced civilization establishing our own civilization in the prehistoric past. For a precise, engaging critique of Velikovsky, see C. Leroy Ellenberger, "A Lesson from Velikovsky", *Skeptical Inquirer* 10.4 (Summer 1986): 380–381, as well as the more extensive work by Gordin, cited above.

95 A good synopsis of this concept of the possible disappearance of ancient advanced civilizations may be found in Garrett G. Fagan and Chris Hale, "The New Atlantis and the Dangers of Pseudohistory", *Skeptic* 9.1 (Spring 2001): 78–87.

96 Robin W. Winks, ed., *The Historian as Detective: Essays on Evidence* (New York: Harper & Row, 1970), xiii.

Some researchers have made attempts to show what earlier civilizations might have been able to accomplish given the technology that they actually possessed. Thor Heyerdahl worked hard to prove his conjectures, and made noble efforts to get historians to think about, for example, the ways that populations might have moved across the vast distances of the Pacific Ocean in ancient times.[97] But Heyerdahl proved only that ancient peoples *could* have crossed the South Pacific with their existing technology — he did not prove that they actually *did*.

Other researchers have penned books that soberly examine ancient technologies. Some works, such as Dick Teresi's *Lost Discoveries: The Ancient Roots of Modern Science*, provide a survey of the relative sophistication of the technological achievements of the past. Others, including K.D. White's *Greek and Roman Technology*, the more recent *Technology and Culture in Greek and Roman Antiquity* by Serafina Cuomo, and Joseph Needham's extensive *Science and Civilisation in China*, serve as solid reference works in the field of ancient science and technology, with research established through careful textual and archaeological investigation. But even these works do not engage directly some of the more peculiar accounts found in ancient texts. Needham talks about early Chinese tales of flight, for example — a topic this book takes on here — but he does not engage in an in-depth interpretation of what the writers of those accounts might have been talking about beyond possible allusions to kites.

In a discussion concerning the dangers of pseudoarchaeology, the historian Garrett G. Fagan notes that archaeology has a very specific and well-established method for looking at evidence. Therefore, he says, a "skeptical approach to new material is entirely appropriate. Challenges to established knowledge have to demonstrate their cogency before they convince people".[98] In the case of this book, the early texts that are presented here may be considered a form of archaeological evidence; however, it should be noted that no particular "challenge" is being presented to established norms. Rather, this book is a call to examine evidence that is often overlooked, or evidence that been examined in only one context. For example, the fact that Ainu folk tales describe mechanical flight in a way that at times is similar to ancient South

---

97 Note the popular Thor Heyerdahl, *Kon-Tiki: Across the Pacific by Raft*, trans. F.H. Lyon (Chicago: Rand McNally, 1950), and idem, *The Ra Expeditions*, trans. Patricia Crampton (Garden City, NY: Doubleday, 1971); also see idem, *Early Man and the Ocean: A Search for the Beginnings of Navigation and Seaborne Civilizations* (Garden City, NY: Doubleday, 1979).

98 See p. 26 of Fagan, "Diagnosing Pseudoarchaeology", in idem, ed., *Archaeological Fantasies* (New York: Routledge, 2006), 23–46.

Asian narratives is not something that has been studied by historians of technology. Generally, academic studies of historical chronicles, myths, and other early source materials, do not tend to focus on the discussion of technological devices that appear in those materials. More profoundly, modern studies — even when they do note the mention of such technologies — do not investigate the deeper contexts in which such mentions are often embedded. For example, as this book shows, mentions of mechanical devices in early sources at times connect those devices to moral corruption and cultural decline.

Such examinations of textual evidence are not always easy. As Fagan points out in reference to archaeological evidence, context plays a vital role, such that "the codified techniques of excavation exist primarily to established as firmly as possible the context of newly revealed evidence."[99] One of his criticisms of pseudoarchaeology, in fact, is the dismissal of context. He notes two particular problems with pseudoarchaeology:

> First, its practitioners adopt an inflexible stance toward the evidence, which is to be shoehorned into their pet conclusions at all costs. Second, pseudoarchaeologists are practically contemptuous of context. It is not unusual for entire monuments, artistic motifs, or myth cycles to be compared with each other (usually in literal-minded fashion) and far-reaching conclusions reached without any regard for their respective contexts.[100]

These are all valid criticisms, and Fagan goes on to critique the approach of writers such as Hancock, mentioned earlier here. One can again take Fagan's discussion of archaeology and apply it to the texts of early writers. Context is very important, but it should be noted that in this case, unlike some archaeological evidence, written works have more complex problems in terms of interpretation: unlike a stone monument, a text can become corrupted in a highly convoluted manner over time, and is subject not only to multiple re-writings and multiple re-interpretations, but also the vagaries of the creative process and even the potential outright mendacity of its writer. Of course, archaeological evidence also gets eroded and rebuilt, but physical materials from the past can be examined chemically and with radiographic methods. Manuscripts can be examined in this way, certainly, but the actual content of the text — its intent, its *meaning* — can be very remote to the modern reader, despite the availability of translations, the methods provided by comparative literature, and the accessibility of commentaries.

99 Ibid., 26.
100 Ibid., 27.

But to Fagan's points: this book certainly does not start with any fixed conclusion other than the idea that early sources contain a number of intriguing discussions of technological knowledge and associated themes, such as civilizational decline. There is no attempt to force the evidence into a "pet conclusion" here. However, this book urges a more perspicacious approach to early sources — that is, it assumes that early writers were no less thoughtful than our present ones, and that they were deeply engaged with questions concerning technology, knowledge, and the path of civilizational development. Their methods, and their way of writing about these themes, were certainly different from modern approaches, of course: it is our job to understand in what way.

In terms of Fagan's second point, then, this book talks less about historical context as it is usually considered in academic studies. Instead, "context" here means how a writer was thinking: when a Chinese chronicler wrote about a mirror with the apparent ability to see into the human body, the contextual examination is about Chinese understanding of mirrors generally and what this particular writer might have known about them. In the case of something like an Ainu folktale, or Hopi discussions of peculiar "flying shields", the context is less certain, since the mode of transmission was oral, and the general conception of technological devices held by those cultures is poorly understood. Finally, although Fagan points out that it is not sound method simply to compare various bits of evidence, in the case of texts it can be useful to see cultures that are widely separated in time and location grappling with the same technological conceptions. The sheer universality of ideas about mechanized flight, as well as themes of technological overreach — stories such as the "City of Brass" and Atlantis, for example — tell us something about the past and preoccupations of the human intellect in these earlier periods. Given this framework, and the appropriate caveats, the reader here can now turn to a selection of early texts, and see what kinds of details the authors themselves recount, and some possible interpretations of these curious descriptions.

CHAPTER 2

Ancient Tales of Flying Machines

1 Two Types of Tales

It may surprise a modern reader that even many centuries ago, there existed a variety of tales about flight and flying machines. In his lengthy essay, *The Prehistory of Aviation*, the anthropologist and sinologist Berthold Laufer surveyed a selection of such technological tales.[1] Laufer attempted to interpret the meaning of some those stories of flying machines; writing about those from South Asia, for example, he commented as follows:

> Did the ancient Indians really navigate the air? ... To my way of thinking this point is irrelevant. The main point is: they had the idea ... The Indians saw two points clearly — that aircraft must operate on the principle of the flight of birds and that a mechanism is required to start the machine, to keep it in mid-air, and to make a descent.[2]

This is a key observation for our discussion of the taxonomies of these texts: the fact that a wide variety of sources — from folktales to romances — employed a mechanical device in their narratives, and at times even included an attempt at a technical description of the machine's construction and operation. This chapter presents those sources, including several not cited by Laufer, and examines how the texts — and in the case of the Ainu and the Hopi, for example, oral traditions — treated the subject of something as technologically sophisticated as mechanized flight.

The history of technology in the West has often been built upon the methodical examination of inventions and inventors, including such figures as Leonardo da Vinci, the Wright brothers, and so on. But technological knowledge exists outside this realm, and in particular outside the realm of Western culture and the frequently studied period of history from the Renaissance to the present. K.D. White, in his famous *Greek and Roman Technology*, showed how a great deal of technological innovation took place in the classical world.[3] This shows, then, that technology can exist in a culture even when it is not considered

1 Berthold Laufer, *The Prehistory of Aviation* (Chicago: Field Museum of Natural History, 1928).
2 Ibid., 51.
3 K.D. White, *Greek and Roman Technology* (Ithaca: Cornell University Press, 1984).

 | DOI:10.1163/9789004352728_003

by that culture explicitly as a separate field of knowledge. In the modern world, technology is a distinct field or set of fields, but such an epistemological rendering is an artifact of our present age, reinforced by our particular belief in the central importance of technology. In other cultures, technology existed, but it was either relegated to a lower status, or bound up in other cultural contexts — everything from folktales to philosophical treatises.

Joseph Needham, whose works are cited a number of times in this book, devoted his scholarly career to studying the wide range of scientific and technological knowledge found in early Chinese sources. There was a great deal of technological and mechanical thinking in "pre-modern" cultures, and one can piece together some of this knowledge through the examination of texts from a variety of ancient sources. It is important, moreover, to reiterate that the stories recounted here in relation to flight make a clear distinction between *individual entities* who flew under their own power — such as the Chinese *fei tian* (飛天), or "flying beings" — and *mechanical devices* that flew, marked by a different term in Chinese: *fei che* (飛車), that is, "flying vehicles".

This same differentiation is found not only in Chinese sources, but also, for example, in stories of flight among peoples of the Massim archipelago in the western Pacific Ocean. In some of those stories, there is a clear cultural and narrative differentiation between flying witches and flying canoes:

> [T]here is a certain inversion in the comparison between flying witch and flying canoes. Being women, the flying witches cannot build canoes or sail them; but they transform themselves and fly through the air which men cannot do. Thus the man's canoe magic is an exteriorized activity performed on a manufactured object.[4]

The two types of flight here are distinguished by cultural gender roles, but also by the fact that in the latter case there is a vehicle — a "manufactured object". This is a difference found in other sources, as well: in a discussion on the theme

4 See p. 194 of S.J. Tambiah, "On Flying Witches and Flying Canoes: The Coding of Male and Female Values", in Jerry W. Leach and Edmund Leach, eds., *The Kula: New Perspectives on Massim Exchange* (Cambridge: Cambridge University Press, 1983), 171–200; also see the early research on these tales in Bronislaw Malinowski, *Archipelagoes of Melanesian New Guinea* (London: Routledge, 2002), 134 et ff. In addition, note the critique in J.P. Singh Uberoi, *Politics of the Kula Ring: An Analysis of the Findings of Bronislaw Malinowski* (Manchester: Manchester University Press, 1971), 77 et ff., and in Richard Joyce, *The Myth of Morality* (London: Cambridge University Press, 2003), 234.

of flight in *The Arabian Nights*, one commentator notes that the stories' "mechanical flying horses are. … non-sentient" and so "distinct from Pegasus".[5]

Certainly, much remains obscure. The discussion of early flying vehicles by the thirteenth-century philosopher and writer on science, Roger Bacon, leaves us with a frustratingly incomplete comment as to his sources. In his mid-thirteenth century[6] *De mirabili potestate artis et naturae*, after discussing devices for flight and other machines, Bacon states:

> These devices have been made in antiquity and in our own time, and they are certain. I am acquainted with them explicitly, except with the instrument for flying, which I have not seen. And I know no one who has seen it. But I know a wise man who has thought out the artifice.[7]

Note that the term Bacon uses for an aerial vehicle here is "instrument for flying" — *instrumentum volandi* in the original Latin. Bacon seems to have made one of the earliest allusions in Europe to a flying machine. He wrote in Latin, but his text led to another quite early term for such a device in English. As one author notes: "Among all the nouns that have been applied to aircraft in English, *instrument* seems to have been the earliest. The first occurrence I have seen of it is in a Renaissance translation of Roger Bacon."[8] In the first known English translation of Bacon's *De mirabili potestate artis et naturae*, the 1597 *Mirror of Alchemy*, the passage on flight is rendered as follows:

> There may bee made instruments of Nauigation without men to rowe in them … yea instruments to flie withall, so that one sitting in the middle of the Instrument, and turning about an Engine, by which the winges being artificially composed may beate the ayre after the maner of a flying bird … And it is certaine that there is an instrument to flie with, which I

5 Marina Warner, *Stranger Magic: Charmed States and the Arabian Nights* (Cambridge, MA: Harvard University Press, 2012), 75–76.

6 The exact date of composition is uncertain.

7 Roger Bacon, *Roger Bacon's Letter Concerning the Marvelous Power of Art and of Nature and Concerning the Nullity of Magic*, trans. Tenney L. Davis (Easton, PA: The Chemical Publishing Company, 1923), 27.

8 Svante Stubelius, *Balloon, Flying-Machine, Helicopter: Further Studies in the History of Terms for Aircraft in English* (Göteborg: Almqvist & Wiksell, 1960), 5; also see idem, *Aeroplane, Aircraft: Studies in the History of Terms for Aircraft in English* (Göteborg: Almqvist & Wiksell, 1958), 16–17 and 63–66.

> neuer sawe, nor know any man that hath seene it, but I full wel know by name the learned man that inuented the same.[9]

Unfortunately, Bacon does not give us any further details. Moreover, the description is unrealistic from a modern perspective, in that the artificial imitation of the beating of a bird's wings is not a good solution to the challenge of manned flight, as Leonardo da Vinci discovered. Regardless, it is interesting that just as in the much earlier accounts of flight described later in this chapter, the *mechanical* nature of the device is clear in this description, from the use of the term "instrument" to the mention of an "engine".

Nowhere else in his writings does Bacon tell us where he has received his information on flying vehicles. This chapter, however, attempts to provide a selection of tales of flight, to demonstrate the wide range of cultures that were interested in the concept of mechanical flight, and how each culture framed the discussion of the vehicles necessary for such a technological endeavor. While it is unlikely that the Chinese, Pacific Islanders, or any other early society witnessed an "instrument for flying", as Bacon puts it, it is surprising to note the number of discussions of the idea in so many different cultures.

A number of early Chinese and Korean writings include references to flying machines of various types. The texts speak of certain groups or individuals who constructed devices so as to travel great distances through the air. The Chinese accounts of *fei che* — again, a term explicitly meaning "flying vehicles" — are especially interesting for the historian of technology, as they provide a fascinating insight into early technological concepts and ambitions. These texts prompt us to create a new category that could be termed "mechanical mythologies": this is because the descriptions are distinctly mechanical or technological in nature, while at the same time possessing a mythical quality in the sense of straining our belief about what was mechanically possible in the past, or in the sense of myth not usually having the language to talk about technical subjects and thus rendering the details rather obscurely.

There are two basic story types related to early discussions of flight. The first type involves humans with magical powers, special anthropomorphic entities, or other creatures — for example, dragons — that were capable of flight on their own.[10] In Chinese culture, for example, popular tales often tell of

---

9 Roger Bacon, *The Mirror of Alchemy*, 64–65, reproduced in J.E. Hodgson, *The History of Aeronautics in Great Britain, from the Earliest Times to the Latter Half of the Nineteenth Century* (London, Oxford University Press, 1924), fig. 2; also see Stubelius, *Balloon, Flying Machine, Helicopter*, 5.

10 For a discussion and analysis of this kind of flight, see Mircea Eliade, "Symbolisme du 'Vol magique'", *Numen* 3.1 (January 1956): 1–13.

adepts, divine entities, or other beings flying through the air. The *xian* (仙) were immortals, for example, who were capable of flight under their own divine power. They were said to be feathered, and a term that has been used for Daoist priests is *yu ke* (羽客), meaning "feathered guest."[11] The term *fei tian* (飛天), which again can mean a kind of "flying being", also appears in early tales.[12] These *fei tian* are zoomorphic, and might refer to technology only insofar as creatures in nature — such as birds — inspired the idea that humans might fly given the "technology" of feathers and wings.

The second type of story is one that has received far less attention, since its apparent technological nature has fallen outside the realm of folklorists and historians. That type comprises stories or accounts that explicitly describe mechanical devices for flying, devices in which a person operates as a pilot and does not fly under their own power with wings or by way of magic. Flight, of course, was for millennia the object of fancy and speculation by writers, and the ambition of inventors in a variety of cultures. The texts presented here show that early writers in different cultures were interested in these more specific discussions of *mechanical* flight, however ambiguous the resulting descriptions might be.

Some work in this area has been done before, but there has been no unified presentation of these texts on flying vehicles. Needham, in his germinal work, *Science and Civilisation in China*, pointed out the need to look at several religious mythologies of flight as a background for some of these stories.[13] Prior to Needham, Laufer presented a selection of ancient stories concerning mechanical flight in his essay, *The Prehistory of Aviation*, cited earlier. Needham believed that the importance of the fanciful tales of flying beings in religious context was simply "that it put ideas into people's heads; ideas (for example) of aerial cars and their makers."[14]

Other factors, such as the development of kites, notes Needham, may also have had a general influence on Chinese ideas on flight. However, the texts that are presented here involve neither flying deities nor kite technology. Rather, the texts discuss vehicles that are constructed specifically for manned flight,

---

11 See Livia Kohn, ed., *Daoism Handbook* (Leiden: Brill, 2000), 641.

12 Note Haiwang Yuan, *The Magic Lotus Lantern and Other Tales from the Han Chinese* (Westport, CT: Libraries Unlimited, 2006), 4 and 217; for a slightly different use of this term, see Kohn, 221. For examples of *fei tian* in art, see Shu-Hong Chang and Cheng-Xian Li, *The Flying Devis of Dunhuang* (Beijing: China Travel and Tourism Press, 1980).

13 See Joseph Needham, *Science and Civilisation in China, Vol. 4: Physics and Physical Technology, Part II: Mechanical Engineering* (Cambridge: Cambridge University Press, 1965), 568–570; this volume hereafter is referred to as "Needham, 4.2".

14 Needham, 4.2, 569–570.

using particular technologies for motive power. Again, such texts represent a distinct category of early knowledge and speculation, one that deals with the mechanical rather than the spiritual or fantastic.

This chapter looks at stories about flying vehicles from a number of different cultures, presenting tales from Chinese, Korean, and Indian sources, as well as stories from the cultures of the Ainu, the Hopi, and the peoples of Oceania. In addition, it examines the traditions of "flying horse" tales, which reveal something about both sharing concepts across cultures, and the framing of technological concepts in stories. Finally, it concludes with a look at what kind of language is used in early describes to articulate the — distinctly modern? — idea of a vehicle constructed for flight.

2 Chinese Stories of Flying Machines

The most detailed story of flight in early Chinese sources involves a people called the Qi Gong (奇肱). The account is found in several sources in various versions. The earliest source is a third-century CE work entitled *Bowuzhi* (博物志 "Record of Investigation of Things"), attributed to Zhang Hua (張華). The Qi Gong people themselves, however, are mentioned in the Chinese compendium of myths and tales of uncertain date, the *Shan Hai Jing* (山海經 "Classic of the Mountains and Seas").[15] There they are described as having only one arm, and the characters 奇肱 literally can be translated as "one arm" or "strange arm". This is the story in the *Bowuzhi* concerning them and flight:

> The Chi Kung [Qi Gong] people were good at making mechanical devices for killing birds. They could also make aerial carriages that, with a fair wind, traveled great distances. In the time of the emperor Thang [Tang],[16] a westerly wind carried such a car as far as Yüchow [Yuzhou], whereupon Thang [Tang] had the car taken to pieces, not wishing his own people to see it. Ten years later there came an easterly wind (of sufficient strength), and then the car was reassembled and the visitors were sent back to their own country, which lays 40,000 *li* beyond the Jade Gate.[17]

15 Ibid., 570. A rough date for the composition of the *Shan Hai Jing* (which had multiple authors) may be the 3rd century BCE; see Michael Loewe, ed., *Early Chinese Texts: A Bibliographical Guide* (Berkeley: Society for the Study of Early China and Institute of East Asian Studies, University of California, Berkeley, 1993), 359–60.

16 This is Cheng Tang (成湯), founder of the Shang dynasty, in the 17th century BCE.

17 This translation is from Needham, 4.2, 570.

In the *Diwang Shiji* (帝王世紀 "Annals of the Emperors and Kings") of Huangfu Mi (皇甫謐), dating from roughly the same period, the story is the quite similar. However, in that version Qi Gong is said to be an individual, not a people as in Zhang Hua's account.[18] The claim in the passage above as to the distance travelled is rather remarkable: traditionally, a *li* can be considered approximately equal to half a kilometer, which means that this "aerial carriage" flew roughly 20,000 kilometers. Certainly, there must be some corruption in the text, or the passage is simply spinning a fantastic — if mechanically contextualized — tale.

The story is repeated in other works through the fifth and sixth century CE.[19] A look at the sixth-century CE *Shuyiji* (述異記 "Records of Strange Things"), a work attributed to Ren Fang (任昉), shows how the basic story remains the same, despite the passage of time and the inevitable narrative changes:

> The Chi Kung [Qi Gong] people were skilled, and able to make flying vehicles which used the wind to fly. In the time of Emperor Thang [Tang], a west wind blew the Chi Kung [Qi Gong] people's vehicle east to the border of Yüchow [Yuzhou]. Then ten years later, a wind let them return home to their country, which is 40,000 *li* from the Jade Gate.[20]

Needham does not put forward any speculations as to the meaning of this tale. Laufer, however, provides some conjectures, stating: "I imagine, therefore, that the Ki-kung's [Qi Gong's] 'flying chariot' was built on the aerostatic principle, being driven by a combination of sails and kites."[21] He bases this conclusion solely on the proficiency of the Chinese in constructing kites; however, the text itself provides no clues as to the details of the vehicle's design or configuration.

Laufer believed that the story is not simply a fanciful tale, but rather, a record — albeit a distorted one — of an actual manned flight by the early Chinese. That conjecture aside, what is clear is that the story appears in a distinctly technological context. In the first version, the sentence preceding the discussion of the aerial vehicle states, "The Chi Kung people were good at

18 Also see the brief mention of the Qi Gong on p. 214 of Joachim Kurtz, "Silken Strands: Making Technology Work in China", in Dagmar Schäfer, ed., *Cultures of Knowledge: Technology in Chinese History* (Leiden: Brill, 2012), 45–74.

19 Needham, 4.2, 570.

20 Fang Ren, *Shuyiji* (述異記 "Records of Strange Things") (Taipei: Taiwan shang wu yin shu guan, 1983), bk. 2, ch. 6., my translation; for more details on this work, see Robert Ford Campany, *Strange Writing: Anomaly Accounts in Early Medieval China* (Albany: State University of New York Press, 1996), 83–85.

21 Laufer, *The Prehistory of Aviation*, 21.

making mechanical devices for killing birds." Perhaps this implies that these people were keen observers of birds and interested, therefore, in the mechanics of flight. The introduction to the story is about mechanical methods, and there is no mention of magical powers, deities, or fanciful means of flight. The passage also strikes the reader as technologically focused in that it deals with the reassembly of the vehicle.

In the *Baopuzi Neipian* (抱朴子内篇 "Master Embracing Simplicity — Inner Chapters") by Ge Hong (葛洪), we find both a technical description of a flying vehicle, and some comments on the nature of the atmosphere. Ge Hong (283–343 CE or 284–363 CE) was a noted Daoist scholar who also wrote on alchemical subjects.[22] Concerning flight, he states the following:

> Some use the inner part of a jujube tree [= Chinese date tree] to make a flying vehicle, using ox leather straps fastened to encircling blades, so as to propel this machine. Some others have the idea of making five snakes, six dragons, and three oxen [this refers to kites in the shape of these animals] encounter the "hard wind" (罡風) and so ride it, rising up to 40 *li*. [That region] is called the *tai qing* (太清). In the *tai qing*, the air is very hard, and can lift people. The Master [it is not clear to whom this refers] says that a hawk flying, spiraling higher and higher, only needs to straighten out its two wings and not flap them any more to move forward, because it is riding on the hard wind. The dragons, when they first rise up, step on the clouds, going to 40 *li*, then fly by themselves. This account comes from the [Daoist] adepts, and is recounted, being handed down to ordinary people. Yet these common people are not really able to understand it.[23]

It might be excessive to assume from this passage that early Chinese thinkers had a clear knowledge of the nature of the upper atmosphere. In fact, the way the atmosphere is described in this passage seems incorrect, for in fact one could argue that the atmosphere becomes "softer" — that is, thinner — rather

22 This translation is a slight reworking of that found in Needham, 4.2, 582–583. A translation of this work is also in Hong Ge and James R. Ware, *Alchemy, Medicine, and Religion in the China of AD 320, the Nei P'ien of Ko Hung* (*Pao-p'u-tzu*) (Cambridge, MA: MIT Press, 1966). Concerning the *Baopuzi Neipian* generally, note Nathan Sivin, "On the Pao p'u tzu nei p'ien and the Life of Ko Hung (283–343)", *Isis* 60.3 (Autumn 1969): 388–391, and Jay Sailey, *The Master Who Embraces Simplicity: A Study of the Philosopher Ko Hung, AD 283–343* (San Francisco: Chinese Materials Center, 1978).

23 Hong Ge, *Baopuzi Neipian* (抱朴子内篇 "Master Embracing Simplicity — Inner Chapters") (Taipei: Taiwan shang wu yin shu guan, 1983), ch. 15, author's translation.

than "harder" as one goes up to higher altitudes. However, it may also be that the "hard wind" simply means a kind of jet stream — that is, a wind of constant velocity that may be encountered at a certain altitudes. Most importantly, the discussion here still shows that the writer had a keen interest in the nature of the atmosphere, and how that might affect flying techniques.

The observation that flying animals do indeed stop beating their wings at times and just glide is accurate, although this behavior is not directly connected with altitude. Needham states that the mention of "hard wind" is evidence of early knowledge of the principle of the airfoil:

> [W]hat is to be said of Ko Hung's [Ge Hong's] "hard wind"? From the examples he gives of the gliding and soaring of birds, it is obviously nothing else that the property of "air-lift", the bearing or rising of the inclined aerofoil subjected to the forces of an airstream, whether natural or artificial.[24]

However, a close look at the original passage only reveals an understanding of the fact that birds can glide; there is no evidence of any understanding as to *why* in terms of basic aerodynamics. The text states simply that, "a hawk flying, spiraling higher and higher, only needs to straighten out its two wings and not flap them any more to move forward, because it is riding on the hard wind." Moreover, this is followed by a statement about dragons and how they fly: "they first rise up, step on the clouds, going to 40 *li*, then fly by themselves." For the modern reader, this is a difficult passage to critique, because it contains at the same time fantastic elements (dragons), natural but not fully explained elements (the flight of birds), and mechanical elements. The epistemological framework is quite different from that in modern texts, where the differentiation between the real and the fantastical — and certainly, the mechanical and the fantastical — is more clearly marked.

Open to speculation and interpretation is the beginning of the passage, with its interesting mechanical description. The other sections of the passage can be traced in some measure to Daoist ideas of the cosmos, but this first section cannot. Needham perhaps puts too strong an interpretation to the account, saying:

> There can be no doubt that the first plan which Ko Hung [Ge Hong] proposes for flight is the helicopter top; returning or revolving blades

24 Needham, 4.2, 590.

> can hardly mean anything else, especially in close association with a belt or strap.[25]

Yet it is not at all clear that this is the case, for the manner in which the vehicle flies is not stated in the passage. In addition, it is not evident that we are dealing here with "returning or revolving blades". Another modern scholar, Yan Hongsen, similarly has tried to reconstruct the device described, even drawing a schematic.[26] However, that drawing shows what is essentially a toy with a spinning, bladed rotor. The information in the original text really does not suffice for such a complete rendering. The original text also does not examine frame the idea of mechanical flight in terms of any kind of "scaling up" from such a toy to a proper flying machine.

The original Chinese text that opens the passage above from the *Baopuzi Neipian* is 牛革結環劍, a phrase that does not reveal a precise picture of the device being described. The term 牛革 means "ox leather," and the third character, 結, means "tied" or "bound". The term 環 may mean "around", "round", or even "ring", while the final character, 劍, literally means a "sword" or "dagger". Thus the image that the reader gets of the device is quite ambiguous, with ox leather straps somehow tied to a circle of blades, perhaps.[27] The motion of these blades, though, is never described explicitly in Ge Hong's passage, nor is the overall configuration of the vehicle. In addition, there is no discussion of motive power. We are left with the nonetheless interesting fact that the text explicitly refers to a mechanical device intended for flying.

Both Needham and Yan attempt to draw a parallel between the device mentioned in this passage and a device known as a "Chinese top", which is a rotary flying toy. But the connection seems valid only in an informal sense. The *Baopuzi Neipian* passage discusses the machine made of jujube wood and leather straps in the context of a discussion on reaching high altitudes; the toy Chinese top could never be employed in this manner. In addition, as Needham

---

25 Ibid., 572.

26 See Yan Hongsen, *Reconstruction Designs of Lost Ancient Chinese Machinery* (Dordrecht: Springer, 2007), 49–50.

27 On the question of the mechanism described here, see p. 164 of Nicole Halsberghe, "Introduction and Development of the Screw in Seventeenth-Century China: Theoretical Explanations and Practical Applications by Ferdinand Verbiest", *East Asian Science, Technology, and Medicine* 34 (2011): 163–193, as well as the discussions in Zhenduo Wang, 葛洪《抱朴子》中飛車的複原 ["The Reconstruction of Flying Vehicles from Ge Hong's *Baopuzi*"], *Zhongguo Lishi Bowuguan Guankan* (中國歷史博物館館刊 "Journal of the Museum of Chinese History") 6 (1984): 48–51, and Xianzhou Liu, *Zhongguo Jixie Gongcheng Faming Shi* (中國機械工程發明史 "The History of Inventions in Chinese Mechanical Engineering") (Beijing: Kexue chubanshe, 1962).

himself notes, we do not know the exact provenance of the toy, with the earliest example of it dating from the late medieval period in Europe.[28] The point, then, is that we can not stray too far from the texts themselves in trying to discern what such passages are telling us about flying vehicles. It is sufficiently intriguing to note that the Chinese texts go beyond simple magical elements, and in fact discuss mechanisms for achieving flight. In addition, the *Baopuzi Neipian* here again presents the theme of technological knowledge being hidden or obscure: "This account comes from the [Daoist] adepts, and is recounted, being handed down to ordinary people. Yet these common people are not really able to understand it."

What sets the Qi Gong story and the note about a device with "encircling blades" found in the *Baopuzi Neipian* apart is the fact that those discussions clearly convey the idea that a person or people *constructed a device for flight*. Again, this is quite different from a simple tale of a person being able to float or fly through the air by means of magical power, the use of a magical object, or in a cart drawn through the air by flying creatures.

As Needham puts it: "Flying cars drawn by birds, griffons, or dragons were a separate tradition."[29] That kind of aerial imagery is found, for example, in the famous poetic work *Qiu Ge* (九歌 "Nine Songs") by Qu Yuan (屈原) who lived ca. 340–278 BCE:

> He is going to rest in the House of Life.
> His brightness is like that of the sun and moon.
> In his dragon chariot, dressed in imperial splendor.
> Now he flies off to wander round the sky.
>
> The god had just descended in bright majesty,
> When off in a whirl he soared again, far into the clouds.
> He looks down on Jizhou and the lands beyond it;
> There is no place in the world that he does not pass over.
> Thinking of that lord makes me sigh
> And afflicts my heart with a grievous longing …
>
> North I go, drawn by my flying dragon,
> Steering my course to the Dongting lake:
> My sail is of fig-leaves, melilotus my rigging,
> And iris my flag-pole, my banner of orchids …

---

28 Needham, 4.2, 583, and note *b*.

29 Needham, 4.2, 572.

> The stream runs fast through the stony shallows,
> And my flying dragon wings swiftly above it …[30]
>
> He drives his dragon chariot with thunder of wheels;
> High up he rides, careering heavenwards.
> But I stand where I am, twisting a spray of cassia …[31]
>
> I ride a dragon car and chariot on the thunder,
> with cloud-banners fluttering in the wind.
> I heave a long sigh as I start the ascent.
> Reluctant to leave and looking back longingly.[32]

These are lovely lines, emphasizing the concept of flight through aerial images of the landscape. But while here flight is achieved with dragons, in the Qi Gong story one needs no such creatures to become airborne — the only power invoked is the wind.

The Qi Gong account can also be compared with another discussion of flight found in the sixth-century CE *Shuyiji*, cited earlier. Indeed, the Qi Gong story that appears in the *Shuyiji* is set alongside what seems to the modern reader to be a very fanciful kind of tale, which also concerns flying:

> Beyond the western sea there is the country of the swan, where the people are only 7 *cun* [about 18 centimeters] in height. In a day, they can travel 1000 *li*. Most animals do not attack them. These people only fear the sea swan [?]. If this bird sees these men, it will definitely swallow them. If the man does not perish in the bird's stomach, then he can fly inside the bird for a 1000 *li*.[33]

This story has many fantastic and peculiar elements, but most important to note is that it does not mention a mechanism or device.

However, another very brief tale of a device constructed for flight is found in the same work:

---

30 John Minford and Joseph S.M. Lau, eds., *Classical Chinese Literature: An Anthology of Translations*, 2 vols. (Hong Kong: The Chinese University Press, 2000), 1.256–1.257; also see the introduction on p. 240 of this same volume.

31 Ibid., 259.

32 Ibid., 260; also note the brief comments concerning "aerial travel" in David Hawkes, John Minford, and Siu-kit Wong *Classical, Modern and Humane: Essays in Chinese Literature* (Shatin, N.T., Hong Kong: Chinese University Press, 1989), 99.

33 Fang Ren, bk. 2, ch. 6.

> On the southern peak of Tian Lao (天姥) mountain, a long time ago, Lu Ban (魯班) carved some wood into a crane that then flew 700 *li*. Later, the bird was placed on the west peak of the northern mountain. Emperor Wu of Han (漢武帝) ordered his people to go take it, but it then flew back to the southern peak. Often, when it looks like it is about to rain, then the bird's wings begin to move, flapping as if it is about to fly.[34]

Wooden birds that are capable of flight appear in some other tales as well, in Chinese, South Asian, and other sources.[35] The story above includes the interesting comment that "when it looks like it is about to rain, then the bird's wings begin to move"; one might conjecture that this implies a device sensitive to changes in barometric pressure.

The passage above notes that Lu Ban constructed a bird made of wood, and that the mechanism was capable of flight. Another story says that the astronomer and engineer Zhang Heng (張衡), who lived 78–139 CE, made a wooden bird (木鳥) that contained a mechanism inside it that allowed it to fly. Heng himself refers to this kind of device in one of his writings.[36] Needham notes that accounts of mechanical birds are also found in Western sources, notably the second-century CE writer Aulus Gellius. In these cases, the wooden birds seem to be small-scale devices, while in other tales they are described as being piloted.

Gellius notes in his *Attic Nights* the apparent construction of a small mechanical bird by the Greek philosopher Archytas of Tarentum (fl. ca. 400–350 BCE).[37] But it is interesting to see exactly in what context Gellius discusses these. After he dismisses some fantastic stories concerning magic invisibility potions, spells carried out with animals, and so on, as being wrongly attributed by Pliny the Elder to the Greek philosopher Democritus, Gellius says:

---

34 Ibid., bk. 2, ch. 3.

35 A Chinese tale of this type is found in connection with the famed Chinese philosopher Mozi (墨子): see Motse, *The Ethical and Political Works of Motse. Translated from the Original Chinese Text by Yi-Pao Mei* (London: Probsthain, 1929), 9. In that story, a rival of Mozi builds a bird of bamboo and wood, and flies in it for three days. Also note Needham, *Science and Civilisation in China, Vol. 2: History of Scientific Thought* (Cambridge: Cambridge University Press, 1956), 54, and see the analysis in Zhaomin Liu, 我國古代有關飛行及航空器的構想 ["The Concept of Flight and Aircraft in Ancient China"], *Zhonghua Keji Shixue Huixuekan* (中華科技史學會學刊 "Journal of the Chinese Society of Science and Technology") 16 (2011): 53–59.

36 Needham, 4.2, 574–575.

37 Ibid., 575–576.

> It seems that many fictions of this kind have been ascribed to the name of Democritus, by those poorly skilled men who employed the shelter of his reputation and authority. But that which Archytas the Pythagorean is said to have devised and made ought to seem just as marvellous, without, however, being similarly absurd. For many reputable Greeks, and Favorinus the philosopher, most diligent in pursuit of ancient records, have written most affirmatively that a model of a dove was made by Archytas out of wood, constructed according to a particular method and mechanical discipline so that it flew. Thus, it is evident that it was suspended using counterweights and propelled by a current of air enclosed and hidden in it.[38] Concerning a matter so difficult to believe, I am disposed to employ the words of Favorinus himself: "Archytas of Tarentum, being among other things mechanically inclined, made a flying wooden dove; when it alighted, it no longer arose."[39]

This passage provides an interesting perspective on how an early author — Gellius, in this case — might judge a source and provide analysis. Gellius rejects the practice engaged in by some authors, such as Pliny the Elder, of attributing ideas to a particular figure, notably Democritus. He is wary of misattribution, especially purposeful misattribution, by writers attempting to gain authority by such a practice.[40]

As for the discussion concerning the invention of Archytas the Pythagorean — the flying wooden dove — Gellius seems to be making a distinction between the "fictions" generally promulgated by lazy or deceitful writers and the description of this mechanical device: the latter, he says, "ought to seem just as marvellous, without, however, being similarly absurd." One scholar carefully dissects Gellius' language to illustrate this point:

> As Gellius moves from Pliny's slander of Democritus to Favorinus' amazing-but-true tale of Archytas' dove (authorized and quoted in full),

38 On this setup, see Carl A. Huffman, *Archytas of Tarentum: Pythagorean, Philosopher and Mathematician King* (Cambridge: Cambridge University Press, 2005), 578.

39 Aulus Gellius, *Attic Nights*, 10.12.8–10; this is author's translation from the Latin in Rolfe's edition, *Attic Nights of Aulus Gellius*, 3 vols., trans. John C. Rolfe (Cambridge, MA: Harvard University Press, 1927), 2.244–2.245.

40 See Irene Peirano, *The Rhetoric of the Roman Fake: Latin Pseudepigrapha in Context* (Cambridge: Cambridge University Press, 2012), 50–51.

> he activitates the double meaning of the word *commentum* ["fiction"], used here to describe the things Democritus is supposed to have written.[41]

Although is he talking about Gellius' use of language, this commentator goes on discuss such word choice in a way that is of particular interest to the historian of technology:

> *Commentum* and *comminiscor* can refer both to the contrivance of a device or scheme and to the fabrication of a lie or the perpetration of a deception. The potential pun is impossible to reproduce in English, although the word "device," in its literal and figurative senses, comes close. Pliny and others ascribed various lies/devices to Democritus, but Archytas made an actual device — a device, moreover, that has the appearance and behavior of a real thing on account of its ingenious construction. One kind of creative ingenuity (coming up with lies to attribute to authorities) is compared to another (crafting a mechanical *simulacrum* of a real thing).[42]

Gellius believes this account of the flying wooden device, citing Greek accounts and especially Favorinus (ca. 80–160 CE), a Roman philosopher and orator, as well as his mentor and friend.[43] A modern classicist notes that this mechanical

> dove is at about the same level of sophistication as and shows some similarities to the automata in the earliest surviving Greek treatises on the subject ... which date to the third century BC. Archytas' device would be a hundred years or more earlier, but there is evidence for the invention of isolated "marvelous devices" even before Archytas.[44]

Unfortunately there is a lacuna in the original text of Gellius after the passage about the flying wooden dove, so Gellius' recounting of Favorinus description is incomplete. It is interesting to see, however, how Favorinus is described: "most diligent in pursuit of ancient records" — *memoriarum veterum*

41 Joseph A. Howley, *Aulus Gellius and Roman Reading Culture: Text, Presence, and Imperial Knowledge in the* Nocetes Atticae (Cambridge: Cambridge University Press, 2018), 138.

42 Ibid., 138.

43 See Leofranc Holford-Strevens, *Aulus Gellius: An Antonine Scholar and his Achievement* (Oxford: Oxford University Press, 2003), 98–130, and Stephen M. Beall, "Homo Fandi Dulcissimus: The Role of Favorinus in the Attic Nights of Aulus Gellius", *American Journal of Philology* 122.1 (Spring 2001): 87–106.

44 Huffman, 571; also see ibid., 82–83 and 572–579.

exequentissimus, in the original Latin. Again, we see an ancient author quite aware of the nature of history and the transmission of knowledge.

Returning to the Chinese sources, recall that in a passage cited above from the sixth-century CE *Shuyiji*, Emperor Wu is said to have ordered his people to seize the wooden bird fashioned in the shape of a crane. We find Emperor Wu mentioned once more in another peculiar tale, again in the *Shuyiji*.[45] This one involves flight as well, but with more magical elements. Certainly, the above tale about the wooden bird could be taken as pure fantasy, but note that the story has no explicit mention of any spirits or magical powers — nor indeed is there any fantastic imagery whatsoever. Contrast that story of the wooden bird, then, with the following:

> Emperor Wu of Han, in the first year of Yuan Ding (元鼎), built a pavilion for beckoning spirits.[46] One time, a female spirit left behind a jade hairpin for the emperor. The emperor bestowed it on Zhao Jieyu.[47] In the Yuan Feng (元鳳) year of Emperor Zhao, the palace residents saw this hairpin shining in a special way.[48] Together, they contrived to break it open [i.e., to see why it shined in this way]. They looked in the hairpin box, and clearly saw a white swallow rise straight [out of the box and] into the sky. Later, the palace residents often made jade hairpins. And so it is that [they are now] called "jade swallow" hairpins.[49]

Another tale of flight, with a similar tone, is found in a poem entitled "Li Sao" (離騷), meaning "Sorrow at Parting" (sometimes rendered as "Encountering Trouble"), attributed to poet Qu Yuan (屈原) (340–278 BCE).[50] In the poem, the narrator flies about in a special vehicle; the original Chinese reads:

> 駟玉虬以乘鷖兮，
> 溘埃風餘上徵。

45 Emperor Wu ruled from 141 to 87 BCE, during the Han dynasty.
46 The Yuan Ding era ran from 116 to 111 BCE.
47 The term *jieyu* (婕妤) refers to a grade of concubine.
48 The Yuan Feng era ran from 110 to 105 BCE.
49 Fang Ren, bk. 2, ch. 6.
50 For an English translation, see Cyril Birch, ed., *Anthology of Chinese Literature: From Early Times to the 14th Century* (New York: Grove, 1965), 51–62; he reprints Hawkes' translation of this poem. Also note the brief comments in Gregory Shushan, *Conceptions of the Afterlife in Early Civilizations: Universalism, Constructivism, and Near-Death Experience* (London: Continuum, 2009), 113.

A literal rendering:

> I rode a phoenix carriage driven by four jade dragons,
> Suddenly, with a dusty wind, I ascended on my journey.[51]

Looking at the passages above, we see the two basic story types: First, there are tales that contain technical or mechanical elements. In addition, there are stories that follow a fantastic format, with magical powers or spirits invoked to bring about flight. In addition, there are combined formats: a poet writes of flying through the air, but does so in a vehicle, however fancifully described. In those latter descriptions, there is a harnessing of what one might call animate entities — from birds to dragons — to achieve flight.

Chinese popular fiction involving flying machines began to appear in the late Qing Dynasty, partly due to inspiration from European sources such as Jules Verne.[52] This period saw the rise of Western technology and invention — and colonization. China, meanwhile, was also witnessing the end of its millennia-old imperial system and experiencing doubts about its place in the world. This created a peculiar, combined attraction-and-repulsion to Western ideas — and even attempts to find Chinese origins for Western technological superiority. One commentator notes that during the late Qing period, "Adherence to the notion that modern western science originated from Chinese legends and fables was not limited to learned circles."[53] But despite the intriguing mentions in early Chinese texts of mechanical devices for flight, there are not enough details to decipher fully these "legends and fables" and properly interpret their historicity.

3 Korean Accounts of Flying Machines

A Korean work entitled *Oju Yŏnmun Changjŏn San'go* (五洲衍文長箋散稿 "Scattered Manuscripts of Glosses and Comments of Oju") also examines

---

51 My translation; also see Birch, 56, and note Hawkes, 48. For further comments, note Minford and Lau, 1.247, n. 17, as well as p. 48 of Fusheng Wu, "*Sao* Poetry: *The Lyrics of Chu* (*Chuci*)" in Zong-qi Cai, ed., *How to Read Chinese Poetry: A Guided Anthology* (New York: Columbia University Press, 2008), 36–56.

52 Pingyuan Chen, "From Popular Science to Science Fiction: An Investigation of Flying Machines", in David Pollard, ed., *Translation and Creation: Readings of Western Literature in Early Modern China, 1840–1918* (Philadelphia: John Benjamin, 1998), 209–240; note especially p. 231 for a brief discussion of the flying machines featured in the *Jinghua Yuan.*

53 Ibid., 230, and see 239 n. 5.

accounts of flying vehicles.[54] This work is an encyclopedia of "some two thousand items about all kind of things, probably written not later than about 1836"; the author, Yi Kyu-gyŏng (李圭景), was born in 1788 and died in 1856.[55] The author includes a look at the story of the Qi Gong, and ends his rendering of the tale with his opinion as to its veracity:

> According to the records of the Qi Gu (奇股) people, and recorded in the ancient book, *Diwang Shiji*, in a place 40,000 *li* from the Jade Gate, people were able to produce a flying machine, which was able to fly with the force of the wind. As it was flying far into the area of the Tang, it met a westerly wind that blew it into Yuzhou. The machine was then destroyed, in order that it would not be discovered by others. Ten years later, when the winds began blowing from the east, they produced another flying machine, and presented it to the people of Yuzhou. This is an unreliable legend handed down from the dawn of civilization, and there is no means of verifying the truth now.[56]

This version indeed seems to be drawn from the *Diwang Shiji* of Huangfu Mi, cited earlier. Here, Qi Gong (奇肱) has become Qi Gu (奇股): again, the characters in Qi Gong (奇肱) mean "one arm" or "strange arm", and Qi Gu (奇股) can be rendered as "one foot" or "strange foot", an unsurprising conflation.[57] The implications in this version of the story again include secrecy concerning technical knowledge: in the earlier version, the emperor did not want his own people to witness the craft, and he has the craft disassembled; in this version, the device is destroyed completely. At the same time, the passage above implies that the knowledge as to how to construct this flying machine was preserved, since a new one is described as having been built "ten years later".

---

54 Kyu-gyŏng Yi, *Oju Yŏnmun Changjŏn San'go* (五洲衍文長箋散稿 "Scattered Manuscripts of Glosses and Comments of Oju"), 2 vols. (Sŏul t'ŭkpyŏl-si: Tong'guk munhwasa, 1959). The title of this work is sometimes translated as *Random Expatiations of Oju*; in the title, the term *Oju* (五洲) literally means "the five continents".

55 Prof. Seong-Rae Park, private communication, 9 February 1995.

56 Yi, vol. 1, *chuan* 2, 33–34. I wish to thank James Ascencio for his assistance with this translation; Prof. Ihnhee Lee of the University of Pennsylvania also provided help with some of the Korean texts. See the brief discussion of Yi's book in Michael Finch, *Min Yông-hwan: A Political Biography* (Honolulu: University of Hawai'i Press, 2002), 19.

57 See Yilu Lu, 《山海經》中的一足形象與圖像探析－兼論奇肱國與奇股國問題 ["Investigating the Images and Graphics of 'One Foot' in the *Shan Hai Jing* and an Analysis of the 'One Arm' Country and the 'One Foot' Country"], *Danjiang Zhongwen Xuebao* (淡江中文學報 "Tamkang Journal of Chinese Literature") 29 (2013): 29–50.

Despite his doubts about the reliability of this tale, Yi goes on to discuss other accounts of flight in the same text. One modern author notes that "Yi Kyu-gyŏng's work is marked by careful methods of empirical scholarship" — and therefore the topic of mechanized flight is treated with both curiosity and skepticism.[58] That curiosity is revealed as the text of the *Oju Yŏnmun Changjŏn San'go* turns to a discussion of another type of mechanical flying device, those constructed in the form of creatures. A chapter entitled *Feiche Bianzheng Shuo* (飛車辨證說 "An Investigation of Flying Machines"), includes the following passage:[59]

> During the reign of the Tang emperor Mu Zong (穆宗), there was a person named Han Zhihe (韓志和), whose ancestors had come from Japan, and who was a skilled woodcarver. He produced the external forms of phoenix-like birds, cranes, ravens, and magpies. Regardless of whether the birds were drinking or eating, the representations were very close to reality. By installing a mechanical shaft inside the stomach of the carving, he was able to start the mechanism, and go up into the sky and fly quickly. He was able to reach a height of one hundred meters, and fly as far as two hundred meters before beginning to descend.[60]

As noted earlier, the *Shuyiji* recounts that Lu Ban, "carved some wood into a crane that then flew 700 *li*", and that this crane moved by flapping its wings.[61] The *Oju Yŏnmun Changjŏn San'go* has this:

> The *Taiping Guangji* (太平廣記 "Extensive Records of the Taiping Era") makes a note of a man named Lu Ban, from the area of Dun Huang, today's Gansu Province. The era in which he lived is not clear, only that he was good at a precise and skillful carving of the things of the earth. In the Liang Zhou area, in today's Gansu Province, he did carving in temples and Buddhist pagodas. He produced a wooden bird, and each time he

58 Michael J. Seth, *A Concise History of Korea: From the Neolithic Period Through the Nineteenth Century* (Lanham, Maryland: Rowman & Littlefield, 2016), 238.

59 A brief discussion (in Korean) of that chapter's material on flying machines is found in a two-part article: Seok-hwan Kim, 해설 실학과 (實學) 비차변증설에 (飛車辨證說) 대한 고찰 (1) ["Review: Korean Classic Pragmatism and Flying Vehicles (1)"], *Journal of the Korean Society for Aeronautical and Space Sciences* 16.4 (1988): 4070, and idem, 해설: 실학과 (實學) 비차변증설에 (飛車辨證說) (2) ["Review: Korean Classic Pragmatism and Flying Vehicles (2)"], *Journal of the Korean Society for Aeronautical and Space Sciences* 17.1 (1989): 1080.

60 Yi, vol. 1, *chuan* 2, 33–34.

61 Fang Ren, bk. 2, ch. 6.

> struck the mechanism three times, he was able to mount it, and fly to the peaceful and happy world in the West.[62]

In the first of these two passages, it is possible to place the woodcarver historically by noting that Emperor Mu Zong reigned from 820 to 824 CE. However, there do not seem to be other extant sources that describe Han Zhihe's flying machines. Both of the passages display a mixture of the pragmatic — a person with skill in carving wood structures extends those skills to carving a wooden bird — and the fantastic: it is not clear where, if anywhere, one might find "the peaceful and happy world in the West" that the pilot reaches. The author adds: "It is said that if Lu Ban's and Han Zhihe's mechanisms and methods were adapted to vehicles, those vehicles could then soar." However, the details of such "mechanisms and methods" are not provided.

In another passage, the *Oju Yŏnmun Changjŏn San'go* provides further comments about flying machines:

> If one wants to copy the production techniques of the flying machine, that person must first make a machine that resembles a bird, install wings, set up the internal mechanism, and people can sit inside. After starting the mechanism, the flying machine will seem like a skilled swimmer swimming, and as natural as the larvae of a moth stretching its wings and agitating the air to fly freely. Within a short period of time, it will fly a thousand *li*.[63]

Note here how the writer has drawn together a discussion of the natural and the artificial: a bird serves as the model, which is then scaled up so that it can carry passengers. From the perspective of modern aerodynamics, "agitating the air" will not work — flapping wings, such as the kind also conjectured by Leonardo da Vinci, cannot create sufficient lift to carry a person into the air. Human-powered ornithopters may be possible nowadays, because we have specialized, lightweight materials. But in an early period, vehicles using materials such as wood, canvas, and leather could not have achieved flight using human power to "agitate" the wings. The Italian physiologist Giovanni Alfonso Borelli (1608–1679) was one of the first to understand this, and commented on the problem in his book *De Motu Animalium*.[64] A modern book on gliding

---

62 Yi, vol. 1, *chuan* 2, 33–34.

63 Ibid., 33–34.

64 An English translation of Borelli's original analysis can be found in Giovanni Alfonso Borelli, *On the Movement of Animals*, trans. Paul Maquet (Berlin: Springer-Verlag,

notes that a "human-powered ornithopter is virtually incapable of flight due to the dramatic difference in the strength-to-weight ratio of birds compared to humans".[65]

The *Oju Yŏnmun Changjŏn San'go* includes a story of a flying machine used in wartime. In other versions of that tale, the machine apparently did not work in the style of bird, but rather like a modern helicopter. A Korean account of bravery in battle, for example, includes an account of how a Korean hero in the sixteenth century created such a flying machine. According to the tale, during the Hideyoshi invasions (1592–1597) of Korea, while the Japanese besieged Chinju castle, an individual named Jŏng Pyŏng-gu rescued a compatriot by means of a flying vehicle. The vehicle allegedly flew for 30 *li*. This story first gained recognition in modern times when it appeared in a Korean language textbook entitled *Chosŏn ŏmun kyŏngwi* (朝鮮語文經緯 "Choson [i.e., Korean] Language Reading Methods"), written by Kwŏn Tŏk-kyu (權悳奎) in 1923.[66] Kwŏn's account of the tale comes in a discussion of nouns that have their origins in proper names:

> There is another story related to these [ways of naming]. Jŏng Pyŏng-gu was the inventor of an airplane in Chosŏn.[67] During the Imjin War, when Chinju fortress was in danger, he rescued his friend by an airplane, flying 30 miles away, and then landing. The spinning top with which children play, [called] "P'inggu," whirls around on the ground as it is whipped by them, and then lifts into the air. This is like Jŏng Pyŏng-gu's riding of the airplane, moving up and down. So the name "Pyŏnggu" was given [to the top], changed to "Pinggu".[68]

1989), 162–182; also note Borelli, *The Flight of Birds*, ed. Thomas O'Brien Hubbard and John H. Ledeboer (London: Printed and Published for the Aëronautical Society of Great Britain by King, Sell & Olding, 1911), 36–40. In addition, see the comments in Richard P. Hallion, *Taking Flight: Inventing the Aerial Age, from Antiquity Through the First World War* (Oxford: Oxford University Press, 2003), 28.

65 Federal Aviation Administration, *Glider Flying Handbook* (Washington, D.C.: Federal Aviation Administration, 2003), 1–1. Also see Benjamin J. Goodheart, "Tracing the History of the Ornithopter: Past, Present, and Future", *Journal of Aviation/Aerospace Education & Research* 21.1 (Fall 2011): 31–44; Goodheart points out (p. 34) that even prior to Borelli, Leonardo da Vinci understood that "that man does not possess strength in adequate supply to sustain flight like a bird".

66 This information is based on notes contained in a private communication from Prof. Seong-Rae Park of the Hankuk University of Foreign Studies, Seoul, Korea, 9 February 1995; all translations are by Chungsoo J. Lee.

67 Chosŏn was a kingdom in what is now Korea.

68 Tŏk-kyu Kwŏn, *Chosŏn ŏmun kyŏngwi* (朝鮮語文經緯 "Choson Language Reading Methods") (Kyŏngsong: Kwangmunsa, 1923), sec. 39, 22–23.

The historian of science Seong-Rae Park notes that this story is available in a few other sources as well.[69] A seven-volume history of Korea has the following passage in a discussion of various early weapons:

> In the Jingu month, during the battle of Kyŏngju, there was the thunder-like sound that shook the heavens. It was produced by a kind of short-fused explosive fired from a large cannon. Also, there was a French cannon … imported at the time of the Imjin War against Japan [1592–1598]. Besides these, King Munjong [reigned 1450–1452] had already ordered his brother, General Yong Im, to invent a fire-car [a term for a kind of military vehicle] …[70] In addition, according to a legend, an airplane was invented … During the Imjin War against Japan, an old fortress in Yŏngnam province fell under a siege.[71] It is said that a friend of the lord of that land rode an airplane, entered into the fortress, carried away the lord, came out, and landed about 30 *li* away. In Chŏnra province, he is called Kim Jae-in [or] Jŏng Pyŏng-gu.[72]

This work uses as its source the *Kosat'ong* (故事通 "Compilation of Old Events") by Ch'oe Nam-sŏn (崔南善) (1890–1957), a Korean historian. In the *Kosat'ong*, in a section entitled "Flying Vehicles", the story runs as follows:

> During the siege, a soldier's household was searching for a way to make contact with the outside. There were no methods other than utilizing the rain or a delaying tactic. According to Shin Kyŏng-jun's plan, during the Imjin War, an old fortress in the Southern province fell under siege and was about to be taken. There was someone who was a very close friend of the lord and who had courage. He made an airplane, entered into the fortress, carried the lord off, and flew up. After about 30 miles, he landed

69 Park, private communication, 9 February 1995.

70 This was likely a *hwa cha* (火車) — literally "fire vehicle" — a kind of rocket launcher mounted on a wooden cart, which was used against the Japanese; see Ki-baik Lee, *A New History of Korea*, trans. Edward W. Wagner (Cambridge, MA: Harvard University Press, 1984), 518, and note George M. Siouris, *Missile Guidance and Control Systems* (New York: Springer, 2004), 1.

71 For more on the Imjin War, see Samuel Hawley, *The Imjin War: Japan's Sixteenth-Century Invasion of Korea and Attempt to Conquer China* (Seoul: Royal Asiatic Society, Korea Branch, 2005).

72 Sang-baek Yi, ed., *Han'guksa* (韓國史 "The History of Korea"), 7 vols. (Seoul: Ŭryu munhwasa, 1961–1971), 4.791–4.792.

> on the ground. In Chŏnra province, he was called Kim Jae-in [or] Jŏng Pyŏng-gu.[73]

The *Oju Yŏnmun Changjŏn San'go*'s story of mechanized flight used in battle, in turn, has several elements that parallel the tale of Jŏng Pyŏng-gu. In the chapter entitled "An Investigation of Flying Machines", the account runs as follows:

> In recent times, there was a person named Sin Seungseon (申丞宣) … who, in a civil service examination article entitled *Cha Je* (車制 "Vehicle Systems") stated that after the Imjin period, the invasion of Japanese pirates became very severe. In Yeongnam (嶺南),[74] an independent city had been entered and surrounded by bandits; the situation was critical, and the city was about to capitulate. At that time, there was a person who was very close to the city's ruler … He built a flying machine, flew into the city and rescued the city's ruler, and then flew out of the isolated city. He flew 30 *li* before landing, escaping the bandits' attack. If this really happened, then China long ago had methods for building a flying machine … But nothing remains today.[75]

The story is oddly similar to the one describing Jŏng Pyŏng-gu's flight — even the distance of "30 *li*" is the same. But the lack of other early sources makes this story difficult to trace. In addition, there survive no illustrations of the device that the Korean hero, Jŏng Pyŏng-gu, used to make his daring rescue. The description seems to indicate a device like helicopter, since the analogy is drawn to a child's top — but there are no other sources to help clarify this description. Indeed, this story of a Korean achievement in attempts to build flying vehicles does not seem to be found elsewhere in that culture's literature.

There is, however, another connection in Korean tradition between military endeavors and flight is in the story of a general named Choe Yeong (崔瑩). In the late fourteenth century, he arrived in the Kingdom of Tamna (耽羅國), on what is today the island of Jeju in South Korea. He was there to put down a rebellion, but upon arrival, he found that the coast was very steep. In one version of the account, Choe Yeong conceived a plan to launch fire-carrying kites from his fleet.[76] A variation on this story says that a soldier was tied to a kite

---

73 Nam-sŏn Ch'oe, *Kosat'ong* (故事通 "Compilation of Old Events") (Kyŏngsong: Samjungdang, 1943), 144–145.

74 A region in the southwest part of the Korean peninsula.

75 Kyu-gyŏng Yi, vol. 1, *chuan* 2, 33–34.

76 Sang-su Choe, *The Survey of Korean Kites* (Seoul: Korea Books Publishing Co., 1958), 2.

to enable him to land on a cliff on the coast.[77] The only other reference to the ability to fly in Korean sources is in *The Legend of Zon U-Czi*, where the hero, Zon, and his men transform themselves into kites so as to engage in a surprise attack on a group of bandits.[78] But perhaps this tale falls into the separate category of the purely magical approach to aerial journeys.

Returning to the *Oju Yŏnmun Changjŏn San'go*, it should be pointed out that the text provides comments concerning flying vehicles that are not found in other East Asian works. It is noteworthy, too, that the *Oju Yŏnmun Changjŏn San'go* provides some opinions on the issues of mechanized flight generally. After speaking of some flying machines that were able to fly a thousand *li* in one day, the author notes:

> If doubts are created because we cannot witness these events with our own eyes, there is no way to answer the question of the truth or falsity of these claims of flying machines. If, for instance, we want to find out the principles and methods of flying-machine manufacture, we are actually only able to ascertain the flying machine's means of producing power, and that is all.[79]

The author does not go on to explain any details of how one ascertains a flying vehicle's motive power. He comments, wistfully perhaps:

> Now if these methods and systems [for a constructing a flying vehicle] really existed, then this machine could ascend with upward wind currents, raising large quantities of dust, and could fly with great momentum. It would travel throughout the world with the ease of moving about in one's home or on the road. One could go wherever that person wanted to go, without obstacle. Isn't this enough to make one feel carefree and leisurely?[80]

But it is important here that author is looking at history from a kind of "meta" perspective, carefully considering the problem of verifiability. Moreover, the wistful tone again implies knowledge that was more sophisticated in the past, and that it subsequently became lost.

---

77 Kyu-t'ae Chŏn, *Korean Heritage* (Seoul: Jeong Eum Sa, 1975); also see Clive Hart, *Kites: An Historical Survey* (Mount Vernon, NY: P.P. Appel, 1982), 34.

78 See In-sŏb Zŏng, ed. and trans., *Folk Tales from Korea*, 3rd ed. (Elizabeth, NJ: Hollym International, 1982), 229.

79 Yi, vol. 1, *chuan* 2, 33–34.

80 Ibid., 33–34.

He ends his chapter on flying machines with the following:

> Looking back on the conjectures and claims [concerning these flying machines], we can retain them, and there is no need for further verification. The book entitled *Haiguo Tuzhi* (海國圖志 "Illustrated Gazetteer of the Maritime Nations") records diagrams of flying machines, and offers verification for future generations.[81]

This book that is said to have "diagrams of flying machines" is discussed below. Interestingly, in this same section of the text, the *Oju Yŏnmun Changjŏn San'go* mentions other works that also concern flying vehicles:

> In the faraway Western areas, there lived a man named Deng Yuhan (鄧玉函). He had a book called *Qiqi Tu* [奇器圖 "Drawings of Peculiar Apparatus"], in which were recorded books regarding people flying. However, the original manuscript is lost. Was this in order to keep the techniques of flying hidden, and only to arouse people's interest and not let them know the secret? Russia and Xiwu (西塢) are in the west, so how can we know whether or not their people were descendants of Qi Gu (奇股) of old? And, how can we know whether or not Deng Yuhan's books of flying men were about the methods handed down from ancient Qi Gu?[82]

The *Qiqi Tu* (奇器圖 "Drawings of Peculiar Apparatus") likely refers to the *Yuanxi Qiqi Tushuo Luzui* (遠西奇器圖說錄最 "Collected Diagrams and Explanations of Wonderful Machines from the Far West"). This work was the first translation into Chinese of European mechanical studies. It was compiled by in China by the Jesuit missionary Johann Terrenz Schreck (1576–1630), whose Chinese name was Deng Yuhan (鄧玉函), and the Chinese scholar Wang Zheng (王徵) in 1627.[83] Perhaps the "the faraway Western areas" refers to Schreck's birthplace, Swabia, in what is today Germany. Despite what the passage above claims, the *Yuanxi Qiqi Tushuo Luzui* does not seem to include anything about flight. Nor is it clear that there was an "original manuscript" that was lost.[84] Again, too,

81 Ibid., 33–34.

82 Yi, vol. 1, *chuan* 2, 33–34.

83 On Wang Zheng and his interests in technology, see Peter J. Golas, *Picturing Technology in China: From Earliest Times to the Nineteenth Century* (Hong Kong: Hong Kong University Press, 2015), 144 et ff.

84 See Baichun Zhang, Miao Tian, and Qiang Liu, "The Evolution of Editions of *Yuanxi Qiqi Tushuo Luzui* (A Record of the Best Illustrations and Descriptions of Extraordinary

notice the mention of the Qi Gu (奇股), as the author tries to piece together the story of early flight here.

As for the location of Xiwu, this likely refers to Xiwuzhumuqin Qi (西烏珠穆沁旗), a location in Inner Mongolia; it is known in English as the "West Ujimqin Banner", with the term *qi* (旗) — literally, "banner" — referring to a county administrative division. However, no other source appears to mention this region in connection with flying machines. The *Oju Yŏnmun Changjŏn San'go* mentions *Xiwu* in another passage, pinpointing more precisely than any other extant text the alleged location of manufacture of flying vehicles:

> Several tens of thousands of *li* from China, in what is today Russia, situated between China and Europe, it is said that the people were able to manufacture wheels for a flying machine. Each flying machine required four wheels in all. Furthermore, legend has it that they were able to fly one thousand *li* in one day. The people of Xiwu (西塢) also produced a flying machine, and utilizing a bellows [usually] used in smelting, a pulley hauling device, and other methods, enabled the flying machine to move. The machine was able to float in the air and move without obstacle on rivers or land. During the dangerous times of war and turmoil, this vehicle could be used to ward off attacks from the enemy.

The passage presents a list of parts, but no indication of what the overall vehicle looked like, or why it would require wheels. Is it supposed to be some kind of balloon, or a powered, winged aircraft? The text here includes the comment that each "flying machine required four wheels in all." An 1832 reprint of a Chinese fantasy novel, *Jinghua Yuan* (鏡花緣 "Flowers in the Mirror") by Li Ruzhen (李汝珍), includes two illustrations by Xie Yemei (謝葉梅) showing flying vehicles. The elegant flying machines in these pictures each have four wheels.[85] However, we can draw no clear connection between the textual description in the *Oju Yŏnmun Changjŏn San'go* of the flying vehicle, and this novel with its illustrations. The vehicles as they appear in this novel are rendered simply as traditional carriages of the period, with the only remarkable feature being the fact that they are shown as floating in the clouds.

Devices of the Far West) and *Xinzhi Zhuqi Tushuo* (Illustration and Descriptions of Several Newly-built Devices)", *The Chinese Journal for the History of Science and Technology* 27.2 (2006): 115–136.

85 See Needham, 4.2, 593, and figs. 709 and 710. Concerning the *Jinghua Yuan*, see An-Chi Wang, *Gulliver's Travels and Ching-Hua Yuan Revisited: A Menippean Approach* (New York: Peter Lang Publishing, 1995); also see Hsin-Sheng C. Kao, *Li Ju-Chen* (New York: Twayne Publishers, 1981).

The peculiar mention of the bellows is repeated in another brief passage later in the same chapter:

> Lie Kou (列寇) rode the wind and soared for fifteen days, traveling a great distance, and like the roc shaking its wings, stirred up a turbulent spray for three thousand *li*. There is nothing that surpasses this. The main reason why he could fly is that he used silk rope horizontally and vertically interconnected, stretching and contracting, linked together on the machine, in order to induce the effect of agitating the bellows to produce wind power. In this way, the two wings moved, and immediately floated on the strong winds. The flying momentum could not be hindered, because [the machine] used wind power as its principal moving force, following the flying methods of birds. This principle is sufficient and reasonable.

This passage again strikes the modern reader as odd in that it combines ancient legend with technical descriptions. The figure cited here is Lie Yukou (列禦寇), a figure in Daoist philosophy who is sometimes considered the author of the Daoist book *Liezi* (列子). In another Daoist text, the *Zhuangzi* (莊子), he is said to be able to ride the wind; however, in that text this ability is not attributed to any kind of flying mechanism. In the passage above, Lie Yukou's ability is clearly connected with a technological apparatus. How that device actually works is not quite clear: it seems almost as if the air from the bellows blows the wings into motion, and then the device flies on the natural "strong winds" of the atmosphere.

However, for the writer of the *Oju Yŏnmun Changjŏn San'go*, Yi Kyu-gyŏng, the idea that the machine works according to the "flying methods of birds" is "reasonable". Again mentioning Xiwu, the author adds:

> The flying machines built by the people of Qi Gu (奇股) and Yeongnam (嶺南) were constructed according to the methods of the people of Russia and Xiwu (西塢). Each time we travel to Yandu (燕都), or have contact with the people of Russia, we do not attempt to understand or ask them to explain their systems or methods. We only want to obtain their glass, mirrors, and ceramic implements and wares. Why is this? I believe it is because we are not skilled at harnessing those methods.[86]

This is an interesting perspective on technology — again, there is the theme of hidden knowledge, but the author describes it in this case as a

86 Yi, vol. 1, *chuan* 2, 33–34.

willful avoidance of learning about mechanical "systems or methods". From a modern perspective, the knowledge is hidden in another sense: we have no corroborating record of any kind of "flying machines built by the people of Qi Gu and Yeongnam", nor any indication of what "methods" for constructing flying vehicles were known in Russia or Xiwu. More particularly, the veracity of the account itself is uncertain — from where is the author drawing this peculiar historical narrative?

In further discussions of flight, the author of the *Oju Yŏnmun Changjŏn San'go* mentions another encyclopedic collection, the *Haiguo Tuzhi* (海國圖志 "Illustrated Gazetteer of the Maritime Nations") by Wei Yuan (魏源) (1794–1857).[87] He adds that it has depictions of *fei che* (飛車) — again the Chinese term literally meaning "flying vehicles" — and one volume of the *Haiguo Tuzhi* indeed includes a brief discussion of such a machine:

> There is a *tian chuan* (天船 lit. "sky ship"), short and small, in the style of a garden pavilion or gazebo. It can hold ten people. Inside, a very ingenious *feng gui* (風櫃 lit. "wind cabinet"), like a *hun tian* (渾天 "armillary sphere"), is placed. With the help of a few men, it is started up. Then it is able to fly, soaring to great heights. There, the wind carries it along. Depending on where you wish to go, you set sail for that direction using *tian chi* [天尺 "celestial feet"] for measurement.[88] In this place, then, one gathers in the sails, and lets it [i.e., the "sky ship"] descend. It is reported that [the vehicle] has been burned and destroyed by the sun's fire, and people have perished, so one does not dare to use it frequently.[89]

87 On the *Haiguo Tuzhi*, see Jane Kate Leonard, *Wei Yuan and China's Rediscovery of the Maritime World* (Cambridge, MA: Council on East Asian Studies, Harvard University Press, 1984).

88 The term *tian chi* (天尺) had different values at different points in history (Nathan Sivin, private communication, 27 November 2008); see the brief mention on p. 13 of Zezong Xi, "Chinese Studies in the History of Astronomy, 1949–1979", *Isis* 72.3 (September 1981): 456–470, and the longer discussions in Shitong Yi, 量天尺考 ["On the 'Foot' Used in Celestial Measuring"], *Wenwu* (文物 "Cultural Relics") 2 (1978): 10–16, and Zhenhuan Han, 我國古代海航用的量天尺 ["The Use of the Celestial Measuring Ruler in Maritime Activities of Pre-Modern China"], *Wenwu Jikan* (文物集刊 "Cultural Relics Collection") 2 (1980): 217–221.

89 Yuan Wei, *Haiguo Tuzhi* (海國圖志 "Illustrated Gazetteer of the Maritime Nations"), 60 *chuan* (Yangzhou: Guweitang, 1847), *chuan* 59. This work was published in a number of editions in the nineteenth century, and each differed slightly in terms of the arrangement and extent of the text, with an enlarged edition in 1847, a further enlarged edition, 1852, and other editions in 1880 and 1887. This citation is from the 1847 edition; in the 1880 and 1887 editions, this same passage concerning flight is found in *chuan* 64.

Given that the author, Wei Yuan, lived from 1794 to 1857, it is possible that he had heard of the British emissary, Lord Macartney (George Macartney, 1737–1806), and his plan to bring a hot-air balloon for demonstration in Beijing.[90] But the passage above lacks all details of the persons involved, and provides no dates, so it is impossible to make a firm connection here.[91] What this means is that the reader must try to reconstruct the epistemological framework in which the writer created this description. In the passages analyzed above, one sees a mixture of the technical and non-technical, although Yi's text here mentions neither birds nor mythical dragons.

The technical parts of the description are reflected in the careful listing of the parts of the machine, the noting of the fact that the device can be steered, and that there is a particular method for bringing the aircraft back to the ground. The non-technical aspect of the description is characterized by the vagueness of the terms used for the components, such as *feng gui*. In addition, it is not clear at all how the craft could be "destroyed by the sun's fire", and this seems to reflect a misunderstanding on the part of the writer.

The overall description, then, is puzzling. The vehicle is referred to literally as a "sky ship", and from the subsequent mention of sails, perhaps indeed we are meant to picture a kind of flying ship or boat. The device is described as "short and small, in the style of a garden pavilion or gazebo". If this flying craft is a balloon, that description might refer to the basket or gondola, but not to the rest of the machine. The specific terms describing the motive power of the vehicle are particularly opaque. The word *feng* (風) found in the passage above means, simply, "wind," but here it is paired with *gui* (櫃), a word that by itself means "cabinet" or "cupboard."[92] The lack of clarity in the description of this "engine" is compounded by the comparison — apparently, in terms of shape — to an armillary sphere. Moreover, if this passage is simply describing a hot air balloon, then this *feng gui* seems superfluous, as do the sails.

90 John Fitzgerald, *Awakening China: Politics, Culture, and Class in the Nationalist Revolution* (Stanford, CA: Stanford University Press, 1996), 46. For a primary source on this interesting subject, see John Barrow, *Some Account of the Public Life and a Selection from the Unpublished Writings of the Earl of Macartney* ... 2 vols. (London: T. Cadell and W. Davies, 1807), 2.364.

91 Certainly in a later period — the late nineteenth century — Chinese visitors to Europe indeed witnessed hot air balloons; see Ke Ren, "Fin-de-Siècle Diplomat: Chen Jitong (1852–1907) and Cosmopolitan Possibilities in the Late Qing World" (PhD diss., Johns Hopkins University, 2014), 216.

92 In Taiwan, there is a coastal locale named "Fenggui Cave" (風櫃洞), where the waves push air through blowholes in the rocks that form the coastline, resulting in a howling sound.

The ambiguity is not unexpected, as East Asian culture was experiencing and absorbing a range of Western technologies in this period, and doing so without a particular protocol. An example is the work of the Japanese author Morishima Chūryō (森島中良) (1756–1810) who wrote fiction and non-fiction, the latter comprising studies in the area of what is known as *rangaku* (蘭学). This is a term literally meaning "Dutch learning" and by extension, "European learning" or "Western learning"; it was a conscious effort by the Japanese scholarly elite in the late eighteenth and early nineteenth centuries to gain Western knowledge in the areas of science, technology, art, and so on. Chūryō published a book in 1787 with the interesting title of *Kōmō Zatsuwa* (紅毛雜話), literally meaning "Red-Hair Miscellany" — that is, a collection of miscellaneous knowledge from the Dutch. This included an image of a manned balloon with a sail, a gondola, and a rudder. Chūryō copied this image from a Dutch source, but as one commentator notes:

> How closely Chūryō copied is unclear, as an original has never been traced; but he showed a water-going boat, with skids, a paddle-wheel, and rudder, flying above puffs of cloud, supported by an inadequately sized balloon, and drive, it seems, by wind caught in a sail. The single passenger is bracing himself against a flagpole.[93]

He also adds: "Chūryō's transcription does not inspire confidence that he had looked very carefully at the original print."[94] Certainly, this example shows both the unsurprising ambiguity mentioned earlier, and the more general issue of trying to transmit technical knowledge across time and cultures.

As it stands, the description in this obscure passage from the *Haiguo Tuzhi* also seems to echo certain earlier Western discussion of flying vehicles. In the thirteenth century, Roger Bacon, in his *Epistola de secretis operibus artis et naturae* ("Letter on Secret Works of Art and of Nature") describes a flying machine, with wings driven by a hand-turned crank mechanism.[95] Bacon also makes the interesting point that such vehicles work with a pilot inside — *homo sedeat in medio*. Moreover, Bacon adds that such devices were made in the past (*sunt antiquitus*) and in *nostris temporibus* — that is, "in our times". The passage that describes this machine is in a longer text entitled *De instrumentis*

93 Timon Screech, *The Lens Within the Heart: The Western Scientific Gaze and Popular Imagery in Later Edo Japan* (Honolulu: University of Hawai'i Press, 2002), 225.

94 Idem, *Secret Memoirs of the Shoguns: Isaac Titsingh and Japan, 1779–1822* (London: Routledge, 2006), 53.

95 For the original Latin text, see J.S. Brewer, ed., *Fr. Rogeri Bacon, Opera quaedam hactenus inedita* (London: Longman, Green, Longman and Roberts, 1859), 533; also see Davis, 27.

artificiosis mirabilibus, and includes marine vehicles, as well as aerial ones.[96] In another work, *Communia mathematica fratris Rogeri*, Bacon again speaks of *instrumenta volandi* in the context of other technological marvels.[97] Bacon's discussion is curious in that it is both mechanical — if vague — and rooted in a kind of fantastical past. In the *Epistola de secretis operibus artis et naturae*, Bacon actually draws from the writer Aethicus Ister, bringing in an old legend that connected Alexander the Great with a flying machine.[98] As we examine elsewhere in this book — in the case of "magic mirrors" and other devices — the technological is often inextricably connected with the historical, pseudo-historical, and magical in these early writings.

The Chinese accounts also match aspects of another Western source. In 1670, Francesco Lana Terzi, a Jesuit priest and mathematician, published a book entitled *Prodromo*, a work with a lengthy subtitle including the phrase *saggio di alcune inventioni nuove* — "essay concerning some new inventions".[99] In this book, there is a picture of a flying boat, complete with sails for steering (fig. 1).

In this design, there are also several large metal spheres. The idea was that the ship would be lifted into the sky by use of these spheres, which were made of copper, and completely hollow.[100] The spheres, Lana Terzi stated, should be evacuated of all air, and thus, weighing less than the atmosphere, would pull the ship upwards. He wrote as follows:

---

96 See the text in Brewer, 532–533.

97 The original Latin text can be found in Robert Steele, ed., *Opera hactenus inedita Rogeri Bacon*, 16 vols. (Oxford: Clarendon Press, 1909–1941), 16.43–16.44.

98 See Alfred George Greenhill, "The Legend of Alexander and His Flying Machine", *Nature: A Weekly Illustrated Journal of Science* 101.2524 (14 March 1918): 25–26; also see the discussion on pp. 24–26 of Saskia Dönitz, "Alexander the Great in Medieval Hebrew Traditions", in Z. David Zuwiyya, ed., *A Companion to Alexander Literature in the Middle Ages* (Leiden: Brill, 2011), 21–40. A century after Bacon, the philosopher Nicole Oresme speculated in a similar fashion about flight; see the engaging historical discussion of early aerostatic concepts in Hallion, cited earlier, 29–34.

99 Francesco Lana Terzi, *Prodromo, ouero, Saggio di alcune inuentione nuoue, premesso all'arte maestra / opera che prepara il p. Francesco Lana Bresciano della Compagnia di Giesu., per mostrare li piu reconditi principij della naturale filosofia, riconosciuti con accurata teorica nelle piu segnalate inuentioni, ed isperienze sin'hora ritrouate da gli scrittori di questa materia & altra nuoue dell'autore medesimo* (Brescia: Per li Rizzardi, 1670).

100 Roger Bacon had also put forward the idea of a flying machine that used hollow spheres of copper; see the interesting description of this and other ideas of Bacon on flight in John Wise, *A System of Aeronautics, Comprehending Its Earliest Investigations and Modern Practice and Art* (Philadelphia: Joseph A. Speel, 1850), 20–21.

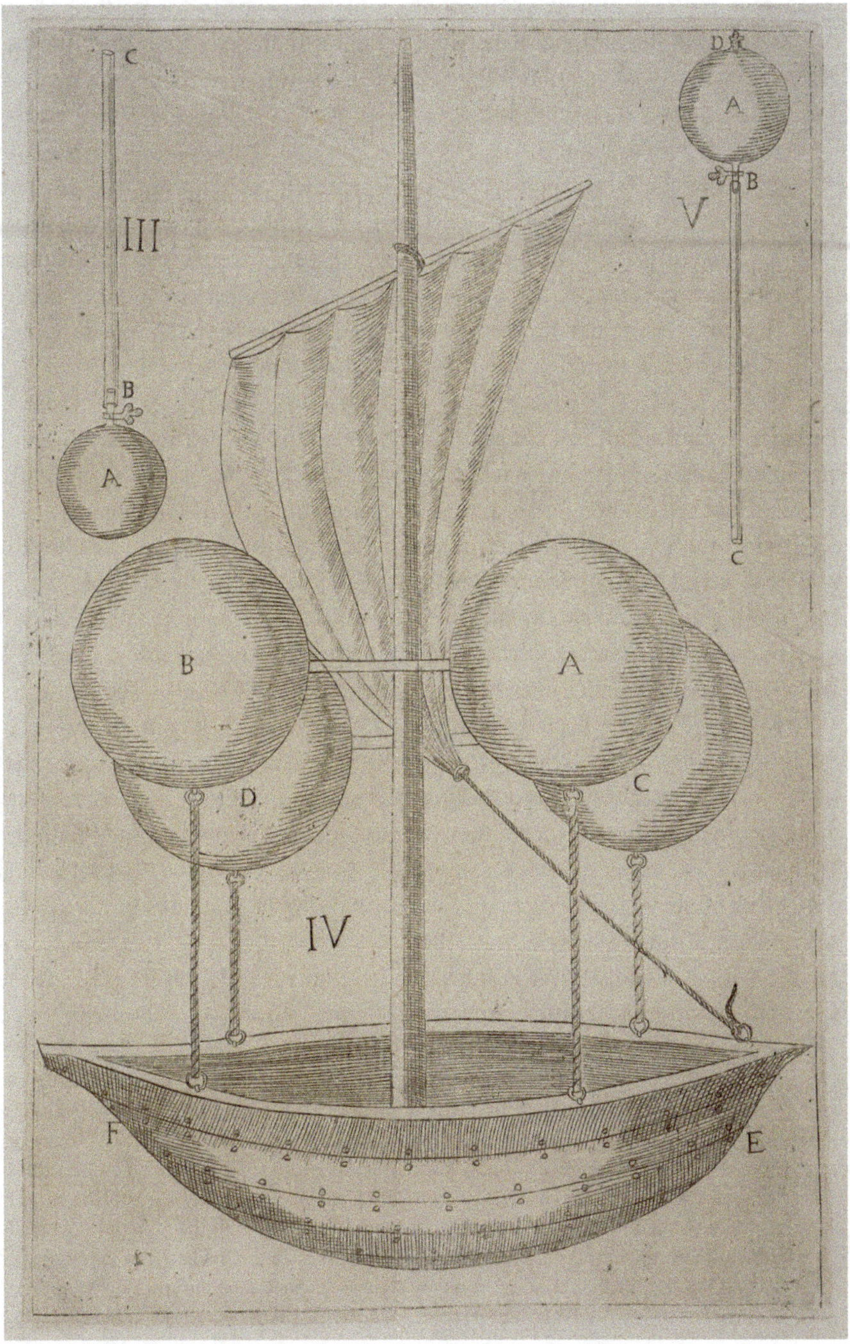

FIGURE 1 Plate of flying machine in Francesco Lana Terzi's 1670 *Prodromo*

> As is the case of a wooden sphere, which ascends to the surface of the water and floats there, because it is lighter in density than the water, so also will a glass vessel full of air float on the water, notwithstanding that glass in itself is heavier than water, but all the substance of the vessel — taking the glass and air together — is lighter than an equal volume of water, which fact is owing to its being, as a whole, lighter in density than water … Presupposing all these things, it is certain that one can construct a vessel of glass or other material which could weigh less than the air contained therein; if, then, one exhausted all the air in the manner before described, this vessel would be lighter in density than the air itself, and, therefore … it would float on the air and ascend.[101]

Perhaps the author of the *Haiguo Tuzhi*, Wei Yuan, is referring to something similar to Lana Terzi's spheres when he describes, "wind cabinets" that are like armillary spheres. When he notes that the flying ship he is discussing has a *feng gui* ("wind cabinet") that is like a *hun tian* ("armillary sphere"), this strikes the reader as a strange comparison. However, what he seems to be saying is simply that these *feng gui* are spherical, as armillary spheres are; more specifically, it may be that Wei Yuan actually is making a comparison to *tianti yi* (天體儀), which are celestial globes and thus truly in the form of a fully closed sphere.

In roughly the same period as Lana Terzi, the Anglican clergyman and natural philosopher John Wilkins wrote a work entitled *Mathematical Magick* that included both traditional descriptions of everyday machines such as the level and the wheel, and more speculative discussions concerning manned flight.[102] Wilkins goes so far as to say that one could construct a "flying chariot, which may be so contrived as to carry a man within it."[103] He includes in his discussion a look at tales from the past that concerned flight, such as those connected with Archytas. What is relevant here, however, is the fact that these speculations and discussions concerning flight — from Wei Yuan, Lana, and

101 See Lana Terzi, *The Aerial Ship*, ed. Thomas O'Brien Hubbard and John Henry Ledeboer (London: Printed and Published for the Aëronautical Society of Great Britain [by] King, Sell & Olding, 1910), 17–18; also note Laufer, *The Prehistory of Aviation*, 21, and Wise, 27–29.

102 John Wilkins, *Mathematical Magick, or, The Wonders that May be Performed by Mechanical Geometry* (London: Printed for Edw. Gellibrand, 1680), 191 et ff. Wilkins' work is divided into two books; the discussion on flying machines (what he calls the "volant or flying automata") is found in chapter 6 of book 2. Also note the connection between literature and technology concerning Wilkins' ideas on flight in Gwin J. Kole, "Johnson's 'Dissertation on Flying' and John Wilkins' *Mathematical Magick*", *Modern Philology* 47.1 (August 1949): 24–31.

103 Wilkins, 209.

Wilkins — wrestle, however obscurely, with actual technology. Moreover, Evangelista Torricelli (1608–1647), Blaise Pascal (1623–1662), and Otto von Guericke (1602–1686) all had developed models of atmospheric pressure and devices that demonstrated that pressure, and Wei Yuan may have understood the concept, at least in an approximate form, and tried to relate it to idea of mechanical flight. These are relatively late sources, but there is no reason to deny that earlier thinkers in East Asia and elsewhere might have similarly combined basic knowledge with speculation to formulate ideas about flying technologies.

The *Oju Yŏnmun Changjŏn San'go* discusses another source concerning flight, employing a rather skeptical tone:

> Some people in the Beijing and Luzhou areas claim that there is a hidden book regarding flying machines, but this is only a legend from Qidong (齊東). Some people think that they have seen a book concerning flying machines concealed by the people of Yuanzhou (原州). The book says that one could use mere nothings to produce a swan shape that could transport four people, and by agitating the abdomen, wind power could be produced. This wind power could carry it aloft, and to a distance of one hundred *zhang* (丈). But while in the air it encountered swirling winds and was not able to proceed, but it was able to descend. If it met with fierce winds, then it could not fly. The production methods and dimensions are very detailed in the book. One 金時讓, of 金州, stated that in the city of 魯 in 湖西, there lived a person named 尹達圭, a descendent of 明齊. He was skilled at producing precision devices, and also had records of the methods for producing and setting up flying machines. However, he concealed these methods, and did not know the complete, detailed contents of the book.[104]

The references here are far from clear. As for 金時讓, the Korean reading of these characters is "Kim Si-yang", and the reference here then might be to a Confucian scholar writer of that name who lived from 1581 to 1643. He is noted as living in 金州, which could be either a place in China, with the characters being read "Jinzhou", or a locale in Korea, with the characters being read "Geumju". It is likely the latter, as the subsequent reference also seems to be

---

104 Yi, vol. 1, *chuan* 2, 33–34.

concerning Korea. In the text, this 尹達圭 (Yun Dalgyu) is said to have lived in 湖西, or Hoseo, a region on the western side of the Korean peninsula.[105]

Again, the author of the *Oju Yŏnmun Changjŏn San'go* is trying to put together a narrative concerning the history of technology of mechanical flight. Moreover, the *Oju Yŏnmun Changjŏn San'go* is an early nineteenth-century work — but treats some of these old stories of flight as if they were recent history. This phenomenon in East Asian sources is not that unusual — in cartography, as late as the eighteenth century, there were traditional Buddhist world maps being made, even as contemporary scientific ones were available from Europe.[106] Overall, the result of this attempt to assemble a technological narrative is that the historical information is fragmentary, and the technical information vague.

This uncertainty is also characterized by how these East Asian texts place the invention of flying vehicles with no specificity in terms of place or time. However, another curious detail is that some of the texts situate the origin of this inventive activity somewhere in the West. Perhaps this is merely an echo of a more general tendency in East Asian history for authors to attribute technological superiority to more distant realms. Historically, the Chinese view of their own culture in relation to foreign cultures "ebbed and flowed with China's might and glory and was more prominent in some periods than in others".[107] In certain periods, such as after the end of the Han dynasty in 220 CE, China had increased contact with South Asia and the Arab world. Much later, by the Qing Dynasty (1644–1912 CE), the Chinese began to view the West as the source of technology, as Jesuits came bringing modern cartography and other scientific methods, for example.[108]

Note the ambiguity in the passage above: the author of the *Oju Yŏnmun Changjŏn San'go* states that "there is a hidden book regarding flying machines", but then comments that this is a legend. Later, he says that there was someone named Yi Dagui who possessed "records of the methods for producing and setting up flying machines", but also that he kept them hidden. This ambiguity about the origins of mechanized flight and the veracity of accounts of their use

105 The only apparent information on this figure and his connection to flight is discussed in Park, 尹達圭의 飛車: 韓國 飛行機의 前史 ["Yun Dalgyu's Flying Machine: Prehistory of Korean Air Flight"], 역사문화연구 ("Journal of History and Culture") 18 (2003): 1–20.

106 See Nobuo Muroga and Kazutaka Unno, "The Buddhist World Map in Japan and its Contact with European Maps", *Imago Mundi* 16 (1962): 49–69.

107 D.E. Mungello, *The Great Encounter of China and the West, 1500–1800* (Lanham, MD: Rowman & Littlefield, 2009), 5.

108 Ibid., 18–19, and 38 et ff.

may be the result of a complex process of a culture taking in new knowledge. In the East Asian context, this can be understood as follows:

> Even when under attack from foreign knowledge, thought and belief ... traditional intellectual resources do not just disappear. Rather, they serve the functions of making analogies and translation to facilitate comprehension of incoming ideas. Traditional knowledge, thought and belief undergo transformation through this process of translation and comprehension, and, at the same time *mutatis mutandis* the intellectual history proceeds ... In China, due to the Chinese propensity to rely on history for support, new changes often appear in the guise of old history. The Chinese employ old terminology to explicate new knowledge, pouring new wine into old bottles.[109]

So, it may be a similar phenomenon in the Korean context, wherein the writer of the *Oju Yŏnmun Changjŏn San'go*, Yi Kyu-gyŏng, is taking this new idea of flight and trying to fit it into the context of earlier books and inventors — but by necessity, of course, those remain obscure. He is struggling to recreate a technological lineage in a situation where that may be impossible.

4 South Asian Tales of Flying Vehicles

Early South Asian texts provide us with abundant and varied references to flight and flying vehicles.[110] These texts contain some notable parallels to the Chinese accounts, particularly in their descriptions of the different types of flying methods and devices. Again, in terms of taxonomic categories, there are two types of stories. First, there are tales of flying gods, similar to the Chinese *fei tian*. The Indian tales also frequently refer to flight achieved by the practice of magic and magical objects, tropes that also appear in Chinese and European sources.[111] A Sanskrit collection of stories compiled by the eleventh-century CE

109 Zhaoguang Ge, *An Intellectual History of China, Volume One: Knowledge, Thought, and Belief before the Seventh Century CE*, trans. Michael S. Duke and Josephine Chiu-Duke (Leiden: Brill, 2014), 40.

110 For a short discussion of mechanical devices, including flying vehicles, in ancient Indian texts, see Benjamin Walker, *The Hindu World: An Encyclopedic Survey of Hinduism*, 2 vols. (London: Allen & Unwin, 1968), 2.55–2.56; also see the entry under *vyomaga* in Arthur A. MacDonell, *A Sanskrit-English Dictionary* (London: Longmans, Green, and Co., 1893), 304.

111 See Laufer, *The Prehistory of Aviation*, 52–55.

writer Somadeva, entitled *Katha Sarit Sagara*, known in English as the "Ocean of Story" or "Ocean of Streams of Stories", enumerates a number of these methods. They include the following: "eating human flesh", "recitation of spells", "the magic power of a sword", and so on. There is even the phenomenon of a person being transformed into a vehicle for flight.[112] In addition, a variety of objects can be invested with the power to become airborne.

But Indian texts also speak of mechanical devices that can go through the air, from flying bird models to flying machines capable of carrying passengers. Laufer notes thc numerous Indian aerial deities of the Vedic period (ca. 1750–550 BCE).[113] Many of them traverse the skies in flying chariots, or "golden ships sailing over the aerial ocean."[114] The *Katha Sarit Sagara* includes a tale entitled "Somaprabha and her Three Suitors" in which the characters fly through the air in an aerial chariot.[115] As is common in these stories, there is a romantic theme, this time with a suitor constructing a flying vehicle to impress the daughter of a Brahmin.[116]

Texts after the Vedic period speak of Garuda, the "lord of the birds" and the conveyor of the god Vishnu. One story recounts how a weaver in love with the king's daughter uses a flying machine to go to the palace and court her. A friend of the weaver constructs the aerial vehicle, which the narrative describes as a wooden machine in the shape of this Garuda. The launching mechanism for this wooden machine is described as a "switch or spring."[117] This mention of a switch we find in other Indian tales as well, and we are reminded of the Chinese accounts of Lu Ban's construction of a wooden bird.[118] Laufer cites another similar tale, also a romance, in which a wooden bird is constructed. The person inside the vehicle, the story tells us, controlled this wooden

---

112 See p. 574 of H.S.V. Jones, "The *Cléomadès* and Related Folk-Tales", *PMLA* (*Publications of the Modern Language Association of America*) 23.4 (1908): 557–598. For a survey and brief discussion of this kind of flight, also see Eliade, "Symbolisme du 'Vol magique'", cited earlier.

113 Laufer, *The Prehistory of Aviation*, 44–57.

114 Ibid., 45.

115 Jones, "The *Cléomadès* and Related Folk-Tales", 575.

116 See Norman Mosley Penzer, ed., *The Ocean of Story Being C.H. Tawney's Translation of Somadeva's Katha Sarit Sagara* (*or Ocean of Streams of Story*), 10 vols. (London: Sawyer, 1924), 6.200–6.203; also note Laufer, *The Prehistory of Aviation*, 48, and Jones, 563 and 575. On the co-occurrence of the idea flight and romantic themes in early tales, note the discussion in Jones, "The *Cléomadès* and Related Folk-Tales", 575–576.

117 Laufer, *The Prehistory of Aviation*, 46; also note Jones, "The *Cléomadès* and Related Folk-Tales", 576–577.

118 Laufer, *The Prehistory of Aviation*, 23.

bird in its flight.[119] The *Katha Sarit Sagara* presents us with an interesting dichotomy here: while it discusses magical ways of achieving flight, as outlined above, it also describes several mechanical devices to achieve the same ability. The text also speaks of other technologies, including robotic devices in the form of human beings and animals, as well as other kinds of mechanical birds.[120] The image of Garuda may have inspired later Arab folklore accounts of the *rukh*, a huge bird that can carry human beings. One commentator notes, in fact, that, "the *rukh* must be conceded a role in the imaginative prehistory of aviation."[121]

Other Indian stories also include descriptions of the construction of mechanical flying machines. In a Buddhist text entitled *Parayanavagga* ("Book of the Way Across"), of uncertain date, we read a story about Katthavahana, a fine carpenter who ran a well-staffed workshop. Bringing lumber from the forests of the Himalayas, the carpenter and his assistants had built houses for the king and the people. Growing weary of that, they constructed flying machines of wood, and flew from Benares to the Himalayas.[122]

Carpenters again are the makers of flying vehicles in a story in an ancient collection of animal fables and other tales from about 300 BCE, the *Panchatantra* ("Five Treatises"). There, we read how a woodworker made "a brand-new mechanical bird, like Garuda, the bird of Vishnu." The carpenter in the story then explains to the recipient of this gift how to operate the flying device with a kind of plug or switch.[123] Similarly, a story from Sri Lanka entitled "The Royal Prince and the Carpenter's Son" features a "wooden peacock machine" that can be "rowed ... in the sky."[124] In the story, the vehicle is flown "very high above the sea."[125] Another story — with a wooden peacock capable of carrying a pilot and passengers — is found in a Tibetan Buddhist tale. In the end, the

---

119 Ibid., 47–48.

120 Jean Philippe Vogel, *The Goose in Indian Literature and Art* (Leiden: Brill, 1962), 48–49.

121 Robert Irwin, *The Arabian Nights: A Companion* (London: Allen Lane 1994), 207.

122 See Nathmal Tatia and Angraj Chaudhary, eds., *The Suttanipata-Atthakatha* (Bihar: Nava Nalanda Mahavihara, 1975), vol. 2, 25; also note Bimala Churn Law, *A History of Pali Literature*, 2 vols. (London: Kegan Paul, Trench, Trubner & Co., 1933), 1.471–1.472, and G.P. Malalasekera, *Dictionary of Pali Proper Names*, 2 vols. (London: J. Murray, 1937–1938), 1.488.

123 *The Panchatantra*, trans. Arthur W. Ryder (Chicago: University of Chicago Press, 1925), 94–95.

124 See Henry Parker, *Village Folk-tales of Ceylon*, 3 vols. (London: Luzac & Co., 1910–1914), 2.18. Also note a second story in that collection, "Concerning a Royal Prince and a Princess", where the "wooden peacock machine" is piloted by these regal characters; see idem, 2.23 et ff.

125 Ibid., 2.20.

remarkable vehicle is destroyed by heavy rains.[126] In these examples, again, the authors wrestle with a technical idea, and it is interesting that the actual technological capability to build a flying machine in this period is attributed to carpenters. This certainly has a logic in epistemological terms for this culture: if some of the most skilled craftsmen were carpenters, then it is logical that they would be the ones said to possess the knowledge to build aerial vehicles. This indicates, in short, an attempt by these early authors to contextualize a technological concept that otherwise might have struck them as anomalous.

Within some of the South Asian stories, we find an interesting clue as to their possible source. In another set of narratives of uncertain date, the *Brihatkatha* ("Great Story"), carpenters again are involved in the attempt to construct a flying vehicle.[127] In the story, Viśvila is said to have constructed a "mechanical cock on which he flew every night from Vārāṇasī to Ujjayanī."[128] Vārāṇasī is modern Benares, while Ujjayanī is modern Ujjain; these cities are about 750 kilometers apart, so this would have been a significant flight, even if it were simply in the writer's imagination. Note again that the device is clearly described as "mechanical", and no magical powers are invoked.

The story then recounts how a king sought information concerning the construction of an aerial chariot. Viśvila, described as a clever craftsman, responds to the ambassadors of the king: "This should not be told to anyone, be it craftsman or someone else. The science of aerial chariots is a difficult one for others than Yavanas." The story further relates that Viśvila flees with his wife through the use of a mechanical bird. Later, a second aerial vehicle in the form of the bird Garuda is again constructed.[129] Also in the *Brihatkatha*, just prior to this story about Viśvila, we find the following passage about a queen named Vāsavadattā:

---

126 *The Life of the Buddha and the Early History of His Order: Derived from Tibetan Works in the Bkah-Hgyur and Bstan-Hgyur*, trans. W. Woodville Rockhill (London: Kegan Paul, Trench, Trübner & Co., 1907), 108; also see the mention in Steven Moore, *The Novel: An Alternative History: Beginnings to 1600* (New York: Continuum, 2010), 449, and C. Rasanayagam, *Ancient Jaffna. Being a Research into the History of Jaffna from Very Early Times to the Portuguese Period* (New Delhi: Asian Educational Services, 1984), 151.

127 The earliest extant reference to the *Brihatkatha* is from the seventh century CE; see R. Vijayalakshmy, *A Study of the Peruṅkatai: An Authentic Version of the Story of Udayana* (Madras: International Institute of Tamil Studies, 1981), 8–11.

128 Eric P. Maten, *Budhasvāmin's Bṛhatkathāślokasaṅgraha: A Literary Study of an Ancient Indian Narrative* (Leiden: Brill, 1973), 18.

129 See Maten, 18, and note pp. 3 and 10–11; also note Laufer, *The Prehistory of Aviation*, 48–49, and Venkatarama Raghavan, *Yantras or Mechanical Contrivances in Ancient India* (*Transactions of the Indian Institute of Culture*, no. 10) (Basavangudi: Indian Institute of Culture, 1952), 15.

> [S]he wants to make a trip in an aerial chariot to see the earth. But the craftsmen invited to do the job maintain that they are not able to make such a vehicle and that only the Yavanas possess such knowledge. A Brahman doubts the sincerity of the craftsmen and accuses them of being unduly secretive.[130]

First, note again the motif of secrecy in relation to technological knowledge. More particularly, the knowledge in question is attributed to a foreign culture. In early Indian literature, the Sanskrit term "Yavana" was a phonetic rendering of the Greek word for "Ionians." The term referred to the Greeks, and later, to Arabs as well, and foreigners generally.[131] In the *Brihatkatha*, there are various indications that the builders of these vehicles — in the text, the machines are called *ākāśa-yantras*, meaning "aether devices" — are foreigners. In the story of Viśvila, when he is caught he beseeches his friends not to tell anyone the secret of these flying vehicles, and the implication is that he has learned the secret by way of outside expertise.[132] Further aerial vehicles are presented in the *Ratnaprabha-lambaka*, and again these vehicles are products of clever craftsman working in wood.[133] Another Indian romance, the *Harshacarita*, or "Deeds of King Harsha," from the seventh century CE, speaks of a flying machine made by a Greek who had been taken prisoner.[134]

In Sanskrit literature there are numerous discussions of mechanical devices, for use in warfare, architecture, and entertainment. These mechanisms collectively are referred to as *yantra* in the texts, a term that refers to a "contrivance" or "engine" — note the terms *ākāśa-yantra* cited above.[135] The flying machines were one category of *yantra*, but the flying machines themselves

130 Maten, 17.

131 Raghavan, 9; also see Maten, 18, n. 1.

132 Ibid., 13–15.

133 Ibid., 18–19.

134 Laufer, *The Prehistory of Aviation*, 50. Also see Panduranga Vamana Kane, ed., *The Harshacarita of Bāṇabhaṭṭa: Text of Uchchhvāsas I–VIII*, 2nd ed. (Delhi: Motilal Banarsidass, 1965), 161, as well as Raghavan, 10–11; Raghavan cites another story where an architect named Mandhata uses an aerial vehicle to fly and visit his son.

135 See Raghavan, 2; also note Mira Roy, "The Concept of *Yantra* in the *Samarāṅgaṇa-Sūtradhāra* of Bhoja", *Indian Journal of History of Science* 19.2 (1984): 118–124, and Dvijendra Nath Shukla, *Samarangana-Sutradhara, Pt. II: Royal Palace and Royal Arts* (Lucknow: Vastuvanmaya-Prakasna-Sala, 1967), pt. 2, 37–52. In addition, note pp. 77–78 of Kali Kumar Dutta Sastri, "A Mediaeval Account of Mechanical Devices in India", *Proceedings of the Indian History Congress* 24 (1961): 74–78. For a highly speculative discussion of this topic, see K.R. Shruthi and Rajani Jairam, "Mechanical Contrivances and Dharu Vimanas Described in 'Samarangana Sutradhara' of Bhojadeva", *IOSR Journal of Humanities and Social Science*, 20.12 (December 2015): 16–20, noting especially pp. 18–19.

also are described through different categories, including wooden birds, flying chariots, and so on. For example, there are *vimanas*, or aerial chariots in which the gods and other entities travel, and *vyomaga*, another kind of aerial vehicle — this latter term literally means "sky-going" or "flying".[136]

Further Sanskrit sources speak of flying machines, again said to be built by Greeks.[137] One source, the eleventh-century CE architectural treatise *Samarāṅganasūtradhāra*, gives an interesting — if peculiar — description of a wooden flying vehicle:

> Make a huge bird-like shape with a wing on each side fully carved out of a light wood with a space in the bottom for fitting up a mechanism full of fire and the *Para* in its proper (the middle) place. Now [with] the traveller on, the switch also on, and with the air produced by the two wings, the *Para* will be in action and with its power so generated, the machine will go to a great distance in the sky, giving amusement both to the man on it, and those below it, the spectators.
>
> And just like this, make a heavier wooden aeroplane in imitation of a *vimana* house. In it, place four strong jars full of *Para* and then also place an iron pot full of fire underneath them. Thus heated, these jars generate power by means of Mercury, *Rasaraja*, the *Para*, and they will make the machine move and by the power of the *Rasaraja* [and] *Para*, it will go up in the air and be the ornament of the sky.[138]

The terms *para* and *rasaraja* are ambiguous — the translator of this text here leaves them untranslated — and their function is not clear. *Rasaraja* was one of several terms used in early Indian medical texts for mercury, and *para* may also refer to the same substance.[139] A "*vimana* house" may refer to some kind of aerial vehicle that has the appearance of a dwelling. An updated translation renders the Sanskrit text into English only slightly more helpfully as follows:

> Having created a huge bird of lighter timber having framework well-knit and stout, (one) may introduce *rasayantra* (i.e., mercurial contrivance) … having the base for fire. Mounted over that the man by the breeze released

136 See Walker, 55, and MacDonell, 304; I wish to thank Robert P. Goldman, Professor of Sanskrit and India Studies at the University of California at Berkeley for his clarification of the definition of *vimanas*.

137 Shukla, 39.

138 Ibid., 49.

139 See S.K. Ramachandra Rao, ed., *Encyclopaedia of Indian Medicine*, 6 vols. (Mumbai: Popular Prakashan, 1985), 1.18; also see the different terms for mercury in 1.82.

> by the flapping locomotion of the wings or flank twain of that one, having inner self cosy [?] by the power of mercury spelling marvel, flies along far. This very way similar in extent [?] to a gods' abode or temple; flies along a heavy aerial car or vehicle (made) of timber or wood. One may provide in an orderly way on the four corners of that one [vehicle], stout ewers mercury laden. By the power of *rasarāja* (quicksilver or *pārada*, i.e., Mercury) heated and exploding quickly ... it becomes an object of decoration in the sky owing to the quality born of the heated pots by slow ignition burnt within the steel or iron potsherds.[140]

Even in this translation, with its parenthetical explications, the description is awkward; details in the passage also are lacking. The reader is left to puzzle as to the exact form and motive power of the flying vehicles described.

It is noteworthy, however, that two kinds of vehicles are described: one in the form of a bird and one in the form of a "house" or "temple". Again, it may be that these latter terms indicate that the writer is conceptualizing a vehicle of large size and square-shaped like a dwelling, thus requiring the four "jars" — one at each corner, perhaps — for propulsion. This is another example where an early text is wrestling to describe an unusual or unfamiliar technology with available terminology. The result is, of course, rather ambiguous. The description itself, with its use of terms such as "temple", is not that surprising in the sense that much of the *Samarāṅganasūtradhāra* deals with architecture, rather than flying vehicles.

One modern commentator interprets the text as indicating that the small, bird-shaped vehicle flies under the power of heated mercury, with additional force supplied by the wings being flapped by the pilot. The larger vehicle is also said to gain power through heated mercury.[141] In addition to its mechanical aspect, mercury has a magical connection to flight in some early South Asian tales: an old story tells of a prince who could fly by holding a "magic gem" in his mouth. One day, by accident, he gasps, and with "live mercury dropping from his mouth" he plummets to the ground.[142]

The large vehicle with the four "jars" described in the *Samarāṅganasūtradhāra* also reminds us again of the flying boat in Francesco Lana's *Prodromo*, with its

---

140 *Samarāṅganasūtradhāra*, Chapter 31, in *Samarāṅgana Sūtradhāra of Bhojadeva (Paramāra Ruler of Dhārā): An Ancient Treatise on Architecture ...* 2 vols., trans. Sudarshan Kumar Sharma (Delhi: Parimal Publications, 2007), 1.381–1.382.

141 Raghavan, 28–29.

142 Gordon H. Luce, *Old Burma — Early Pagán*, 3 vols. (Locust Valley, NY: J.J. Augustin, 1969–1970), 1.119.

metal spheres designed to lift the vehicle. But the writer of the Indian text gives no further details except to say cryptically:

> If the art of construction is not unfolded, it is not for lack of knowledge, but for the simple reason to keep it a secret, otherwise if made public, these yantras would not bring the desired result.[143]

This appeal to secrecy is a theme encountered in the Chinese texts, discussed earlier. It may reflect a genuine belief that such technology should be left to experts, although in this case, the phrase "would not bring the desire result" seems to imply something else. The allusion to secrecy could also be a convenient way for the writers of these various texts to avoid providing more accurate descriptions and details due to their own lack of understanding, or their lack of detailed sources. Nonetheless, despite both the vagueness and the appeal to secrecy, one thing is clear: both the Chinese and South Asian texts here place their discussions of flight in a distinctly mechanical context.

Other Indian texts speak of flying vehicles in more poetic forms. In the classical Sanskrit epic of India, *The Rāmāyaṇa of Vālmīki*, likely dating from the third century BCE, there is the following:

> It was a golden, jewel-encrusted chariot that flew where one desires. Harnessed to it were asses decked out with the trappings of gold ... The majestic overlord ... younger brother of Kubera, bestower of riches, ascended the chariot; with the rumble of a storm cloud he set out for the ocean, lord of streams and rivers.[144]

Further on, there is mention of another flying vehicle, called a *puṣpaka*. This was a large flying "palace" constructed by the architect and engineer of the gods, Vishvakarman. It was given to the god of wealth, Kubera, and then seized by his half-brother, a demon king named Ravana; later, it is inherited from him by the good demon Vibhishana. He, in turn, loans the vehicle to Rama,

---

143 Shukla, 49–50. Note, however, that there is some implication that the author of the *Samarāṅganasūtradhāra*, the eleventh-century King Bhoja, had witnessed aerial vehicles himself; see Raghavan, 29.

144 Sheldon I. Pollock, trans. and Robert P. Goldman, ed., *The Rāmāyaṇa of Vālmīki: An Epic of Ancient India*, vol. 3, *Araṇyakāṇḍa* (Princeton: Princeton University Press, 1991), 156; also note William K. Mahony, *The Artful Universe: An Introduction to the Vedic Religious Imagination* (Albany: SUNY Press, 1998), 25.

who flies it from Lanka to Ayodhya.[145] The text of the *Rāmāyaṇa* describes the vehicle as follows:

> *Puṣpaka*, the sunlike aerial chariot I took by force from my brother ..., is spacious, lovely, and swift as thought. Use it to your heart's content, shapely Sītā, in my company.[146]

This vehicle is further represented in the *Rāmāyaṇa* as capable of carrying a large number of passengers. Some mechanical details are lacking, such as its mode of propulsion: in the text, the vehicle is said simply to be "yoked to *haṃsas*", a species of goose. However, as one scholar of Sanskrit and Indian studies, Robert Goldman, has noted, "medieval commentators on the text do not take the reference to be actual goosepower [and] instead some regard the *haṃsas* merely as a decorative feature."[147] A Western tradition, by comparison, has Apollo's chariot being pulled through the sky by swans — another interesting conflation of natural and artificial flight.[148] It should be noted that in the South Asian context, the *haṃsa* is indeed used as a decorative element and is understood as a symbol of the soul, along with other religious and philosophical elements.[149]

So, how to decipher these Indian tales? Or, at the very least, how to frame our reading of these stories? As noted at the beginning of this chapter, Laufer was interested in the interpretation of early tales of flying machines, but his key point was that the writers of such tales possessed a clear conception of the idea of mechanized flight. Early texts from many cultures discuss mechanical devices for flying from place to place. Of course, this does not necessarily mean that the stories contain elements of fact, or that they are records of some now-lost technology. Yet the authors of these tales were interested in the

---

145 Goldman, private communication, 10 April 2007.

146 Pollock and Goldman, 204.

147 Goldman, private communication, 30 September 2007; also see the discussion concerning the *haṃsa* in Robert P. Goldman, Sally Sutherland Goldman, and Barend A. van Nooten, trans., *The Rāmāyaṇa of Vālmīki: An Epic of Ancient India*, vol. 6, *Yuddhakāṇḍa* (Princeton: Princeton University Press, 2009), 471, 1480, and 1520, and in Robert P. Goldman and Sally Sutherland Goldman, trans., *The Rāmāyaṇa of Vālmīki: An Epic of Ancient India*, vol. 7, *Uttarakāṇḍa* (Princeton: Princeton University Press, 2017), 655, n. 8.

148 See Robert L. Fowler, *Early Greek Mythology*, 2 vols. (Oxford: Oxford University Press, 2013), 2.149, as well as A. Lindsay Price, *Swans of the World: In Nature, History, Myth & Art* (Tulsa, OK: Council Oaks Books, 1994), 44.

149 See the discussion and examples in Vogel, *The Goose in Indian Literature and Art*, cited earlier.

phenomenon of flight, and there remains the question of why this was the case. As Laufer states:

> The Indians saw two points clearly — that aircraft must operate on the principle of the flight of birds and that a mechanism is required to start the machine, to keep it in mid-air, and to make a descent ... [T]he stories cited ... are sufficient evidence of the fact that they devoted considerable thought to problems of the air and aeronautics.[150]

A modern commentator notes the following about the flying machines in the *Samarāṅganasūtradhāra*:

> Being the only unequivocal ancient Indian source on building aeroplanes ... [the *Samarāṅganasūtradhāra*] has attracted considerable attention in recent times (a casual internet search will make this clear). However, I would like to shift the query from the obvious, but strictly speaking unanswerable questions (*did they try? did they work?*) to another set of considerations, that bear upon the entirety of the text.
>
> The author states something that should not pass unobserved: the instructions to make flying machines are, purposely, *incomplete*. Due to ethical considerations, the descriptions are not enough to effectively make any real aeroplane. The construction of the machines has not been explained, for the sake of secrecy, and not due to lack of knowledge. In this respect, that should be known as the reason, i.e. that the details are not fruitful when disclosed. Nonetheless, the lengthy chapter was felt worthy of inclusion in this large Vāstu [i.e., architectural] treatise. If at all one wished to put it into practice, extra information (perhaps to be provided in person) would have to be acquired.[151]

Whatever the veracity of the existence of machines capable of flight in South Asia in such an early period, what is important is how the early authors deal with the technical knowledge — it is rendered as obscure:

> Although this may be the case due to the esoteric nature of the knowledge involved in the specific chapter, what bears upon the entirety of the

150 Laufer, *The Prehistory of Aviation*, 51.

151 See pp. 41–42 of Mattia Salvini, "The *Samarāṅganasūtradhāra*: Themes and Context for the Science of Vāstu", *Journal of the Royal Asiatic Society*, series 3, 22.1 (January 2012): 35–55.

> work is that a gap between textual and extra-textual knowledge is explicitly mentioned and nowhere seen as a flaw. The list of qualifications of someone who would be able to actually make the machines is worth mentioning:
>
> Connection to a traditional lineage, skill, direct instructions,
> Practice of the *śāstra* [i.e., rules, precepts], exertion in the activities of Vāstu [i.e., architecture], and intelligence.
>
> At least two of the requirements, it will .be noticed, indicate the necessity of additional information compared to what is found in the text.[152]

In general terms, this suggests either of two things: first, there is the possibility that an early author would cloak the details in obscurity to mask the fictional nature of these flying machines — that is, obscurity is written into the text to provide a sense of false erudition and spare the author having to invent greater detail. Alternatively, a writer of one of these texts may have heard a fanciful story of flight, but taken it as true. The original tale would have lacked details, and so the author's written account, in turn, would have communicated to readers that further details were unavailable to him.

More specifically, in the cases of these early written accounts of flying vehicles, there is indeed "a gap between textual and extra-textual knowledge". What these early authors tell us is incomplete, and the implication is that there is some "extra-textual knowledge" — available perhaps through a connection "to a traditional lineage" or "direct instructions". As noted earlier in this book, this is an idea still familiar to us nowadays, since one learns about things ranging from bicycle repair to cooking, for example, both from texts and direct instruction, demonstration, and participation. In the case of these South Asian texts, however, the "traditional lineage" — if it indeed existed in the case of flying vehicles — was broken long ago, and the "direct instructions" are lost. All that remain are the texts, and their implied claim that there was indeed a parallel chain of unwritten knowledge; however, the texts are difficult to decipher, and the existence of a chain of such knowledge is impossible to verify.

It is not known if South Asian and East Asian cultures devoted the same degree of interest and attention to actual experiments in manned flight as they did to their intriguing tales. The early South Asian tales of flight have led to some claims — claims with a nationalist tinge — that India in ancient times somehow had advanced aeronautical knowledge.[153] In the 1950s, an insightful

152 Ibid., 42.

153 Note, for example, Kanjilal, *Vimana in Ancient India*, and Bharadvaja, *Vymaanika-Shaastra Aeronautics*, cited earlier — the latter, as noted, works from a very peculiar

Western journalist in India provided a wonderful critique of these beliefs in a brief essay cleverly entitled "Science by Sutras".[154] The belief that these tales of flight in the South Asian texts reflect, in completely literal terms, some kind of ancient, advanced technology goes back at least to the nineteenth century.[155] A similar nationalistic claim in the invention flight is found on behalf of the Ethiopians in a 1926 book, noting that they "may have invented flying machines, for the 'flying horse Pegasus' and the 'ram of the golden fleece' may not have been mere fairy tales."[156] The archaeological record, of course, has given us no evidence of ancient flying machines in Ethiopia, South Asia, or anywhere else.

The references in South Asian stories about flight to the Greeks naturally lead us to the question of Greek sources. Laufer notes that various Indian stories recount other Greek mechanical innovations, such as automata. He suggests that the Indian authors may have reasoned that since the Greeks were such wonderful mechanics and artisans in other technological fields, they also could make flying vehicles.[157] Indeed, there are not many other suppositions one can make — Greek sources themselves recount virtually nothing concerning flying vehicles or mechanical flight of the kind found in South Asian texts. The works of Heron and others discuss various machines and automata, but none mention aerial chariots, for example. The one clear reference on this subject is in the work of Aulus Gellius, cited earlier, which briefly speaks of a flying wooden dove, supposedly fashioned by Archytas. But again, there are no discussions in extant Greek sources of mechanical machines carrying people into the air, the type of description found in the South Asian texts.[158]

---

premise and offers outlandish conjectures concerning these early stories of aerial vehicles. There have even been absurd claims of evidence of nuclear warfare in South Asia in the remote past; see David W. Davenport and Ettore Vincenti, *2000 a.C., Distruzione Atomica* (Milan: Sugarco, 1979). For a critique of *Vymaanika-Shaastra Aeronautics*, see H.S. Mukunda, et al., "A Critical Study of the Work *Vymanika Shastra*", *Scientific Opinion* (May 1974): 5–12. For a critical examination of South Asian nationalist interpretations of history, see Ramaswamy, *The Lost Land of Lemuria*, cited earlier.

154 See Jean Lyon, *Just Half a World Away: My Search for the New India* (New York: Thomas Y. Crowell Company, 1954), 197–202.

155 One of the earliest Western citations is in J.B. Tolentino Ferrão, "Barcos e trens a vapor, balões aerostaticos, e outras invenções entre os antigos Hindus", *Vasco da Gama* 1 (1872): n.p.

156 Drusilla Dunjee Houston, *Wonderful Ethiopians of the Ancient Cushite Empire* (Oklahoma City: The Universal Publishing Co., 1926), 5.

157 Laufer, *The Prehistory of Aviation*, 52–53.

158 In Greek texts, the most detailed description of flying equipment is the account of the construction of wings by Daedalus in Ovid; see *The Metamorphoses of Ovid*, trans. Michael Simpson (Amherst: University of Massachusetts Press, 2001), 338–339.

Needham notes that accounts of mechanical birds are also found in Western sources, citing Aulus Gellius, and Laufer discusses the mechanical eagle attributed to Regiomantus (1436–1476 CE) as well as the seventeenth-century text of Francesco Lana, which also treats the subject of mechanical birds.[159] But Western sources are quite limited in terms of references to flying vehicles that can actually carry a pilot. Many such accounts of flight are fanciful or allegorical tales, such as the story of Daedalus. Lucian, the satirist of the second century CE, touches upon the theme of manned flight: in his fantastic and ironically titled *True History*, he writes of men riding enormous birds. Missing in the Graeco-Roman stories are any mechanical details such as those found in the Indian tales, or in the Chinese and Korean accounts. In the Western classical texts, specific discussions of technology appear rather limited, other than specialized sources such as Heron of Alexandria and Vitruvius.

In particular, since the discovery of the Antikythera Mechanism, one is struck by the contrast between the complexity that ancient Mediterranean cultures apparently were capable of in technological terms and what they described in extant texts. Alexander Jones, in his study of the Antikythera Mechanism cited earlier, *A Portable Cosmos*, indeed argues that the device had some textual context. He writes that one can even reconstruct some kind of history, if fragmented:

> We are obviously a long way from being able to put together a coherent story of the evolution and eventual degeneration of the ancient tradition of astronomical mechanisms, but there is enough evidence to suggest that complex and scientifically ambitious mechanisms were being made at least through the three centuries from about 100 BC to AD 200.[160]

But certainly there is no way to reconstruct a history in a similar manner for ancient ideas about flight from Western classical texts.

In other non-Asian sources, there are several accounts of clever artisans constructing wings and attempting flight; Laufer cites a tenth-century CE Arab narrative and an eleventh-century English one, as well as others.[161] None of these accounts, however, are as early as those found in the Indian texts, and they do not involve the construction of complete flying vehicles. Again, Greek sources do not seem to contain material relating to any kind of sophisticated flying machines, despite the claims of the Indian texts. Moreover, if the Greeks were the

159 For the reference to Aulus Gellius, see Needham, 4.2, 575–576.

160 Jones, *A Portable Cosmos*, 242.

161 Laufer, *The Prehistory of Aviation*, 68.

inventors of aerial vehicles, one might expect to see more Arab development of the idea, since they were the inheritors of so much Greek knowledge.[162] But Arab sources seem largely silent on the matter. What we do find is that the Indian theme of the romantic rescue by air is, perhaps not surprisingly, paralleled in certain Western tales. A commentator notes that the "theme of winning or rescuing a girl by means of aerial journey ... is widely current in folk tales."[163] One may recall, for example, that the Indian stories often involved the use of the flying vehicle to win the king's daughter.

In his popular book, *The Prehistory of Flight*, the scholar Clive Hart gives us a chronological guide to references in various Western texts to flying machines.[164] It is notable that most of the early references to flight in these sources involve the use of man-made wings. These references include a variety of sources, including chronicles, histories (such as that of Suetonius), and the writings of Roger Bacon.[165] However, again, there are no early discussions of the more complex type of man-carrying aerial vehicles as are found in the Indian tales, or for that matter in the Chinese and Korean texts.

Some early Western histories and chronicles include mentions of aerial vehicles, but lacking in any mechanical detail. In *The Irish Version of the Historia Britonum of Nennius*, dating from the fourteenth or early part of the fifteenth century, there is the following peculiar entry:

> Congalach [king of Ireland, 944–956 CE], son of Mailmithigh, was at the fair of Taillten on a certain day, and he perceived a ship in the air. He saw one of them [i.e., the ship's crew] cast a dart at a salmon. The dart fell down in the presence of the fair, and a man came out of the ship after it. When his head came down it was caught by a man from below. Upon which the man from above said, "I am being drowned" ... "Let him go", said Congalach; and he was allowed to come up, and he went away from them, swimming in the air, afterwards.[166]

162 Note, for example, Arab development of Ptolemy's work.

163 Jones, "The *Cléomadès* and Related Folk-Tales", 563.

164 Hart, *The Prehistory of Flight* (Berkeley: University of California Press, 1985), 195–197 et ff.

165 Ibid., 195–197.

166 James Henthorn Todd, ed. and trans., *The Irish Version of the Historia Britonum of Nennius* (Dublin: Irish Archaeological Society, 1848), 212–213; a similar mention of an aerial ship is found in the *Annals of Ulster*: "Ships with their crews were seen in the air above Cluain Moccu Nóis [the monastery of Clonmacnoise]"; see William M. Hennessy, ed., *Annala Uladh: Annals of Ulster, otherwise, Annala Senait, Annals of Senat; a chronicle of Irish affairs from AD 431, to AD 1540 ...* 4 vols. (Dublin: Printed for H.M. Stationery Office by A. Thom & Co., 1887–1901), 1.213.

This kind of account of a mysterious aerial vehicle is also found in *The Five Strange Wonders, in the North and West of England*, a seventeenth-century English report to members of Parliament. The writer of this work reports a coffin-shaped airborne object, hovering in the air accompanied by another object with the form of a dart or arrow, near Nottingham.[167]

But unlike the Indian tales, or the Chinese and Korean stories, in these early Western accounts there is no clear indication of an inventor or a craftsman, nor any specifics in terms of mechanical operation. In terms of Western invention, it is only in the works of Leonardo da Vinci that we begin to see extensive Western descriptions of flying vehicles, along with mechanical details. There seems to be no direct lineage from the early South Asian and East Asian tales and accounts to Western ideas of mechanized flight. What remains are several separate traditions, and a frustratingly incomplete picture.

5 Ainu Stories of Flying Machines

Oddly, an echo of the Indian tales of flying vehicles, and deities using such conveyances, appears in folklore from the far northern reaches of Asia. In the mythological narratives of the Ainu people of Japan and northern Russia, a character identified as the creator of the world, known as *samaikur*, has to journey through the celestial realm to retrieve the soul of his kidnapped princess. After a series of battles, *samaikur* is able to bring the soul back and return it to her body. The princess herself often is described in these tales as residing on a high mountain.[168]

Most interesting here is the fact that *samaikur* rides a vehicle to make his trip. Thus there are the combined themes — as in the Indian tales — of romance, rescue, flight, and mechanical devices. Whereas in some of the Asian tales of flight the vehicle takes the form of an artificial bird, in the Ainu tale the device is called a *sinta* or *shinta*, which is usually the term for a baby's cradle.[169] However, among the Ainu this word is also used to refer to an aerial vehicle — one story even describes a character who "flew around the sky and fought from a *sinta* (air conveyance) like a goddess", a scene rather reminiscent

167 See *The Five Strange Wonders, in the North and West of England: as they were communicated to divers honourable members of Parliament* ... (London: Printed for W. Thomas, 1659).

168 Note p. 306 in Kan Wada, "Some Shamanistic Features of Ainu Religion", in Juha Pentikäinen, ed., *Shamanism and Northern Ecology* (Berlin: Walter de Gruyter & Co., 1996), 305–312.

169 The term is usually transcribed into Japanese *katakana* as シンタ [*shi-n-ta*], or in a mixture of *hiragana* and *katakana* as しンタ [*shi-n-ta*].

of the Indian accounts.[170] Other stories recount the riding of a *sinta* through the air to travel from place to place, as a kind of aerial "sledge".[171]

For the Ainu, the connection between a "cradle" and an aerial vehicle is not that peculiar, and even a traditional lullaby equates the two, with the cradle referred to as a "sleeping ship".[172] One commentator says that

> Ainu gods are said to fly through the sky riding aboard vehicles called *shinta*, which is the common word for "cradle." The Ainu cradle consists of a wooden frame with rails at the sides. Cords are attached to it at four places, and it is suspended from a rafter or a branch. The cradle is swung back and forth like a swing to put the baby to sleep. The "chariots" in which the gods fly through the sky, emitting fierce rumblings, are clearly patterned after these cradles.[173]

Another study notes:

> The *shinta* is also the spaceship of the gods who visit the Ainu land from the upper world. In a *kamuy yukar* called "Shower of Embers" a deity, son of the thunder god or dragon god in high heaven, descends to the land of human beings riding a *shinta*.[174]

---

170 Katsuichi Honda, *Harukor: An Ainu Woman's Tale*, trans. Kyoko Selden (Berkeley: University of California Press, 2000), 297; also see Emiko Ohnuki-Tierney, "A Northwest Coast Sakhalin Ainu World View", 2 vols. (PhD diss., University of Wisconsin, 1968), 1.55 and 1.122.

171 See p. 494, as well as n. 367 on p. 717 of Bronisław Piłsudski, "On the Bear Festival of the Ainu on the Land of Sakhalin", in Alfred F. Majewicz, ed., *The Collected Works of Bronisław Piłsudski — Volume 1: The Aborigines of Sakhalin* (Berlin: Mouton de Gruyter, 1998), 438–561. Also note the comments in Mashiho Chiri, *Ainu bungaku* (アイヌ文学 "Ainu Literature") (Tokyo: Gengensha, 1955), 90 and 96, and in idem, *Chiri Mashiho chosakushū* (知里真志保著作集 "Collected Works of Chiri Mashiho"), 6 vols. (Tokyo: Heibonsha, 1973–1976), 1.340 and 1.388.

172 Note p. 658 in Sakurako (Sherry) Tanaka, "Ainu Shamanism", in Eva Jane Neumann Fridman and Mariko Namba Walter, eds., *Shamanism: An Encyclopedia of World Beliefs, Practices, and Culture*, 2 vols. (Santa Barbara, CA: ABC-CLIO, 2004), 1.657–1.665.

173 Donald L. Philippi, *Songs of Gods, Songs of Humans: The Epic Tradition of the Ainu* (Tokyo: University of Tokyo Press, 1979), 78; also see the discussion in Tanaka, "The Ainu of Tsugaru: The Indigenous History and Shamanism of Northern Japan" (PhD diss., University of British Columbia, 2000), 216–217.

174 Tanaka, "The Ainu of Tsugaru", 217. The term *kamuy yukar* can be rendered "epic concerning deities", with *kamuy* (カムイ) being the Ainu word for a deitiy, or a spiritual or supernatural entity, and *yukar* (ユカラ) referring to the Ainu epic poem format; see Masayoshi Shibatani, *The Languages of Japan* (Cambridge: Cambridge University Press, 1990), 4–5,

For the Ainus,

> thunder is believed to be hurled by the Thunder Deity, *Kanna Kamui* (literally, the Upper Deity"), for punishing disrespectful people. The deity is believed to drive *shinta*, a kind of divine vehicle, and hurl thunder.[175]

One Ainu *kamuy yukar* concerns a *kane shinta* (カネ シンタ), or "metal *shinta*", that flies.[176] In another story, a god in a flying vehicle visits the Ainu "Spider Goddess":

> One day
> from far out at sea
> a god was heard coming this way
> with a load roaring
> and rumbling.
> After a while
> he stopped his chariot [*shinta*]
> over
> my house.[177]

Even in this brief mention, the technological aspects of flight are implied: the roaring sound, and the fact that a particular landing spot can be chosen.

In addition to such tales, it should be noted that the Ainu refer to themselves generally as "Sky People", and they have a legend that says that their ancestors came from the skies on one of these *sinta*.[178] There are similar claims of

---

as well as Joseph M. Kitagawa, *On Understanding Japanese Religion* (Princeton: Princeton University Press, 1987), 60–62.

175 Takako Yamada, *The World View of the Ainu: Nature and Cosmos Reading from Language* (London: Kegan Paul, 2001), 60.

176 Tanaka, "The Ainu of Tsugaru", 217, and see the original text in Toshi Ueda, *Ishikari o mamoru kumo-shin no yumewomita paroatte* (石狩を守るクモ神の夢を見たパロアッテ "Paroatte Dreamed and Saw the Spider Goddess Who Protected Ishikari") (Hokkaido: Ainu Museum, 15 June 1999), 189. Also note pp. 176, 179, 180, and 183 of "Nabesawa-3 kamuyyukar (2)「ツブドルマケ ランラン」" in Shiho Endō and Hiroshi Nakagawa, eds., *Motozo Nabesawa's Ainu-language Notes (1928–1959) at the National Museum of Ethnology* [= *Senri Ethnological Reports* 134] (Osaka: National Museum of Ethnology, 2016), 173–186.

177 Philippi, 78; on the Spider Goddess, also see Michael Ashkenazi, *Handbook of Japanese Mythology* (Santa Barbara, CA: ABC-CLIO, 2003), 294.

178 See the comments in M. Inez Hilger, *Together with the Ainu, a Vanishing People* (Norman, OK: University of Oklahoma Press, 1971), and the brief mention in Robert Moes and

such origins in other cultures, although not fully documented — for example, the Torajan people of Indonesia apparently view their origins as being from the sky.[179] One account notes that the

> nobility of the southern Toraja districts traces its origin to the descent from heaven of the mythological noble Puang Tamborolangi. He is said to have come to earth in a magical Toraja house that descended atop sacred Mount Kandora.[180]

Here, as in the account from the *Samarāṅganasūtradhāra*, there is an odd connection between architectural structures and the idea of flight. In turn, the Japanese — perhaps through their historical connection with the Ainus — have tales of deities descending from the sky in the oddly-named *ama no iwafune* (天の岩船), literally meaning "stone boat of the cosmos".[181]

Again as in the Indian tales, the Ainu gods fly and even engage in combat on the *sinta*. As one author notes:

> For the Ainu, the beginning of the universe is characterized by the activities of the deities and the demons. In the beginning, the deities fought one another on air conveyances (*sinta*).[182]

In his dictionary of the Ainu language, the famous Ainu anthropological linguist Chiri Mashiho has the following entry:

Anne Pike Tay, *Mingei: Japanese Folk Art from the Brooklyn Museum Collection* (New York: Universe Books, 1985), 175.

179 Note the discussion in Lawrence Blair, *Ring of Fire: An Indonesian Odyssey* (Singapore: Didier Millet, 2010), 53 and 62.

180 See p. 14 of Eric Crystal, "Myth, Symbol and Function of the Toraja House", *Traditional Dwellings and Settlements Review* 1.1 (Fall 1989): 7–17.

181 See Tanaka, "The Ainu of Tsugaru", 496; also note Ian Reader and George J. Tanabe, Jr., *Practically Religious: Worldly Benefits and the Common Religion of Japan* (Honolulu: University of Hawai'i Press, 1998), 171, and Robin D. Gill, *Mad in Translation: A Thousand Years of* Kyōka, *Comic Japanese Poetry in the Classic* Waka *Mode* (Key Biscayne, FL: Paraverse Press, 2009), 42.

182 See p. 489 of Ohnuki-Tierney, "Concepts of Time among the Ainu of the Northwest Coast of Sakhalin", *American Anthropologist*, new series, 71.3 (June 1969): 488–492; also see Tanaka, "Ainu Shamanism", in Fridman and Walter, 1.658.

> § 24. Baby Cradles
> (1) sinta [šin-ta シンタ] …
> Note — In Karahuto [i.e., South Sakhalin], except the northwestern part, the Ainu did not use cradles. However, in [Ainu] ancient epics, e.g., in Hokkaido, this term appears [in reference to] the airship of the deities.[183]

The Japanese linguist Kindaichi Kyōsuke collected and studied the *yukar*, the ancient epics of the Ainu.[184] One such epic that he studied has descriptions of deities fighting in the air, and another sacred work tells of a god who comes down on a *sinta* and destroys a village.[185]

A similar scene is recounted in an Ainu tale entitled "A Crow Baby". A young man, who is also described as a deity, leaves his home to go and fight in a war:

> He prepared for war. To be well-prepared … he took out a sword and fastened it [to his waist]. Then he took out a bow and … arrows; he then took out a spear and went outside to engage in war. He went out through the skylight. The wife was in the house. As she was in the house, she heard her husband going up in the sky to fight in the war. She heard him riding on the aircraft [*sinta*] going up in the sky. Therefore she went outside.
>
> After getting outside, she looked about. As she looked toward to the sky, she saw her husband flying in the aircraft in all directions and fighting. Then she went into the house. After getting inside, she stayed there. Then there was a very loud sound from the war in the sky. Then there was a loud sound from the aircraft which was flying around and fighting.
>
> Even though she stayed at home, the sound of the war became louder and louder; then in the calm late afternoon, there was a very loud sound as though something had fallen in the back of the house. Therefore, the

183 Chiri, *Bunrui Ainugo jiten* (分類アイヌ語辞典 "Classified Dictionaries of the Ainu Language"), 3 vols. (Tokyo: Nihon Jōmin Bunka Kenkyūjo: 1953–1954; 1962), 3.7; also see Nana Sato-Rossberg, "Chiri Mashiho's Performative Translations of Ainu Oral Narratives", *Japanese Studies* 28.2 (2008): 135–148.

184 Kyōsuke Kindaichi, ed., *Kitaezo koyō ihen* (北蝦夷古謡遺篇 "Ancient Folklore of Northern Ezo") (Tokyo: Kōin Sōsho Kankōjo, 1914). Also see the references to *sinta* in idem, *Ainu bunka — minzokugaku* (アイヌ文化・民俗学 "Ainu Culture — Folklore") (Tokyo: Sanseidō, 1993), 130, 318, and 323, and idem, Ainu seiten (アイヌ聖典 "Ainu Scriptures") (Tokyo: Kaizōsha, 1930), 68, 70–71, 95–96, etc.

185 See Shigeru Kayano, *Honoo no uma: Ainu minwashū* (炎の馬：アイヌ民話集 "Horse of Fire: A Collection of Ainu Folklore") (Tokyo: Suzusawa Shoten, 1977), 264–266.

> woman went to the back of the house and looked about; it looked as though the young deity husband [*sic*] had been defeated in the war. He was killed in the air conveyance, and it looked [as if] he had fallen in the air conveyance from the sky.[186]

Emiko Ohnuki-Tierney, the anthropologist who collected and studied this tale, makes the following comment concerning the passage above:

> The *sinta* is the air conveyance which the Ainu believe the deities use for travel in the sky. They believe that the deities often fought one another in the sky at the beginning of the universe. Chiri[187] also encountered the same use of this term both in Sakhalin and Hokkaido Ainu tales, but he points out that in the recent Hokkaido Ainu dialects, the same term *sinta* means a "cradle for a baby." Chiri suggests that earlier the term meant "air conveyance," but it has changed among the Hokkaido Ainu.[188]

Although the conflation of the terms for an aerial vehicle and a cradle may seem strange to us, this can be understood as culturally equivalent to the "flying carpets" of ancient Persia or — closer to the Ainu tradition, perhaps — shamans riding their drums into the sky.[189] Given a technological concept, such as flying, the question for a storyteller is how to contextualize that concept. Something familiar must be used to make the story accessible to readers (or listeners). The storyteller may have heard the story in a form where the flying vehicle was presented with a different cultural iconography and changed it

186 Ohnuki-Tierney, *Sakhalin Ainu Folklore* (Washington, D.C.: American Anthropological Association, 1969), 92–94.

187 This refers to the anthropological linguist Chiri Mashiho (1909–1961), cited earlier; that author notes other Ainu tales of *sinta* on p. 19 of Chiri, 樺太アイヌの説話 ["Folktales of the Sakhalin Ainu"], *Karafuto-chō hakubutsukan ihō* (樺太庁博物館彙報 "Karafuto Prefecture Museum Bulletin") 3.1 (1944): 1–146, and on pp. 54–55 of idem, 樺太アイヌの説話 ["Folktales of the Sakhalin Ainu"], *Minzoku-gaku kenkyū* (民族学研究 "Ethnology Research") 12.4 (1948): 46–56.

188 Ohnuki-Tierney, *Sakhalin Ainu Folklore*, 106.

189 Concerning flying drums, see the mention on p. 6 of Ter Ellingson-Waugh, "Musical Flight in Tibet", *Asian Music* 5.2 (1974): 3–44, as well as p. 538 of J.W. Layard, "Shamanism: An Analysis Based on Comparison with the Flying Tricksters of Malekula", *The Journal of the Royal Anthropological Institute of Great Britain and Ireland* 60 (July–December 1930): 525–550. It is interesting to note that the Chinese also have a tradition of "flying carpets" — see p. 304 of Rosalie Hankey, "Ghosts and Shamanism in Kwangtung", *California Folklore Quarterly* 2.4 (October 1943): 303–308. On "flying carpets" in the context of the mythology of the Middle East, see Warner, 63–70.

to suit the audience of his or her own culture: therefore, as tales get transmitted, a flying carpet might become a wooden bird, and then perhaps appear as a aerial cradle.

In fact, the Ainu also have a tradition of a flying horse made of gold — reminiscent of Western tales of a flying horse of brass, discussed later in this chapter. Part of one Ainu story runs as follows:

> A man had lost his wife, and was searching for her everywhere, over hill and dale, forest and sea-shore. At last he came to a wide plain, on which stood an oak-tree. Going up to it he found it to be not so much an oak-tree as a house, in which dwelt a kind-looking old man. Said the old man: "I am the god of the oak-tree. I know of your loss, and have seen your faithful search. Rest here awhile, and refresh yourself by eating and smoking. After that, if you hope to find your wife again, you must obey my orders, which are as follows: Take this golden horse, get on his back, fly up on him to the sky, and, when you get there, ride about the streets, constantly singing."
>
> So the man mounted the horse, which was of pure gold. The saddle and all the trappings were of gold also. As soon as he was in the saddle, the horse flew up to the sky. There the man found a world like ours, but more beautiful. There was an immense city in it; and up and down the streets of that city, day after day, he rode.[190]

Whether by using a kind of "cradle" or a metal horse, notice that even in the Ainu tradition, there is a clear differentiation: individuals do not simply fly about, but do so in *vehicles*. In fact, this concept of flying with the aid of a device — rather than simply through magical levitation or angel-like wings — is found in other indigenous cultures, as well.

---

190 Basil Hall Chamberlain, *Aino Folk-Tales* (London: Folk-Lore Society, 1888), 21; also see Carl Etter, *Ainu Folklore: Traditions and Culture of the Vanishing Aborigines of Japan 1949* (Chicago: Wilcox & Follett, 1949), 155, and p. 10 of Juha Janhunen, "Tracing the Bear Myth in Northeast Asia", *Acta Slavica Iaponica* 20 (2003): 1–24, as well as Tanaka, "The Ainu of Tsugaru", 217.

6 Hopi Lore about Flying Vehicles

Similar to the Ainu, Hopi culture also has figures flying about on vehicles. In the Hopi tales, there is the *paatuwvota* — a "magic flying shield". Other Hopi stories speak of flying devices that are described as "trays" and "baskets":[191]

> In Kawaíhkaa, a Pueblo village in New Mexico, the people were living. North of the plaza, at the house where there was a long ladder, lived two maidens. They were sisters, and persistently refused to marry any of the young men in the village. Finally the Night (Tókila) concluded to try to marry the two maidens, and came to the house. He came there in the evening and asked them to marry him. They said they would lay the matter before their parents, and if they were willing they would marry him. The parents were willing, and so the two sisters waited for their suitor. The next evening he came to fetch his two brides.
>
> Leaving the village they went through a narrow passage. Outside of the village they found a large tray (póta), which the Night had left there. "This we shall enter," he said. So they all took a place on the tray, whereupon they were lifted up and carried through the air to Kawaíhka Nuwátoaovi, where they entered a deep canyon or gulch.[192]

Another Hopi tale, in fact, has a *kachina* or "spirit being" on a flying shield chase the woman whom he intends to marry.[193] A story of a newlywed couple travelling on one of these vehicles describes the flight:

---

191 See the discussion in Ekkehart Malotki, Ken Gary, and Karen Knorowski, *Hopi Stories of Witchcraft, Shamanism, and Magic* (Lincoln, NE: University of Nebraska Press, 2001), xl, and Ekkehart Malotki, Michael Lomatuway'ma, Lorena Lomatuway'ma, and Sidney Namingha, *Hopi Tales of Destruction* (Lincoln, NE: University of Nebraska Press, 2002), 65. Also see Harold Courlander, *Hopi Voices: Recollections, Traditions, and Narratives of the Hopi Indians* (Albuquerque, NM: University of New Mexico Press, 1982), 200–202, as well as p. 105 of Armin W. Geertz, "Tsu'ngyam Tradition: Men, Women and Snakes in Hopi Theology", in Cora Bender, Thomas Hensel, and Erhard Schüttpelz, eds., *Schlangenritual: Der Transfer der Wissensformen vom Tsu'ti'kive der Hopi bis zu Aby Warburgs Kreuzlinger Vortrag* (Berlin: Akademie Verlag, 2007), 103–118.

192 H.R. Voth, *The Traditions of the Hopi* (Chicago: Field Columbian Museum, 1905), 143; also note Malotki, Gary, and Knorowski, 66–67.

193 Robert S. McPherson, *Viewing the Ancestors: Perceptions of the Anaasází, Mokwič, and Hisatsinom* (Norman, OK: University of Oklahoma Press, 2014), 45; also note p. 39.

> [T]he young man came out with his new wife and said to her, "All right, let's climb aboard. Then I want you to close your eyes tightly again. And you're not to open them until I tell you to."
>
> This time on their way home, the two were going to travel by flying shield. Together they climbed onto the shield, and she firmly shut her eyes. As the shield lifted off, the *kachinas* all gave out a boisterous yell. The spectacle was incredible; every sort of *kachina* was present. All of a sudden, as the couple flew along, flashes of lightning were visible in the air, and the rumble of thunder could be heard. When the shield rose higher, drizzle began to fall. The *kachinas* were now accompanying them. They actually followed the pair in the form of clouds.
>
> Thus the entire entourage traveled along until it neared Mishongnovi. Everyone landed on the west side of the village. The young man and his new wife, who had not yet opened her eyes, were the first to descend, followed by the *kachinas*. At this point the husband turned to his young wife and said, "You can open your eyes now." When she did, she discovered that they had landed at Kyeletipkya, a place west of Toriva. A host of *kachinas* had accompanied them, bearing a variety of gifts. The youth prompted his wife, "All right, you can climb down. We've arrived."[194]

As in the South Asian tales of flight, there is a peculiar combination here of the themes of aerial vehicles and romance. Despite this rather non-technical setting, Hopi storytelling makes clear that the *paatuwvota* is a device for flying — that is, as in the tales from other cultures examined here, the deities are not themselves winged nor able to magically fly on their own. Rather, they possess these flying shields for travelling. Again, the technology sits within the cultural context of the society that is telling the story:

> In [Hopi] stories, powerful beings, such as gods, are bound to own a *paatuwvota*, or "flying shield," that they employ as a mode of transportation. With it, they reach any destination in no time because they can fly about with the shield. The shield is made from cotton and is woven in the manner of a Hopi wedding robe. The owner of such a vehicle needs only to climb aboard, tug on something and utter a command, whereupon the

194 Ekkehart Malotki and Michael Lomatuway'ma, *Earth Fire: A Hopi Legend of the Sunset Crater Eruption* (Walnut, CA: Kiva, 2005), 43–44.

> shield rises in the air and takes him wherever he says. To descend, he goes through the same procedure.[195]

Despite the obscurity of the mechanism, it is still clear that there is a machine in operation here, and that the flying shield is used as a piece of technology to carry out tasks. A Hopi story entitled "The Destruction of Awat'ovi" describes in a very matter-of-fact fashion a *kachina* named Pavayoykyasi using a flying vehicle, as if the technology is something the listener of the story would not find at all surprising:

> Pavayoykyasi, meanwhile, had walked a little ways southeast where he had left his … magic flying shield. The shield had two parts, with the lower one spinning and the upper one remaining still. Climbing aboard, Pavayoykyasi rose up into the air and flew off. He was traveling to his home, in the Buttes Area, where he lived with his grandmother at a place called Siipa.[196]

Note a mechanical description is included here, however brief. Pavayoykyasi goes on to pilot his flying shield "to the top of the mountain, where he found the village he was looking for." We then read that he "sat down his flying shield right by its edge [i.e., the edge of the mountain], climbed out, and after chewing his medicine headed for the village."[197]

In terms of contextualizing the idea of flight as a whole, Patrick Pérez, a specialist in Native American architecture and landscape, notes:

> From the height and the plunging view over the valleys and the interior of canyons, the Hopi, as all the modern or ancient Pueblos, derive their habit of describing and representing space from "up above". This modality is most evident in the geometry of their architecture, but it is also present in the tales and myths describing territories from atop a flying shield.[198]

---

195 Ekkehart Malotki, Michael Lomatuway'ma, Lorena Lomatuway'ma, and Sidney Namingha, *Hopi Ruin Legends: Kiqötutuwutsi* (Lincoln, NE: University of Nebraska Press, 1993), 429.

196 Ibid., 307.

197 Ibid., 317.

198 See p. 87 of Patrick Pérez, "Elements of an Amerindian Landscape: The Arizona Hopi", in Arnar Árnason, Nicolas Ellison, Jo Vergunst, and Andrew Whitehouse, eds., *Landscapes Beyond Land: Routes, Aesthetics, Narratives* (New York: Berghahn, 2012), 83–97.

In addition to the flying shield, the Hopi shamans also have a flying "gourd", which is also described in a surprisingly mechanical fashion: it is activated by the twisting of a cord, and makes a "humming noise" as it flies about. Hopi stories also mention a "hissing" round aerial vehicle that transports people through the air.[199]

7 Tales from Oceania about Flying Vehicles

Tales that combine aerial vehicles and romance are found in South Asian texts, Hopi traditions, and in stories from Oceania. One anthropologist, William Lessa, collected a number of stories from the people of the Caroline Islands in the western Pacific that concern flying vehicles. In one story, "a special tree characterized by its light wood is sought by a man who wants to retrieve his wife, and in a dream, the tree itself shows the man its possibilities for being made into a flying machine."[200]

Another story from the same cultural tradition runs as follows:

> Jol was a spirit who went to Puluwat and there married a woman named Jat. When the people of Puluwat, who thought that he was a human, stopped giving his wife fish, he decided to go fishing himself and made highly successful catches with his trap and with hooks tied to his hair. One day before submerging under the water to place his trap he gave his wife a coconut ... A spirit came and took her away to Yap. When Jol was through fishing he surfaced and looked about for his wife but she did not answer his calls. He paddled to land, felled a breadfruit tree, and fashioned a frigate bird from it. He got into it and flew away, first to all the islands east as far as Ponape and then to Hok ... and all the islands west of Puluwat, but although he searched for his wife he did not see her.
>
> However, when he flew over Yap he saw her. The spirit who had taken his wife to Yap to live there with him was called Haujap. The women told Haujap that she craved fish and pestered him to catch some for her, even though he said he did not know how to fish. Haujap took a fish and set out to sea to catch bonito. The frigate bird espied the woman, swept down, and flew up with her. When Haujap saw the frigate bird, he had already

199 Malotki, Gary, and Knorowski, xl.

200 William A. Lessa, *More Tales from Ulithi Atoll: A Content Analysis* (Berkeley: University of California Press, 1980), 72.

> caught hundreds of fish. He tossed first one fish then another into the air and each time the bird swept down and caught them.
>
> Then Jol, who sat in the frigate bird with his wife, told her that all the fish were gone and that she should now tell Haujap that she was going away to Puluwat. The woman leaned out of the frigate bird, which was like a canoe, and cried out, "Haujap! I am leaving you to go to Puluwat!" Haujap looked up and saw his wife and was so startled that he collapsed and died.[201]

Lessa discusses a number of variants of this tale, and related stories of birds made from wood that can carry people.[202] Note the mention that the flying vehicle is "like a canoe"; this comparison is discussed further below. In another version of the tale above, there is a discussion about the technique of one particular craftsman, Maholbule:

> After the tree had been felled he cut a section about three out-stretched arm spans long. From the section he built a bird like a frigate bird. After having finished making the frigate bird he put some feathers [of wood] all over the bird. He also made a door under the wing of the bird ... He entered the frigate bird and flew away.[203]

While the stories are fanciful, again details are provided to make the tale have a distinctly technological aspect — and the actual feat of flying, then, is by way of machine, rather than simple magic.

In terms of a broad analysis of technology and myth, some of the early stories concerning flight may tell us something about how each culture chooses its vehicle. The South Asians have wooden birds and horses, but a Native American tale from the Seneca people has "two women [who] appeared from the west, sailing in a canoe through the air". In another story, there is this description: "[T]he old man shoved the canoe out of the doorway, and at once it rose into the air, through which it passed with great rapidity."[204] Similar flying canoes are found in stories from Oceania, such as this one:

201 Ibid., 69.

202 Ibid., 67–73.

203 Ibid., 68.

204 See J.N.B. Hewitt, ed., *Thirty-Second Annual Report of the Bureau of American Ethnology to the Secretary of the Smithsonian Institution 1910–1911* (Washington, D.C.: Smithsonian Institution, 1918), 224 and 407; also note p. 445 of Philip S. Lesourd, "The Passamaquoddy 'Witchcraft Tales' of Newell S. Francis", *Anthropological Linguistics* 42.4 (Winter 2000): 441–498.

> A well-known legend tells of the Princess of Vanuakula being kidnapped by the Prince of Tonga, and her being rescued by a chief of Kadavu in his magic flying canoe made of *kīkī* (*Flagellaria indica*, a cane used for weaving baskets).[205]

Further tales from this region describe various adventures using flying canoes.[206] A story from Namonuito, in Micronesia, notes that a person

> who knew how to do many things built two canoes that did not sail in water but flew through the air so fast that he was able to get from Satawan to Truk in two hours after sunrise. Two people on Truk saw the canoe arriving and called out, "O! A canoe is coming!" People on Satawan desired to know from him how it was possible to sail so quickly through the air with a canoe. But the man, who alone was able to fly it, intended to keep the secret for himself and did not want to show them how.[207]

Note how in these accounts the *fantastic* — flight — is contextualized technologically in the *familiar*: canoes. However, even the stories of flight from these remote cultures talk about it in terms of a technological object; that is, the stories make clear that the flying device is something *constructured*, and done so with special and even hidden knowledge. Once more, the theme of technological prowess is coupled with the theme of secrecy.

8 A Synthesis of Traditions in the "Flying Horse" Tales

Stories of flying horses illustrate two interesting points. First, there is the distinction similar to the one mentioned earlier in this chapter between the *fei tian*, or "flying beings", and *fei che*, or "flying vehicles": the former is mythical and the latter is mechanical, even if the latter is often placed in a mythical or fabulous context. In addition, the history of these stories of flying horses at times indicates cultural crossover and common sources.

As for the first point, one author notes:

---

205 See pp. 375–376, n. 30 of Paul Geraghty, "Pulotu, Polynesian Homeland", *The Journal of the Polynesian Society* 102.4 (December 1993): 343–384.

206 Note, for example, Tambiah's study, cited earlier.

207 Lessa, 72.

> Flying horses have a distinguished literary history. At least since Plato's "team of winged steeds and their winged charioteer" in the Phaedrus and the ancient myths of Pegasus and Phaethon, horses with wings have figured the human imagination and the unruly power of the individual will. To these themes the marvelous steeds of medieval romance (and their heirs in such Renaissance narratives as those of Ariosto and Cervantes) add technical mastery and the human penchant for deceiving and controlling other people with the help of sophisticated machinery.[208]

There are flying horses — such as Pegasus, a creature with wings — but there are also flying machines in the *form* of horses. Those latter devices are a separate category, and appear in various stories from a range of cultures; they illustrate the second point, concerning the transmission of stories between cultures. Tales clearly reflecting an idea of "sophisticated machinery" were transmitted and shared over great stretches of history and geography, indicating a strong interest in technology and its potential, well before the modern "technological age".

In terms of such an exchange of ideas, one can begin with Geoffrey Chaucer (ca. 1343–1400), and work backwards. Chaucer wrote of a special flying horse, and it seems that he was working from both Old French and Arabic sources.[209] Chaucer's "The Squire's Tale" is a story in the famed collection known as *The Canterbury Tales*. In the story, a fictionalized Genghis Khan is celebrating with a feast; a knight appears with gifts, one of which is a large horse made of brass. The knight states:

> The king of Arabia and India, my Liege lord, on this solemn day, salutes you as best he knows how, and sends you by means of me, who is already ready at your behest, in honor of your festival, this steed of brass. This horse can, easily and well, in the space of a day — that is to say, in four and twenty hours — bear your body unto every place your heart wishes, whether in drought or in torrents, through fair winds or foul, without harm to you. Or, if you wish to fly as high in the air as does an eagle when

208 See p. 287 of Craig A. Berry, "Flying Sources: Classical Authority in Chaucer's 'Squire's Tale'", *ELH* 68.2 (Summer 2001): 287–313.

209 See Carol F. Heffernan, *The Orient in Chaucer and Medieval Romance* (Cambridge: D.S. Brewer, 2003), 64–65. Also note the discussion of Chaucer's work in relation to other tales of flying horses on pp. 279 et ff. of W.A. Clouston, "On the Magical Elements in Chaucer's 'Squire's Tale,' with Analogues," in Frederick J. Furnivall, ed., *John Lane's Continuation of Chaucer's "Squire's Tale"* (London: Published for the Chaucer Society by K. Paul, Trench, Trübner & Co., 1888, 1890), 263–471.

> he soars, this self-same steed shall carry you evermore, in safety, till you arrive where you please. And even if you choose to sleep upon his back or rest, with the simple turning of this pin, you shall return unto your home again. He who wrought this magnificent horse knew of full many an ingenious contrivance. The craftsman bided his time and observed many differing configurations of the stars ere he completed this operation, and he knew full well many a magical seal and many a mystical spell.[210]

Note that although magic is said to have been involved in its creation, this steed is mechanical. In addition, there is the "pin" to operate the device, similar to the switch described in Indian tales of flying machines. This idea of a pin or switch might imply a system for the rapid release of stored or potential energy that would make the machine take flight.[211]

Perhaps the description of such a setup might help a reader of Chaucer's period understand that what the author sought to portray. But while it certainly is true that a switch or spring could imply potential energy being released, as in a wind-up mechanism, the description is quite vague in both Chaucer and the Indian sources. Nonetheless, it is interesting that what is essentially something like a modern "on" switch is being described in such an early text — although something similar existed, of course, as release mechanisms in clocks and other such devices. Regardless, what strikes the reader is a very modern, even anomalous, concept — that is, the idea of turning on a machine, especially one that then takes flight. This modern tone is reinforced by another aspect of the description of this mechanical horse: "And even if you choose to sleep upon his back or rest, with the simple turning of this pin, you shall return unto your home again." In short, this device can function on what one would call today an autopilot mechanism.

In Chaucer's story, the mechanical horse inspires a great deal of wonder, and people are described as comparing the device to Pegasus, and even to the Trojan horse — a kind of interesting and rather modern implication of suspicions about technological power. The story notes:

---

210 Gerald J. Davis, *The Canterbury Tales: The New Translation* (Bridgeport, CT: Insignia, 2016), 207–208.

211 I would like to thank Steven A. Walton of Michigan Technological University for making this point.

> Of sundry doubts did they thus chatter and debate, as ignorant people are wont to do about things that are crafted more cunningly than they can comprehend in their ignorance, and they usually expect the worst.[212]

Once again, it is clear that this is a case of a writer who understands technology and how it can be written about in terms that differ from simple tales of magic — and how technology can seem magical and dangerous.[213] Later in the tale, the knight gives the king a description of the horse's operation:

> Sir, there is no more to say than, when you desire to ride anywhere, you must turn this pin in his ear, which I shall tell you how to do just betwixt us two. You must also name unto him to what place, or to what country, that you wish to ride. And, when you come to where you wish to abide, bid him descend, and turn another pin, for therein lies the secret of the engine, and he will down descend and do your will, and in that place will he abide still.... Or, if you wish to bid him from thence to be gone, turn this pin and he will vanish anon.[214]

Again, the description is at the same time fantastical — the horse can take you anywhere — and pragmatic, with take-off and landing requiring the movement of two corresponding pins.[215] Invocation and incantation are not sufficient: there is also mechanical operation.

Chaucer's story, in turn, draws from earlier sources.[216] Mechanical flying horses appear in versions of *The Arabian Nights*, in *Li roumans de Cléomadès*, by the thirteenth-century French writer Adenet le Roi (Adenés li Rois), and in

---

212 Davis, 209.

213 See the comments in Minsoo Kang, *Sublime Dreams of Living Machines: The Automaton in the European Imagination* (Cambridge, MA: Harvard University Press, 2011), 68, as well as in Paul Beekman Taylor, *Chaucer Translator* (Lanham, MD: University Press of America, 1998), 11. Also see the discussion of the relationship between the technological and magical qualities of Chaucer's brass horse in Susan Crane, *Animal Encounters: Contacts and Concepts in Medieval Britain* (Philadelphia: University of Pennsylvania Press, 2013), 147–150.

214 Davis, 211.

215 See the discussion concerning the operating instructions for this horse on p. 540 of Kathryn L. Lynch, "East Meets West in Chaucer's Squire's and Franklin's Tales," *Speculum* 70.3 (July 1995): 530–551.

216 On the complex connection between these sources, see, for example, the discussion in Heffernan, 64–66, and in Robert M. Correale and Mary Hamel, eds., *Sources and Analogues of the Canterbury Tales*, 2 vols. (Cambridge: D.S. Brewer, 2002–2005), 1.182–1.186. Also see Crane, *Gender and Romance in Chaucer's "Canterbury Tales"* (Princeton: Princeton University Press, 1994), 138, n. 6.

another thirteenth-century French tale, *Meliacin ou le Cheval de fust*, by Girart d'Amiens.[217] In the *Cléomadès*, which may have been known to Chaucer, there is again a special horse.[218] It is a gift from Crompart, one of three African kings, for a Spanish princess named Marine. The horse is made of ebony, and can fly. Cléomadès, the brother of Marine, who is not happy about the romantic proposal of Crompart, tests the horse. Although the story is an adventure, it is interesting that there is a strong technological element here; not only is there the ebony horse, but the other gifts of the kings also include mechanical devices: chickens made of gold that can walk and sing, and a man of gold who wields a golden trumpet — in short, automata.[219]

In this tale, too, it is made clear that this "horse" is a mechanical device. Cléomadès launches the flying vehicle with the turn of a pin, and subsequently learns how to operate the mechanism ably.[220] He then goes on a series of adventures using this horse, and all ends happily.[221] In the *Méliacin*, the story is quite similar, with four pins — *chevilles* — noted as being used to control the mechanical beast.[222] The *Méliacin* gives some details concerning the mechanism of the horse; first, there is again a switch or pin that starts the mechanism:

> He turned a pin
> which was placed in the neck of the horse
> and the horse took up its way
> faster than a bird flies,
> so that I think nothing goes as fast
> as that horse was going,

217 See Adenet le Roi and André Henri Constant van Hasselt, *Li roumans de Cléomadès*, 2 vols. (Bruxelles: Victor Devaux et cie., 1865–1866), and Girart d'Amiens, *Meliacin ou le Cheval de fust*, ed. Antoinette Saly (Aix-en-Provence: CUERMA, 1990).

218 See pp. 221 and 225 of H.S.V. Jones, "The Cléomadès, the Méliacin, and the Arabian Tale of the 'Enchanted Horse'", *The Journal of English and Germanic Philology* 6.2 (January 1907): 221–243, as well as the longer discussion in idem, "The *Cléomadès* and Related Folk-Tales", cited earlier. Also see idem, "Some Observations upon the Squire's Tale", *PMLA* (*Publications of the Modern Language Association of America*) 20.2 (1905): 346–359, as well as Aurélie Houdebert, "L'histoire du cheval d'ébène, de Tolède à Paris: propositions sur les modalités d'une transmission", *Actes du colloque des 26–29 novembre 2012 "Byzance et l'Occident: Rencontre de l'Est et de l'Ouest"* (Budapest: Collège Eötvös József ELTE, 2013): 143–156.

219 See the brief mention of these in Heffernan, 65.

220 Jones, "The Cléomadès, the Méliacin ...", 225–226.

221 See the summary in ibid., 225–229, as well as Albert Henry, *Les Œuvres d'Adenet le Roi, Tome V: Cléomades* (Geneva:, Slatkine Reprints, 1996), 586 et ff.

222 Jones, "The Cléomadès, the Méliacin ...", 229–230.

> everywhere its master wished
> for an hour.[223]

Then, the text turns to further details:

> For now I wish to tell you how
> this very noble magic was done.
> The philosopher, who puts his pleasure in ingenious things,
> had placed there four pins
> which were made through necromancy,
> and so subtly designed
> that, if it was not by chance,
> no one might ever learn the construction.
> The first was placed in the neck;
> and the second, behind, in the croup;
> and the third pin was in the left flank;
> and the fourth was in the right.
> They were attached and fixed in the horse of wood,
> so that no one, through art or knowledge, could move the horse
> so as to lead it as much as a foot,
> if beforehand he did not know how to control
> and place the pins correctly.
> But they were in such awkward places
> to find or seek out
> that a man could scarcely locate them
> if he were not helped,
> unless Fortune brought it about.[224]

In the original French text, the term "noble magic" and "necromancy" are actually both rendered with *nigremance*. As examined in the discussion below, there is the odd idea here of magic — indeed, sorcery — being connected to what is clearly a mechanical device. It is as if the storyteller believes not in magic *per se*, but rather that those who practice such arts are, more fundamentally, possessors of secret *technological* knowledge.

This combination of mechanical knowledge and secrecy, in turn, is not that strange for this period, if one considers figures such as Leonardo da Vinci, who sometimes veiled his explorations of mechanics in arcane language. Folio 34v

223 Correale and Hamel, 1.179.
224 Ibid., 179–180.

of the Codex Arundel (Arundel 263 34v), for example, is filled with drawings of mysterious machines — investigations of an idea popular in Leonardo's time, perpetual motion. Next to one of his machine drawings there, in Leonardo's hand is the phrase, "Fanne modello, uno segreto rinchiuso, e di fori un verisimile da mostrare"; he seems to be describing a device that has a secret mechanism inside.[225] In the description of the horse device in the *Méliacin*, the details surrounding the location of the pins and their being so carefully hidden are aligned with the obscurity of their actual operation — and the even more obscure question as to what apparatus is actually inside the horse.

In a tale in *The Arabian Nights*, an ebony horse operated by a pin or screw plays a key role in helping a couple escape.[226] In the story, a man presents this mechanical horse, which a prince then mounts to see if he can get it to work:

> [T]he prince came up and mounted it, but, however much he moved his legs, the horse would not budge from where it was.[227]

Note how the story clearly lays out the idea that someone is faced with a piece of technology that is beyond their understanding: the prince is treating the mechanical horse as if it were a real horse, trying to get it to move with a squeeze or tap of the legs, as a trained equestrian would expect to do. The prince then addresses the man who has presented this horse:

> "Where is this speed that you claimed for it, wise man?" the prince asked, but at that the man came up to him and showed him a screw that would make it rise in the air. "Turn this," he said, and when the prince turned it, the horse started to move and then flew up with him into the sky, going on and on until it was out of sight.[228]

The tale also presents a contrast between religious belief or traditional thinking and technology and its mastery:

---

225 See pp. 30–31 of Benjamin B. Olshin, "Sophistical Devices: Leonardo da Vinci's Investigations of Perpetual Motion", *ICON: Journal of the International Committee for the History of Technology* 15 (2009): 1–39.

226 For a discussion of the three versions of *The Arabian Nights* version of this tale, see Correale and Hamel, 1.170–1.171; Correale and Hamel also look at the story as it appears in the *Cléomadès* and the *Méliacin*.

227 *The Arabian Nights: Tales of 1001 Nights*, 3 vols., trans. Malcolm C. Lyons (London: Penguin, 2008), 2.128.

228 Ibid., 2.128.

> The prince was startled, and regretted having mounted it, telling himself that this was a trick that the man had played on him in order to get him killed. "There is no might and no power except with God, the Exalted, the Omnipotent!" he exclaimed, and he then started to examine all the parts of the horse's body. While he was doing this, he noticed a protuberance like a cock's comb on its right shoulder and another on its left. As these were the only projections that he saw, he rubbed the one on the right shoulder of the horse, but as this made it go further up into the sky, he took his hand away. Then he looked at the knob on the left shoulder, and when he rubbed it, the horse's motion changed from a climb to a descent. It continued to come slowly down to earth, with the prince taking what care he could.[229]

In other words, the prince's conceptualization of this technological contraption goes from shock and suspicion to invocation of religious belief, and then from curiosity to — perhaps fearfully — mastery of the device and its operation. The technological setting and tone of this *Arabian Nights* tale is clear: the horse is indicated in the story as being one of several mechanical devices presented to a king, including an artificial peacock that "flaps its wings as each hour passes" and an artificial trumpeter with a special brass trumpet that can serve as an automatic alarm.[230] Interestingly, too, at the conclusion of the tale in some versions, the mechanical horse is completely disassembled.[231]

The technical description of how the mechanism flies is not explicitly given in these tales, but all the versions of the story are careful to portray this as the operation of a machine, with little reference to magic spells or incantations, contrary to what a reader might expect to find in a story from an essentially pre-technological age. A commentator notes the following about the Chaucer version of the flying horse tale:

> One of the first items of information that the "strange knyght" provides about the horse is that it can promptly be returned to earth by the simple "writhyng of a pyn" ... Later, when the knight offers more precise instructions about the horse's operation, he does so matter of factly and explicitly. The text even suggests that the knight may literally be pointing out

229 Ibid., 2.128.

230 See p. 294 of Zahra Faridany-Akhavan, "All the King's Toys", *Muqarnas* 10 (1993): 292–298.

231 See Irwin, 64, who summarizes the connections between the *Arabian Nights* story concerning the ebony horse and the variants from other sources. Also see the discussion in Ulrich Marzolph and Richard van Leeuwen, *The Arabian Nights Encyclopedia*, 2 vols. (Santa Barbara: ABC-CLIO, 2004), 1.172–174.

> the relevant objects to the king as he speaks, since, when he offers advice about how to make the horse vanish, he suggests, "Trille *this* pyn, and he wol vanysshe anoon" ... Nowhere are the instructions particularly hard, and even the two bits of information that the knight at first withholds he promises to supply before flight. In fact, they are both necessary *to* flight, involving the precise location of the ignition pin in the horse's ear ... and advice about how to retrieve the horse from invisibility ... Not only is the information offered relatively complete as regards navigating and landing the horse; the description of the horse's flying functions seems easily understood.[232]

The Chaucer version includes this explicit reference to technology, and at the time, as one scholar notes, offers "the startling juxtapositions of wonder and demystifying explanations" about these strange devices.[233] That same scholar points out that the

> clarity of the instruction to "Trille this pyn, and he wol vanysshe anoon" contends with the narrator's "The hors vanysshed, I noot in what manere" ... as if to insist on a residual mystery within the masterful construction.[234]

Specifically, then, this story type is not completely fantastical — like a story of a Pegasus might be — but rather one in tension, between describing a mechanical device and the inability to truly understand how it works.

Another study of these tales of flying devices makes the following point:

> As a means of aerial locomotion, in all of these stories, a horse not infrequently appears, and in several notable instances a horse of wood. One should note, too, that the trip through the air is made ... by means of a wonderful mechanical contrivance ... Furthermore, the striking motive of aerial locomotion and the dominant interest in wonderful objects and accomplishments must have served to associate all of the above groups [of tales] in the minds of story-tellers.[235]

The point here, again, is that the authors of a number of tales recounted in this book are clearly interested in the mechanical operation of devices — even if

232 Lynch, 540.

233 Crane, *Gender and Romance in Chaucer's "Canterbury Tales"*, 137.

234 Ibid., 139.

235 Jones, "The *Cléomadès* and Related Folk-Tales," 593–594.

the language is never sufficiently detailed to tell us how they work in a truly technical sense. But there *is* sufficient specificity for the reader to know that a machine — rather than a magical creature or deity — is being discussed, and this should make us consider the question of how the author came to formulate or repeat what is in some sense rather strange: a discussion from centuries ago of mechanized flying vehicles.

9 Terms and Types

As far as the terminology for flying vehicles is concerned, it is clear that cultures must formulate appellations for technologies that are novel to them. One study notes that there "are two ways of denoting a new referent: either existing words may be used or new formations may be made".[236] A device that flies may be referred to using the existing word "ship" or even just "machine" in a slightly modified sense. But through history, terms for flying devices became subject to new formations, such as "flying machine", "aircraft", and "airliner", especially in English.[237] In the case of early accounts, however, there is the same pattern in terminology: On the one hand, existing terms might be used — the *sinta* or "cradle" used by the Ainu, perhaps through an analogy in shape or the fact that the traditional Ainu cradle was indeed suspended above the ground, to allow it to be rocked back and forth. On the other hand, an existing term might be subject to a new formation or clarification through compounding — the Hopi "flying shield" and the Chinese *fei che*, or "flying vehicle".

More generally, however, what should one make of these early tales of mechanized flight? The images of flying beings and winged spirits in East Asian and other sources are engaging but do not surprise the modern reader. Clearly, they have roots in various elements of early mythology. Considering early belief systems such as animism, it is not surprising that the flying abilities of animals such as birds and bats might readily be attached to anthropomorphic beings.[238] However, it is necessary to recall the taxonomic distinction between two kinds of stories here. The first type of tale includes such anthropomorphic beings and winged deities. This category might also be said to include flight achieved through the practice of magic, or the use of magical objects.

236 Stubelius, *Balloon, Flying Machine, Helicopter*, 290.

237 See the discussion in ibid., 291–296.

238 On this and a number of other related mythic constructs concerning flight, see Bayla Singer, *Like Sex with Gods: An Unorthodox History of Flying* (College Station, TX: Texas A&M University Press, 2003).

A separate category comprises stories where humans fly in some kind of mechanism or aerial vehicle specifically designed for manned flight. Of course, there is some overlap in these categories, as we read of various deities who do not fly under their own power, but instead ride in some kind of aerial chariot.

The latter category of stories, where various individuals construct and fly in aerial vehicles is of particular interest. This is *not* because they lead one to conclude that such vehicles were actually constructed; rather, it is that these stories may represent a proto-technological interest in the idea of mechanized flight itself. Historically, some of the solutions in the quest for flight not surprisingly involved the imitation of birds, a trend that continued into the late nineteenth century. The author of the *Oju Yŏnmun Changjŏn San'go*, cited earlier, states simply:

> If one wants to copy the production techniques of the flying machine, that person must first make a machine that resembles a bird, install wings, set up the internal mechanism, and people can sit inside. After starting the mechanism, the flying machine will seem like a skilled swimmer swimming, and as natural as the larvae of a moth stretching its wings and agitating the air to fly freely.[239]

The Indian tales, of course, also speak of vehicles in the form of birds. There seems to be an assumption in these stories and accounts that for a vehicle to fly, it must take the form of something in nature that flies. This may reflect a synthesis of "magical thinking" — the principle of homomorphism, for example — with proto-technological musings.

The vehicles described in these texts perhaps are fanciful, but the concepts represented in the stories — to construct a mechanical device and fly from place to place — are not. The stories may represent ambitions to create such a practical flying machine, and a realization that the kinds of flights believed to have been undertaken by the winged spirits and deities of the past would not always remain outside the realm of human activity. Moreover, the tales of flying vehicles — the *fei che*, *vimana*, and so on — exhibit an understanding that humans might be able to fly only with the assistance of a mechanical device. A hymn written in the second century BCE speaks of a deity appearing in chariots drawn by flying dragons; by the third century CE, we read of people — ordinary mortals — constructing a mechanical flying vehicle.[240]

239 Yi, vol. 1, *chuan* 2, 33–34.
240 Laufer, *The Prehistory of Aviation*, 19–20.

Notice the change: no longer is the vehicle drawn by any animal, be it a magical phoenix or dragon — instead, it is self-propelled.

What caused the evolution of this separate type of story, one that was more technological in nature and of a type that could be called a "mechanical myth", is not entirely clear. One what might well ask why the authors of various stories included discussions of machines at all. Why didn't stories that required flight simply continue portraying winged beings and levitating immortals? Goldman, cited earlier, comments on flying vehicles in Indian sources:

> It is difficult to say what the references to flying machines, vehicles and palaces in the ancient Indian literature "mean." The power of flight is ... widely shared by various types of beings in this literature. The gods, in particular, are said to be possessed of "vimanas" or flying vehicles, and the foremost example of these is the great "Pushpaka palace cum flying machine" described in the *Ramayana*. Perhaps the idea of the poet is to magnify the grandeur and power of a titanic figure like Ravana, the demon king, through endowing him with this marvelous conveyance.[241]

This chapter has examined some of these *vimanas*, but there still remains the key question of why gods or demons would be described as using a vehicle rather than flying under their own power. Clearly, there was a shift in the nature of the mythical narrative. Perhaps divine beings came to be seen as more like human beings, with their divine powers defined by advanced technological capabilities.

In the Chinese case, on a more mundane level, perhaps there is a connection to the development of kites, which occurred at a period in China before many of these tales of "aerial carriages". Kites gave people a view of flight made practical: structures made of bamboo, wood, cloth, and paper — man-made devices that actually became airborne. There are even stories of kites large enough to hold individuals, a not implausible scenario.[242] From kites, it was perhaps not an unreasonable step to imagine independent flying machines, untethered and flying from place to place. Needham himself points out some connections between kites and early conceptions of aircraft.[243]

However, Needham does not argue that kite-flying in ancient Chinese history might have inspired the specific tales of aerial vehicles presented in this study. The texts examined here, with the variety of flying machines that they

241 Goldman, private communication, 29 May 2007.

242 See Needham, 4.2, 588–591 for a discussion of early Chinese kite technology.

243 Ibid., 4.2, 568.

present, do not seem to suggest too strong a connection either. The Chinese stories speak of aerial vehicles traversing great distances, and explicitly refer to machines built with leather and metal parts, or carved wooden animals that ascend in to the air by means of internal mechanisms. These do not seem related to kites in any way. Furthermore, the stories we have examined include mentions of motive power or engines, something also lacking in a kite. The Korean tale of Jŏng Pyŏng-gu speaks of a craft that was able to undertake a controlled flight to and from a fortress, in a manner quite different from the limited aerial abilities of a kite. Finally, the Indian narratives depict a variety of flying craft, involved in rescues and combat, and these are described, too, in ways that seem to have no relation to the structure or function of kites.

In the Chinese and Indian sources, one can trace some patterns in these stories — for example, the wooden birds that carry a pilot and passengers appear in more than one source. Yet an overall pattern remains elusive; certain items, like the propulsion system fabricated "from the inner part of a jujube tree," are mentioned in only one extant text. Further research into ancient sources in various parts of the world may reveal other descriptions of early flying vehicles, and allow the assembling of a more complete understanding of this intriguing aspect of writing about technology.

Looking at what is commonly understood as the actual beginnings of powered, manned aircraft — the work of Otto Lilienthal and the Wright brothers — it is easy to accept that the "invention of flight represented the culmination of centuries of thought and desire", as one author puts it.[244] But the perspective of the early writers themselves — the writers examined in this chapter — was that their predecessors had knowledge that had somehow been lost, and that this knowledge was of a very specific, technical kind. Moreover, it is important to note — as we will see in as different a context as Plato's description of a place called Atlantis — that early writers were not simply passive figures. They provided their own critiques concerning the veracity of various tales, and understood the processes concerning both the transmission of knowledge and the loss of knowledge through the ages.

244 Hallion, xix.

CHAPTER 3

Magic Mirrors and Early Televisions

1 Mirrors in History[1]

Mirrors appear throughout human history in many forms. They have been used for scientific investigation, decoration, and in magical rituals. Passages discussing mirrors are found in Chinese, Indian, Greek, Roman, and many other sources. But as with ancient accounts of flight, in some of these texts on mirrors one finds peculiar accounts that concern apparently lost technical knowledge — for example, certain early Chinese sources present devices that are labelled as "mirrors", but have the power to see inside the human body. To explore these accounts, it is useful to have a brief history of mirrors and related optics. In looking at such a history, the narrative does not follow the linear model that one might expect. This "uneven" history makes the anomalous accounts of Chinese mirrors even more striking. Where do those accounts really "fit" in this complex historical pattern?

Mirrors appear throughout human history in many forms. These pieces of polished metal or metal and glass have been used in many cultures, in locales as diverse as North America, Africa, and Siberia.[2] The word "mirror" itself has an interesting history; it first appears in English in the early thirteenth century, coming from the Old French term *mireor*. This term, in turn, came from the Latin verb *mirari*, meaning, "to wonder at, admire". Already in Latin, the words for the common mirror and for something special or magical were intertwined: *speculum* was the general term for a mirror, while *speculatio* had a broad range of meanings including, "watching", "contemplation", and "speculation". In Greek, the term for mirror was κάτοπτρο; in the seventeenth century, there appeared in English the term "catoptromancy" — meaning divination by means of a mirror.[3]

1 A more abbreviated version of part of this study appeared in Olshin, "A Revealing Reflection: The Case of the Chinese Emperor's Mirror", *ICON: Journal of the International Committee for the History of Technology* 18 (2012): 122–141.

2 See, for example, the interesting mention of shamanic uses of such objects in Nicholas J. Saunders, "Stealers of Light, Traders in Brilliance: Amerindian Metaphysics in the Mirror of Conquest", *RES: Anthropology and Aesthetics* (33: *Pre-Columbian States of Being*) (Spring 1998): 225–252, noting especially p. 237.

3 For some background on the uses of mirrors throughout history, and the complex set of ideas that has developed around them, see Jurgis Baltrušaitis, *Le miroir: essai sur une légende*

 | DOI:10.1163/9789004352728_004

Abundant literary and archaeological evidence reveals the existence of metal mirrors in antiquity. Such mirrors were most often small, circular, hand-held affairs, with a highly polished mirrored surface on one side and a decorative scene on the reverse. Examples of these have been found in tombs and other sites throughout the Graeco-Roman world. There is also evidence, mostly literary, of larger mirrors, gracing the walls of places as diverse as houses, barbershops, and palaces. However, one also encounters references to mirrors that were set up in various "trick" arrangements, yielding enlarged, diminished, or otherwise distorted images. Ancient texts not only describe these systems of mirrors, but also comment on their uses and abuses. Not surprisingly, vanity is the most common target of these literary digressions. The sin of Narcissus, who became enamored of his image in the most primitive of mirrors, a pool of water, is the earliest example. Even in these classical Western accounts, there is a mixture of the technical and the literary. In Chinese accounts, detailed below, the technical descriptions of mirrors become particularly mysterious, and it is in those accounts that the reader again encounters the issue of lost knowledge. However, before moving to those accounts, it is worth noting the degree of sophistication of knowledge concerning mirrors and optics that the early sources reveal.

One of the earliest known mirrors was found in a tomb in Egypt, and is made of copper. It dates from almost 5,000 years ago.[4] Mirrors in Egypt had both practical and symbolic value: they were used for the simple reflection of a person's face, and for reflecting sunlight, but also as a method of allowing the soul of the dead to reach the gods and as symbols of the Sun.[5] In ancient Greece, mirrors were fabricated from thin discs of bronze, with one side polished to provide a reflective surface. Mirrors made of glass coated with tin or silver — similar in construction to modern mirrors — have been found at Pompeii. As with the Egyptians, the Greeks saw mirrors in both a practical and metaphorical way; the most famous "mirror myth" is the Greek tale mentioned above — that of Narcissus, who falls in love with his own reflection.

Classical sources recount details concerning the making of mirrors, and the materials required in the process. The first-century AD Roman writer Pliny the Elder, in his encyclopedic *Natural History*, speaks of glass mirrors made at Sidon, a Phoenician city in what is today Lebanon. The Greek philosopher

---

scientifique: révélations, science-fiction et fallacies (Paris: Elmayan, 1978), and G.F. Hartlaub, *Zauber des Spiegels: Geschichte und Bedeutung des Spiegels in der Kunst* (Munich: R. Piper, 1951).

4 Richard Gregory, *Mirrors in Mind* (New York: W.H. Freeman, 1996), 48.

5 Ibid., 48.

Alexander Aphrodisias mentions glass mirrors coated with tin; the most common material, however, seems to have been bronze. In the time of the Roman Republic, says Pliny, silver mirrors were in vogue — in the Imperial period, he goes on to note, they were even used by maid-servants.

The famous Roman architect Vitruvius speaks of mirrors as well; his famous text on architecture, *De Architectura,* mentions large reflective surfaces on walls. In a digression about the Greek engineer Ctesibius of Alexandria, he also notes large mirrors:

> Ctesibius, born at Alexandria, was the son of a barber. Preëminent for natural ability and great industry, he is said to have amused himself with ingenious devices. For example, wishing to hang a mirror in his father's shop in such a way that, on being lowered and raised again, its weight should be raised by means of a concealed cord.[6]

It is interesting to see that — as with some of the other ancient writers examined in this book — Vitruvius, too, is quite conscious of his place in the transmission of knowledge. In his introduction to Book VII, he says:

> It was a wise and useful provision of the ancients to transmit their thoughts to posterity by recording them in treatises, so that they should not be lost, but, being developed in succeeding generations through publication in books, should gradually attain in later times, to the highest refinement of learning. And so the ancients deserve no ordinary, but unending thanks, because they did not pass on in envious silence, but took care that their ideas of every kind should be transmitted to the future in their writings. If they had not done so, we could not have known what deeds were done in Troy, nor what Thales, Democritus, Anaxagoras, Xenophanes, and the other physicists thought about nature, and what rules Socrates, Plato, Aristotle, Zeno, Epicurus, and other philosophers laid down for the conduct of human life; nor would the deeds and motives of Croesus, Alexander, Darius, and other kings have been known, unless the ancients had compiled treatises, and published them in commentaries to be had in universal remembrance with posterity.[7]

---

6 Vitruvius, *De Architectura* 9.8.2, in Vitruvius, *The Ten Books on Architecture,* trans. Morris Hicky Morgan (Cambridge, MA: Harvard University Press, 1914), 273.

7 Vitruvius, *De Architectura* 7. Introduction. 1–2, trans. Morgan, 195.

The sources show us that early writers of all kinds — poets, philosophers, historians, and so on — were interested in what is clearly technical knowledge, and that technical writers, such as Vitruvius, were interested in the passage of knowledge through history.

Vitruvius, and Greek commentators on optics and mirrors, certainly had an understanding of the technologies involved. However, what is notable is the fact that one finds non-technical writers also discussing mirrors and their properties.[8] The second-century CE writer Lucian of Samosata, like Vitruvius, mentions barbershop mirrors. In his satirical tale, "The Ignorant Book Collector", the narrator says:

> Consider the barbers and you will observe that the master-craftsmen among them have only a razor and a pair of shears and a suitable mirror, while the unskilled, amateurish fellows put on view a multitude of shears, and huge mirrors; but for all that, they cannot keep their ignorance from being found out. In fact, what happens to them is as comical as can be — people have their hair cut next door [i.e., in the good barbershops with the small mirrors] and then go to their mirrors to brush it.[9]

Mirrors even appear in discussions of oratory. The first-century Roman rhetorician Quintilian writes:

> On the other hand, if gesture and the expression of the face are out of harmony with speech ... our words will not only lack weight, but will also fail to carry conviction. Gesture and movement are also productive of grace. It was for this reason that Demosthenes used to practise his delivery in front of a large mirror, since, in spite of the fact that its reflexions are reversed, he trusted his eyes to enable him to judge accurately the effect produced.[10]

The famous second-century writer Apuleius, in his humorous *Apologia*, provides an extensive defense of the use of the mirror. In one part of this work,

8 For a look at some of the writers on optics in antiquity, see Jones, "On Some Borrowed and Misunderstood Problems in Greek Catoptrics", *Centaurus* 30 (1987): 1–17.

9 Lucian, "The Ignorant Book Collector", 29, in *Lucian, with an English Translation*, 8 vols., trans. A.M. Harmon, K. Kilburn, and M.D. Macleod (London: William Heinemann, 1913–1967), 3.211.

10 Quintilian, *Institutio Oratoria* 11.3.67–11.3.68, in *The Institutio Oratoria of Quintialian, with an English Translation,* 4 vols., trans. H.E. Butler (London: William Heinemann, 1920–1922), 4.281.

Apuleius claims that he has been unfairly attacked by a certain Pudens, who has apparently accused the writer of a marked degree of vanity with the mirror. The response:

> Come now, let me admit that I *have* looked into it. Is it a crime to be acquainted with one's own likeness and to carry it with one wherever one goes ready to hand within the compass of a small mirror, instead of keeping it hidden away in some one place? Are you ignorant of the fact that there is nothing more pleasing for a man to look upon than his own image?[11]

He adds:

> For all we mould in clay or cast in bronze or colour with paint, in a word, every attempt at artistic representation by the hand of man after a brief lapse of time loses its truth and becomes motionless and impassive like the face of a corpse. So far superior to all pictorial art in respect of truthful representation is the craftsmanship of the smooth mirror and the splendour of its art.[12]

Again, note that even in a work that is not technical in nature, the author takes the time to talk about a mechanical device and its function. Moreover, there is again the theme of the danger of a technology — in this case, mirrors leading to vanity.

Later, Apuleius, like Quintilian, cites the story of Demosthenes:

> Again, who is ignorant of the fact that Demosthenes, the greatest master of the speaking, always practised pleading before a mirror as though before a professor of rhetoric? When that supreme orator had drained deep draughts of eloquence in the study of Plato the philosopher, and had learned all that could be learned of argumentation from the dialectician Eubulides, last of all he betook himself to a mirror to learn perfection of delivery.[13]

---

11 Apuleius, *Apologia* 14, in *The Apologia and Florida of Apuleius of Madaura*, trans. H.E. Butler (Oxford: Clarendon Press, 1909), 38.

12 Apuleius, *Apologia* 15, trans. Butler, 39.

13 Apuleius, *Apologia* 15, trans. Butler, 40.

Apuleius also provides us with some more technical speculation about the functioning of mirrors:

> Is Epicurus right when he asserts that images proceed forth from us, as it were a kind of slough that continually streams forth from our bodies? These images when they strike anything smooth and solid are reflected by the shock and reversed in such ways as to give back an image turned to face its original ... all agree that when these emanations strike any dense, smooth, and shining surface, they return to the surface from which they proceeded in such a manner that the angle of incidence is equal to the angle of reflection, and as a result that which they approach and touch outside the mirror is imaged within the mirror.[14]

Here in Apuleius' work, there is a combination of history, technical knowledge of the workings of mirrors, and an urging for further inquiry. Not only does Apuleius cite the ideas of Epicurus, but he also comments on the nature of reflections; there is a very similar discussion of reflections, in fact, in another "non-scientific" writer, Plutarch, who discusses angles of incidence and reflection in his work entitled *Concerning the Face which Appears in the Orb of the Moon*, cited earlier in this book.[15] Both writers seem to have been familiar with the pseudo-Euclidean *Catoptrics*, which deals with mirrors and their properties.

As already noted, Apuleius begins his whole treatment of mirrors in a philosophical tone, defending their use and stating: "So far superior to all pictorial art in respect to truthful representation is the craftsmanship of the smooth mirror and the splendour of its art." Towards the end of his discussion, he shifts to a series of highly technical questions, prefacing it with the statement, "Should not philosophers make all these problems subjects of research and inquiry and in solitary study look into mirrors of every kind, solid and liquid?"[16] He goes on:

> [W]hy is it that in flat mirrors all images and objects reflected are shown in almost precisely their original dimensions, whereas in convex and spherical mirrors everything is seen smaller, in concave mirrors on the

---

14 Apuleius, *Apologia* 15, trans. Butler, 41.

15 See Plutarch, "Concerning the Face which Appears in the Orb of the Moon", 929f–930b, trans. Cherniss, et al., 12.104–12.107; also see footnote *e* in ibid., 12.105, concerning the early understanding of the angle of reflection equaling the angle of incidence.

16 Apuleius, *Apologia* 16, trans. Butler, 42.

> other hand larger than nature? Why again and under what circumstances are left and right reversed? When does one and the same mirror seem now to withdraw the image into its depths, now to extrude it forth to view? Why do concave mirrors when held at right angles to the rays of the sun kindle tinder set opposite them? What is the cause of the prismatic colours of the rainbow, or of the appearance in heaven of two rival images of the sun, with sundry other phenomena treated in a monumental volume by Archimedes of Syracuse, a man who showed extraordinary and unique subtlety in all branches of geometry, but was perhaps particularly remarkable for his frequent and attentive inspection of mirrors.[17]

Apuleius' synthesis of morality, history, and technical detail deserves careful consideration. In examining ancient texts from a modern perspective, there is the bias towards believing that there must be separate categories: technical works, such as Vitruvius' treatise on architecture, and then philosophy, poetry, myth, and so on. But in fact, as already shown in the analysis of flying vehicles in Chapter 2, there are many technological and scientific discussions — or at the very least technological and scientific asides — in early philosophical and literary texts. Moreover, of course, even the rigid differentiation between topics such as philosophy, history, and myth did not exist in the ancient world, an issue examined here in the chapter on Atlantis.

The Greek and Roman references to mirrors are abundant; in addition to the sources cited above, there discussions of mirrors in the work of Ulpian, the great writer on Roman law, who speaks several times of large mirrors, and in the writings of the famed Plutarch, who uses mirrors as a metaphor in his treatise "Advice to Bride and Groom" in his *Morals*. In his "On Listening to Lectures", also in the *Morals*, there is again a look at the simple mirrors in a barbershop.

Aulus Gellius, in his *Attic Nights*, a collection of miscellaneous texts on literary, historical, and philosophical subjects, also touches upon the subject of mirrors. However, here the mirrors are of a different kind. For these are not the simple hand-held *specula*, nor even the full-length mirrors of Ulpian. Rather, Aulus Gellius, in a brief passage on optics, states that this science of reflection

> effect[s] many surprising things, such as the appearance in one mirror of several images of the same thing; also that a mirror placed in a certain position shows no image, but when moved to another spot gives reflections;

17 Apuleius, *Apologia* 16, trans. Butler, 42.

> also that if you look straight into a [certain type of] mirror, your reflection is such that your head appears below and your feet uppermost.[18]

Pausanias, in his description of Greece, briefly mentions a mirror of the second type cited by Gellius, which "placed in a certain position shows no image". He describes a peculiar setup in a Greek temple in Arcadia:

> On the right as you go out of the temple, there is a mirror fitted into the wall. If anyone looks into this mirror, he will see himself very dimly or not at all, but the actual images of the gods and the throne can be seen quite clearly.[19]

The pseudo-Ptolemaic treatise *De speculis* (2.11–12) seems to describe a similar arrangement: "And how can one not think it wonderful when they see by a mirror neither themselves nor anyone else, but only whatever someone has chosen in advance?"[20]

Pliny the Elder's description of reflecting surfaces is quite extensive.[21] He discusses convex and concave mirrored surfaces, as well as concave surfaces that offer multiple reflections. Seneca the Younger describes something similar:

> [J]oin mirrors together: the images will not merge into one, but each will embrace within itself a likeness of the observed object. There are some mirrors composed of many tiny mirrors. If you place one person in front of them, a nation is visible, as each segment generates its own shape; although the mirrors are joined together and contiguous, nevertheless they keep their images separate, and from one person they create a crowd.[22]

---

18 Aulus Gelllius, *Attic Nights* 16.18, trans. Rolfe, 3.186–3.189.

19 Pausanias, *Description of Greece* 8.37.7, in Pausanias, *Description of Greece*, 4 vols. trans. W.H.S. Jones and H.A. Ormerod (Cambridge, MA: Harvard University Press, 1918–1935), 4.89. Also see the interpretation of the mirror setup described in this passage in Verity Platt, *Facing the Gods: Epiphany and Representation in Graeco-Roman Art, Literature and Religion* (Cambridge: Cambridge University Press, 2011), 222.

20 See Jones, "Pseudo-Ptolemy, *De Speculis*", *SCIAMVS: Sources and Commentaries in Exact Sciences* 2 (2001): 145–186.

21 See Pliny the Elder, *Natural History* 34.128–130, in Pliny the Elder, *Natural History*, 10 vols, trans. D.E. Eichholz, W.H.S. Jones, and H. Rackham (Cambridge, MA: Harvard University Press, 1938–1963), 9.96–9.99.

22 Seneca, *Natural Questions* 1.5.5, in Lucius Annaeus Seneca, *Natural Questions*, trans. Harry M. Hine (Chicago: University of Chicago Press, 2010), 148–149.

Pliny also speaks of "trick" arrangements of mirrors.[23] Like those cited by Pausanias, these are found in a temple. Apuleius, again in his *Apologia*, speaks about mirrors that yield distorted reflections:

> There is also over and above these questions further matter for discussion. For instance, why is it that in flat mirrors all images and objects reflected are shown in almost precisely their original dimensions, whereas in convex and spherical mirrors everything is seen smaller, in concave mirrors, on the other hand, larger than nature? Why again and under what circumstances are left and right reversed? When does one and the same mirror seem now to withdraw the image into its depths, now to extrude it forth to view?[24]

A long discussion of mirrors and their properties, as well as special arrangements of mirrors, also appears in Lucretius, in Book 4 of his well-known first-century BCE work, *On the Nature of Things*.[25] Again, these are authors who in modern terms are certainly not technical writers, but who nonetheless take the time and care to describe mechanical devices — and in the end preserve and transmit some rather sophisticated knowledge. The inclusion of these technical matters also suggests that readers of the time were interested in such details.

These discussions of technology are often accompanied by moral considerations; in one of the Chinese sources, in fact, there is mention of a mirror that can detect an "evil heart". Seneca the Younger, in turn, felt that mirrors should not serve the causes of vanity and perversity. Rather, he says:

> Mirrors were invented in order that man may know himself, destined to attain many benefits from this: first, knowledge of himself; next, in certain directions, wisdom. The handsome man, to avoid infamy; the homely man, to understand that what he lacks in physical appearance must be compensated for by virtue. The young man, to be reminded by his youth that it is a time of learning, and of daring brave deeds; the old man, to set aside actions dishonourable to his grey hair, to think some thoughts

23 Pliny the Elder, *Natural History* 33.126, trans. Eichholz, Jones, and Rackham, 9.96–9.97.

24 Apuleius, *Apologia* 16, trans. Butler, 41–42.

25 See Lucretius, *On the Nature of Things* 4.90 et ff., in Lucretius, *On the Nature of Things*, trans. Martin Ferguson Smith (Indianapolis, IN: Hackett, 2001), 103 et ff.

> about death. This is why nature has given us the opportunity of seeing ourselves.[26]

Seneca was a philosopher, a statesman, and a writer; it is significant that although not a "scientifically-trained" figure, he used a piece of a technology — the mirror — to talk about human character. In fact, the passages above show how non-technical writers used the knowledge of mirrors and their functions in various ways. Even though technology was not as integrated into society as it is in the present age, there clearly was an understanding about technological devices, their meaning, and their consequences. In terms of information transmission, this means that the knowledge of technology was preserved not only in more formal sources, such as Ptolemy or Heron of Alexandria, but also in the informal network of writers who were penning everything from amusing stories to philosophical treatises.

2 Two Chinese Diagnostic Mirrors

In the Western classical world, then, mirrors — even when used metaphorically — were understood as actual devices, a technology with the specific function of created reflected images in various ways through various configurations. But mirrors in Chinese history were more intimately connected with the ideas of seeing or perceiving, as well as unmasking and revealing. Indeed, one early Chinese text speaks of a mirror with a particularly peculiar power to unmask. But before presenting that text, it is helpful to examine an entry in a Chinese work entitled *Kaiyuan Tianbao Yishi* (開元天寶遺事 "Anecdotes of the Kaiyuan and Tianbao Periods") by Wang Renyu (王仁裕), who was active in the first part of the tenth century CE. The entry is entitled *zhao bing jing* (照病鏡) — literally, "illuminating [*zhao*] illness [*bing*] mirror [*jing*]":

> *A Mirror to Illuminate Illness*
> Ye Fashan [葉法善] had an iron mirror that could reflect objects in the same manner as water. Whenever a person had an illness, the mirror could be used to illuminate and see completely any obstructions in his

26 Seneca the Younger, *Natural Questions* 1.17.4, cited in Rabun Taylor, *The Moral Mirror of Roman Art* (Cambridge: Cambridge University Press, 2008), 19; also see the comments in Tobias Foster Gittes, *Boccaccio's Naked Muse: Eros, Culture, and the Mythopoeic Imagination* (Toronto: University of Toronto Press, 2008), 326, n. 40.

> internal organs. Then, using medicines, he would be treated until he recovered.[27]

This is a puzzling and rather ambiguous description: the mirror is described as reflecting in the same manner as water, yet we are told in the next line that mirror was able to do more than reflect — it could also reveal what was *inside* a person's body.[28] Certainly, early Chinese sources spoke of mirrors in a fantastical way — capable of "miraculous deeds" — but the passage here provides a somewhat more "technical" description.[29]

Another brief mention of a special mirror, similar to the one above, is found in a work known as the *Gujing Ji* (古鏡記 "Record of an Ancient Mirror"), dating from the Tang dynasty (618–907 CE).[30] In a passage there, a monk says the following concerning a certain kind of mirror:

> Unfortunately ... I do not have a certain essential medication; otherwise I could make the mirror that illuminates the *fu* [腑] and *zang* [臟].[31]

The *fu* and *zang* in the passage above refer to the system of internal organs in traditional Chinese medicine, where certain organs are classed as *yang* (陽) and others as *yin* (陰). The *fu* (腑) are *yang* organs, and comprise the large intestine, small intestine, stomach, gall bladder, urinary bladder, and

27 Renyu Wang, *Kaiyuan Tianbao Yishi* (開元天寶遺事 "Anecdotes of the Kaiyuan and Tianbao Periods") (Taipei: Yiwen Yinshuguan Yinxing, 1965–1970); my translation. 開元 ("Kaiyuan") and 天寶 ("Tianbao") refer to two periods of the reign of the Tang Dynasty emperor Xuanzong (唐玄宗): 713 to 741 CE and 742 to 756 CE, respectively. The *Kaiyuan Tianbao Yishi* includes a number of tales of peculiar phenomena; see p. 45 of Leo Tak-hung Chan, "Text and Talk: Classical Literary Tales in Traditional China and the Context of Casual Oral Storytelling", *Asian Folklore Studies* 56.1 (1997): 33–63.

28 See the mention of this passage in Laufer, *The Prehistory of Aviation*, 88; however, Laufer does not provide an interpretation.

29 See p. 33 of Jue Chen, "The Mystery of an 'Ancient Mirror': An Interpretation of *Gujing Ji* in the Context of Medieval Chinese Cultural History", *East Asian History* 27 (June 2004): 33–50; also note Pauline Bentley Koffler, "The Story of the Magic Mirror (*Gujingji*) by Wang Du", in Jean-Pierre Diény, ed., *Hommage à Kwong Hing Foon, Études d'histoire culturelle de la Chine* (Paris: Collège de France, Institut des hautes études chinoises, 1995), 165–214.

30 Chen, 33; also note idem, "Wang Ji's Travel Route in the South — An Itinerary Historical or Fictional?", in Mechtild Leutner and Klaus Mühlhahn, eds., *Reisen in chinesischer Geschichte und Gegenwart: Erfahrungen, Berichte, Zeugnisse* (Wiesbaden: Harrassowitz, 2008), 53–69. For a comprehensive study of the *Gujing Ji*, see Chen, *Record of an Ancient Mirror: An Interdisciplinary Reading* (Wiesbaden: Harrassowitz, 2010).

31 Translation adapted from Chen, *Record of an Ancient Mirror: An Interdisciplinary Reading*, 19.

the "triple-warmer' (the openings of the stomach, small intestine, and the bladder). The *zang* (臟) are *yin* organs, and comprise the heart, liver, spleen, lung, and kidneys.[32] The implication is that this medication or special substance — the Chinese term used here is *yao* (藥), which is the typical word for medicine, drug, remedy, or cure — is to be used in conjunction with the mirror. As with many of these references, very little detail is provided, although it is sufficient to see that the actual use of the mirror described is at the same time both fantastical and pragmatic. What to make of such descriptions?

The *Kaiyuan Tianbao Yishi*, where the passage above — concerning a "Mirror to Illuminate Illness" — is found, has been described as an "informal" text, "comprised of very short and titled anecdotes in no apparent order". Moreover, the author does not cite sources for these entries.[33] What we do know is that Ye Fashan (d. 720 CE), who is mentioned in the text, was an alleged "Daoist" master who lived during the Tang dynasty.[34] Such masters were often said to have special abilities, but here we see something more than fanciful magic powers: there is the explicit mention of a technology — the illuminating mirror — an element that makes the text seem more than a bit of whimsy.

The second mention of a mirror cited above, from the *Gujing Ji*, is perplexing for a modern reader because — like many early texts — the work generally cannot be readily classified as "fact" or "fiction". One sinologist points out that in this text, "truth and fiction are presented simultaneously".[35] Some of the historical details are verifiable and "the descriptions in fiction [i.e., the apparently fictional parts of the text] here accord with the historical records".[36] Textual interpretation aside, the present chapter provides the reader a few accounts of these peculiar Chinese mirrors, and suggests that the descriptions in some cases may derive, in fact, from a simple but subtle illusion — one that was produced through mirror technology developed by the Chinese centuries

32 On this classification, see Daniel P. Reid, *Chinese Herbal Medicine* (Boston: Shambhala, 1987), 32.

33 Richard D. McBride, II, "A Koreanist's Musings on the Chinese *Yishi* Genre", *Sungkyun Journal of East Asian Studies* 6.1 (April 2006): 31–59; note especially pp. 41–42.

34 For more on Ye Fashan, see Russell Kirkland, "Tales Of Thaumaturgy: T'ang Accounts of the Wonder-Worker Yeh Fa-shan", *Monumenta Serica* 40 (1992): 47–86. On the problematic use of the term "Daoist' for such figures as Ye Fashan, see Nathan Sivin, "On the Word 'Taoist' as a Source of Perplexity. With Special Reference to the Relations of Science and Religion in Traditional China", *History of Religions* 17.3–4 (1978): 303–330, and idem, "Taoism and Science", in Sivin, ed., *Medicine, Philosophy and Religion in Ancient China: Researches and Reflections* (Aldershot: Variorum, 1995), 1–72.

35 See p. 182 of Chen, "History and Fiction in the *Gujing Ji* (Record of an Ancient Mirror)", *Monumenta Serica* 52 (2004): 161–197.

36 Ibid., 180 and 181.

ago. The authors of the *Kaiyuan Tianbao Yishi* and the *Gujing Ji* may have been recounting — knowingly or not — examples of that technology.

3 A Mirror to Locate Illness and a Mirror to "Illuminate the Bones"

A further account of a special mirror of this kind appears in a Chinese text entitled *Xijing Zaji* (西京雜記 "Miscellaneous Records of the Western Capital"), another collection of anecdotes, of uncertain authorship and dating from around 500 CE.[37] The description here is more detailed, and begins with a discussion of other curious mechanical devices:[38]

> When Gaozu [高祖][39] first entered the palace at Xianyang [咸陽],[40] he toured the treasury storehouse, which was filled with unfathomable amounts of gold, jade, and other precious treasures. They were beyond description, and particularly surprising were five green columns seven *chi* [尺] five *cun* [寸] in height, made of jade, [supporting] a lamp.[41] Below them, there was a coiled hornless dragon holding a lamp in its mouth.

---

37 See the discussion of this work in W.H. Nienhauser, Jr., *The Indiana Companion to Traditional Chinese Literature*, 2 vols. (Bloomington: Indiana University Press, 1986–1998), 1.406–1.407. 西京 ("Xijing") here literally means "Western Capital", and refers to the city of Chang'an (長安) and the surrounding area; today it is known as Xi'an (西安).

38 A brief mention of this passage is found in Needham, *Science and Civilisation in China, Vol. 4: Physics and Physical Technology, Part I: Physics* (Cambridge: Cambridge University Press, 1962), 91–92; this volume hereafter is referred to as "Needham, 4.1"; also note Laufer, *Prehistory of Aviation*, 11; and Gregory, 52. See, as well, A.G. Wenley, "A Chinese Sui Dynasty Mirror", *Artibus Asiae* 25.2–3 (1962): 141–148, noting especially 141–142; there, the author notes that this passage is referred to by two other Chinese sources, including a poem inscribed on a mirror now in the Freer Gallery of Art. A short discussion is found in Chen, "The Mystery of an 'Ancient Mirror' …", 44–45.

39 Emperor Gaozu (256–195 BCE) was the first emperor of the Han Dynasty. He was one of the leaders of an insurrection in the late Qin Dynasty. Although this has been characterized as comprising "peasant uprisings", one author argues that instead it was a true rebellion left by "men eager … to establish new socio-political identities for themselves"; see p. 315 of Jack L. Dull, "Anti-Qin Rebels: No Peasant Leaders Here", *Modern China* 9.3 (July 1983): 285–318.

40 This was the royal palace of the state of Qin at Xianyang, the capital; when Qin Shihuang unified China, it became his palace.

41 These are traditional Chinese units of measurement; one *chi* (尺) equals approximately one-third of a meter, and one *cun* (寸) equals approximately one-thirtieth of a meter, but the precise definitions of these units has varied during the course of Chinese history. See Klaas Ruitenbeek, *Carpentry and Building in Late Imperial China: A Study of the Fifteenth-Century Carpenter's Manual* Lu Ban Jing (Leiden: Brill, 1993), 1.

> When the lamp was lit, the scales [of the dragon] all moved and shone like stars, lighting up the whole room.
>
> There were also twelve seated figures made of cast bronze. They were all the same height, three *chi*, and were seated on bamboo mats, playing the *qin* [琴], *zhu* [筑], *sheng* [笙], etc.[42] All [the figures] were decorated with multi-coloured designs. They almost looked like real people. Underneath the bamboo mats, there were two bronze pipes, the top openings of which were several *chi* in height beyond the mats.[43] One of the pipes was empty, and inside the other there was a rope as thick as a finger. If one blew into the empty pipe while another person twisted the rope, all the instruments started playing, and [the effect] was no different from real music.
>
> There were also lutes, six *chi* in length, with thirteen strings and twenty-six frets; each was completely decorated with the seven treasures[44] and an inscription that read "music of precious jade".
>
> There [also] was a wind instrument of jade, two *chi* three *cun* in length, with twenty-six holes. When it was played, one would see vehicles, horses, and mountain forests, one right after another.[45] When one ceased playing [the instrument], one stopped seeing these images. [This instrument] was inscribed "jade tube of brightness and beauty".

The text then turns to another device in the storehouse, a large mirror:

> There was [also] a rectangular mirror, four *chi* wide and five *chi* nine *cun* high. The outside and inside [of the mirror] were luminous. If one were coming directly to face the mirror, then their image would be reversed. If one touched one's heart with the hand, and approached the mirror, then the colon, stomach, and the five *zang* were clearly visible [in the mirror]. If one had an internal illness, and covered their heart and faced the mirror, then the location of the illness would be known. And if a woman

42 The *qin* is a Chinese instrument similar to a zither, the *zhu* is a type of Chinese lute, and the *sheng* is a kind of panpipe.

43 The description here is not particularly clear; Needham translates this line as follows: "Under the mat there two bronze tubes, the upper opening of which was several feet high and protruded beyond the end of the mat." See Needham, 4.2, 158.

44 The "seven treasures" refer to gold, silver, lapis lazuli, crystal, coral, agate, and pearl.

45 The meaning here is uncertain; the phrase "one would see vehicles, horses, and mountain forests" may mean that one would see *images* of these when playing this jade wind instrument. Perhaps some kind of "magic lantern" device is being referred to here, or a mechanism powered by a current of air.

> who had an evil heart was facing the mirror, her gall would swell and her heart would palpitate. Qin Shihuang [秦始皇] often used [the mirror] to check his concubines.[46] If their galls swelled and hearts were agitated, he had them executed. Knowing this, Gaozu had [the storehouse] closed and waited for Xiang Yu [項羽], who took it all [i.e., the contents of the storehouse] eastward.[47] Afterwards, [its] whereabouts were unknown.[48]

In this text, it is interesting to see how the discussion of the mirror with special powers is part of a larger discussion of mechanical devices or automata. The technological context makes the description of the mirror's features seem less fantastic, although still quite peculiar. Recall that in Chapter 2, in the Chaucer tales and their antecedents, strange mechanical devices — such as flying horses made of wood or brass — are also set in a technological context with other devices, including automata.

In the passage above, there are again the five *zang*, comprising the heart, liver, spleen, lung, and kidneys. One modern commentator has noted that "the *zang* and *fu* weren't anatomically conceived" in early Chinese medicine but were understood in a more "functional" sense.[49] However, this does not mean that we should interpret the description above as not referring to actual human

46 Emperor Qin Shihuang (259–210 BCE) is perhaps most famous today for the Great Wall and the terracotta warriors of his mausoleum. He was born as Ying Zheng (嬴政) and was the son of the king of the State of Qin. Ying Zheng's ambition was to subjugate and then unify the other states such as Han, Zhao, Wei, Chu, Yan and Qi. He realized this goal by 221 BCE, building the first centrally-governed empire in Chinese history, and commencing what is now called the Qin Dynasty.

47 Xiang Yu (232–202 BCE) was a famous military leader and and a head of the coalition against the Qin (see Dull, "Anti-Qin Rebels …", 307ff). Gaozu — at that time known as Liu Bang (劉邦) — was a member of this coalition, but he and Xiang Yu subsequently became enemies. In 202 BCE, Gaozu, as emperor, defeated Xiang Yu; for a brief summary of this conflict, see Matthew Bennett, *The Hutchinson Dictionary of Ancient & Medieval Warfare* (Chicago: Fitzroy Dearborn, 1998), 139.

48 Xin Liu, *Xijing Zaji* (西京雜記 "Miscellaneous Records of the Western Capital") (Taipei: Shangwu, 1979), ch. 3, 3a; another edition of this text, with commentary, is found in Xinyang Xiang and Keren Liu, eds., *Xijing Zaji Jiaozhu* (西京雜記校注 "Annotated Edition of Miscellaneous Records of the Western Capital") (Shanghai: Guji chubanshe, 1991); the translation here is by the author with the assistance of Thomas Radice, of the University of Pennsylvania, and Lin Li-Chuan. The final sentence of this text is somewhat unclear; the connection between Xiang Yu and the fate of the mirror is obscure, although it is known that after the victory of Gaozu over the Qin, Xiang Yu arrived at Xianyang — the location of the palace and the storehouse with the mirror. Also see Chen, *Record of an Ancient Mirror: An Interdisciplinary Reading*, 20.

49 See Shigehisa Kuriyama, *The Expressiveness of the Body and the Divergence of Greek and Chinese Medicine* (New York: Zone Books, 1999), 266.

anatomy. The early and well-known Chinese medical work *Huangdi Neijing* (黃帝內經 "The Yellow Emperor's Classic of Internal Medicine") includes "descriptions of the [human] body ... largely based on dissections".[50]

That work preceded the *Xijing Zaji* and so knowledge of human anatomy potentially was available to the writer of the passage above. This suggests that the author of the *Xinjing Zaji* account has taken the idea of a simple mirror — a reflecting device — and somehow attributed to it the more fantastic ability to see inside the body. Perhaps such a conflation of technological capacities already existed in the writer's source. Regardless, this is a peculiar, if intriguing, way of framing a technology.

The passage above also mentions the use of the mirror by the emperor "to check his servants", to see if "their galls swelled". At first, this might seem rather strange; however, a Tang dynasty mirror, currently at the Cleveland Museum of Art, bears an inscription noting that it can reveal a person's gall, and, by extension, reveal their emotions.[51] An early eighth-century CE bronze mirror in the Shanghai Museum has a similar inscription that seems to echo directly the passage from the *Xijing Zaji*: "It is often said that the bright mirror of the King of Qin illuminates the gall and the heart. It possesses the efficacy unobtainable even by good craftsmen."[52] In traditional Chinese medicine, the gall bladder is an organ with a key role in determining a person's initiative and resolve.[53] A story in the *One Thousand and One Nights*, "The Tale of Zayn Al-Asnam", similarly mentions a prince who uses a mirror that can determine whether or

50 See p. 636 of Catherine Despeux, "The Body Revealed: The Contribution of Forensic Medicine to Knowledge and Representation of the Skeleton in China", in Francesca Bray, Vera Dorofeeva-Lichtmann, and Georges Métailié, eds., *Graphics and Text in the Production of Technical Knowledge in China: the Warp and the Weft* (Leiden: Brill, 2007), 635–684.

51 This mirror is labeled "Mirror with Six Circular Flowers" due to a design on one side (Cleveland Museum of Art, accession number 1995.341).

52 See pp. 47–48 and fig. 6 of Eugene Yuejin Wang, "Mirror, Moon, and Memory in Eighth-Century China: From Dragon Pond to Lunar Palace", *Cleveland Studies in the History of Art* 9 (2005): 42–67.

53 See Ning Yu, "Metaphor, Body, and Culture: The Chinese Understanding of Gallbladder and Courage", *Metaphor and Symbol* 18.1 (2003): 13–31, and Giovanni Maciocia, *The Foundations of Chinese Medicine: A Comprehensive Text for Acupuncturists and Herbalists* (Edinburgh: Churchill Livingstone, 1989), 116. An earlier article on Chinese mirrors also notes that in Chinese culture a mirror could be understood as a "a thought and crime detector, a precursor of X-ray or a light-emitting object" — see p. 188 of A.R. Hall, "The Early Significance of Chinese Mirrors", *Journal of the American Oriental Society* 55.2 (June 1935): 182–189. Also see pp. 49–50 of Wang, "Mirror, Moon, and Memory in Eighth-Century China", cited above.

not a woman is still a virgin.[54] A separate Chinese tradition speaks of a 業鏡 (*ye jing*) — a "mirror of karma" or "mirror of deeds" — as a Buddhist device to examine the conduct of an individual.[55]

But what is it exactly that rings odd about the account in the *Xijing Zaji* of this mirror found in a storehouse? Some of the other devices described — the seated figures and the musical instruments — fit with our contemporary notion of technological possibilities for a pre-modern civilization. It is the other devices in the text that defy easy historical placement or explanation; in particular, there is the strange "wind instrument of jade" which allows the player to see "vehicles, horses, and mountain forests, one after another" and hear "the hidden rumble of the vehicles" — and, of course, there is the rectangular mirror which could reveal the internal organs. Again, the reader witnesses the apparent conflation — as we see in a number of early sources — of descriptions of apparatus at the same time both magical and technological. In the case of the mirror described here in the the *Xijing Zaji*, the account is given a kind of further veracity by claims that it was displayed in front of a palace during the Three Kingdoms period (220–280 CE).[56]

A similar account, perhaps deriving from the in the *Xijing Zaji*, is found in a ninth-century CE work, the *Youyang Zazu* (酉陽雜俎 "Miscellaneous Morsels from Youyang") by Duan Chengshi (段成式). In this account, the mirror is also described as being large in size and able to display the five viscera.[57] This text also notes that this mirror was known as the *zhao gu bao* (照骨寶), meaning the "treasure [*bao*] that illuminates [*zhao*] the bones [*gu*]".[58]

54 See the mention of this tale on p. 68 of Russell McDougall, "Blind Man's Bluff in a Sandstorm? Literary Education in the Late Condominium Sudan," in Annalisa Oboe and Shaul Bassi, eds., *Experiences of Freedom in Postcolonial Literatures and Cultures* (London: Routledge, 2011), 59–70, and in Clouston, 330.

55 See B.J. ter Haar, *The White Lotus Teachings in Chinese Religious History* (Honolulu: University of Hawai'i, 1999), 167–169. Also note Stephen F. Teiser, *The Scripture on the Ten Kings: And the Making of Purgatory in Medieval Chinese Buddhism* (Honolulu: University of Hawai'i, 1994), 175 and p. 50 of W. Gilbert Walshe, "Some Chinese Funeral Customs", *Journal of the North China Branch of the Royal Asiatic Society* 35 (1903–1904): 26–64, where the device is called a "mirror of retribution".

56 See Chen, *Record of an Ancient Mirror: An Interdisciplinary Reading*, 20.

57 On the mention of mirrors in this work, note Wenley, 141–142, and Chen, 45, n. 52. On this work generally, see Carrie E. Reed, *A Tang Miscellany: An Introduction to the Youyang Zazu* (New York, 2003), as well as idem, "Motivation and Meaning of a 'Hodge Podge': Duan Chengshi's *Youyang zazu*", *Journal of the American Oriental Society* 123.1 (January–March 2003): 121–145.

58 See the brief comment on this in Chen, *Record of an Ancient Mirror: An Interdisciplinary Reading*, 20, n. 74.

How should one read these texts? They could be read as pure fiction, fabrications of the author or his source; but these texts — particularly the passage in the *Xijing Zaji* with its context of mechanical devices — have details that go beyond the framework of a simple, fanciful tale. Certainly, the Chinese had standard hand mirrors of polished metal, similar in design to those of the Greeks and Romans.[59] Research into this field has been quite extensive, with examinations of mirrors carried out by a number of sinologists in the nineteenth and twentieth centuries. Friedrich Hirth, who compiled an extensive study and bibliography on the subject, notes that the invention of mirrors in China was attributed to the legendary emperor, Huang Di (皇帝), who is alleged to have reigned around 2700 BCE.[60]

4 Looking into Chinese Mirrors

Even before the Han dynasty, there were already established terms in Chinese writings to refer to mirrors: *jian* (鑑 / 鑒) and *jing* (鏡). The discussion at beginning of this chapter notes that in the West, terms for mirrors included broader meanings, such that *speculatio*, for example, had the definition of "contemplation". In Chinese, *jian* (鑑 / 鑒) also includes the more metaphorical meaning of "reflect" and is even found in book titles such as *Zizhi Tongjian* (資治通鑒), translated as "A Comprehensive Mirror for the Aid of Government", a historical work written by an eleventh-century CE Chinese historian.[61]

59 See, for example, the brief history and sample images in Stephen Little and Shawn Eichman, *Taoism and the Arts of China* (Chicago: The Art Institute of Chicago, 2000), 140–141.

60 Friedrich Hirth, "Chinese Metallic Mirrors, With Notes on Some Ancient Specimens of the Musée Guimet", in Berthold Laufer and H.A. Andrews, eds., *Boas Anniversary Volume: Anthropological Papers Written in Honor of Franz Boas, Professor of Anthropology in Columbia University, Presented to Him on the Twenty-Fifth Anniversary of his Doctorate* (New York: G.E. Stechert & Co., 1906), 208–256, noting especially 208–212; also see Schuyler Cammann, "Chinese Mirrors and Chinese Civilization", *Archaeology* 2.3 (September 1949): 114–120; R.W. Swallow, *Ancient Chinese Bronze Mirrors* (Peiping, 1937); and Milan Rupert and Oliver Julian Todd, *Chinese Bronze Mirrors: A Study based on the Todd Collection of 1,000 Bronze Mirrors Found in the Five Northern Provinces of Suiyuan, Shensi, Shansi, Honan, and Hopei, China* (Peiping: San Yu Press, 1935). For a brief survey of the development of mirrors in various cultures, see Jay M. Enoch, "History of Mirrors Dating Back 8000 Years", *Optometry and Vision Science* 83.10 (October 2006): 775–781, which notes the possible foreign origin of Chinese mirrors.

61 On the term *jian* (鑑 / 鑒), see Wang, "Mirror, Death, and Rhetoric: Reading Later Han Chinese Bronze Artifacts", *The Art Bulletin* 76.3 (September 1994): 511–534; note especially p. 511, n. 1. Also see the interesting discussion of mirror symbolism in seventeenth-century

FIGURE 2 A traditional Chinese bronze mirror; maker unknown, 8th century CE, bronze

More pragmatically, both of the Chinese characters for mirror, 鑑 and 鏡, include the radical *jin* (金), here meaning "metal". Chinese metallurgy in the crafting of mirrors developed over a long period of time; however, with very sophisticated casting techniques already in evidence by the fourth-century BCE, the Chinese had the ability to make mirrors of bronze, an alloy of copper, tin, and lead. The alloy may be gold or silver in appearance, depending on the ratio of tin in the mixture (fig. 2).[62]

Chinese literature in Jing Zhang, "In His Thievish Eyes: The Voyeur/Reader in Li Yu's "The Summer Pavilion", *Southeast Review of Asian Studies* 34 (2012): 25–42. Of course, the mirror also has a long history as a concept in Eastern philosophy; a well-known discussion is found in Paul Demiéville, "Le miroir spirituel", *Sinologica: Zeitschrift für chinesische Kultur und Wissenschaft* 1.2 (1947): 112–137.

62 For a history of Chinese bronze mirrors and possible outside origins, see Li Jaang, "Long-Distance Interactions as Reflected in the Earliest Chinese Bronze Mirrors", in Lothar von

When polished, bronze is highly reflective, and it was used in many cultures — before the development of glass mirrors — for the creation of mirrored surfaces for various purposes.

In China, bronze mirrors came into wide circulation by the time of the Warring States period (ca. fifth-century to third-century BCE).[63] These mirrors often had auspicious designs on the back, and some had inscriptions as well.[64] Hirth notes that mirrors were said by the Chinese to have a number of magical powers, including the ability to ward off of evil spirits.[65] But the texts here describe mirrors that are not simply for reflection — nor are they for magical or shamanistic practices. In the *Kaiyuan Tianbao Yishi* account, we have a mirror that had the ability, the text tells us, to "illuminate" the *inside* of a patient's body. Again the language is strange to us, because we tend to think of mirrors simply as objects that show us a reflected image, not a penetrating one.

The two Chinese texts that talk about these mirrors with penetrating abilities give very specific details, and again one should ask the question: if the account is a complete fabrication, why the specificity of detail? One of the particular details is that the mirror is rectangular, rather than disk-shaped. This is curious: Hirth notes that almost "all the Chinese metallic mirrors we

---

Falkenhausen, ed., *The Lloyd Cotsen Study Collection of Chinese Bronze Mirrors: Volume II: Studies* (Los Angeles: Cotsen Occasional Press, UCLA Cotsen Institute of Archaeology Press, 2011), 34–49.

63 See the introductory remarks in Zhongfu Zhou, Shuyun Sun, Rubin Han, and T. Ko, "A Study of the Lacquerish Patina (*Qi Gu*) on Ancient Bronze Mirrors", in Alan K.L. Chan, George K. Clancey, and Hui-Chieh Loy, eds., *Historical Perspectives on East Asian Science, Technology and Medicine* (Singapore: Singapore University Press, 2002), 465–469, and Toru Nakano, Tseng Yuho Ecke, and Suzanne E. Cahill, *Bronze Mirrors from Ancient China: Donald H. Graham, Jr. Collection* (Honolulu: Donald H. Graham, Jr., 1994). Also see the discussion in Kimpei Takeuchi, "Ancient Chinese Bronze Mirrors", *The Burlington Magazine for Connoisseurs* 19.102 (September 1911): 311–313 and 316–319, which includes excellent photographs of some Han and Tang dynasty mirrors. A brief survey can also be found in Alfred Salmony, "Chinese Metal Mirrors: Origin, Usage, and Decoration", *Magazine for the Buffalo Museum of Science* 25 (1945): 96–104.

64 On the decoration of mirrors, see Sueji Umehara and Jiro Harada, "The Late Mr. Moriya's Collection of Ancient Chinese Mirrors", *Artibus Asiae* 18.3–4 (1955): 238–256, as well as the extensive discussion in Anneliese Bulling, *The Decoration of Mirrors of the Han Period: A Chronology* (Ascona, Switzerland: Artibus Asiae Publishers, 1960).

65 Hirth, 229–230. Concerning different types of early Chinese mirrors, see Chen, "The Mystery of an 'Ancient Mirror' …", 35 et ff. For a longer discussion on mirrors, magic, and Daoism, see Max Kaltenmark, "Miroirs magiques", in *Mélanges de sinologie offerts à Monsieur Paul Demiéville*, 2 vols. (Paris: Presses universitaires de France, 1966–1974), 2.151–2.166, as well as Edward H. Schafer, "A T'ang Taoist Mirror", *Early China* 4 (1978–1979): 56–59.

know of are disk-shaped" and the archaeological record from China seems to support this.[66] However, Hirth goes on to comment:

> Very large mirrors, a number of feet long and broad, probably rectangular in shape, and reflecting the whole human body, are also mentioned in the post-Christian literature, some of them being imported from abroad. Several passages are on record which tell of the importation of such mirrors under the Emperor Wu-ti, who hoarded up many precious things he had sent for from western Asia; but they are mostly blended with legendary matter, which makes it difficult to treat them seriously. Such large metallic mirrors, aside from the possibility of their having been originally constructed in China itself, could easily be accounted for as importations, or imitations, of Western specimens, certainly during the first century CE, when Seneca the Younger mentioned mirrors of sizes equal to those of human bodies.[67]

However, Western accounts of mirrors, which include discussions by Seneca the Younger, Pliny the Elder, Pausanias, and others, do not attribute to them any penetrative powers. These sources have discussions of various kinds of "trick" mirrors — for example, convex and concave mirrors — that can distort reflections, but nothing exactly like the descriptions found in the Chinese texts presented above.[68]

5 Mirrors, Meaning, and Context

It seems as if the authors of these Chinese texts are struggling to explain some kind of mechanical device, but with their language not being sufficient to the task. This does not mean that the Chinese language itself is to blame, and in fact the belief that Chinese is somehow inherently unable to communicate

66 Ibid., 223. It may be simply the case, however, that the small round mirrors are more commonly found because the Chinese used them as grave goods; therefore, it is difficult to say what sizes and shapes of mirrors the Chinese actually used in daily life. I would like to thank Nathan Sivin for making this important point.

67 Ibid., 224; there have also been finds of early Chinese square mirrors — see Hall, 182. Square mirrors are also discussed in Doris Dohrenwend, "The Early Chinese Mirror", *Artibus Asiae* 27.1–2 (1964): 79–98, and see the comments on large, square mirrors in Chinese lore in Chen, "The Mystery of an 'Ancient Mirror' …", 45, n. 52.

68 For some early Western descriptions of distorting mirrors, see, for example, Seneca the Younger, *Natural Questions* 1.16, in Seneca the Younger, *Natural Questions*, 2 vols., trans. Thomas H. Corcoran (Cambridge, MA: Harvard University Press, 1971–1972), 1.82–1.83.

scientific and technological concepts has been roundly refuted.[69] However, the authors of the Chinese texts under examination here were not necessarily technical experts; they were chroniclers and collectors of anecdotal information, and they did not have the "language" — in the more abstract sense of that word — to describe fully the matter at hand.

The word "mirror" here — and that is the precise term used in these Chinese texts — might be metaphorical, standing in for some other metallic or glass-like material or a contraption utilizing such material. Even in our contemporary technological age, we often use metaphorical language to describe mechanical devices: a space "ship", a "computer" (although it does much more than simply compute), and so on. The term "horseless carriage" was used to refer to automobiles in the beginning of the twentieth century, as a way of explaining a new technology by using existing language. The earlier chapter here on flying vehicles mentions the similar phenomenon of using familiar objects — such as canoes, cradles, and shields — as ways of describing or naming a device that does something quite novel in technological terms, that is, go up into the air.

One cannot assume from these ancient texts that describe mirrors that can see inside the human body an unreasonably advanced technological knowledge on the part of the early Chinese.[70] However, there may be a few scattered clues in the history of Chinese mirrors to help us interpret these peculiar descriptions, with two key considerations. The first concerns the connection between medicine and mirrors, notable because the Chinese texts presented here have mirrors serving a medical function. Hirth points out several accounts where mirrors were used to prevent or cure sickness, with the methods never sufficiently articulated — at least from a modern perspective — by the Chinese authors.[71] The second considerations concerns the existence in Chinese history of so-called "magic mirrors", described below.

The passages presented above are somewhat more explicit than the ones cited by Hirth in relation to medicine, in that they at least state specifically what these mirrors were said to do: reveal the organs of the human body. Despite evidence for a number of uses of mirrors in ancient Chinese sources — including simple reflection, light intensification, performance of rituals, and keeping away evil spirits — the precise medical utilization of

---

69 Note pp. 303–305 of Sivin, review of *Science and Civilisation in China. Volume 7, The Social Background. Part 2, General Conclusions and Reflections*, by Joseph Needham, *China Review International* 12.2 (2005): 297–307.

70 Hall, 188, in passing, notes wryly how a mirror might be credited as "a protector against evil … a disc for divination like a crystal gazer's glove, a thought and crime detector, [or] a precursor of X-ray or a light emitting object" in early Chinese texts.

71 Hirth, 230–232.

mirrors recounted in the *Kaiyuan Tianbao Yishi* and the *Xijing Zaji* does not seem to appear in many other Chinese accounts.[72] Again, too, there is no similar account — where mirrors or other devices are used to display what is inside the human body — in any Western text.

What *is* found in early Chinese writings is an odd miscellany of medical uses for mirrors, even including bits of them being ground up and consumed orally as a prescription.[73] There is also what one writer calls "mirror lore", particular in the Six Dynasties period, which ran from the early third century to the late sixth century CE. This "lore" involved a number of different kinds of mirrors with special powers: the *zhaoyao jing* (照妖鏡, "Mirror that Reveals Demons'), the *zhao gu jing* (照骨鏡, "Mirror that Reveals Bones'), the *wu ji jing* (無疾鏡, "No Illness Mirror'), the *zhi shi jing* (知事鏡 "Mirror that Knows [Future] Matters"), and the *jian shen jing* (見神鏡 "Mirror that Sees the Spirits").[74]

In some sense, then, these strange "X-ray" mirrors sit in a larger cultural context, a context that included both practical technologies for making metal mirrors and fantastical powers attributed to mirrors or mirror-like devices. The Chinese in this period certainly had an understanding of mirrors in technical terms — that is, the principles of reflection, inversion, and so on.[75] But those technical aspects of mirrors noted in Chinese texts somehow become conflated with the odd accounts found here in the *Kaiyuan Tianbao Yishi* and the *Xijing Zaji*.

6 Another Diagnostic Device

In another Chinese source, there again is the idea of using a device or artifact to see *inside* a body. This time, the item in question is not a mirror, and the description and context are less technical — indeed, they are rather fabulous. The late ninth-century CE collection of anecdotes entitled *Duyang Zabian*

---

72 On the range of uses of mirrors in ancient China, see ibid., 225 ff.

73 Note the comments on these kinds of uses of mirrors in medical practices in Shizhen Li, *Bencao gangmu* (本草綱目 "Compendium of Materia Medica"), 6 vols. (Shanghai: Shangwu yinshuguan, 1959), *juan* 9, 22–23; also see Hirth, 231–232, and Mark Pendergast, *Mirror Mirror: A History of the Human Love Affair with Reflection* (New York: Basic Books, 2003), 20.

74 Chen, "The Mystery of an 'Ancient Mirror' ...", 41–42; for the sake of precision, I provide here more literal translations than those given in Chen's paper; see the list and discussion in idem, *Record of an Ancient Mirror: An Interdisciplinary Reading*, 15.

75 Note, A.C. Graham and Nathan Sivin, "A Systematic Approach to Mohist Optics", in Shigeru Nakayama and Nathan Sivin, eds., *Chinese Science: Explorations of an Ancient Tradition* (Cambridge, MA: MIT Press, 1973), 105–152.

(杜陽雜編 "Compilation of Miscellanea from Duyang"), by the scholar Su E (蘇鶚), includes the mention of a huge, shining rock from a country called *Rilin* (日林) — apparently, Japan.[76] This rock had the power to reveal the internal organs of a patient, aiding the work of a physician in healing them.[77] The passage on this subject in the *Duyang Zabian* frames it with an odd description:

> In the *Dali* [大曆] period,[78] the *Rilin* country [日林國] sent as presents luminous beans and dragon-horn hairpins. This country lays forty thousand *li* [里] northeast across the sea.[79] In the southwest of the country, there is a strange rock several hundred *li* square, luminous and clear; it can reflect a person's five "solid" organs and six "hollow" organs.[80] It is designated the "Mirror of the Immortals".[81] If a person in this country has an illness, then [the rock] is used to illuminate his body, so as to discover [the problem] in a particular organ. Then, *shen cao* [神草] is taken, and the person without exception is healed.[82] The size of the beans resembles that of the Chinese green bean, [but] the colour is a dark red. Moreover, their rays [i.e., the rays of light that the luminous beans emit] can reach several *chi* in length.[83]

Apparently, the story of this interesting rock was in circulation earlier; it also is found in almost identical form in the early sixth-century Shuyiji (述異記 "Record of Strange Things"), and that seems to have been

76 E Su, *Duyang Zabian* (杜陽雜編 "Compilation of Miscellanea from Duyang") (Beijing shi: Zhonghua shuju, 1985).

77 See Schafer, *The Golden Peaches of Samarkand: A Study of T'ang Exotics* (Berkeley: University of California Press, 1963), 37.

78 The *Dali* period was 766 to 779 CE, during the Tang Dynasty.

79 The value of the *li* varied through various periods of Chinese history, but it can be understood as approximately equivalent to half a kilometer here; this yields, however, the fantastic distance of 20,000 kilometers in the text.

80 This again refers to the system of organs as found in traditional Chinese medicine, based on the principle of *yin* and *yang*. The five "solid' or *yin* organs are the heart, liver, spleen, lung, and kidneys, and the six "hollow' or *yang* organs are the large intestine, small intestine, stomach, gall bladder, urinary bladder, and the "triple-warmer' (the openings of the stomach, small intestine, and the bladder); see Reid, *Chinese Herbal Medicine*, 32.

81 The *xian ren* (仙人) are the Daoist immortals or transcendent beings of Chinese tradition.

82 The term *shen cao* (神草) literally means "spirit herb" — this is a traditional way of referring to ginseng.

83 See the mention of this passage in E.D. Edwards, *Chinese Prose Literature of the T'ang Period, AD 618–906*, 2 vols. (London: Arthur Probsthain, 1937–1938), 1.84–1.85; also note Chen, *Record of an Ancient Mirror: An Interdisciplinary Reading*, 20–21.

Su E's source.[84] Again, despite the strangeness of the story, the word used here for mirror is simply the technical term *jing* (鏡).

7 Mirrors and Medicine

At this point, it may be useful to provide a brief comment on the connections between these Chinese mirrors that are said to be able penetrate the body, and the more general idea in early Chinese medicine of seeing inside the body to make a diagnosis and a prognosis, as well as to heal. One author has noted that especially in terms of dissection — although it was carried out — "anatomy in China never gained dominance as a way of understanding the body."[85] Nonetheless, as noted earlier, there indeed was anatomical knowledge, and more critically, there was a discussion of what was going on *inside* the body in functional terms.

One story recounts how the Chinese physician Bian Que (扁鵲) warned Duke Huan about a "disease which lies in the blood vessels", and eventually in the bone marrow.[86] The story does not make clear how the physician knows this, but it clearly does discuss what one might term "levels" of an illness, deeper and deeper within the body. Bian Que says to the sceptical Duke Huan — who subsequently dies, having delayed treatment too long — the following:

> When a disease lies in the pores, it can be treated by poultices. When it lies in the blood vessels, it can be treated with needles. When it lies in the stomach and intestines, it can be treated with medicines. But when the disease lies in the bone marrow, not even the God of Life can do anything about it.[87]

---

84 Schafer, *The Golden Peaches of Samarkand*, 288, n. 241 and n. 243.

85 Kuriyama, 155.

86 Ibid., 163. The identity of Bian Que is unclear; his real name may have been Qin Yueren (秦越人), with Bian Que being a title coming from the legendary physician of that name from the age of Huang Di (黃帝) — see *The Yellow Emperor's Classic of Internal Medicine*, trans. Ilza Veith (Berkeley: University of California Press, 2002), 3, n. 10. Also note H.A. Giles, *A Chinese Biographical Dictionary* (London: Bernard Quaritch, 1898), 155 (= entry 396, "Ch'in Yüeh-jen 秦越人"), as well as Jürgen Kovacs and Paul U. Unschuld, *Essential Subtleties on the Silver Sea: The* Yin-Hai Jing-Wei*: A Chinese Classic on Ophthalmology* (Berkeley: University of California Press, 1998), 120–121 and n. 3. In addition, see W.N. Whitney, "Notes on the History of Medical Progress in Japan", *Transactions of the Asiatic Society of Japan* 12.4 (July 1885): 245–470, noting especially p. 281, n. 54.

87 Kuriyama, 163.

Even though a medical disorder might be internal, as deep as the viscera of the body, Chinese traditional medicine had (and still has) a method of "reading" that disorder on the body's surface, so to speak, through a complex series of pulse readings at the wrist.[88]

The question here is how this relates to the stories of penetrating mirrors. For Chinese physicians, a key question was how to "contemplate a body organized by depth", the "levels" or layers that we saw in passage about Bian Que, above.[89] On the one hand, as one commentator puts it, "the skin is an occluding screen" and blocks the diagnostician from understanding the underlying dysfunction.[90] On the other hand, in traditional Chinese medicine the skin could also reveal what was going on inside, as could the pulse. Chinese physicians used their "gaze" (*wang* 望) to make diagnoses, with this "gaze" not only including seeing *per se*, but also diagnosing through the colour or hue of the skin, smelling the patient's body, and so on.[91] One scholar has noted that in early Chinese medicine, "it was sufficient to deduce the interior functioning of the body from signs observed externally", with diagnosis and subsequent treatment "based [not only] on vision but also on other senses."[92]

But the real coup, of course, would be the ability to see *directly* into the body, and perhaps the story of these penetrating mirrors arose in the context of early scholars contemplating such an idea. That is, the "pinnacle of medical acumen" was *wang er zhi* (望而知) — literally, "to gaze and to know".[93] So, while the Chinese medical model clearly comprised levels or layers of the body, and external aspects could reveal those, what if there was a device that could allow the physician's "gaze" to enter into the body unhindered? Musing on this topic might have led to the technological speculations described in the texts above.

In early Chinese medicine, there arose a model of illness that was based, in fact, on the concept of "dysfunctions within the systems of vessels [*mai* 脈]" and a "deeper physiological dysfunction".[94] Of course, this then suggests

---

88 Ibid., 166; this Chinese diagnostic method is unrelated to the Western method of taking the pulse to measure heart rate. See the discussion in Elisabeth Hsu, "Pulse Diagnostics in the Western Han: How *mai* and *qi* Determine *bing*", in idem, ed., *Innovation in Chinese Medicine* (Cambridge: Cambridge University Press, 2001), 51–91.

89 Kuriyama, 167.

90 Ibid., 167.

91 See ibid., 167 ff., where Kuriyama provides an interesting discussion of this idea of the "gaze", as well as the importance of the hues of the skin, in diagnosis.

92 Despeux, 637.

93 Kuriyama, 179.

94 Donald Harper, "Iatromancy, Diagnosis, and Prognosis in Early Chinese Medicine", in Hsu, *Innovation in Chinese Medicine*, 99–120. Also see the discussion of *mai* (脈) in Vivienne Lo, "The Influence of Nurturing Life Culture on the Development of Western Han Acumoxa

that the physician would have desired the best possible methods or technology for perceiving or revealing such an internal dysfunction. It is interesting that we find an oblique connection here to another aspect of mirrors. Chinese medicine concerned itself not only with the body itself and its functions, but also with divination and physiognomy, a connection still found today in places such as Taiwan. This makes sense, given the framework of traditional medicine — "reading" a patient as part of the formulation of a medical diagnosis, aligned with "reading" a person by looking at their face, their astrological data, and so on. Mirrors in the West were used for catoptromancy; while the Chinese do not seem to have used their "medical mirrors' in this way, they nonetheless linked other forms of divination, through such vehicles as the *Yijing* (易經), to medicine.[95] In a more general sense, too, the idea of "reading a face" to reveal a person's character is not unlike the use of the mirror to reveal the internal organs — and thus the inclinations and character — of one's concubines, a practice noted in the *Xijing Zaji*.

8 Jīvaka's Diagnostic Device

A description concerning a rock with powers similar to the penetrating capacity of the strange Chinese mirrors is found in an account concerning Jīvaka, the physician to the Buddha. Jivaka is reported to have come into the possession of a magic gem that — apparently like a modern fluoroscope — could light up the inside of a person's body.[96] An old tale notes that while traveling, Jīvaka came upon

> a man carrying a load of wood to the city, of whom nothing was left but skin and bone, and the whole of whose body was dropping sweat; he said to him, "O friend, how came you into such a plight?" The man replied, "I know not. But I have got into this state since I began to carry this load." Jīvaka carefully inspected the wood, and said, "Friend, will you sell this wood?" "Yes!" "For how much money?" "For five hundred Kārshāpaṇas."

Therapy", in Hsu, 19–50; note especially pp. 20 et ff. Lo also elaborates on the idea of how Chinese medicine included a range of practices, from breath-cultivation to advice on sexual practices.

95 See the discussion in Harper, 100; also see idem, "Physicians and Diviners: The Relation of Divination to the Medicine of the *Huangdi neijing*", *Extrême-Orient, Extrême-Occident* 21 (1999): 91–110.

96 A brief mention of this is found in Laufer, *The Prehistory of Aviation*, 11.

> Jīvaka bought the wood, and when he had examined it, he discovered the gem which brings all beings to belief. The virtue of the gem is of this kind: when it is placed before an invalid, it illuminates him as a lamp lights up all the objects in a house, and so reveals the nature of his malady.
>
> When Jīvaka had gradually made his way to the Udumbara land,[97] he found there a man who was measuring with a measure, and who, when he had finished measuring, inflicted a wound upon his head with the measure. When Jīvaka saw this, he asked him why he behaved that way. "My head itches greatly." "Come here and I will look at it."
>
> The man lay down and Jīvaka examined his head. Then he laid on the man's head the gem which brings all beings to belief, and it immediately became manifest that there was a centipede inside.[98] Thereupon Jīvaka said, "O man, there is a centipede inside your head." The man touched Jīvaka's feet and said, "Cure me." Jīvaka promised to do so ... Next day Jīvaka ... opened the skull with the proper instrument, touched the back of the centipede with the heated pincers, and then, when the centipede drew its arms and feet together, he seized it with the pincers and pulled it out.[99]

As with the Chinese account, this passage contains both fantastical elements — the magical gem — and more technical language: the description of the surgical procedure, not to mention the diagnostic function of the strange jewel. Another early South Asian tradition speaks of the *bhaiṣajya rāja* tree — the name means "king of healing" or "king of medicine" in Sanskrit. This tree was said to contain a gem with the ability to reveal the internal organs

---

97 Literally, "land of the fig", as uḍumbara in Pali and in Sanskrit refers to the *Ficus racemosa*; however, it is not clear what geographical location the text here is indicating.

98 This may refer to a tumour (see Laufer's *The Prehistory of Aviation*, 11, cited earlier), or to *Echinococcus granulosus*, a kind of tapeworm that can lodge in the brain; see R.M. Pujari, Khle Pradeep, and N.R. Kumar, *Pride of India: A Glimpse into India's Scientific Heritage* (New Delhi: Samskrita Bharati, 2006), 150.

99 F. Anton von Schiefner, *Tibetan Tales, Derived from Indian Sources: Translated from the Tibetan of the Kahgyur. Done into English from the German by W.R.S. Ralston* (London: Kegan Paul, Trench, Trübner & Co., Ltd., 1906), 99–100. In another version of this story, Jivaka uses a kind of magical wood to see into a patient's body; see Laufer, *The Prehistory of Aviation*, 11–12, and the detailed discussion on pp. 197–201 of C. Pierce Salguero, "The Buddhist Medicine King in Literary Context: Reconsidering an Early Medieval Example of Indian Influence on Chinese Medicine and Surgery", *History of Religions* 48.3 (February 2009): 183–210. Also note the brief mention in P.K. Doshi, "History of Stereotactic Surgery India", in A.M. Lozano, P.L. Gildenberg, and R.R. Tasker, eds., *Textbook of Stereotactic and Functional Neurosurgery* (Berlin: Springer, 2009), 155–169. For more on accounts of Jivaka's medical knowledge, see Salguero, cited above.

of a patient.[100] However, just as with other accounts presented here, the precise nature of this penetrating visual device is left unexplained.[101]

Similarly, the ninth-century *Zhou Qin Xing Ji* (周秦行記, "A Journey through Zhou and Qin") includes a description of an "imperial-consort [who] is said to have worn a luminous jade ring which showed the bone of her finger".[102] In another Chinese story, there are no special gems or other such devices, but rather a certain drug, water, or dew which allowed the physician named Bian Que, mentioned earlier, to see the "five viscera and the obstructions and knots of the abdomen".[103]

9 A Magic Mirror Trick?

Is there a possible explanation for the supposed ability of these early devices to see inside the human body? Perhaps at least in the case of the Chinese accounts there indeed is a conjectural model that can be put together. Returning to the two original Chinese texts presented here, a hint might come from the so-called "magic mirrors". "Magic mirrors" were special mirrors created by Chinese craftsmen; these devices employed various optical phenomena in order to cast particular kinds of images. The Japanese also created such mirrors, calling them *makyō* (魔鏡) — literally, "demon mirrors".[104] A modern

---

100 K.M. Choksey, *Dentistry in Ancient India* (Bombay: Popular Book Depot, 1953), 9.

101 For some brief mentions of the Jīvaka account, see H. Bergemann, "Jivaka — Schädelchirurg und 'Röntgenarzt' im alten Indien", *Blätter für Zahnheilkunde / Bulletin Dentaire* 29.7 (July 1968): 112–115, and Kenneth G. Zysk, *Asceticism and Healing in Ancient India: Medicine in the Buddhist Monastery* (Oxford: Oxford University Press, 1991), 55. Concerning other accounts of magical stones, see Sydney H. Ball, "Luminous Gems, Mythical and Real", *The Scientific Monthly* 47.6 (December 1938): 496–505; however, none of the stories that Ball recounts speak of gems that have this "x-ray" quality.

102 Edwards, 46–47.

103 Salguero, 201.

104 An early but rather extensive discussion of the similar Japanese devices may be found in William E. Ayrton and John Perry, "The Magic Mirror of Japan", *Proceedings of the Royal Society of London* 28 (1878–1879): 127–148. Also note Ayrton, "The Mirror of Japan and Its Magic Quality", *Nature* 19.493 (10 April 1879): 539–542, and the report in *Notices of the Proceedings at the Meetings of the Members of the Royal Institution of Great Britain ...* 9 (1879–1881): 25–36. There is an even earlier account by the physicist François Arago in *Comptes Rendus de l'Académie des Sciences* 19 (1844), 234. On Western interest in such mirrors in the nineteenth century, see Graeme J.N. Gooday and Morris F. Low, "Technology Transfer and Cultural Exchange: Western Scientists and Engineers Encounter Late Tokugawa and Meiji Japan", *Osiris*, 2nd series, 13 (1998): 99–128, esp. 118–119.

FIGURE 3 A Japanese version of the standard Chinese "magic mirror"

study of these rare devices notes that they "have the uncanny ability to project patterns from the back when light is shining on the front."[105]

Typical "magic mirrors" have cast bronze designs on the back, at times accompanied by written characters (fig. 3).[106]

105 Julia K. Murray and Suzanne E. Cahill, "Recent Advances in Understanding the Mystery of Ancient Chinese 'Magic Mirrors': A Brief Summary of Chinese Analytical and Experimental Studies", *Chinese Science* 8 (1987): 1–8. For technical discussions of Chinese "magic mirrors", see Michael V. Berry, "Oriental Magic Mirrors and the Laplacian Image", *European Journal of Physics* 27.1 (January 2006): 109–118; G. Saines and M.G. Tomilin, "Magic Mirrors of the Orient", *Journal of Optical Technology*, 66.8 (August 1999): 758–765; and D.A. Seregin, A.G. Seregin, and M.G. Tomilin, "Method of Shaping the Front Profile of Metallic Mirrors with a Given Relief of its Back Surface", *Journal of Optical Technology* 71.2 (February 2004): 121–122.

106 For a brief discussion of such inscriptions, see Bernhard Karlgren, "Early Chinese Mirror Inscriptions", *Bulletin of the Museum of Far Eastern Antiquities* 6 (1934): 9–74, and W.P. Yetts, "Two Chinese Mirrors", *The Burlington Magazine for Connoisseurs* 74.430 (January 1939): 22–24 and 26–28.

FIGURE 4 A Japanese *makyō* (魔鏡), with its projected pattern

The reflecting side is convex and made of highly polished bronze. In normal light, this kind of mirror reflects in the typical manner. However, in bright light, one can actually see *through* the reflecting side. Therefore, these mirrors are able to reflect outwards from the polished side whatever designs and characters are on their reverse side (fig. 4).[107]

107 See Needham, 4.1, 94–97; also note Benjamin Goldberg, *The Mirror and Man* (Charlottesville: University of Virginia Press, 1985), fig. 6, and Gregory, 53–54.

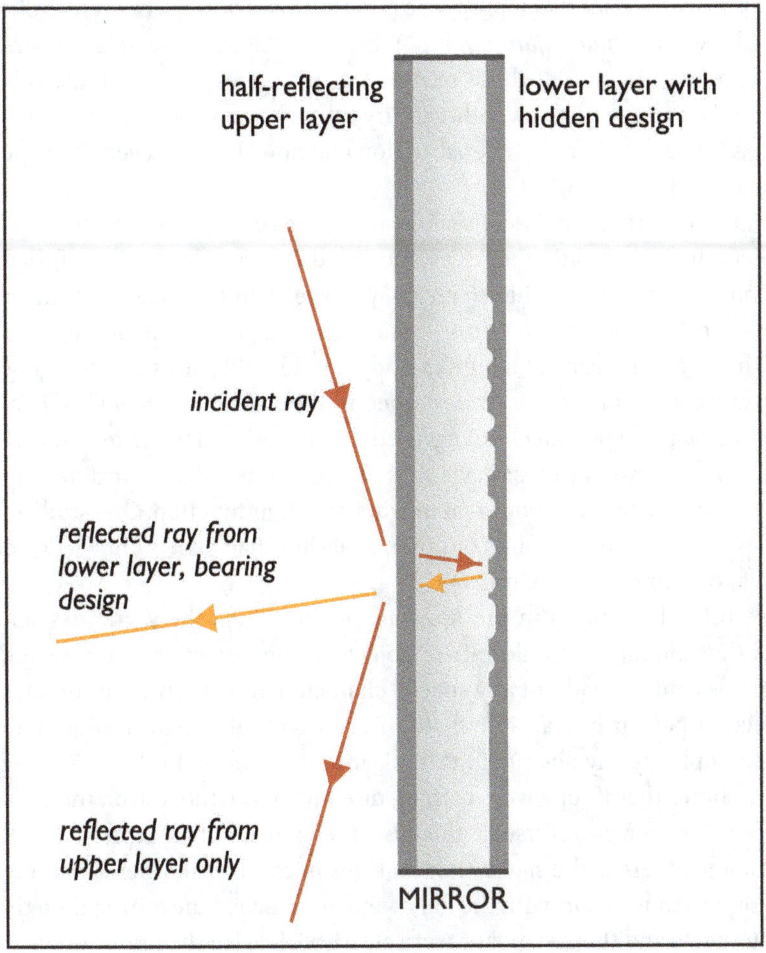

FIGURE 5 Diagram indicating the reflection of a hidden design in a "magic mirror"

In other words, these mirrors usually have two layers: the first is a lower reflecting surface bearing a design, while the second or upper layer is a polished, half-reflecting surface that also lets light through (fig. 5).[108]

---

108 See Se-yuen Mak and Din-yan Yip, "Secrets of the Chinese Magic Mirror Replica", *Physics Education* 36.2 (March 2001): 102–107. There are actually two techniques to hide the images. In one, an additional layer of material on the back hides the engraved designs of the mirror. In the other technique, marks can be made on the reflective surface itself, and those marks are visible only when the mirror is used to cast light on a wall; see Gregory, 54–55 (caption to fig. 3.3), as well as a the conjectures made in the first part of the nineteenth century by the Scottish physicist Sir David Brewster, in John Francis Davis, *The*

Because of these kinds of properties, these types of devices were called by the Chinese *tou guang jian* (透光鑑), meaning "light-permeable mirrors". The design and function of these mirrors was of interest even to those who did not make them — the eleventh-century CE statesman and polymath Shen Kua (沈括) speculated on their construction and how they achieved their special effect.[109]

Perhaps there is some connection here to the idea of mirrors as both reflecting devices and "penetrating" devices. Indeed, the image cast by one of these "magic mirrors" is eerily similar to an X-ray image.[110] In the case of a Chinese "magic mirror", of course, the mirror is simply casting its own image, and not that of the internal organs of a human body. Yet this still leaves us with a possible interpretation of the mirror described in the *Kaiyuan Tianbao Yishi* and the *Xijing Zaji* as a type of technology used for deception. The writers of early Chinese texts — even of those texts that included rather fantastical descriptions — may have known about real mirrors and their function. One study on the Tang dynasty *Gujing Ji* points out that its author "had expert knowledge of the design of Chinese bronze mirrors."[111]

In the brief description in the *Kaiyuan Tianbao Yishi*, the writer explicitly states that the mirror "could reflect objects in the same manner as water." This implies a mirror without any special characteristics. But the author adds: "Whenever a person had an illness, the mirror could be used to illuminate and see completely any obstructions in his internal organs." In the *Xijing Zaji*, the writer notes that if "one were coming directly to face the mirror, then one would see their image reversed"; this also seems to imply a typical mirror. Again, though, there is the suggestion that this mirror has another capability: "If one touched one's heart with the hand, and approached the mirror, then the colon, stomach, and the five organs were clearly visible [in the mirror]."

It might be possible to explain these statements in the context of the "magic mirrors" described above. Imagine a large, rectangular version of the Chinese "magic mirror": the top layer has the highly polished surface of a typical mirror,

---

Chinese: a General Description of the Empire of China and Its Inhabitants, 2 vols. (London: Charles Knight & Co., 1836), 2.236–2.237.

109 See p. 43 of Sivin, "Shen Kua: A Preliminary Assessment of His Scientific Thought and Achievements", *Sung Studies Newsletter* 13 (1977): 31–56.

110 It is interesting to note that in aboriginal art in Australia (and in other locales as well), one finds what is termed "X-ray art" — that is, figures of humans and animals where the bones and internal organs are rendered in some detail; see W.W. Newcomb, Jr., *The Rock Art of Texas Indians* (Austin: University of Texas Press, 1967), 31, as well as Kalyan Kumar Chakravarty and Robert G. Bednarik, *Indian Rock Art and Its Global Context* (Delhi: Motilal Banarsidass, 1997), 183.

111 Chen, *Record of an Ancient Mirror: An Interdisciplinary Reading*, 6.

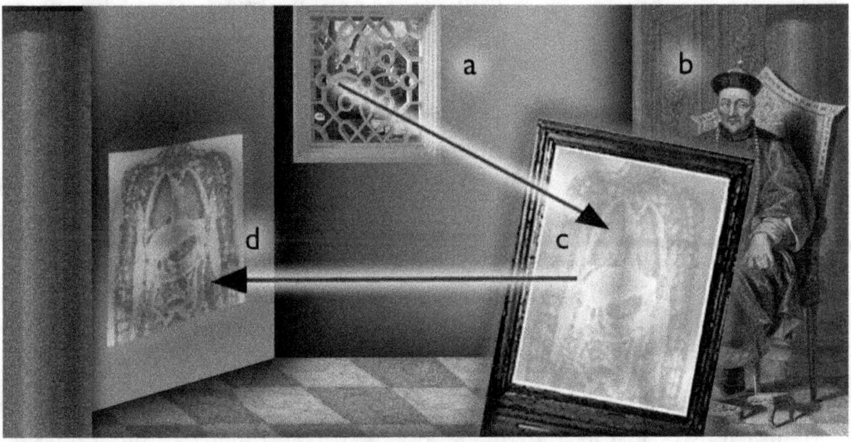

FIGURE 6 A possible configuration using a Chinese "magic mirror"

while the hidden layer comprises a subtle relief rendering of the torso of a human body with the internal organs depicted. Again, the way such a "magic mirror" works is that while the front appears as a smooth reflecting surface, when sunlight or other kind of bright light is reflected off that surface onto a wall, the design from the hidden layer appears in the resulting reflection.

As a supposed diagnostic device, the mirror would have been placed in front of the patient, with a strong light source projected on the reflecting surface (fig. 6).

In the figure, light comes in through a window (a), to a mirror that has been placed in front of a patient (b). The mirror casts a reflection of its hidden design of internal organs (c) to yield an image on the wall (d). In this way, the resultant image would have *appeared* to be revealing the internal organs of the patient. In actuality, the image revealed would be that from the hidden layer of the mirror itself. Printed editions of the *Huangdi Neijing* — the early Chinese medical text mentioned above — include reproductions of rough drawings of the various human organs, and the image cast on the wall may have looked something like those figures.[112] The mirrors described in the *Xijing Zaji* and the *Youyang Zazu* — supposed to reveal the mysterious inner workings of the human body — may have been no more than well-crafted optical tricks.

112 For some examples of the drawings of human organs in the *Huangdi Neijing*, see the reproductions in Veith, 26–27, 28, 31–33, etc. Note especially fig. 10, "Position of the Five Viscera, the Stomach, the Large Intestines, and the Small Intestines', on p. 38, and fig. 13, "The Internal Organs', on p. 41.

The case of these peculiar Chinese mirrors illustrates several key points in terms of the history of technology. Most importantly, it is an example of descriptions of mechanical devices appearing in a broad variety of literature, from ancient Chinese narratives to Buddhist tales. How those descriptions should be treated is a complex matter, since the contexts in which these technologies are discussed are different from the ones we use: again, these are not textbooks or technical manuals. One commentator on the *Gujing Ji* tale concerning a mirror thoughtfully describes this issue of context as follows:

> The history of bronze mirrors from the Han dynasty to the Tang is preserved not only in museums and the works done by mirror specialists, but also in various literary, philosophical, religious, sociological and political texts. Just as a textual landscape can be different from a physical landscape, so can textualized mirror lore differ from the physical history of real mirrors. The study of mirror lore is, to a certain degree, independent of the study of mirror casting history. While the mirror in *Gujing Ji* is typologically a mirror of "nowhere," it is ideologically a mirror of "somewhere".[113]

The amazing mirror in the *Gujing Ji*, a device that can reveal the inner workings of the body, is indeed a mirror of "nowhere" — it may never have existed. But it is also certainly "somewhere" — in the imagination of an author who understood how mirrors worked and could conjecture about a device possibly having penetrating optical ability, a device at the same time both magical and technological.

So, even without putting the template of the present onto the past, it is possible to reconstruct an earlier conceptualization of mechanical devices through looking at extant technologies. It is also certain that writers and storytellers across the globe, in the Americas, Asia, Europe, Africa, and Oceania engaged technology, through recounting what they had observed, heard, or read about. These same figures contextualized the technological ideas that inspired them — flying machines, mirrors, or other devices — in a variety of ways, embedding them in everything from historical chronicles to fanciful tales.

More particularly, early authors also, consciously or unconsciously, encoded technological discussions in synthetic forms, combining genres in a manner largely unfamiliar to modern readers. It is for this reason that some of the descriptions — for example, a mirror or a special gem that function like an X-ray device — strike us as so odd. For the writers and storytellers themselves,

113 Chen, *Record of an Ancient Mirror: An Interdisciplinary Reading*, 15.

however, given their perspective that sophisticated technologies had existed in the past, or that they existed in the present but were shrouded in secrecy, these accounts may have seemed less fantastical. Understanding our sources through their own conception of history — history, perhaps, as cyclical rather than linear — can help us to answer the question raised by these early texts — what *were* these authors describing? These authors may have believed that there was nothing particularly strange about these reports of remarkable technologies, since they could more easily conceptualize the idea of lost or hidden knowledge.

10 Traditions of Transmitted Images in Central American and Persian Cultures

In 1928, Berthold Laufer — who also wrote about early tales of flying machines — published an article with the delightful title, "The Prehistory of Television".[114] Television was quite new even in the period in which Laufer was writing, but he was keen enough to see that this new technology had a long lineage in terms of an ancient human desire to see and hear at a distance. As Laufer put it: "Thus television ... has a prehistory in the domain of oriental folklore."[115] Once more, one finds that early sources of all kinds, including folktales, include discussions of fantastic technologies. This chapter cites Laufer's work, but also provides more details as to these early sources on the idea of transmission and reception of images and sounds. Again, the idea here is that earlier cultures had concepts that on the one hand seem rooted in pure fantasy, but on the other hand involved distinct, technological conceptualizations. As one modern author notes, it is indeed possible to "track the idea of live moving-image transmissions back to the distant past."[116]

Television itself is a very particular technology, with at least three distinct forms in its history. At present, television is based on digital technologies, but in an earlier generation, it used cathode ray tubes to generate images. Prior to that, for a few years, the technology comprised a rapidly spinning disk generating images; a set of apertures set in a spiral on the disk allowed the scanning and subsequent re-assembly of a picture. This is important, because if

114 Laufer, "The Prehistory of Television", *The Scientific Monthly* 27.5 (November 1928): 455–459.

115 Ibid., 455.

116 See p. 289 of William Uricchio, "Television's First Seventy-Five Years: The Interpretive Flexibility of a Medium in Transition," in Robert Kolker, ed., *The Oxford Handbook of Film and Media Studies* (Oxford: Oxford University Press, 2008), 286–305.

one considers the early days of television, it is easy to understand that the disk setup — the "Nipkow disk" is the technical name — is a mechanical device, and even ancient societies had an understanding of wheels, apertures, and so on.

This is not to say that some ancient inventor would have built such a disk, but it means that in the general discussion of technological concepts presented in this book, one can consider television as something closer to shadow images — which early cultures indeed created — and moving magic lanterns and phenakistoscopes, which those same cultures also may have constructed. While one commenter notes that it was the "telephone [that] sparked an anticipatory interest in visual systems that could share the instrument's ability to link distant locations point to point in real time", the general idea of the transmission of images and sounds occurs in a number of early cultures.[117] For example, Laufer cites a Persian story of pictures of distant events seen in a divining cup, as well as a tale of a special "mirror" in Alexandria that could display images of incoming ships, even if they were "twenty days' voyage off".[118] Thus Laufer's title has a resonance even beyond the folkloric connection. The main point here is that technological *ideas* existed before the *actual technology* to construct them existed. Relatively simple technologies, such as shadow puppets, along with the basic concept of signaling at a distance by light reflected with a mirror, may have led to early speculations — or, again, "mechanical mythologies" — about transmitting actual images across significant distances.

Hindu nationalism has given rise to some extreme claims on this subject, with a 2014 book entitled *Tejomay Bharat* (meaning "Shining India") claiming television to be an invention from ancient India.[119] While that claim is extreme, one certainly can see the idea of viewing at a distance in Batra's source, the story of the character Sanjaya in the early Indian epic, the *Mahābhārata*. In the tale, Sanjaya receives the divine gift of being able to observe a battle remotely, and then recount events to King Dhritarashtra. One author goes so far as to say, "Conceptually, this was the equivalent of a TV system being set up for transmitting a particular event."[120] However, the story in the *Mahābhārata*

117 Uricchio, 289.

118 Laufer, 456 and 458.

119 On this subject, see Ravi Meena, *Saffronising Education: "Acche Din"!* (Lucknow: Onlinegatha, 2015), 78; also note Sally Anne Steindorf, "Walking Against the Wind: Negotiating Television and Modernity in Rural Rajasthan" (PhD diss., Syracuse University, 2008), 101–102.

120 Varadaraja V. Raman, *Indic Visions in an Age of Science* (New York: Metanexus, 2011), 194.

implies no technological apparatus helping Sanjaya accomplish his feat.[121] The "panoptic sight" is the story is highly effective, but its source is the sage Vyāsa, not a mechanical device.[122]

Perhaps not surprisingly, in some sources the idea of the transmission of images is connected with the tradition of scrying. Scrying is the practice of looking into a translucent ball, or the shiny surface of some other material to see the past or future, or to communicate with various spiritual entities. Highly polished stone, as well as iron or other ores were used.[123] Many cultures practiced this form of divination and prophecy, including those in Mesoamerica. One of the most famous accounts of such reflective surfaces in this way involves the famed ruler of Tenochtitlan, Moctezuma II.[124] In a mirror, so the story goes, he saw the ultimate fall of his empire. The account comes from the *Historia de las cosas de Nueva España* by the Franciscan friar and missionary, Bernardino de Sahagún (1499–1590):

> The seventh sign or omen is that hunters of waterfowl caught a dark bird the size of a crane, and then went to Moctezuma, who was in a hall that they called Tlillan calmecac. This was after midday. This bird had in the middle of its head a round mirror in which appeared the sky and the stars, especially Castor and Pollux, which move close to the Pleiades.

---

121 *The Bhagavad-Gita: Krishna's Counsel in Time of War*, trans. Barbara Stoler Miller (New York: Bantam Dell, 2004), 5.

122 Diana L. Eck, *India: A Sacred Geography* (New York: Three Rivers Press, 2012), 107.

123 An obsidian mirror from Central America ended up in possession of John Dee, the astrologer to Elizabeth I; it is now in the British Museum. On these mirrors, note the interesting early account in John Lloyd Stephens, *Incidents of Travel in Central America, Chiapas, and Yucatan ... Revised from the Latest American Edition, with Additions by Frederick Catherwood* (London: Arthur Hall, Virtue, & Co., 1854), 311 et ff. Modern studies on these mirrors include, for example, Emiliano Gallaga and Marc G. Blainey, eds., *Manufactured Light: Mirrors in the Mesoamerican Realm* (Boulder: University Press of Colorado, 2016), as well as Saunders, "A Dark Light: Reflections on Obsidian in Mesoamerica", *World Archaeology* 33.2 (October 2001): 220–236. Also see Karl A. Taube, "The Iconography of Mirrors at Teotihuacan", in Janet C. Berlo, ed., *Art, Ideology, and the City of Teotihuacan* (Washington, D.C.: Dumbarton Oaks Research Library and Collections, 1993), 169–204, and John B. Carlson, "Olmec Concave Iron-Ore Mirrors: The Aesthetics of a Lithic Technology and the Lord of the Mirror", in Elizabeth P. Benson, ed., *The Olmec and Their Neighbors: Essays in Memory of Matthew W. Stirling* (Washington, D.C.: Dumbarton Oaks Research Library and Collections, 1981), 117–148.

124 See Olivier Guilhem, *Mockeries and Metamorphoses of an Aztec God: Tezcatlipoca, "Lord of the Smoking Mirror"* (Boulder: University Press of Colorado, 2003), and Olivier Guilhem and Leonardo López Lujan, "Images of Moctezuma and His Symbols of Power", in Colin McEwan and Leonardo López Lujan, eds., *Moctezuma: Aztec Ruler* (London: British Museum Press, 2009), 78–91.

> When Moctezuma saw this, he became frightened. The second time that he looked in the mirror on the bird, a bit later, he saw a crowd of people together all coming, armed and mounted in horses. Then Moctezuma ordered his augurs and diviners to be summoned, and asked them, "Do you not know what this is that I have seen? That many people are coming?" And before the diviners could respond, the bird disappeared, and they said nothing in reply."[125]

Naturally, this prophecy was fulfilled upon the arrival of the Spanish *conquistadores*.[126]

One can see in this account, too, how a mirror no longer is simply a reflecting object, but serves — like a television — as a kind of portal to other places and other times. This is, in short, a proto-technological concept: the idea that an object can have a particular technical capability, however fantastical it may seem. The assumption of great powers for communication by these mirrors in Mesoamerican cultures may have its roots in the simple observation of what reflection can do. One study argues for the possibility that in addition to their magical uses, early mirrors such as those made of pyrite could have been used for a long distance communication system via reflected sunlight.[127]

Tales of mirrors or similar devices for communication at a distance appear in various cultures. The Persian story entitled the *Shahnameh* ("The Book of Kings") tells of a king, Khosrow, who is able to use the divining cup to see where another character, Bizhan, is located. The cup in the tale is called the "world-revealing cup that shows the seven climes".[128] The "seven climes" are a reference to an ancient way of rendering the entire globe, with the *climata* being latitudinal divisions of the Earth.[129]

A story concerning the city of Labtayt (Toledo) in *The Arabian Nights* presents a mirror with similar abilities. The setup is like that in the *Xijing Zaji*

---

125 See James Lockhart, trans. and ed., *We People Here: Nahuatl Accounts of the Conquest of Mexico* (Berkeley: University of California Press, 1993), 55–56. The manuscript of the *Historia de las cosas de Nueva España*, commonly called the *Florentine Codex*, is in the Biblioteca Medicea Laurenziana in Florence.

126 On this passage, see pp. 241–243 of Stephen A. Colston, "'No Longer Will There Be a Mexico': Omens, Prophecies, and the Conquest of the Aztec Empire", *American Indian Quarterly* 9.3 (Summer 1985): 239–258.

127 See the discussion in José J. Lunazzi, "On How Mirrors Would Have Been Employed in the Ancient Americas", in Gallaga and Blainey, 125–142.

128 Abolqasem Ferdowsi, *Shahnameh: The Persian Book of Kings*, trans. Dick Davis (New York: Penguin, 2016), 349.

129 See, for example, Dimitry A. Shcheglov, "Eratosthenes' Parallel of Rhodes and the History of the System of *Climata*", *Klio* 88.2 (February 2006): 351–359.

(西京雜記, "Miscellaneous Records of the Western Capital"), examined earlier. In the *Xijing Zaji*, Emperor Gaozu enters the treasury storehouse in the palace at Xianyang and finds various technological marvels there. In *The Arabian Nights* tale, in turn, a king opens a long-locked tower in the city, and comes upon statues of Arabs on horseback, a book with information on herbs, minerals, talismans, and other subjects, a map of the Earth, Solomon's mirror, and other treasures.[130] This device of Solomon is described as being a "huge circular mirror of remarkable composition that had been made for the prophet Solomon, son of David". It is a magical technology that, like the cup from Persian *Shahnameh*, allows one to "see the seven climes" of the world".[131] In traditional medieval cartography, the world was divided into five latitudinal zones or *climata* (from the Greek κλίματα), including a northern frigid zone, a northern temperate zone, and so on. The system goes back to Ptolemy, who created a seven-climate system that was picked up by later Islamic cosmographers and cartographers. The "seven climes" thus refers to the entire world — meaning that this fantastic mirror allows one to see any part of the planet desired.

The cup that can access the whole world is connected to the Persian tradition as the "Cup of Jamšid" — a device belonging to a mythical Persian king "in which he saw the reflection of all events and phenomena."[132] A related story speaks of "Sikander's [that is, Alexander's] mirror", which allowed Alexander the Great to see into the enemy's camp. Again, the magical technology is said to have come from a lost civilization: a tradition surrounding this story notes that the origin of this mirror is from a time before Adam.[133]

Perhaps related to this same Persian tradition, in a Manichaean religious text there is a reference to the idea of seeing at a distance.[134] The *Rex honoris* (literally, "King of Honor" or "King of Ten Heavens") is a deity in the highly complex Manichaean cosmology, and as one study notes, "Several texts associate the Great King of Honor with some sort of device through which he

---

130 See Marzolph and van Leeuwen, 1.265–1.266, as well as the citation in Victor Chauvin, *Bibliographie des ouvrages arabes ou relatifs aux Arabes, publiés dan l'Europe chrétienne de 1810 à 1885*, 12 vols. (Liège: H. Vaillant-Carmanne, 1892–1922), 6.91.

131 Lyons, 1.889.

132 Seyyed Hossein Nasr, *Sufi Essays* (Albany: SUNY Press, 1972), 33, n. 17.

133 Richard J.H. Gottheil, *The Literature of Persia*, 2 vols. (New York: Colonial Press, 1900), 1.375.

134 Manichaeism was a religion that was founded by the third-century CE prophet Mani, known in Latin as Manichaeus or Manes. There are a number of work on this faith; see, for example, John Kevin Coyle, *Manichaeism and Its Legacy* (Leiden: Brill, 2009), and the bibliographic discussion in Nicholas J. Baker-Brian, *Manichaeism: An Ancient Faith Rediscovered* (London: T. & T. Clark, 2011), vii–ix.

monitors the firmaments."[135] This device is obscurely described in the extant Manichaean literature; one description is found in a Manichaean text written in Sogdian, an Eastern Iranian language that was spoken in Central Asia:

> Thereunder they formed ten Firmaments, [and] set up one magic twelve-faced lens. There they seated a Son of God as watcher, so that in all the ten Firmaments the demons could do no harm.[136]

As the translator of this text notes, this "magic lens" is an interesting term:

> Only one such object is known from other sources: the "wheel that lies in front of the *Rex honoris*" ... All that happens in the firmaments can be seen in it by the *Rex honoris*, seated on a throne in the seventh heaven; it is his magic mirror.[137]

That "wheel" appears in another Manichaean text, known as the *Kephalaia of the Teacher* and dating from the late third century CE. In Chapter 36 of that text — a chapter entitled "Concerning the Wheel that exists in front of the King of Honour" — there is the following:

> Again the apostle speaks about the wheel that exists in front [of the King of Honour who dwells in] the seventh firmament ... [I]f he [i.e., a traitorous or rebellious ruler] should wish to cause a trick in his bond, [it will be] apparent and known in [that] wheel ... [T]he spheres and stars and [the] leader of all the powers exist there, and can be known by it in front of the King of Honour. The twelfth seal that is in it is its five shining men that are conjoined and enmeshed through the entire zone; because every

---

135 See pp. 78–79 of Zsuzsanna Gulácsi and Jason BeDuhn, "Picturing Mani's Cosmology: An Analysis of Doctrinal Iconography on a Manichaean Hanging Scroll from 13th/14th-Century Southern China", *Bulletin of the Asia Institute* 25 (October 2015): 55–105.

136 See p. 312 of W.B. Henning, "A Sogdian Fragment of the Manichaean Cosmogony", *Bulletin of the School of Oriental and African Studies, University of London* 12.2 (1948): 306–318. This passage is also noted on p. 392 of Yutaka Yoshida, "Southern Chinese Version of Mani's Picture Book Discovered?" in Siegfried G. Richter, Charles Horton, and Klaus Ohlhafer, eds., *Mani in Dublin: Selected Papers from the Seventh International Conference of Manichaean Studies in the Chester Beatty Library, Dublin, 8–12 September 2009* (Leiden: Brill, 2015), 389–398, and on p. 66 of Gábor Kósa, "Translating the *Eikōn*. Some Considerations on the Relation of the Chinese Cosmology Painting to the *Eikōn*", in Jens Peter Laut and Klaus Röhrborn, eds., *Vom Aramäischen zum Alttürkischen. Fragen zur Übersetzung von manichäischen Texten* (Berlin: Walter de Gruyter, 2014): 49–84.

137 Henning, 315–316.

> place where they are cramped, and every place where they have space, is displayed by that wheel. For the root of all the firmaments above and below, and that of their rulers, is ensnared and fixed in it. The observation is set right at all times opposite the great King of Honour.[138]

The chapter concludes:

> This is also the case for all the rulers who are in the firmaments, and every power of the zone: If they [i.e., demons] should wish to escape, they shall be recognized and revealed by that wheel. [As] the wheel is like a great mirror, for the discrimination of all is things [... is] in it.[139]

The translator of this text summarizes this as follows:

> The King of Honour is enthroned in the seventh firmament from where he is able to control and judge all the evil powers fettered in the heavens. The wheel is here presented as a divination device by which he can know the plans and tricks of these demons as they seek to escape; but also the location of the five shining men, that is the soul mixed throughout the material universe.[140]

The Manichaean texts themselves do not elaborate on this device any further, but it certainly strikes the reader as being some kind of mysterious mechanical panopticon.

Working from textual descriptions to try to visualize such a "lens", "wheel", or "mirror" is not easy. However, one modern study looks to a Chinese Manichaean cosmological painting to find a connection.[141] The painting, entitled *Yuzhou Tu*

---

138 See Iain Gardner, *The Kephalaia of the Teacher: The Edited Coptic Manichaean Texts in Translation with Commentary* (Leiden: Brill, 1995), 91–92 (= *Kephalaia* 87.30–88.21).

139 Ibid., 92 (= *Kephalaia* 88.29–33); also note Gulácsi and BeDuhn, 79.

140 Ibid., 91.

141 See Kósa, "Translating the Eikōn"; also note pp. 10–11 of idem, "The Sea of Fire as a Chinese Manichaean Metaphor: Source Materials for Mapping an Unnoticed Image", *Asia Major*, 3rd series, 24.2 (2011): 1–52, and idem, "The Sun, the Moon and Paradise — An Interpretation of the Upper Section of the Chinese Manichaean Cosmology Painting", *Journal of Inner Asian Art and Archaeology* 6 (2011): 179–203. For a further discussion, see Abolqasen Esmailpour, *Manichaean Gnosis and Creation Myth* (Philadelphia: University of Pennsylvania, Department of East Asian Languages and Civilizations, 2005), 66; a longer treatment of artwork with Manichaean themes is found in Gulácsi, "Searching for Mani's Picture Book in Textual and Pictorial Sources", *Transcultural Studies* 1 (2011): 233–262, and idem, "The Central Asian Roots of a Chinese Manichaean Silk Painting in

(宇宙圖 "Diagram of the Cosmos"), includes a rendering of the *Rex honoris* holding a large disc. The author of the study believes that this disc is indeed an artist's interpretation of the "lens … to detect imminent dangers caused by the demons in the firmaments".[142]

Another Manichaean manuscript, entitled *Moni Guangfu* (摩尼光佛, "Mani, Buddha of Light"), also mentions this device:

> In this firmament, there is a precious mirror with twelve sides; the upper side observes *nirvana*, the lower side illuminates the netherworld, and the ten [other] sides inspect the various hidden betrayals and other such deeds in the ten firmaments.[143]

The device here is labeled as a 鏡 (*jing*) — as noted above, this is a traditional term in Chinese for a mirror. But the description makes it out to be more of a mechanism for observation, rather than reflection; the verbs used in the original text are 觀 (*guan*), 照 (*zhao*), and 鑒 (*jian*). The first means to observe, look at, or watch; the second means to illuminate, shine, or reflect; and the third means to inspect or reflect. A text from the gnostic religion of Mandaeanism also includes a brief account of a similarly ambiguous device — a spirit figure named Ruha possesses "a magic mirror in which [her son] beholds the upper and lower worlds".[144]

In the Manichaean passage above, there is some ambiguity. First of all, the text above uses the term 面 (*mian*), which can mean both "face" — literally and figuratively — and "side". But the visual rendering in the *Yuzhou Tu*, cited above, simply shows a disc with actual faces. Moreover, the disc is not depicted

---

the Collection of the Yamato Bunkakan, Nara, Japan", in Jacob Albert Berg, et al., eds., *"In Search of Truth": Augustine, Manichaeism and other Gnosticism — Studies for Johannes van Oort at Sixty* (Leiden: Brill, 2011), 315–338.

142 Kósa, "Translating the Eikōn", 66.

143 此天內有十二面寶鏡，上面觀於涅槃，下面照于陰司地府，十面鑒於十天諸庇背叛等事. This Chinese text is from p. 379 of Jinguo Chen and Jun Lin, 明教的新發現–福建霞浦縣摩尼教史跡辨析 ["New Manichaean Discoveries — An Analysis of the Relics of Manichaeism in Xiapu County, Fujian"], in Shaowen Li, ed., *Bu Zhi Yu Yi: Zhongyang Mei Yuan "Yiwen Ketang" Mingjia Jiangyan Lu* (不止於藝–中央美院"藝文課堂"名家講演錄 "More than Art — Famous 'Arts and Literature' Lectures, Central Academy of Fine Arts") (Beijing: Beijing daxue chubanshe, 2010), 343–389; see Kósa, 66–67. I wish to thank Gábor Kósa for pointing out this passage, and sharing it with me in a draft of his paper entitled, "Who is the King of Honour and What Does He Do? Gleanings from the New Chinese Manichaean Sources."

144 Jorunn Jacobsen Buckley, *The Mandaeans: Ancient Texts and Modern People* (Oxford: Oxford University Press, 2002), 41.

as configured to view above and below clearly, as suggested in the textual description.[145] Perhaps this is due to the vagaries of artistic interpretation, for the textual description of the device as a whole indeed outlines a panopticon.

A mirror that can see the whole world is found in other early sources, as well, including Islamic accounts.[146] One story, from the Persian polymath Muhammad ibn Jarir al-Tabari (839–923 CE), speaks of a mirror with a similar connection to religious ideas, as in the Manichaean account. The story recounts how, "God bestows on Adam, the first man, a magic mirror that allows Adam to apprehend all things."[147] The English poet Edmund Spenser, in *The Faerie Queen*, has a "looking glass" that is crafted by Merlin and possesses a similar power.[148] Of particular note here, in looking back at the Manichaean account in particular, is the fact that a technical device mentioned in a religious text. Magic or pure spiritual power is not invoked, but rather a device is used — another bit of "mechanical mythology", which strikes the modern reader as a curious combination. By contrast, for example, Norse mythology attributes to the god Odin a panoptic ability, but through the use of two birds, Huginn and Munnin, who survey the world each day and bring back information to him.[149]

145 See the brief discussion in Kósa, "Translating the Eikōn", 67, n. 94.

146 See Aḥmad ibn al-Rashīd Ibn al-Zubayr, *Book of Gifts and Rarities (Kitab al-Hadāyā wa al-Tuḥaf): Selections Compiled in the Fifteenth Century from an Eleventh-Century Manuscript on Gifts and Treasures*, trans. Ghāda al Hijjāwī al-Qaddūmī (Cambridge, MA: Center for Middle Eastern Studies of Harvard University, 1996), 175–176, as well as Alicia Walker, *The Emperor and the World: Exotic Elements and the Imaging of Middle Byzantine Imperial Power, Ninth to Thirteenth Centuries CE* (Cambridge: Cambridge University Press, 2012), 103.

147 Steven M. Wasserstrom, *Between Muslim and Jew: The Problem of Symbiosis Under Early Islam* (Princeton: Princeton University Press, 1995), 109.; also see p. 157, n. 47 of Moshe Gil, "The Babylonian Encounter and the Exilarchic Hosue in the Light of Cairo Geniza Document and Parallel Arab Sources", in Norman Golb, ed., *Judaeo-Arabic Studies: Proceedings of the Founding Conference of the Society for Judaeo-Arabic Studies* (Amsterdam: Harwood Academic Publishers, 1997), 135–174, and Amikam Elad, *The Rebellion of Muḥammad al-Nafs al-Zakiyya in 145/762* (Leiden: Brill, 2016), 95.

148 A brief discussion of Spenser's mention of this device is found in Douglas Brooks-Davies, *The Mercurian Monarch: Magical Politics from Spenser to Pope* (Manchester: Manchester University Press, 1983), 44–45; also see the contextual discussion in Taylor, *The Moral Mirror of Roman Art*, 2–4.

149 See the discussion in Margaret Clunies Ross, *A History of Old Norse Poetry and Poetics* (Cambridge: D.S. Brewer, 2005), 94, and in Richard North, *Pagan Words and Christian Meanings* (Amsterdam: Rodopi, 1991), 104 et ff.

11 Prester John and Western Traditions of Long-Distance Mirrors

An obvious way to "transmit" images is to bounce them from a mirror. This is a fairly simple method — but also rather limited, since the light of the image will dissipate with distance, and will be distorted and obscured by atmospheric effects. Despite this fact, one finds in early texts this idea that a simple mirror could be used to see things from afar. One such mirror is found in the famous "Letter of Prester John". Prester John was a legendary Christian figure, said to be residing in the East, who appeared in popular European accounts for hundreds of years, beginning in the twelfth century.[150] Around this time, an apocryphal letter appeared that was alleged to have been sent to the Byzantine Emperor Manuel I Comnenus (1118–1180) by Prester John.[151] The text came to be translated into many languages, and circulated widely. The letter recounted various marvels that would have resonated with a European audience; one passage in the letter discusses in elaborate detail the setting for a special mirror:

> In front of the gates of our palace … is a mirror of exceptional height, to which one must climb 125 steps. Indeed, the steps of the lower third are made of porphyry, and partly of serpentine and alabaster. For the middle third, the steps are made of crystal, jasper, emerald, and sardonyx. Indeed, the upper third are of amethyst, jasper, and leopard-skin. Indeed, the mirror is supported by a single column. Upon this is set a base, upon the base are two columns, upon which likewise lies another base, and upon that lie four columns, upon which likewise lies another base, and upon that lie eight columns, upon which likewise likes another base, and upon that lie 16 columns, upon which likewise lies another base, upon which like 32 columns, upon which likewise lies another base, and upon that lie 64 columns, upon which likewise lies another base, upon which lie 32 columns. And in this way the columns diminished by descending just as they increase by ascending, al the way down to one. The columns and bases are also of the same kind of stones as the steps by which one ascends to them. Indeed, at the top of the highest column is the mirror that is consecrated by such art that all machinations and

150 On Prester John, see Keagan Brewer, *Prester John: The Legend and its Sources* (Farnham: Ashgate, 2015) and Vsevolod Slessarev, *Prester John: The Letter and the Legend* (Minneapolis: The University of Minnesota Press, 1959). Also see the discussion in Christopher Eric Taylor, "Unknowing the Middle Ages: How Middle English Poetics Rewrote Literary History" (PhD diss., University of Texas at Austin, 2014), 222 et ff.

151 Brewer., 11 et ff. On this letter, also see O.R. Dathorne, *Imagining the World: Mythical Belief Versus Reality in Global Encounters* (Westport, CT: Bergin & Garvey, 1994), 56–64.

> all things that happen for us and against us in the provinces adjacent and subject to us can be seen most clearly and known by those looking upon it. It is guarded by twelve thousand soldiers during the day just as at night, so that it may not be broken or struck down by some calamity.[152]

This idea of a mirror capable of observing the activities of friends and foes — "all things that happen for us and against us in the provinces adjacent and subject to us" — is an old one, and may have entered Europe from foreign sources. One commentator notes that, "Towers with magic mirrors occur frequently in Arabic tales of wonders and sorcery", adding that this passage in the tale of Prester John "represents the earliest occurrence in Western literature of this technical wonder, which is obviously a borrowing from some Eastern tale."[153]

Roughly a century later, the famed thirteenth-century English philosopher and Franciscan friar Roger Bacon wrote about this idea, first commenting that a mass of "dense air" in the atmosphere might be sufficient to reflect an image.[154] Bacon comments that, using this method, "by reflected vision all things hidden in secret places in cities, armies, and the like can be brought to light."[155] Peculiarly, Bacon adds that Socrates used this method to detect a dragon! Bacon goes on to say:

> Similarly mirrors might be erected on an elevation opposite hostile cities and armies, so that all that was being done by the enemy might be visible. This can be done at any distance we desire, since, according to the book on mirrors, one and the same object can be seen by means of as many mirrors as we wish, if they are placed in the manner required. Therefore they can be placed more closely and more remotely, so that we might see an object as far off as we pleased.[156]

---

152 This translation is from Brewer, 83–84, with slight modifications.

153 Slessarev, 49–50; also see the extensive listing of sources in Hermann Thiersch, *Pharos, Antike, Islam und Occident; ein Beitrag zur Architekturgeschichte* (Leipzig: B.G. Teubner, 1909), 39 et ff. and 91–96. In addition, see the discussion in G. Hartwell Jones, *Celtic Britain and the Pilgrim Movement* (London: Honourable Society of Cymmrodorion, 1912), 481–484. There is also a tradition concerning magic mirrors used for warning, a tradition which stems from a misattribution to Virgil: see p. 94 of Juliette Wood, "Virgil and Taliesin: The Concept of the Magician in Medieval Folklore", *Folklore* 94.1 (1983): 91–104.

154 For a careful examination of the activities of Roger Bacon in the context of historical sources, see A.G. Molland, "Roger Bacon as Magician", *Traditio* 30 (1974): 445–460.

155 See Roger Bacon, *The Opus Majus of Roger Bacon*, 2 vols., trans. Robert Belle Burke (New York: Russell & Russell, 1962), 2.581.

156 Ibid., 2.581.

Bacon's concept is logical, if impractical. Again, strangely, Bacon includes the following:

> For in this way Julius Caesar, when he wished to subdue England, is said to have erected very large mirrors, in order that he might see in advance from the shore of Gaul the arrangement of the cities and camps of England.[157]

It is not clear what Bacon's source is here; one commentator states drily: "This is a piece of information about Caesar's British expedition that does not appear to be recorded elsewhere."[158]

The idea of using mirrors to protect a city by observing enemies at a distance is also found in a thirteenth-century French romance, *Cléomadès*, mentioned earlier in this book in relation to its inclusion of a tale about a mechanical flying horse. The *Cléomadès* drew from Arabic sources, so the appearance of a mirror story in the text is not surprising.[159] There was, in fact, a long tradition of the alleged use of mirrors for defensive purposes that dated back to the ninth century.[160] Returning to Bacon's account, his description — odd as it is — nonetheless is placed in an otherwise rather sober discussion of reflections and mirrors.[161]

The idea of viewing images from afar is followed up — and perhaps distorted — in the late fourteenth century in the work of Peter of Trau. He wrote about various Franciscans, and in his writing, he attributes some odd inventions to *frater Rogerius dictus Bachon Anglicus*. He describes a mirror assembled at Oxford that would allow one to see *quid agebant homines in quantumcumque remotis ... partibus* — that is, one could see what people where doing in any remote location. Apparently this invention led to students spending a great deal of time checking in on their relatives and friends who might be dying or unwell; this distraction led the university to destroy the device.[162]

---

157 Ibid., 2.581; also see Molland, 452–453 and 456–457.

158 Robert Bartlett, *The Natural and the Supernatural in the Middle Ages* (Cambridge: Cambridge University Press, 2008), 124–125.

159 See Jones, "The *Cléomadès* and Related Folk-Tales", cited earlier, and Heffernan, 64–66.

160 For a detailed description of this lineage, see pp. 186–195 of Vincent DiMarco, "The Squire's Tale", in Correale and Hamel, 1.169–1.210.

161 On the broader content and context of Bacon's *Opus Maius*, see Chapter 2 (and especially pp. 113–115) of Amanda Power, *Roger Bacon and the Defence of Christiandom* (Cambridge: Cambridge University Press, 2013).

162 See Correale and Hamel, 1.195; also see the comments in Power, 24–25, and p. 69 of idem, "Seeking Remedies for Great Danger: Contemporary Appraisals of Roger Bacon's Expertise", in Joseph Canning, Edmund J. King, and Martial Staub, eds., *Knowledge,*

Like other technological ideas, this concept of seeing at a distance through a device is featured in a range of types of early sources. Even a medieval story entitled "Reynard the Fox" includes it — the clever fox in the tale describes the device this way:

> Now ye shall hear of the mirror. The glass was of such virtue that what was done a mile off by man or beast was seen therein.[163]

It is interesting to note that this passage goes on to includes mention of another technological marvel: the flying wooden horse from the Cléomadès tale.[164] The fox says that the wood used in the frame of this special mirror is similar to the obviously very remarkable wood used to make the flying horse.[165] There is, then, a kind of interweaving of a vaguely magical quality with technological ideas. Concepts that would not be realized for many centuries — mechanized flight and the transmission of images over long distances — found a crude mechanical realization in these tales.

In a discussion of catoptromancy, the French magistrate Pierre de Rosteguy L'Ancre (1553–1631) writes of another mirror that could access images from a distance.[166] L'Ancre states this special mirror allowed the people of Paris to see what was happening in Milan, during a battle in Lombardy between Francis I and Charles V in the 1520's. L'Ancre ascribes this feat to a magician whom he does not name. However, it may have been Henricus Cornelius Agrippa (1486–1535), who served at the court of Francis I as personal physician and astrologer. Agrippa wrote a work entitled *Philosophia Occulta*, and a passage in that book describes this same idea:

Discipline and Power in the Middle Ages: Essays in Honour of David Luscombe (Leiden: Brill, 2011), 63–78. In addition, note A.G. Little, *Franciscan Papers, Lists, and Documents* (Manchester: Manchester University Press, 1943), 73.

163 Henry Morley, ed., *Early Prose Romances* (London: George Routledge and Son, 1889), 129; also see Correale and Hamel, 1.190–1.191. In addition, note the discussion of Claxton's translation in Kiyokazu Mizobata, "Caxton's Revisions: the *Game of Chess*, the *Mirror of the World*, and *Reynard the Fox*", in Takashi Suzuki and Tsuyoshi Mukai, eds., *Arthurian and Other Studies: Presented to Shunichi Noguchi* (Cambridge: D.S. Brewer, 1993), 257–262.

164 See the discussion of flying mechanical horses in Chapter 2 here.

165 See Morley, 129–130, and p. 64 of Paul Wackers, "Medieval French and Dutch Renardian Epics: Between Literature and Society", in Kenneth Varty, ed., *Reynard the Fox: Cultural Metamorphoses and Social Engagement in the Beast Epic from the Middle Ages to the Present* (New York: Berghahn Books, 2000), 55–72.

166 Pierre de Rostéguy L'Ancre, *L'incredulité et mescreance du sortilege plainement convaincue, où il est amplement et curieusement traicté de la vérité ou illusion du sortilège et d'une infinité d'autres rares et nouveaux subjects* (Paris: N. Buon, 1622), 252.

> And there is another sleight, or trick yet more wonderful. If any one shall take images artificially painted, or written letters, and in a clear night set them against the beams of the full Moon, whose resemblances being multiplied in the Air, and caught upward, and reflected back together with the beams of the Moon, any other man that is privy to the thing, at a long distance sees, reads, and knows them in the very compass, and circle of the Moon, which Art of declaring secrets is indeed very profitable for towns, and cities that are besieged, being a thing which Pythagoras long since did often do, and which is not unknown to some in these days, I will not except my self.
>
> And all these, and many more, and greater then these, are grounded in the very nature of the Air, and have their reasons, and causes declared in mathematics, and optics. And as these resemblances are reflected back to the sight, so also sometimes to the hearing, as is manifest in the Echo. But there are more secret arts then these, and such whereby any one may at a very remote distance hear, and understand what another speaks, or whispers softly.[167]

One of the peculiar things about this passage is that as fantastical as it sounds, it is immediately preceded in the same text by something much more sober:

> And it is well known, if in a dark place where there is no light but by the coming in of a beam of the sun somewhere through a little hole, a white paper, or plain looking glass be set up against that light, that there may be seen upon them, whatsoever things are done without, being shined upon by the Sun.[168]

This is a relatively clear description of the mechanics of a *camera obscura*. However, Agrippa is rather less clear in describing the method — which he oddly ascribes to Pythagoras — for the quite unconventional idea of bouncing messages off the Moon. Regardless, it is again interesting to see this elaborate concept of long-distance communication embedded among more reasonable and straightforward technical discussions of optics.

167 Heinrich Cornelius Agrippa von Nettesheim, *Three Books of Occult Philosophy*, Donald Tyson, ed. and James Freake, trans. (St. Paul, MN: Llewellyn, 1993), 18. This passage is from Book 1, Chapter 6 of the *Philosophia Occulta*; for the original Latin text, see V. Perrone Compagni, ed., *De Occulta Philosophia* (Leiden: Brill, 1992), 98.

168 Freake and Tyson, 18.

12 Remote Communication in the Works of Paracelsus and Francis Bacon

In the same period that Agrippa was penning his thoughts on these subjects, Paracelsus (1493–1541) wrote even more obscurely concerning the ability to communicate remotely. In his *Astronomia Magna*, he speaks of the mysterious *ars cabalistica*

> which can allow one to hear a voice beyond sea, or enable long-distance simultaneous communication, such as that between those in the Occident and those in the Orient.[169]

He goes on to say:

> Then, what nature is able to hear at a hundred steps away ... this kind [of magic] may hear at a hundred German miles.[170]

It is not clear what Paracelsus is referring to here, nor is any particular apparatus mentioned. This is magic — but a magic that confers to the user what one might label a technological capability. Moreover, for Paracelsus, these abilities to communicate at a distance do "not spring from spirits or sorcery," but instead come from the "natural way of subtle nature."[171] As with the story of the mechanical flying horse in the *Méliacin* described in Chapter 2, this is not magic in the more fantastical sense of the word, but instead "magic" defined as a special kind of knowledge of the natural world.

The polymath Francis Bacon (1561–1626) also alluded to ways in which a signal could be transmitted. In his futuristic novel, *New Atlantis*, a number of passages are devoted to various technologies. One such passage discusses the transmission of sound:

---

169 "Also vermag auch die Magica zu handlen, das einer mag hören ein Stimme jenseits dem Meer. Also auch das einer in Occident mag reden mit einem der in Orient wohnet." See Paracelsus, *Astronomia Magna: oder die ganze Philosophia Sagax der grossen und kleinen Welt*, in Karl Sudhoff, ed., *Sämtliche Werke* (Munich: R. Oldenbourg, 1929), 12.84; I wish to thank Prof. Dane T. Daniel for this translation.

170 See p. 102 of Urs Leo Gantenbein, "Converging Magical Legends: Faustus, Paracelsus, and Trithemius", in James M. Van der Laan and Andrew Weeks, eds., *The Faustian Century: German Literature and Culture in the Age of Luther and Faustus* (Rochester: Camden House, 2013), 93–124.

171 Ibid., 102.

> We have also sound-houses, where we practice and demonstrate all sounds and their generation. We have harmonies, which you have not, of quarter-sounds and lesser slides of sounds. Divers instruments of music likewise to you unknown, some sweeter than any you have, together with bells and rings that are dainty and sweet. We represent small sounds as great and deep, likewise great sounds extenuate and sharp; we make divers tremblings and warblings of sounds, which in their original are entire. We represent and imitate all articulate sounds and letters, and the voices and notes of beasts and birds. We have certain helps which set to the ear do further the hearing greatly. We also have divers strange and artificial echoes, reflecting the voice many times, and as it were tossing it, and some that give back the voice louder than it came, some shriller and some deeper; yea, some rendering the voice differing in the letters or articulate sound from that they receive. We have also means to convey sounds in trunks and pipes, in strange lines and distances.[172]

The last line in particular — with its reference to the ability "to convey sounds in trunks and pipes, in strange lines and distances" — reminds us of a later technological age. Certainly, the use of a tube to carry sound was a relatively simple technology that long pre-dated the *New Atlantis*, but one wonders if Bacon was referring to something more complex here, or trying to foresee such acoustic technology put into the context of a broader technological system.

In another passage in the *New Atlantis*, there is a more obscure description of technologies, this time in relation to images:

> We have also perspective-houses, where we make demonstrations of all lights and radiations; and of all colours: and out of things uncoloured and transparent, we can represent unto you all several colours; not in rainbows (as it is in gems, and prisms), but of themselves single. We represent also all multiplications of light, which we carry to great distance, and make so sharp as to discern small points and lines. Also all colourations of light; all delusions and deceits of the sight, in figures, magnitudes, motions, colours all demonstrations of shadows. We find also divers means, yet unknown to you, of producing of light originally from divers bodies. We procure means of seeing objects afar off; as in the heaven and remote

172 Francis Bacon, *New Atlantis and The Great Instauration*, ed. Jerry Weinberger (Arlington Heights, IL: Harlan Davidson, 1989), 78–79; on this passage, see Jonathan Sterne, *The Audible Past: Cultural Origins of Sound Reproduction* (Durham: Duke University Press, 2003), 72.

> places; and represent things near as afar off; and things afar off as near; making feigned distances. We have also helps for the sight, far above spectacles and glasses in use. We have also glasses and means to see small and minute bodies perfectly and distinctly; as the shapes and colours of small flies and worms, grains and flaws in gems, which cannot otherwise be seen, observations in urine and blood not otherwise to be seen. We make artificial rainbows, halos, and circles about light. We represent also all manner of reflexions, refractions, and multiplications of visual beams of objects.[173]

This passage seems to be discussing multiple devices: prisms, which are mentioned explicitly, lenses ("multiplications of light"), and distorting mirrors ("delusions and deceits of the sight"). The phrases alluding to a method of "seeing objects afar off" perhaps refer to telescopes, a technology that was available in Bacon's own time. The passage also notes microscopes: "glasses and means to see small and minute bodies perfectly and distinctly; as the shapes and colours of small flies and worms, grains and flaws in gems".

These comments, unlike ones noted in the other texts examined here, do not seem to refer to the explicit *transmission* of images, however. In short, Bacon suggests the transmission of sound, but not image, in his utopian novel. Nonetheless, this is an interesting example of how technological ideas are embedded in what is to us an unfamiliar, non-technical context. Bacon's work is a novel, but it is not pure fantasy:

> Although Bacon's text was Utopian, there is some indication that he was influenced by technological experiments of the day. For example, Salomon de Caus (1576–1626), a French hydraulics engineer, demonstrated in England such sonic curiosities as musical fountains, hydraulically activated songbirds, and even a kind of player piano. Bacon's well-known interest in new technologies … and his inductive method of scientific research established a basis for imaginary solutions of the physical. This is presumably why his imaginings are at once so precise and so prophetic.[174]

Another commentator notes the sophistication of Bacon's treatment of technological concepts:

---

173 Ibid., 77–78.

174 Andrew Hugill, "Imaginary Music Technologies: A Study" (paper presented at the *Technologie imaginaire: l'approche pataphysique de la musique*, Université de Paris-IV Sorbonne, 2005).

> Men had been making predictions for a long time. Bacon distinctively embeds his predictions, however, in a workshop which, while much more visionary than practicable … reflects great foresight. Although Bacon does not know how to bring about such feats, he has the capacity to envision a world that would require many generations of scientists, engineers, and technicians to bring about.[175]

If Bacon had such foresight, it can be assumed that some of the other writers examined in this book — including those in ancient China and other cultures — had a similar "capacity to envision" technologies for the transmission of sound and image, mechanized flight, and so on.

Again, it is important to understand that technological descriptions are bound to appear in works — even fictional ones — produced by any society that is making devices and experimenting with technology, even in an informal way. Bacon's *New Atlantis* represents a "world … produced by his famous project for modern science and technology".[176] Given Bacon's context of an already wealthy and sophisticated culture, to imagine such a world perhaps was not that difficult for him. But even remote cultures in Oceania could conceive, as we have seen, the idea of a technological solution to a challenge such as flight.

13 Traditions Concerning Special Mirrors and Telescopes

In some early texts concerning long-distance viewing, the term "mirror" is used, but not as a device for simple reflection. Instead, the "mirror" takes on other powers, and the term seems to serve as some kind of metaphor for a fantastical device to view scenes remotely. An example of this, discussed earlier, is the device described in the Manichaean text, but there are others, as well. The *Gesta Romanorum* ("Deeds of the Romans") is a collection of anecdotes dating from the late thirteenth or early fourteenth century. The work seems to have been written for the purpose of providing morality tales, as each anecdote or story in the text includes a heading referring to a particular moral virtue or vice — for example, *de invidia*. One of the tales in this book deals with the idea of viewing images at a distance in this context. A knight has married a woman

175 Robert P. Ellis, *Francis Bacon: The Double-Edged Life of the Philosopher and Statesman* (Jefferson, NC: McFarland & Company, 2015), 170.

176 See p. 196 of Weinberger, "On the Miracles in Bacon's *New Atlantis*", in Bronwen Price, ed., *Francis Bacon's* New Atlantis*: New Interdisciplinary Essays* (Manchester: Manchester University Press, 2002), 106–128.

who is unfaithful, but he is unaware of that fact; moreover, he is away from home, travelling in the city of Rome:

> In the meantime, while the knight was passing through the main street of Rome, a wise master met him in the way, and observing him narrowly, said, "My friend, I have a secret to communicate."
>
> "Well, master, what would you please to say?"
>
> "This day, you are one of death's children, unless you follow my advice: your wife is a harlot, and contrives your death." The knight, hearing what was said of his spouse, put confidence in the speaker.[177]

The story then turns to the device:

> Then putting into his hand a polished mirror, [the wise master] said, "Look attentively upon this, and you will see wonders." He did so, and the meanwhile the master read to him from a book. "What see you?" he asked. "I see," said the knight, "a certain clerk in my house."[178]

The story goes on to recount how the knight is able to see that this clerk has murderous intentions. When the knight returns home, his wife denies that there is anything going on, but of course, he is able to describe the events in detail, as he has seen them in the magic mirror.

Again note that a technical idea — however fancifully expressed it may be — is embedded in a moral narrative, a pattern that occurs a number of times in early sources. In fact, the idea of viewing at a distance through the use of some device is a trope found in several early tales. The Aarne-Thompson index lists a tale, traditionally labeled "AT 653A", entitled "The Rarest Thing in the World":

> A princess is offered to the one bringing the rarest thing in the world. Three brothers set out and acquire magic objects: a telescope which shows all that is happening in the world, a carpet (or the like) which transports one at will, and an apple (or other object) which heals or

177 *Gesta Romanorum*, trans. Charles Swan (London: George Bell & Sons, 1906), 174–175.

178 Ibid., 175; for further information on these *Gesta Romanorum* accounts, see DiMarco, "The Squire's Tale", in Correale and Hamel, 1.192. Also note the discussion concerning the "specchio maraviglioso" in Domenico Comparetti, *Virgilio nel medio evo*, 2 vols. (Firenze: Bernardo Seeber, 1896), 2.84 et ff., and Killis Campbell, ed., *The Seven Sages of Rome: Edited from the Manuscripts, with Introduction, Notes, and Glossary* (Geneva: Slatkin Reprints, 1975), xciv–xcvii.

> resuscitates. With the telescope it is learned that the princess is dying or dead. With the carpet they go to her immediately and with the apple they cure or restore her to life. Dispute as to who is going to marry her.[179]

This is known as a "dilemma tale", but particularly relevant here is the interesting quest for devices, especially the odd "telescope".[180] This idea appears not only in Indo-European tales, but also in Africa.[181] In a Wolof tale, the device again is used to see a girl who has died, but will be brought back to life.[182] In another story, a magic mirror provides the three brothers with the ability to see that the chief's daughter has passed away.[183]

A similar story of three brothers appears in *The Arabian Nights*, and includes an ivory tube or telescope.[184] One of the brothers, Prince Ali, is offered this device by a vendor while strolling through the bazaar in Shiraz, Iran. The vendor shows him that the ivory tube, which has glass at either end, allows the prince to see anything he wishes, even if it is very far away.[185] A different story, entitled "Jûdar and His Brethren", also concerns three brothers, and again there is a device — described as a planisphere — to observe any place in the world:

> And as to the celestial planisphere, whoso possesseth it, if he desire to behold all the countries from the east to the west, he will behold them, and divert himself with viewing them, while he sitteth: whatsoever quarter

179 William R. Bascom, *African Dilemma Tales* (The Hague: Mouton Publishers, 1975), 7; concerning the Aarne-Thompson classification system for folktales, see Alan Dundes, "The Motif-Index and the Tale Type Index: A Critique", *Journal of Folklore Research* 34.3 (September–December 1997): 195–202.

180 On this type of tale, see Bascom, 7–8.

181 For a discussion of the origins of this type of tale, see Steven S. Jones, "'The Rarest Thing in the World': Indo-European or African?", *Research in African Literatures* 7.2 (Autumn 1976): 200–210.

182 The Wolof people are an ethnic group living in Senegal, The Gambia, and Mauritania.

183 Bascom, 46 and 48.

184 Note the entry entitled "Magic clairvoyant telescope" in Stith Thompson, *Motif-Index of Folk-Literature*, 6 vols. (Bloomington: Indiana University Press, 1955–1958), 2.180, and see entries on the "Magic clairvoyant mirror" and "Magic clairvoyant sphere" in Hasan M. El-Shamy, *A Motif Index of The Thousand and One Nights* (Bloomington: Indiana University Press, 2006), 53.

185 For this passage in the *Arabian Nights* in English, see Robert L. Mack, ed., *Arabian Nights' Entertainments* (Oxford: Oxford University Press, 1995), 826–827, and Richard Francis Burton, *A Plain and Literal Translation of the Arabian Nights' Entertainments, Now Entituled* [sic] *The Book of The Thousand Nights and a Night*, 10 vols. (Benares: Kamashastra Society, 1885–1888), 10.254–10.255. Also note Laufer, 456, as well as Christine Goldberg, "Dilemma Tales in the Tale Type Index: The Theme of Joint Efforts", *Journal of Folklore Research* 34.3 (September–December 1997): 179–193.

> he desireth to see, he will turn the face of the planisphere towards it, and, looking in the planisphere, he will see that quarter and its inhabitants, as though all were before him.[186]

What is interesting is that this device not only allows one to see any place on the planet — it also allows the destruction of that place:

> Moreover, if he be incensed against a city, and turn the face of the planisphere towards the sun's disk, desiring to burn that city, it will be burnt.[187]

The writer of this tale seems to have "created" this device from a conceptual combination of a traditional planisphere of a kind that was familiar in the Islamic world, with a magical crystal ball.[188]

The idea of long-distance communication using a special mirror device is also found in the works of Chaucer. It is important to understand that Geoffrey Chaucer (1343–1400) is not only famous in the world of literature, but also has been noted in the field of astronomy.[189] Not surprisingly, then, themes related to astronomy and optics appear in his fiction. "The Squire's Tale", also examined elsewhere in this book, recounts the arrival of a mysterious knight. This knight rides a strange brass horse, wears a magic ring that allows one to understand the speech of birds, has at his side a sword that can both wound and heal, and carries a special mirror. The power of the mirror is described as follows:

> This mirour eek, that I have in myn hond,
> Hath swich a might, that men may in it see
> Whan ther shal fallen any adversitee
> Un-to your regne or to your-self also;
> And openly who is your freend or foo.

---

186 *The Thousand and One Nights, Commonly Called, in England, The Arabian Nights Entertainments*, 3 vols., trans. Edward William Lane (London: Charles Knight and Co., 1841), 3.193–3.194.

187 Ibid., 3.194.

188 See the brief summary of this tale on pp. 87–89 of El-Shamy, "Siblings in *Alf layla wa-layla*" in Ulrich Marzolph, ed., *The Arabian Nights in Transnational Perspective* (Detroit: Wayne State Press, 2007), 83–102, and in Marzolph and van Leeuwen, 1.244–1.245, where the object is referred to as a "magic clairvoyant sphere".

189 See, for example, Hamilton M. Smyser, "A View of Chaucer's Astronomy", *Speculum* 45.3 (July 1970): 359–373; James Winny, "Chaucer's Science", in Maurice Hussey, A.C. Spearing, and James Winny, *An Introduction to Chaucer* (Cambridge: Cambridge University Press, 1965), 153–184; and Walter Clyde Curry, *Chaucer and the Mediaeval Sciences* (New York: Barnes & Noble, 1960).

And over al this, if any lady bright
Hath set hir herte on any maner wight,
If he be fals, she shal his treson see,
His newe love and al his subtiltee
So openly, that ther shal no thyng hyde.[190]

The mirror, in short, has the ability to reveal actions happening at a distance. The device is also mentioned again a while later:

And slepte hire firste sleep, and thanne awook.
For swich a joye she in hir herte took
Bothe of hire queynte [= marvelous] ring and hire mirour,
That twenty tyme she changed hir colour [= blushed];
And in hire sleep right for impressioun [because of the effect]
Of hire mirour, she hadde a visioun.[191]

Here, the mirror also seems to have the power to cause visions in one's sleep. Again, the descriptions are poetic, but it is important to note that the mirror itself is portrayed as being part of a set of devices — including the brass horse and the ring — that are both magical and mechanical.

The story draws from many sources, and it is clear that this mirror, too, was inspired by similar accounts in other texts.[192] There are, for example, some parallels between the description in Chaucer and the tale of the mirror in the tower from the "Letter of Prester John", discussed earlier.[193] Writing in the same period as Chaucer, the author Jean Froissart offered a slight variant on this kind of special mirror. In his text, the mirror serves as a communication device in a very particular way: a "knight when parted from his lady puts beneath his pillow a mirror that apparently causes him to see his lady in a dream."[194] The tradition in Western literature of the mirror as an object for viewing at a distance

190 "The Squire's Tale", line 132 et ff. See Mark Allen and John H. Fisher, eds., *The Complete Poetry and Prose of Geoffrey Chaucer* (Boston: Wadsworth, 2012), 193.

191 Ibid., line 367 et ff. See Allen and Fisher, 197.

192 See Jones, "Some Observations upon the Squire's Tale", 352–358.

193 Concerning the parallels, see pp. 7–9 of John Livingston Lowes, "The Squire's Tale and the Land of Prester John", *Washington University Studies* 1 (October 1913): 3–18.

194 Donald C. Baker, ed., *The Variorum Edition of the Works of Geoffrey Chaucer*, vol. 2 *The Canterbury Tales*, part 12, *The Squire's Tale* (Norman, OK: University of Oklahoma Press, 1990), 14. Also see the discussion on p. 111 of Douglas Kelly, "Imitation, Metamorphosis, and Froissart's Use of the Exemplary *Modus tractandi*," in Donald Maddox and Sara Sturm-Maddox, eds., *Froissart Across the Genres* (Gainesville: University of Florida Press, 1998), 101–118.

continued into the sixteenth century. For example, the English playwright Anthony Munday (1560?–1633), in his play *John a Kent and John a Cumber*, has a magic mirror or "glasse" that again allows people to see others at a distance.[195]

14 Chinese Tales of Image Transmission

The *Xijing Zaji* (西京雜記 "Miscellaneous Records of the Western Capital"), cited earlier, describes how Emperor Gaozu (256–195 BCE) enters the storehouse of the palace at Xianyang. At one point, the text notes that among other mechanical devices in the storehouse, there is the following item:

> There [also] was a wind instrument of jade, two *chi* three *cun* in length, with twenty-six holes. When it was played, one would see vehicles, horses, and mountain forests, one right after another. When one ceased playing [the instrument], one stopped seeing these images. [This instrument] was inscribed "jade tube of brightness and beauty".

As noted earlier, the meaning here is not entirely clear. Particularly curious is the phrase "one would see vehicles, horses, and mountain forests"; as the passage states, this indicates that one would see *images* of these things when playing the instrument. This may refer to some kind of moving magic lantern, or even a type of zoetrope or phenakistoscope.[196] Perhaps the power of one's breath entering into the wind instrument also provided some motive force to a wheel that had the images on it.

The phenakistoscope dates back to the beginning of the nineteenth century, but it is likely that an understanding of the "link between rotary motion and persistence of vision" dates from a much earlier period.[197] Euclid (fl. 300 BCE)

---

195 See Anthony Munday, *John a Kent and John a Cumber: A Comedy* (London: Shakespeare Society, 1851), 29 and 64, note to p. 29, line 31. Also see Adam Max Cohen, *Shakespeare and Technology: Dramatizing Early Modern Technological Revolutions* (New York: Palgrave Macmillan, 2006), 164–167 and 210, n. 47, and note the references in Herbert Grabes, *The Mutable Glass: Mirror-imagery in Titles and Texts of the Middle Ages and English Renaissance*, trans. Gordon Collier (Cambridge: Cambridge University Press, 2009), 127.

196 For some background on the magic lantern in later European culture, see Koen Vermeir, "The Magic of the Magic Lantern (1660–1700): On Analogical Demonstration and the Visualization of the Invisible", *The British Journal for the History of Science* 38.2 (June 2005): 127–159.

197 See p. 178 of Richard J. Leskosky, "Phenakiscope: 19th Century Science Turned to Animation", *Film History* 5.2 (June 1993): 176–189; Leskosky provides a good overview of the complex development of this device.

wrote concerning the related phenomenon of induced motion: if a series of objects move together in one direction while another object remains stationary, then the latter object ends up appearing to move in the opposite direction from the former.[198] Lucretius, in the first century BCE, seems to have understood the phenomenon of apparent motion, a facted noted by the Belgian scientist Joseph Plateau, who developed the phenakistoscope.[199] Lucretius writes as follows:

> It should be added that there is nothing remarkable in the fact that images walk and rhythmically move their arms and other limbs. It is indeed true that images seen in sleep seem to do this, and the reason for it is this: when one image fades away and is succeeded by another in a different position, it looks as though the former image has changed its posture. Of course we must assume that this happens extremely rapidly: so immense is the velocity of the images, so immense the store of them, and so immense is the store of particles emitted at any single perceptible point of time, to ensure that the supply of images is continuous.[200]

It seems clear that this is an explanation of how apparent movement is caused by the succession of images. Whether this was ever put into practical use in Lucretius' time is an open question.

Perhaps the Chinese also understood mechanical animation centuries ago, as the passage from the *Xijing Zaji* implies. The historical evidence certainly reveals that there were early technologies for image creation, such as shadow puppetry — which has a long history throughout Asia, from China to Java.[201] The actual history of Chinese experimentation with shadow images is somewhat obscure.[202] Regardless, although Chinese sources do not imply any

198 Nicholas J. Wade, *A Natural History of Vision* (Cambridge, MA: MIT Press, 1998), 203.

199 Ibid., 207.

200 See Lucretius, *On the Nature of Things*, 4.769–776, trans. Smith, 121.

201 On this history, see Fan-Pen Li Chen, *Chinese Shadow Theatre: History, Popular Religion, and Women Warriors* (Montréal: McGill-Queen's University Press, 2007), especially pp. 14–59. Also see Alvin P. Cohen, "Documentation Relating to the Origins of the Chinese Shadow-Puppet Theater", *Asia Major*, 3rd series, 13.1 (2000): 83–108; David Currell, *Shadow Puppets & Shadow Play* (Ramsbury: Crowood Press, 2007); and Amin Sweeney, *Malay Shadow Puppets: The Wayang Siam of Kelantan* (London: British Museum, 1972).

202 See p. 28 of Chen, "Shadow Theaters of the World", *Asian Folklore Studies* 62.1 (2003): 25–64; Chen also examines possible roots of shadow puppetry in Turkish, Indian, and other cultures. A survey of sources concerning the Chinese use of shadow images and related optical devices is found in Needham, 4.1, 122 et ff. For the role of shadow theatre in the Arabic world, see Dalya Cohen-Mor, *A Matter of Fate: The Concept of Fate in the Arab World*

transmission of images, they indicate an understanding of mechanical devices that could create images. To be precise, references to any idea concerning the transmission of images in Chinese sources are few. An early folktale concerns a princess of the Tang Dynasty named Li Wencheng, or Wencheng Gongzhu (文成公主). She was married off to the Tibetan king Songtsän Gampo.[203] The story recounts how the princess' father gave her a special mirror that would allow her to see Chang'an (modern Xi'an).[204]

A late ninth-century Chinese collection of anecdotes entitled *Duyang Zabian* (杜陽雜編 "Compilation of Miscellanea from Duyang") was cited earlier, in the discussion on mirrors. In addition to its description of a special rock that could reveal the internal organs of a patient, the work also describes a special pillow (!) sent from a distant country.[205] The description is brief, yet oddly detailed:

> In the eighth year of Yuanhe Era,[206] the *Dazhen* [大軫] state paid tributes [to the Emperor], including a *zhong ming* pillow [重明枕], a *shen jin* quilt [神錦衾], blue-green wheat, and purple rice. The envoy from Dazhen said that his state is located in the sea thirty thousand *li* to the southeast. Because the state was located in the *zhen su* [軫宿 "chariot mansion"],[207] it was called the *Dazhen* [i.e., "big chariot"] state. The state ran through the mountains of Heqiu [合丘] and Yugao [禹稾]. The *zhong ming* pillow has a length of one *chi* [尺] and two *cun* [寸] and a height of six *cun*. Its pure brightness surpasses crystal. Inside the pillow, there are architectural structures in the shape of a tower and a platform. On the four sides of the tower and platform are ten Daoist priests, holding incense sticks and books in their hands. The ten Daoist priests keep circling around the tower and platform and never stop, and hence are

as *Reflected in Modern Arabic Literature* (Oxford: Oxford University Press, 2001), 21–23, and the discussion in Shmuel Moreh, "The Shadow Play ('Khayāl al-Zill') in the Light of Arabic Literature", *Journal of Arabic Literature* 18 (1987): 46–61.

203 See Jennifer W. Jay, "Li, Princess Wencheng", in Lily Xiao Hong Lee and Sue Wiles, eds., *Biographical Dictionary of Chinese Women: Tang Through Ming, 618–1644* (Armonk, NY: M.E. Sharpe, 2014), 204–205; also note the discussion in Hugh Richardson, "Mun Sheng Kong Co and Kim Sheng Kong Co: Two Chinese Princesses in Tibet", *The Tibet Journal* 22.1 (Spring 1997): 3–11.

204 Note Jianqiang Liu, *Tibetan Environmentalists in China: The King of Dzi*, trans. Ian Rowen, Cyrus K. Hui, and Emily T. Yeh (Lanham, MD: Lexington Book, 2015), 146.

205 See the brief citation of this passage in Schafer, *The Golden Peaches of Samarkand*, 37.

206 AD 814; this was during the Tang dynasty, under Emperor Xianzong.

207 The *zhen su* (軫宿), or "chariot mansion", is one of the twenty-eight "mansions" in the traditional Chinese system of constellations.

> Daoist masters, practicing *xing dao* [行道 "walking the way"].[208] The roof tiles, wood, and paintings of the tower and platform, and the clothing, hairpins, and capes worn by the Taoist priests — none are lacking finest detail. The great transparency and clearness of the whole artifact allows viewers to observe every detail as if they were watching things placed under the water.[209]

The description of this *zhong ming* pillow suggest another fantastical viewing device. A literal translation of the name of the device is not particularly helpful. The Chinese characters are: 重明枕 (*zhong ming zhen*); the first character (重 *zhong*) means "serious" or "to attach importance to", the second character (明 *ming*) means "bright" or "clear", and the last character (枕 *zhen*) indeed means "pillow". The translation of the term as a whole, then, would be "a pillow which values brightness".

The idea of an entire scene visible in a pillow is vaguely echoed in a mention — in a very different kind of work — concerning a moving image observable in a special mirror. The device is recounted in a text entitled *De abditis rerum causis* ("On the Hidden Causes of Things") by the physician Jean Fernel (1497–1558).[210] This work takes the form of a dialogue among three interlocutors, and examines philosophical questions — rather than physiology, which was the focus of Fernel's other investigations. A passage in the text runs as follows:

> I have seen a person use the power of words to divert various phantoms into a mirror, images that promptly displayed there whatever he

---

208 *Xing dao* (行道) literally means, "walking the Way", where *dao* — that is, "way" (道) — also denotes the "Way" of Daoism. Thus the term here refers to both the actual walking of the priests around the architectural structure and the practice of Daoism.

209 [唐宪宗]元和八年，大軫國貢重明枕、神錦衾、碧麥、紫米。云其國在海東南三萬里，當軫宿之位，故曰大軫國，經合丘、禺稾之山。。。重明枕，長一尺二寸，高六寸，潔白逾於水精，中有樓台之狀，四方有十道士，持香執簡，循環無已，謂之行道真人。其樓台瓦木丹青、真人衣服簪帔，無不悉具，通瑩焉如水睹物。I wish to thank Xin Lu, Victor H. Mair of the University of Pennsylvania, and Cheng Fangyi of Tsinghua University for their work on this text and the translation; the Chinese text is from p. 14 of E Su's *Duyang Zabian*, cited earlier.

210 See Forrester and Henry, eds., *Jean Fernel's* On the Hidden Causes of Things*: Forms, Souls, and Occult Diseases in Renaissance Medicine*, cited earlier, as well as Hiro Hirai, *Medical Humanism and Natural Philosophy: Renaissance Debates on Matter, Life and the Soul* (Leiden: Brill, 2011), 47 et ff.

> requested, either in writing or in genuine images, so lucidly that everything was being rapidly and readily recognised by those present.[211]

This passage appears in a longer discussion about what Fernel calls *daemones* ("demons") — that is, spirits that affect human affairs. The language is obscure, and nothing in the rest of the text clarifies what Fernel is talking about here.[212] Moreover, this seems to be part of the general tradition of catoptromancy, the use of mirrors in the context of the practice of magic.[213] However, from Fernel's description, the modern reader might imagine something like a computer screen responding to commands — "images that promptly displayed there whatever he requested, either in writing or in genuine images". Certainly, a vivid *technological* imagination was at work here, even with the inclusion of "phantoms" in the description.

15 Technology in Context

The passages describing devices for viewing images certainly are not purely technical descriptions, but at the same time they cannot be read as pure fantasies, either. Indeed, the attempt to place the various texts that discuss such devices into contemporary categories of fiction, non-fiction, technological treatises, and so on, is anachronistic. These earlier writers viewed their sources and their ideas in a much more holistic manner, and thus the more rational (to us) parts of the texts — such as the mention of basic optical principles — are embedded in what a modern reader would consider more fanciful material.

This embedding was not done consciously — the texts, in short, are what they are. But looking at them collectively, some conclusions might be drawn: the most conservative conclusion might be that these writers were interested in the possibilities of technology. In the periods when these texts were written, there obviously was an intersection between craft, existing technological capabilities, and speculations about potential technological capabilities, along

211 See Forrester and Henry, 392–393; the editors note (p. 393, n. 224) that the "sources of this [passage] ... have not been traced".

212 However, Fernel's engagement — despite his generally analytical approach to medicine — with the idea of "demons" was not unusual in the period in which he was writing; see Forrester and Henry, 54.

213 On this passage in Fernel's text, see the brief mentions on p. 32 of H. Carrington Bolton, "A Modern Oracle and Its Prototypes: A Study in Catoptromancy", *The Journal of American Folklore* 6.20 (January–March 1893): 25–37, and Ernest Bosc de Vèze, *Petite encyclopédie synthétique des sciences occultes* (Nice: Bureau de la Curiosité, 1904), 29–30.

with beliefs in magic, animism, and so on. Moreover, in a later period when particular inventions for obtaining images from afar — such as telescopes — were a reality, the actual discovery of such devices at times seem to have been "backdated" and attributed to ancient sources.[214]

Technological discussions — perhaps much more than today — were contextualized in all kinds of writing, as illustrated elsewhere in this book. In the case of these "mirrors", and the less well-defined devices that supposedly allowed one to observe scenes at a distance, this was also the case.[215] The practice of including these devices in a tale also seems to be very old; concerning tale type AT653A, where a special "telescope" allows one to see what is happening far away, one study notes:

> What appears to be the most satisfactory answer to the origin of AT653A is the proposal that, as an essential narrative pattern, its history may extend back over three or four thousand years. In essence, this suggests that AT 653A is a prehistorical tale known both to Africans and Indo-Europeans that became a part of both their folkloristic heritages. This conclusion demonstrates the ultimate limitations of origins scholarship when trying to identify the authorship of possibly prehistorical tales.[216]

This implies that even thousands of years ago, in a wide range of cultures, oral storytellers — and later, writers — pondered the possibilities of technological devices, even if the contexts seem peculiar to us. As pointed out elsewhere in this book, such "mechanical mythologies" included stories of machines for flight, "mirrors" to see into the human body, and special devices to see images across great distances.

---

214 Note the discussion in Eileen Reeves, *Galileo's Glassworks: The Telescope and the Mirror* (Cambridge, MA: Harvard University Press, 2008), 6–7.

215 See p. 61 of DiMarco, "The Dialogue of Science and Magic in Chaucer's *Squire's Tale*", in Thomas Kuhn and Ursula Schaefer, eds., *Dialogische Strukturen/Dialogic Structures: Festschrift für Willi Erzgräber zum 70. Geburtstag* (Tübingen: Gunter Narr, 1996), 50–68.

216 Jones, "'The Rarest Thing in the World': Indo-European or African?", 208.

CHAPTER 4

The Missing Land of Atlantis

1 A Question of Identity

The land of Atlantis is not lost; it is simply missing. The land may not have existed in reality, but it exists in text, and so perhaps it has not been sought in the proper way. Atlantis certainly is there, at least in Plato's narrative, and a careful look at his words may help in understanding the nature of this strange land, the Atlantean idea, and what role it has in the broader concept of "lost knowledge". As with the other passages from early texts examined in this book, it may be necessary to look at Plato's tale not as modern readers might *wish* to do — hoping to find a clear, simple piece of fiction — but rather as a perhaps purposefully opaque and ambiguous narrative. The story of Atlantis can also be understood as a technological tale: in part, this is because the text uses specific descriptive language to outline aspects of the island configuration, but more particularly because it repeats the theme of a sophisticated civilization that existed in the remote past and that subsequently disappeared.

Of course, few subjects have generated as much idle and ill-informed historical musing and fantastic archaeological speculation as Plato's tale. The literature concerning the story is massive, with commentators placing Plato's mysterious continent everywhere from Antarctica to the Mediterranean.[1] Modern speculations concerning Atlantis began with Ignatius Donnelly's *Atlantis: The Antediluvian World*, which has seen numerous reprints following

1 See Edwin S. Ramage, ed., *Atlantis: Fact or Fiction?* (Bloomington: Indiana University Press, 1978), 30; further speculations and fantasies concerning Atlantis may be found in a number of books, including Andrew Collins, *Gateway to Atlantis: The Search for the Source of a Lost Civilization* (New York: Carroll & Graf Publishers, 2000); Colin Wilson, *The Atlantis Blueprint: Unlocking the Ancient Mysteries of a Long-Lost Civilization* (New York: Delacorte Press, 2001); Rand Flem-Ath and Rose Flem-Ath, *When the Sky Fell: In Search of Atlantis* (New York: St. Martin's Press, 1995); and Ivar Zapp and George Erikson, *Atlantis in America: Navigators of the Ancient World* (Kempton, IL: Adventures Unlimited Press, 1998). In terms of more analytical approaches, note the work by the French scholar Pierre Vidal-Naquet, *L'Atlantide: Petite histoire d'un mythe platonicien* (Paris: Les Belles Lettres, 2005) and the brief study in Heinz-Günther Nesselrath, *Platon und die Erfindung von Atlantis* (Munich: K.G. Saur, 2002).

 | DOI:10.1163/9789004352728_005

its nineteenth-century debut.[2] Bizarre musing by the Theosophists followed, with W. Scott-Eliot's *The Story of Atlantis and the Lost Continent of Lemuria*.[3]

Further back, in the period during and after the voyages of Columbus to the New World, there was a re-examination of Plato's narrative, along with other Latin and Greek stories of early voyages.[4] Such readings led to conjectures that these ancient cultures had perhaps reached the Americas centuries before the ships of Columbus. Diodorus Siculus, a Greek writer who flourished some time before 30 BCE, wrote what is known as the *Historical Library*, a work covering the history of Greece, Egypt, and other parts of the ancient world. In a discussion of islands, he writes:

> But now that we have discussed what relates to the islands which lie within the Pillars of Heracles [i.e., in the Mediterranean], we shall give an account of those which are in the ocean. For there lies out in the deep off Libya an island of considerable size, and situated as it is in the ocean it is distant from Libya a voyage of a number of days to the west. Its land is fruitful, much of it being mountainous and not a little being a level plain of surpassing beauty. Through it flow navigable rivers which are used for irrigation, and the island contains many parks planted with trees of every variety and gardens in great multitudes which are traversed by streams of sweet water; on it also are private villas of costly construction, and throughout the gardens banqueting houses have been constructed in a setting of flowers, and in them the inhabitants pass their time during the summer season, since the land supplies in abundance everything which contributes to enjoyment and luxury …
>
> In ancient times this island remained undiscovered because of its distance from the entire inhabited world, but it was discovered at a later period for the following reason. The Phoenicians, who from ancient times on made voyages continually for purposes of trade, planted many colonies throughout Libya and not a few as well in the western parts of

2 Ignatius Donnelly, *Atlantis: The Antediluvian World* (New York: Harper, 1882); on Donnelly's ideas, see pp. 124–125 of Paul Jordan, "Esoteric Egypt", in Fagan, *Archaeological Fantasies*, 109–128.

3 W. Scott-Eliot, *The Story of Atlantis and the Lost Continent of Lemuria* (London: Theosophical Publishing House, 1925). For a thorough cultural study of the concept of Lemuria, see Ramaswamy, *The Lost Land of Lemuria*, cited earlier.

4 For a look at some of the Greek sources, and a brief discussion of these in the context of the Atlantis story, see Nesselrath, "'Where the Lord of the Sea Grants Passage to Sailors through the Deep-Blue Mere No More': The Greeks and the Western Seas", *Greece & Rome*, 2nd series, 52.2 (October 2005): 153–171.

> Europe. And since their ventures turned out according to their expectations, they amassed great wealth and essayed to voyage beyond the Pillars of Heracles into the sea which men call the ocean ... The Phoenicians, then, while exploring the coast outside the Pillars for the reasons we have stated and while sailing along the shore of Libya, were driven by strong winds a great distance out into the ocean. And after being storm-tossed for many days they were carried ashore on the island we mentioned above, and when they had observed its felicity and nature they caused it to be known to all men. Consequently the Tyrrhenians, at the time when they were masters of the sea, purposed to dispatch a colony to it; but the Carthaginians prevented their doing so, partly out of concern lest many inhabitants of Carthage should remove there because of the excellence of the island, and partly in order to have ready in it a place in which to seek refuge against an incalculable turn of fortune, in case some total disaster should overtake Carthage. For it was their thought that, since they were masters of the sea, they would thus be able to move, households and all, to an island which was unknown to their conquerors.[5]

A similar story is found in the works of the famous Greek writer Plutarch, who offers an account concerning Roman soldiers who return from

> Atlantic islands ... two in number, separated by a very narrow strait; they are ten thousand furlongs distant from Africa, and are called the Islands of the Blest. They enjoy moderate rains at long intervals, and winds which for the most part are soft and precipitate dews, so that the islands not only have a rich soil which is excellent for plowing and planting, but also produce a natural fruit that is plentiful and wholesome enough to feed, without toil or trouble, a leisured folk.[6]

The same tale appears in the pseudo-Aristotelian *On Marvellous Things Heard*; that version runs as follows:

> In the sea outside the Pillars of Hercules, they say that a desert island was found by the Carthaginians, having woods of all kinds and navigable rivers, remarkable for all other kinds of fruits, and few days' voyage away; as the Carthaginians frequented it often owing to its prosperity, and

---

5 Diodorus Siculus, *Library of History* 5.19–5.20, in Diodorus Siculus, *Library of History*, 12 vols., trans. R.M. Geer, et al. (Cambridge, MA: Harvard University Press, 1933–1967), 3.144–3.151.

6 Plutarch, *Life of Sertorius*, 8.2, in Perrin, 8.20–8.21.

> some even lived there, the chief of the Carthaginians announced that they would punish with death any who proposed to sail there, and they massacred all the inhabitants, that they might not tell the story, and that a crowd might not resort to the island, and get possession of it, and take away the prosperity of the Carthaginians.[7]

These ancient stories were in circulation during the period of early Atlantic exploration by the Spanish and Portuguese; the tales are discussed, for example, in the writings of Ferdinand Columbus.[8] They supported the idea, moreover, that the New World had been visited in antiquity — thus further supporting the more fundamental belief that the ancients had carried out sophisticated feats in the distant past, with the resulting knowledge then lost but for the traces found in these tales.

In both the popular and academic literature, many authors have used Plato's text primarily as a launching point for speculative — and actual — archaeological investigations in various locales around the world, from Bimini to Bolivia to Antarctica.[9] More serious studies of Plato's text also have led to a similar obsession to "place" Plato's account somewhere in actual geographical terms, but even these analyses have usually led to reading into the text what simply is not there. As one author has put it, "Plato's story of Atlantis takes on

---

7 Pseudo-Aristotle, *On Marvellous Things Heard*, 836b–837a, in *Minor Works: On Colours. On Things Heard. Physiognomics. On Plants. On Marvellous Things Heard. Mechanical Problems. On Indivisible Lines. The Situations and Names of Winds. On Melissus, Xenophanes, Gorgias*, trans. W.S. Hett (London: W. Heinemann, 1936), 270–271.

8 See Ferdinand Columbus, *The Life of the Admiral Christopher Columbus by his Son Ferdinand*, ed. and trans. Benjamin Keen (New Brunswick, NJ: Rutgers University Press, 1959), 25. Also note p. 79 of James Romm, "New World and '*novos orbes*': Seneca in the Renaissance Debate over Ancient Knowledge of the Americas", in Wolfgang Haase and Meyer Reinhold, eds., *The Classical Tradition and the Americas*, vol. 1, *European Images of the Americas and the Classical Tradition*, Part 1 (Berlin: Walter de Gruyter, 1994), 77–116. An earlier discussion of the connection of classical works with the discovery of the Americas is found in Peter de Roo, *History of America Before Columbus According to Documents and Approved Authors*, vol. 1, *American Aborigines* (Philadelphia: J.B. Lippincott, 1900), 117 et ff.

9 Concerning the "search for Atlantis" in Bimini, see David Zink, *The Stones of Atlantis* (New York: Prentice Hall, 1990); for a skeptics reaction, see Eugene A. Shinn, "A Geologist's Adventure with Bimini Beachrock and Atlantis True Believers", *Skeptical Inquirer* 28.1 (January–February 2004): 38–44, as well as Marshall McKusick and Eugene A. Shinn, "Bahamian Atlantis Reconsidered", *Nature* 287.5777 (September 1980): 11–12. For the quest to find Atlantis in Bolivia, see John M. Allen, *Atlantis: The Andes Solution* (New York: St. Martin's Press, 1999), reviewed in Michael Ceasar, "Atlantis of the Altiplano: The Latest Theory Regarding an Ancient Mystery", *Mercator's World* 4.2 (March–April 1999): 24–31. Finally, for speculations that Antarctica is Plato's Atlantis, see Flem-Ath and Flem-Ath, *When the Sky Fell: In Search of Atlantis*, cited earlier.

the characteristics of a Rorschach test, with readers encountering whatever they were hoping to find."[10]

Rather sober investigation by classicists and archaeologists has focused on the Mediterranean, with a number of scholars arguing that in his Atlantis narrative, Plato was referring to Minoan Crete. The island of Thera (Santorini), some 120 kilometers from Crete, suffered a volcanic eruption in the fifteenth century BCE, and this has led to a conjecture that the decline of Minoan culture was a direct result. Further, it is argued, Plato had this information somehow, and went on to "mythologize" it in the form of the Atlantis account.[11] Another book argues that Plato was referring to an area of what is now modern Turkey, while a study by a geophysicist posits the idea that Plato and even earlier authors were discussing an area in what is today Andalusia.[12]

But all of these interpretations are speculative readings of the text, and in making speculations, what the text actually says is — if not ignored — often gently set aside. This is unfair to the account we find in Plato, which is, in short, the origin of all other ruminations on Atlantis. As one commentator says, the

> Thera proponents must face a battery of discrepancies between the Platonic account and the allegedly 'real events" at Thera: Plato's island was of huge dimensions ... outside the Pillarts of Heracles in the Atlantic Ocean; it and its city were fully ten times larger than any possible dimension at either Knossos or Thera ... [and in Plato's account] the Atlantic

---

10 Kenneth L. Feder, "Diagnosing the Atlantis Syndrome", *Skeptical Inquirer* 26.3 (May–June 2002): 53–54.

11 Christopher Gill, *Plato: The Atlantis Story* (*Timaeus 17–27 / Critias*) (Bristol, England: Bristol Classical Press, 1980), viii–ix. A book continuing the arguments for the Atlantis story being an account of Minoan culture is Charles R. Pellegrino's *Unearthing Atlantis: An Archaeological Odyssey* (New York: Random House, 1991); also see the work of the Danish geologist, Walter L. Friedrich, *Fire in the Sea: The Santorini Volcano: Natural History and the Legend of Atlantis* (Cambridge: Cambridge University Press, 2000). For a rebuttal to the idea that Plato was referring to Santorini, see Kevin Christopher, "Atlantis Behind the Myth: No Way, No How, No Where", *Skeptical Inquirer* 26.1 (January–February 2002): 44–45.

12 Peter James, *The Sunken Kingdom: The Atlantis Mystery Solved* (London: Pimlico, 1996); for an engaging broader look at the possible connection between Plato's Atlantis tale and actual geological events in the ancient Mediterranean world, as well as the importance of myth in such investigations, see James W. Mavor, Jr., "Atlantis and Catastrophe Theory", *Oceanus* 28.1 (Spring 1985): 44–51. Concerning the Andalusia conjecture, see Stavros P. Papamarinopoulos, "Atlantis in Homer and Other Authors Prior to Plato", in S.A. Paipetis, ed., *Science and Technology in Homeric Epics* (Dordrecht: Springer, 2008), 469–508.

> isle was totally submerged forever, whereas Thera, or at least half of it, and all of Crete, remain.[13]

Only a few scholars — such as Diskin Clay, Christopher Gill, and Harold Tarrant — have tackled Plato's account directly, and this chapter will cite those studies extensively.

Besides providing the reader a sober look at Plato's narrative, the other purpose of this chapter is to place that text in parallel with some of the other texts examined in this book. Since Plato's Atlantis story contains an odd array of technical detail, to a certain degree it can be considered in a similar way to, for example, the Chinese tales of flight. That is, the question can asked as to what Plato *means* by including this strange story in his work — what was he saying? And what is the reason for Plato's making the narrative so ambiguous? Or does the ambiguity arise from our modern approach to the text? Perhaps readers of Plato's own time understood the story in a very different way than we do: those readers would have approached the text without the restrictive categories of "fact" and "fiction", and with the capacity to understand the strange technical detail in some quite different context.

2 Atlantis in Plato's *Timaeus*

The account of Atlantis appears in two places in Plato, the *Timaeus* and *Critias*. Later mentions of Atlantis appear in the works of Diodorus Siculus and Strabo, among others. Plato's accounts are given below; first, it is helpful to look at the story in the *Timaeus*, where the text itself gives us some of the background. The interlocutors in the *Timaeus* are Timaeus, about whom we have no historical information; Hermocrates, a statesman from Syracuse, who was involved in the defeat of the Athenians' military expedition against Sicily around 415 BCE; and Critias (ca. 460–403 BCE), who was one of the rulers set up by the Spartans when they defeated the Athenians in 404 BCE. It is Critias who relates the story of Atlantis in this dialogue, noting that he had heard this tale from his grandfather, who was also named — somewhat confusingly — Critias. The grandfather, in turn, had heard it through Solon, an Athenian statesman and poet.

The account in the *Timaeus* comes in the context of a discussion about an ideal state, an issue that comprises Plato's *Republic*. Socrates essentially reviews the structure of the ideal state as outlined in the *Republic*, but also

13 See p. 83 of S. Casey Fredericks, "Plato's Atlantis: A Mythologist Looks at Myth", in Ramage, 81–99, and note Richard Ellis, *Imagining Atlantis* (New York: Alfred A. Knopf, 1998), 236.

distinguishes the present discussion — that is, the one in the *Timaeus* — from preceding discussions.[14] Then Socrates requests that his interlocutors provide an example of an ideal state. But what kind of example? This is the first question as to the nature of the story itself: Is this a made-up example? A historical account? A combination of the two? One modern commentator notes that Socrates

> seems to have in mind an imaginary story; for he regards the task as a 'poetic' or literary one. But he wants the story to be told by narrators who (unlike most poets) genuinely understood the nature of their subject, who combine knowledge of philosophy and politics; and he claims his interlocutors are such men.[15]

Critias responds to the request of Socrates, and in his very first sentence, he attests to the veracity of the tale he is about to tell. Below is the text in full, so that one can better appreciate the context in which the Atlantis story appears:

> CRITIAS: Let me tell you this story then, Socrates. It's a very strange one, but even so, every word of it is true. It's a story that Solon, the wisest of the seven sages once vouched for. He was a kinsman and a very close friend of my great-grandfather Dropides. Solon himself says as much in many places in his poetry. Well, Dropides told the story to my grandfather Critias, and the old man in his turn would tell it to us from memory. The story is that our city had performed great and marvelous deeds in ancient times, which, owing to the passage of time and to the destruction of human life, have vanished. Of all these deeds one in particular was magnificent. It is this one that we should now do well to commemorate and present to you as our gift of thanks. In so doing we shall also offer the goddess a hymn, as it were, of just and true praise on this her festival.
>
> SOCRATES: Splendid! Tell me, though, what was that ancient deed our city performed, the one that Solon reported and old Critias told you about? I've never heard of it. They say it really happened?
>
> CRITIAS: I'll tell you. It's an ancient story I heard from a man who was no youngster himself. In fact, at the time Critias was pretty close to ninety years old already — so he said — and I was around ten or so. As

14 Gill, *Plato: The Atlantis Story*, 31–32.
15 Ibid., 35.

it happened, it was the day of the presentation of children during the Apaturia.[16] On this occasion, too, we children got the customary treatment at the feast: our fathers started a recitation contest. Many compositions by many different poets were recited, and many of us children got to sing the verses of Solon, because they were new at the time. Now someone, a member of our clan, said that he thought that Solon was not only the wisest of men in general, but that his poetry in particular showed him to be the most civilized of all the poets. (The man may have been speaking his mind, or else he may have just wanted to make Critias feel good.) And the old man — how well I remember it — was tickled. He grinned broadly and said, "Yes, Amynander, it's too bad that Solon wrote poetry only as a diversion and didn't seriously work at it like the other poets. And too bad that he never finished the story he'd brought back home with him from Egypt. He was forced to abandon that story on account of the civil conflicts and all the other troubles he found here when he returned. Otherwise not even Hesiod or Homer, or any other poet at all would ever have become more famous than he. That's what *I* think, anyhow." "Well, Critias? What story was that?" asked the other. "It's the story about the most magnificent thing our city has ever done," replied Critias, "an accomplishment that deserves to be known far better than any of her other achievements. But owing to the march of time and the fact that the men who accomplished it have perished, the story has not survived to the present." "Please tell us from the beginning," said the other, "What was this 'true story' that Solon heard? How did he get to hear it? Who told him?"

"In Egypt," Critias began, "in that part of the Delta where the stream of the Nile divides around the vertex there is a district called the Saïtic. The most important city of this district is Saïs.[17] (This is in fact also the city from which King Amasis came.) This city was founded by a goddess whose name was 'Neith' in Egyptian and (according to the people there) 'Athena' in Greek.[18] They are very friendly to Athens and claim to be related to our people somehow or other. Now Solon said that when he arrived there the people began to revere him. Furthermore, he said that when he asked those priests of theirs who were scholars of antiquity about ancient times, he discovered that just about every Greek, including himself, was

16 The Apaturia (Ἀπατούρια) was an ancient Ionian autumnal festival.
17 This was an Egyptian town situated on a branch of the Nile.
18 Neith was an Egyptian goddess of creation.

all but completely ignorant about such matters. On one occasion, wanting to lead them on to talk about antiquity, he broached the subject of our own ancient history. He started talking about Phoroneus — the first human being, it is said — and about Niobe, and then he told the story of how Deucalion and Pyrrha survived the flood. He went on to trace the lines of descent of their posterity, and tried to compute their dates by calculating the number of years which had elapsed since the events of which he spoke. And then one of the priests, a very old man, said, 'Ah, Solon, Solon, you Greeks are ever children. There isn't an old man among you.' On hearing this, Solon said, 'What? What do you mean?' 'You are young,' the old priest replied, 'young in soul, every one of you. Your souls are devoid of beliefs about antiquity handed down by ancient tradition. Your souls lack any learning made hoary by time. The reason for that is this: There have been, and there will continue to be, numerous disasters that have destroyed human life in many kinds of ways. The most serious of these involve fire and water, while the lesser ones have numerous other causes. And so also among your people the tale is told that Phaethon, child of the Sun, once harnessed his father's chariot, but was unable to drive it along his father's course. He ended up burning everything on the earth's surface and was destroyed himself when a lightning bolt struck him. This tale is told as a myth, but the truth behind it is that there is a deviation in the heavenly bodies that travel around the earth, which causes huge fires that destroy what is on the earth across vast stretches of time. When this happens all those people who live in mountains or in places that are high and dry are much more likely to perish than the ones who live next to rivers or by the sea. Our Nile, always our savior, is released and at such times, too, saves us from this disaster. On the other hand, whenever the gods send floods of water upon the earth to purge it, the herdsmen and shepherds in the mountains preserve their lives, while those who live in cities, in your region, are swept by the rivers into the sea. But here, in this place, water does not flow from on high onto our fields, either at such a time or any other. On the contrary, its nature is always to rise up from below. This, then, explains the fact that the antiquities preserved here are said to be the most ancient. The truth is that in all places where neither inordinate cold nor heat prevent it, the human race will continue to exist, sometimes in greater, sometimes in lesser numbers. Now of all the events reported to us, no matter where they've occurred — in your parts or in ours — if there are any that are noble or great or distinguished in some other way, they've all been inscribed here in our temples and preserved from antiquity on. In your case, on the other hand, as in that of others, no sooner have you achieved literacy and all the other resources that cities require, than there again, after the usual number of years, comes the heavenly flood. It sweeps upon you like a plague, and leaves only your illiterate and uncultured people behind. You become infants all over again, as it were, completely unfamiliar with anything there was in ancient times, whether here or in your own region. And so, Solon, the account you just gave of your people's lineage is just like a nursery tale. First of all, you people remember only one flood, though in fact there had been a great many before. Second, you are unaware of the fact that the finest and best of all the races of humankind once lived in your region. This is the race from whom you yourself, your whole city, all that you and your countrymen have today, are sprung, thanks to the survival of a small portion of their stock. But this has escaped you, because for many generations the survivors passed on without leaving written record. Indeed, Solon, there was a time, before the greatest of these devastating floods, when the city that is Athens today not only excelled in war but also distinguished itself by the excellence of its laws in every area. Its accomplishments and its social arrangements are said to have been the finest of all those under heaven of which we have received report.'

"When Solon heard this he was astounded, he said, and with unreserved eagerness begged the priests to give him a detailed, consecutive account of all that concerned those ancient citizens. 'I won't grudge you this, Solon,' the priest replied. 'I'll tell you the story for your own benefit as well as your city's, and especially in honor of our patron goddess who has founded, nurtured and educated our cities, both yours and ours. Yours she founded first, a thousand years before ours, when she had received from Earth and Hephaestus the seed from which your people were to come. Now our social arrangement, according to the records inscribed in our sacred documents, is eight thousand years old. Nine thousand years ago, then, did these fellow citizens of yours live, whose laws and whose finest achievement I'll briefly describe to you. At another time we'll go through all the details one by one at our leisure and inspect the documents themselves.

"'Let's compare your ancient laws with ours today. You'll discover many instances that once existed among you, existing among us today. First, you'll find that the class of priests is marked off and separated from the other classes. Next, in the case of the working class, you'll find that each group — the herdsmen, the hunters and the farmers — works independently, without mixing with the others. In particular, I'm sure you've noticed that our warrior class has been separated from all the others. It's been assigned by law to occupy itself exclusively with matters of war. Moreover, the style of armor used is that of shields and spears, which we were the first among the peoples of Asia to use for arming ourselves. The goddess instructed us just as she first instructed you in the regions where you live. Moreover, as for wisdom, I'm sure you can see how much attention our way of life here has devoted to it, right from the beginning. In our study of the world order we have traced all our discoveries, including prophecy and health-restoring medicine, from those divine realities to human levels, and we have also acquired all the other related disciplines. This is in fact nothing less than the very same system of social order that the goddess first devised for you when she founded your city, which she did once she had chosen the region in which your people were born, and had discerned that the temperate climate in it throughout the seasons would bring forth men of surpassing wisdom. And, being a lover of both war and wisdom, the goddess chose the region that was likely to bring forth men most like herself, and founded it first. And so you came to live there, and to observe laws such as these. In fact your laws improved even more, so that you came to surpass all other peoples in every excellence, as could be expected from those whose begetting and nurture were divine.

"'Now many great accomplishments of your city recorded here are awe-inspiring, but there is one that surely surpasses them all in magnitude and excellence. The records speak of a vast power that your city once brought to a halt in its insolent march against the whole of Europe and Asia at once — a power that sprang forth from beyond, from the Atlantic Ocean. For at that time this ocean was passable, since it had an island in it in front of the strait that you people say you call the 'Pillars of Heracles.' This island was larger than Libya and Asia combined, and it provided passage to the other islands for people who traveled in those days. From those islands one could then travel to the entire continent on the other side, which surrounds that real sea beyond. Everything here inside the strait we're talking about seems nothing but a harbor with a narrow entrance, whereas that really is an ocean out there and the land that embraces it all the way around truly deserves to be called a continent. Now on this Isle of Atlantis a great and marvelous royal power established itself, and ruled not only the whole island, but many of the other islands and parts of the continent as well. What's more, their rule extended even inside the strait, over Libya as far as Egypt, and over Europe as far as Tyrrhenia. Now one day this power gathered all of itself together, and set out to enslave all of the territory inside the strait, including your region and ours, in one fell swoop. Then it was, Solon, that your city's might shone bright with excellence and strength, for all humankind to see. Preeminent among all others in the nobility of her spirit and in her use of all the arts of war, she first rose to the leadership of the Greek cause. Later, forced to stand alone, deserted by her allies, she reached a point of extreme peril. Nevertheless she overcame the invaders and erected her monument of victory. She prevented the enslavement of those not yet enslaved, and generously freed all the rest of us who lived within the boundaries of Heracles. Some time later excessively violent earthquakes and floods occurred, and after the onset of an unbearable day and a night, your entire warrior force sank below the earth all at once, and the Isle of Atlantis likewise sank below the sea and disappeared. That is how the ocean in that region has come to be even now unnavigable and unexplorable, obstructed as it is by a layer of mud at a shallow depth, the residue of the island as it settled.'"

What I've just related, Socrates, is a concise version of old Critias' story, as Solon originally reported it. While you were speaking yesterday about politics and the men you were describing, I was reminded of what I've just told you and was quite amazed as I realized how by some supernatural chance your ideas are on the mark, in substantial agreement with what Solon said. I didn't want to say so at the time, though. Because it had been so long ago, I didn't remember Solon's story very well. So I realized that I would first have to recover the whole story for myself well enough, and then to tell it that way. That's why I was so quick to agree to your assignment yesterday. The most important task in situations like these is to propose a speech that rewards people's expectations, and so I thought that we would be well supplied if I gave this one. And that's how — as Hermocrates has already said — the moment I left here yesterday, I began to repeat the story to him and to Timaeus as it came back to me. After I left them I concentrated on it during the night and recovered just about the whole thing. They say that the lessons of childhood have a marvelous way of being retained. How true that is! In my case, I don't know if I'd be able to recall everything I heard yesterday, but I'd be extremely surprised if any part of this story has gotten away from me, even though it's been a very long time since I heard it. What I heard then gave me so much childlike pleasure — the old man was so eager to teach me because I kept on asking one question after another — that the story has

> stayed with me like the indelible markings of a picture with the colors burnt in. I told the whole story to Timaeus and Hermocrates first thing this morning, so that not just I, but they, too, would have a supply of material for our speech.
>
> I've said all this, Socrates, to prepare myself to tell Solon's story now. I won't just give you the main points, but the details, one by one, just the way I heard it. We'll translate the citizens and the city you described to us in mythical fashion yesterday to the realm of fact, and place it before us as though it is ancient Athens itself. And we'll say that the citizens you imagined are the very ones the priest spoke about, our actual ancestors. The congruence will be complete, and our song will be in tune if we say that your imaginary citizens are the ones who really existed at that time. We'll share the task among us, and we'll all try our best to do justice to your assignment. What do you think, Socrates? Will this do as our speech, or should we look for another to replace it?[19]

That is the account of Atlantis in the *Timaeus*; how should it be approached?

At the beginning, Critias responds to Socrates' discussion of the state with a narrative. It is interesting to note that Critias begins by saying that this tale is "a very strange one, but even so, every word of it is true." This at least makes us ask why such an assertion would be necessary. If Critias is just putting together a story to illustrate some points concerning the concept of the ideal state, why the need to assert that it is true? Moreover, why chose a "strange" story and not something simpler, if the purpose of the story is just to illustrate a point?

In addition, Critias does not simply recount the story, but first gives its provenance. That provenance is somewhat complex — in the text, Critias says that Solon heard the story from Egyptian priests of Neith at Sais. Solon was a "a kinsman and a very close friend" of Dropides.[20] Dropides was the great-grandfather of Critias, the text tells us. Dropides told the story to another figure named Critias, who (we are told) was the grandfather of the storyteller in the text here. The chain of communication is, in short, as follows: Egyptian priests → Solon → Dropides → Critias (the grandfather) → Critias (the narrator here). So, the story is being related more or less fifth-hand, but nothing in the account

19 Plato, *Timaeus* 20d–26e, trans. Donald Zeyl, in John M. Cooper and D.S. Hutchinson, eds., *Plato: Complete Works* (Indianapolis, IN: Hackett, 1997), 1228–1233.

20 Dropides may also have been Solon's brother; see Diskin Clay and Andrea Purvis, *Four Island Utopias: Being Plato's Atlantis, Euhemeros of Messene's Panchaia; Iamboulos' Island of the Sun; Sir Francis Bacon's New Atlantis* (Newburyport, MA: Focus Publishing, 1999), 60, n. 17.

implies that as a result the story has become corrupted or embellished. Indeed, the account in the text is labeled as "a concise version of old Critias' story, as Solon originally reported it". Moreover, the penultimate paragraph gives an honest discussion of memory:

> They say that the lessons of childhood have a marvelous way of being retained. How true that is! In my case, I don't know if I'd be able to recall everything I heard yesterday, but I'd be extremely surprised if any part of this story has gotten away from me, even though it's been a very long time since I heard it.

This is an idea familiar even to a modern reader: that as we get older, short-term memory weakens, but memories from our younger years become ever more clear.

Admittedly there is no evidence, other than what we read in Plato's text, that Solon actually heard this Atlantis story. It is clear that Solon (ca. 639–ca. 559 BCE) was a real person; he was a statesman and there is Plutarch's biography of him. Herodotus also records that Solon visited Sais.[21] Dropides was an *archon* (that is, magistrate) of Athens, and Solon's brother. Our narrator, Critias, was Plato's second cousin; he was a political leader and a poet.[22] These are all historical figures — though that of course does not mean that Plato couldn't have placed them in an account that has no historical veracity.

3 Plato and the Idea of History

Moreover, Plato can and does use historical figures as mouthpieces for various themes and ideas in his writings. Yet here these are not just the historical figures, but rather a clearly delineated "chain" of people through whom the story is passed down. There seems to have been some contraction of this "family tree", with a generation between the elder Critias and the younger Critias left out.[23] However, once again it is important to point out that there was no need to give the story a provenance at all, yet it is there in the text.

---

21 Herodotus, *Histories*, 1.30, in *The History: Herodotus*, trans. David Grene (Chicago: University of Chicago Press, 1987), 45.

22 Clay and Purvis, 58, n. 15.

23 See Gill, *Plato: The Atlantis Story*, 39.

Furthermore, there are other places in the text where the "historicity" of the account is emphasized, *and the idea of history and historical events in general* is highlighted.[24] Near the beginning of the account, Critias says:

> Now Solon said that when he arrived there the people began to revere him. Furthermore, he said that when he asked those priests of theirs who were scholars of antiquity about ancient times, he discovered that just about every Greek, including himself, was all but completely ignorant about such matters. On one occasion, wanting to lead them on to talk about antiquity, he broached the subject of our own ancient history. He started talking about Phoroneus — the first human being, it is said — and about Niobe, and then he told the story of how Deucalion and Pyrrha survived the flood. He went on to trace the lines of descent of their posterity, and tried to compute their dates by calculating the number of years which had elapsed since the events of which he spoke.[25]

In the context of what comes later in this story, this part should make the reader stop and consider: here Solon discusses history, and in particular the "prehistory" of his own culture. In addition, although we might view Phoroneus and the other figures merely as characters in Greek mythology, note that Solon goes so far as to try to calculate years for these ancient events; in other words, he is talking to the Egyptian priests about *history* — not myth — and about the historical "location" of events.

The response of the Egyptian priests is an even more direct reference to the idea of history and the recording of historical events:

> And then one of the priests, a very old man, said, 'Ah, Solon, Solon, you Greeks are ever children. There isn't an old man among you.' On hearing this, Solon said, 'What? What do you mean?' 'You are young,' the old priest replied, 'young in soul, every one of you. Your souls are devoid of beliefs about antiquity handed down by ancient tradition. Your souls lack any learning made hoary by time.'[26]

---

24 Plato's use of and view of history in general has been the subject of various analyses — see, for example, R.G. Bury, "Plato and History", *The Classical Quarterly*, new series, 1.1/2 (January–April 1951): 86–93; Bury sees Plato's interest in history as a philosophical one, and argues that his history primarily is that of human affairs.

25 Plato, *Timaeus* 21e–22b, trans. Zeyl, in Cooper and Hutchinson, 1229–1230.

26 Plato, *Timaeus* 22b–22c, trans. Zeyl, in Cooper and Hutchinson, 1230.

The priest clearly talks about the transmission of tradition and, by extension, the transmission of knowledge, and does not mention myth at all. Moreover, he adds the comment about "learning". The Greek term that Plato uses here is *mathema* (μάθημα), meaning, "that which is learnt", and thus, knowledge. There is also a connection here to mathematical knowledge, and the Greek term provides the root of that modern English word. The point is that the Egyptian priest is making an explicit comment about knowledge of a certain type and its transmission over time.[27] Again, in this passage the stage is being set for a discussion of history, not for the construction of a fanciful tale.

The Egyptian priest continues in his answer to Solon:

> The reason for that is this: There have been, and there will continue to be, numerous disasters that have destroyed human life in many kinds of ways. The most serious of these involve fire and water, while the lesser ones have numerous other causes. And so also among your people the tale is told that Phaethon, child of the Sun, once harnessed his father's chariot, but was unable to drive it along his father's course. He ended up burning everything on the earth's surface and was destroyed himself when a lightning bolt struck him. *This tale is told as a myth, but the truth behind it is that there is a deviation in the heavenly bodies that travel around the earth,* which causes huge fires that destroy what is on the earth across vast stretches of time. When this happens all those people who live in mountains or in places that are high and dry are much more likely to perish than the ones who live next to rivers or by the sea. Our Nile, always our savior, is released and at such times, too, saves us from this disaster. On the other hand, whenever the gods send floods of water upon the earth to purge it, the herdsmen and shepherds in the mountains preserve their lives, while those who live in cities, in your region, are swept by the rivers into the sea. But here, in this place, water does not flow from on high onto our fields, either at such a time or any other. On the contrary, its nature is always to rise up from below. This, then, explains the fact that the antiquities preserved here are said to be the most ancient.[28]

Here, particularly in the phrase marked with italics, the priest makes as clear a statement as one could ask for concerning the relationship between myth

---

27 As to why this connection to the deep past is put within an Egyptian framework, see Thomas K. Johansen, *Plato's Natural Philosophy: A Study of the* Timaeus-Critias (Cambridge: Cambridge University Press, 2004), 39.

28 Plato, *Timaeus* 22c–22e, trans. Zeyl, in Cooper and Hutchinson, 1230–1231.

and knowledge. In discussing the story of Paethon, the priest says: "Now this has the form of a myth, but really signifies a declination of the bodies moving in the heavens around the earth." In other words, here is a clear example of an ancient text making explicit the concept that myth can be the conveyer of *mathema* — knowledge, and indeed, technical knowledge such as astronomy.

In addition, this passage shows us on a more basic level Plato's clear understanding of the difference between mythical metaphor and empirical knowledge. Plato makes a similar point in his *Republic*, where he has his interlocutors emphasize the difference between a metaphorical and aesthetic view of the heavens and actual astronomy.[29] An even more precise parallel is found in *Statesman*, where the character known as the "Visitor" addresses some stories concerning the rising and setting positions of the Sun and planets:

> Well, all these things together are consequences of the same state of affairs, and besides these thousands of others still more astonishing than they; but through the great lapse of time since then some have been obliterated, while others have been reported in a scattered way, each separate from one another.[30]

Again, there is the concept of historical events becoming embedded in mythical contexts, resulting in the fragmentary nature of knowledge.

Returning to the *Timaeus*, the Egyptian priest outlines for Solon some very explicit ideas concerning the transmission of historical knowledge:

> The truth is that in all places where neither inordinate cold nor heat prevent it, the human race will continue to exist, sometimes in greater, sometimes in lesser numbers. Now of all the events reported to us, no matter where they've occurred — in your parts or in ours — if there are any that are noble or great or distinguished in some other way, they've all been inscribed here in our temples and preserved from antiquity on. In your case, on the other hand, as in that of others, no sooner have you achieved literacy and all the other resources that cities require, than there again, after the usual number of years, comes the heavenly flood. It sweeps upon you like a plague, and leaves only your illiterate and uncultured people behind. You become infants all over again, as it were,

29 Plato, *Republic* 529c–530b, trans. G.M.A. Grube, rev. C.D.C. Reeve, in Cooper and Hutchinson, 1145–1146.

30 Plato, *Statesman* 269b–269c, trans. C.J. Rowe, in Cooper and Hutchinson, 310.

> completely unfamiliar with anything there was in ancient times, whether here or in your own region.[31]

Here again the reader is presented with a very articulate discussion of the passing down of knowledge. The priest makes several points; first, that human society survives through the ages, wherever the climate might permit. Next, he notes that historical events ("noble or great") are written down — in other words, there is a conscious recording of history, and an effort to preserve those records in the temples, which are icons of a civilization. Then, the Egyptian priest goes on to talk about how societies "forget": those with an education, the record-keepers, are wiped out in a cataclysm, leaving only those who are "illiterate and uncultured". The priest then concludes with the comment that through the ages, civilizations become "completely unfamiliar with anything there was in ancient times", and so "become like infants all over again".

Plato makes a very similar point concerning the idea of periodic cataclysms disrupting the transmission of knowledge in his dialogue entitled *Laws*, a work concerning the laws of a civilization, ethics, rights, and other related topics.[32] In *Laws*, there is following exchange:

> ATHENIAN: [W]hat about political systems? How are we to suppose they first came into existence? I feel sure that the best and easiest way to see their origins is this.
>
> CLINIAS: What?
>
> ATHENIAN: We take an indefinitely long period of time and study the changes that occur in it.
>
> CLINIAS: How do you mean?
>
> ATHENIAN: Look, do you think you could ever grasp how long it is that states have existed and men have lived under some sort of political organization?
>
> CLINIAS: No, not very easily.

31 Plato, *Timaeus* 22e–23b, trans. Zeyl, in Cooper and Hutchinson, 1230–1231.

32 Note that the *Laws* were written after the *Timaeus*.

ATHENIAN: But at any rate you realize it must be an enormously long time?

CLINIAS: Yes, I see *that*, of course.

ATHENIAN: So surely, during this period, thousands upon thousands of states have come into being, while at least as many, in equally vast numbers, have been destroyed? Time and again each one of them has adopted every type of political system. And sometimes small states have become bigger, and big ones have grown smaller; superior states have deteriorated and bad ones have improved.

CLINIAS: Inevitably.

ATHENIAN: Let's try to pin down just why these changes took place, if we can; then perhaps we shall discover how the various systems took root and developed.

CLINIAS: Admirable! Let's get down to it. You must do your best to explain your views, and we must try to follow you.

ATHENIAN: Do you think there is any truth in tradition?

CLINIAS: What sort of tradition do you mean?

ATHENIAN: This: the human race has been repeatedly annihilated by floods and plagues and many other causes, so that only a small fraction of it survived.

CLINIAS: Yes, of course, all that sort of thing strikes everyone as entirely credible.[33]

Note first the assertion that the concept of states — that is, organized civilizations — have been around for an "enormously long time". The view here is that Greek civilization is nothing new, but rather is one of a chain of many civilizations that have come and gone. Again, this is a cyclical rather than a linear view of history, and allows for the idea that development, including technological development, might appear and disappear through the millennia. The

33 Plato, *Laws* 676b–677a, trans. Trevor J. Saunders, in Cooper and Hutchinson, 1365.

interlocutors in the Platonic dialogue also assert the idea of "truth in tradition", along with a cataclysmic view of history that "strikes everyone as entirely credible".

The speakers continue:

> ATHENIAN: Now then, let's picture just one of this series of annihilations — I mean the effect of the flood.
>
> CLINIAS: What special point are we to notice about it?
>
> ATHENIAN: That those who escaped the disaster must have been pretty nearly all hill-shepherds — a few embers of mankind preserved, I imagine, on the tops of mountains.
>
> CLINIAS: Obviously.
>
> ATHENIAN: Here's a further point: such men must have been in general unskilled and unsophisticated …
>
> CLINIAS: Quite likely.[34]

As in the Atlantis account, as the dialogues goes on, Plato discusses the destructions of mankind. Note that Plato's *Laws* has phrases here that match those in the *Timaeus* concerning the fragile chain of transmission in human cultures. In *Laws*, the Athenian says, "That those who escaped the disaster must have been pretty nearly all hill-shepherds — a few embers of mankind preserved, I imagine, on the tops of mountains." Similarly, the *Timaeus* notes that, "whenever the gods send floods of water upon the earth to purge it, the herdsmen and shepherds in the mountains preserve their lives, while those who live in cities, in your region, are swept by the rivers into the sea."[35] In terms of the broader discussion of technological knowledge here, this is a clear reference to the loss of sophisticated, technical knowledge over time, since those who possess such knowledge are the ones who perish in these civilization-ending disasters.

The tone in this dialogue in the *Laws* is not fantastical, and instead seems to reflect a genuine understanding of actual natural disasters and their effect on cultures:

34 Plato, *Laws* 677a–677b, trans. Saunders, in Cooper and Hutchinson, 1365–1366.

35 Plato, *Timaeus* 22d–22e, trans. Zeyl, in Cooper and Hutchinson, 1230.

> ATHENIAN: And we can take it, can't we, that the cities that had been built on the plains and hear the sea were destroyed root-and-branch?
>
> CLINIAS: Yes, we can.
>
> ATHENIAN: So all their tools were destroyed, and any worthwhile discovery they had made in politics or any other field was entirely lost?[36]

Here, the Athenian speaker adds an interesting line: "You see, my friend, if their discoveries had survived throughout at the same level of development as they have attained today, it is difficult to see what room there can ever have been for any new invention."[37] This seems a brief, even flippant, aside — but again reflects an understanding of the cumulative nature of knowledge and innovation. There is also an understanding here in technological terms: the loss of both the tools of a civilization and the knowledge garnered from discoveries.

Clinias responds with words that reflect, in turn, a comprehensive view of a civilization's place in the context of the many millennia of human existence:

> CLINIAS: The upshot of all this, I suppose, is that for millions of years these techniques remained unknown to primitive man. Then, a thousand or two thousand years ago, Daedelus and Orpheus and Palamedes made their various discoveries, Marsyas and Olympus pioneered the art of music, Amphion invented the lyre, and many other discoveries were made by other people. All this happened only yesterday or the day before, so to speak.[38]

Here, Clinias lists in this passage the various figures from Greek culture who were said to have brought mankind various kinds of knowledge: the famed craftsman Daedelus, the poet and musician Orpheus, Palamedes — credited with inventing, among other things, counting and coinage — and so on. Whether these figures were historical or not is less important here than the exposition of the idea of the place of technical knowledge in the context of the "deep time" of human existence. These passages also reflect the important point that the themes of prehistory, cataclysm, and the transmission and loss of knowledge are not found solely in the Atlantis passages of Plato's works.

---

36 Plato, *Laws* 677c–677d, trans. Saunders, in Cooper and Hutchinson, 1366.

37 Plato, *Laws* 677c, trans. Saunders, in Cooper and Hutchinson, 1366.

38 Plato, *Laws* 677d, trans. Saunders, in Cooper and Hutchinson, 1366.

Later in the dialogue in *Laws*, the speakers expound again on the idea of the passage of long spans of time, and the fate of knowledge:

> ATHENIAN: Perhaps we can describe the state of mankind after the cataclysm like this: in spite of a vast and terrifying desolation, plenty of fertile land was available, and although animals in general had perished it happened that some cattle still survived, together with perhaps a small stock of goats. They were few enough, but sufficient to maintain the correspondingly few herdsmen of this early period.
>
> CLINIAS: Agreed.
>
> ATHENIAN: But at the moment we are talking about the state, and the business of legislation and political organization. Is it conceivable that any trace at all of such things survived — even, so to speak, in the memory?
>
> CLINIAS: Of course not ...
>
> ATHENIAN: So, it was only as time went on, and the number of the human race increased, that civilizations advanced and reached its present stage of development?
>
> CLINIAS: Exactly.
>
> ATHENIAN: The process was probably not sudden, but gradual, and took a considerable time.[39]

The dialogue goes on to make a number of statements, in fact, about knowledge, technological discovery, and cultural memory, all in the context of a discussion on the nature of human civilization. Right after the passage above, *Laws* has the following:

> ATHENIAN: I imagine men were all numbed with fear at the prospect of descending from the hills to the plains.
>
> CLINIAS: Naturally enough.

39 Plato, *Laws* 677e–678b, trans. Saunders, in Cooper and Hutchinson, 1366–1367.

ATHENIAN: And what a pleasure it must have been to see each other, there being so few of them at that time! However, pretty well all vehicles they might have used to visit each other by land or sea had been destroyed, and the techniques used to construct them had been lost, so that I suppose they found getting together none too easy. They suffered from a scarcity of timber, because iron, copper and mineral workings in general had been overlaid with sludge and had been lost to sight, so that it was virtually impossible to refine fresh supplies of metal. Even if there was the odd tool left somewhere on the mountains, it was quickly worn down to nothing by use. Replacements could not be made until the technique of mining sprang up again among men.

CLINIAS: True.

ATHENIAN: And how many generations later did that happen, on our calculation?

CLINIAS: A good many, obviously.

ATHENIAN: Well then, during that period, or even longer, all techniques that depend on a supply of copper and iron and so on must have gone out of use?

CLINIAS: Of Course.[40]

Note the specificity of the discussion; it is clear again that the Greeks had a keen sense not only of history, but also of its patterns and cycles. More particularly, there is an analysis here — in the works of Plato — of technology and technological knowledge: first of all, there is the understanding the statement that "all vehicles they might have used to visit each other by land or sea had been destroyed", implying that the kind of far-reaching but unified culture enjoyed by the Greeks of Plato's time was dependent on such technologies. The Athenian adds that, "the techniques used to construct them had been lost", too, reflecting an understanding of the necessary pairing of technological devices with the methods to construct them. There is also mention of materials, including wood and metals, again showing that even in this non-technical dialogue, such issues are considered and understood.

40 Plato, *Laws* 678c–678e, trans. Saunders, in Cooper and Hutchinson, 1367.

As the dialogue in *Laws* continues, it focuses on a society growing from a few scattered members to a fully developed civilization. It also returns a number of times to the issue of knowledge transmission, with discussions, for example, of societies having "forgotten to quite a remarkable degree" events the "far removed in time".[41] One can see here Plato's keen sense of the vast scope of time and human history, along with an appreciation of technology's role in that history. One modern study points out that however limited their methods might be, "Hellenistic world historians … attempted to determine a suitable starting point from which to commence a historical and chronological narrative", and in that way place and better interpret their own culture and time.[42] Again, these passages in *Laws* — and the whole passage at the beginning of the Atlantis story in the *Timaeus* — show a thorough understanding of "deep time", and the fate of technical and other forms of knowledge.

Returning to the entire "preamble" to the Atlantis story in the *Timaeus*, one can see that it places that story in the context of human history, knowledge, and the transmission of knowledge. Nowhere in that preamble is the reader led to think that the interlocutors are talking about myth, at least myth in the modern sense of a fabrication. Certainly there are details in the story that do not match our historical records, most notably the time scale — that is, events being recounted as having happened some nine thousand years before the dialogue takes place. This figure, on the face of it, immediately makes us return to the traditional modern view of the Atlantis story as a simple fabrication. However, it is important to understand that just because this time scale of history seems excessive to *us*, that does not mean that it was excessive to the original recipients of this story. An ancient reader of this passage in Plato might not have accepted the nine thousand-year figure as an accurate reckoning *per se*, but they perhaps would not have read it as a random choice, either. The figure, or at least the scale of that figure — which, again, seems grotesque to a modern reader — might have made perfect sense to Plato's audience.

The priest makes clear to Solon, first of all, that "all the events reported to us, no matter where they've occurred — in your parts or in ours … [have] all been inscribed here in our temples and preserved from antiquity on."[43] He then goes on speak of "records inscribed in our sacred documents … eight thousand years old"; he adds:

---

41 Plato, *Laws* 682b–682c, trans. Saunders, in Cooper and Hutchinson, 1371.

42 William Adler, *Time Immemorial: Archaic History and Its Sources in Christian Chronography from Julius Africanus to George Syncellus* (Washington, D.C.: Dumbarton Oaks Research Library and Collection, 1989), 108.

43 Plato, *Timaeus* 23a, trans. Zeyl, in Cooper and Hutchinson, 1230.

> Nine thousand years ago, then, did these fellow citizens of yours live, whose laws and whose finest achievement I'll briefly describe to you. At another time we'll go through all the details one by one at our leisure and inspect the documents themselves.[44]

In addition to the vast number of years, the Egyptian priest twice refers to written historical records, the "sacred documents". He does not simply ask Solon to accept the forthcoming account of Atlantis on faith, or to hear it simply as an entertaining tale. Of course, this does not tell us, unfortunately, anything about the veracity of the Atlantis tale; many narratives in antiquity were supported by "bogus sources":

> The problem of verifying references brings us back to the baffling question of bogus sources. I am not here concerned with imaginary works like the unfinished poem of Solon in which Plato claimed to have found the story of Atlantis, or the inscribed dedication on the island of Panchaia where Euhemerus claimed to have learned the truth about King Zeus and his family, or even the bronze tablets supposedly dug up by his father in the family house that Acusilaus cited as authority for his own *Genealogies*, the first prose work of Greek mythography ... The devices of the long-lost manuscript and hidden tablets were to have a rich future in the history of fiction and forgery.[45]

But stepping back in terms of perspective, the modern reader can see that the discussion in Plato's text is framed throughout by the idea of *narrative* — historical events for the ancient Greeks were part of a narrative continuum that could be documented. That is, Plato understood — even if he put it into an apparently fantastical format — that his society occupied a spot in a very long historical timeline, and that situating an account firmly in that context of "deep time" would be of interest to readers. The Atlantis tale is not about using "bogus sources" to spin a yarn, nor was it designed to deceive a reader into taking fiction as fact. Rather, the Atlantis tale is a narrative about the nature of history, about memory, about the fate of civilizations, and about the transmission of knowledge — and the reader of Plato's day would have understand as that.

44 Plato, *Timaeus* 23e–24a, trans. Zeyl, in Cooper and Hutchinson, 1231.

45 Alan Cameron, *Greek Mythology in the Roman World* (Oxford: Oxford University Press, 2004), 124.

4 The Geography of Atlantis

After the introductory passages, the *Timaeus* account of Atlantis turns to a presentation of the setting:

> Now many great accomplishments of your city recorded here are awe-inspiring, but there is one that surely surpasses them all in magnitude and excellence. The records speak of a vast power that your city once brought to a halt in its insolent march against the whole of Europe and Asia at once — a power that sprang forth from beyond, from the Atlantic Ocean. For at that time this ocean was passable, since it had an island in it in front of the strait that you people say you call the 'Pillars of Heracles.' This island was larger than Libya and Asia combined, and it provided passage to the other islands for people who traveled in those days. From those islands one could then travel to the entire continent on the other side, which surrounds that real sea beyond. Everything here inside the strait we're talking about seems nothing but a harbor with a narrow entrance, whereas that really is an ocean out there and the land that embraces it all the way around truly deserves to be called a continent.[46]

This single passage has led to countless speculations as to where a modern archaeological investigation might yield traces of this lost land. But what is most important here simply is the fact that the account *does* give a location — a geographical location in relation to real places, not an abstract or fantastical one. Plato's narrative already has fitted the Atlantis account with an historical preamble; now it gets its geographical setting. The land of Atlantis is clearly placed outside the area of the Mediterranean; indeed, the setting is marked explicitly as the Atlantic Ocean. Furthermore, the narrator says that the island of which he is about to speak "was larger than Libya and Asia combined".

An English historian, P.B.S. Andrews, in a brief article entitled "Larger than Africa and Asia?", argued that this statement in Plato is the result of a (conjectural) transcription error.[47] However, there is no firm evidence for such an error, and Andrews' statement that "to say something is larger than two other things whose size is quite unknown is to say nothing" is not correct — in fact, the Greeks had their own understanding of the full extent of the *oikoumene*, that is, the inhabited landmass of Asia, Africa, and Europe. Andrews also says

46 Plato, *Timaeus* 24d–25a, trans. Zeyl, in Cooper and Hutchinson, 1232.

47 P.B.S. Andrews, "Larger than Africa and Asia?", *Greece & Rome*, 2nd series, 14.1 (April 1967): 76–79.

that "the use of 'island' for a land-mass of such an alleged size is as improper in Greek as in English", but actually the conception of the entire *oikoumene* itself as an island is found as far back as Homer, as well as in later Greek sources such as Strabo.[48] The classicist Duane W. Roller dismisses the Atlantis account simply as a continuation of general Greek ideas of various mythical and semi-mythical islands and lands beyond the ocean.[49] However, Plato's text presents a much more complex narrative than other Greek accounts of such places.

It is worth giving a brief look at the term that Plato uses in his discussion. The original Greek is ἡ Ἀτλαντὶς νῆσος, with the literal meaning of "the Island of Atlas".[50] The word νῆσος means island, although in early Greek texts there are other uses, including "peninsula": a line in Sophocles' *Oedipus at Colonnus* reads ἐν τᾷ μεγάλᾳ Δωρίδι νάσῳ Πέλοπος, "in the great Dorian island of Pelops". Although the Greek language indeed literally labels that geographical area ἡ Πελοπόννησος, "the Peloponnese" — meaning the "island of Pelops" — it is, in fact, a peninsula. Typically, though, the Greek term for a peninsula is χερόνησος, formed from the words χέρσος, meaning "dry", and νῆσος, meaning "island". This ambiguity of terms has led to some debate, in both the academic and non-academic literature, concerning the locational or geomorphic aspect of the Atlantis story.[51] But the term that Plato uses for this land of Atlantis is less important, in a way, than the fact that the account gives a geographical location in relation to real places rather than invented ones. As noted above, Plato's narrative gives the Atlantis account both an historical and geographical context.

As for Plato's mention of "Libya", in ancient Greek geography "Libya" referred to the African continent to the extent that it was known in antiquity, comprising the northern parts of Africa above the Sahara; likewise "Asia" meant Asia Minor, not the lands of India, China, and so on.[52] None of the discussion of Atlantis in Plato's text, then, is out of keeping with traditional Greek geographical concepts of the period. The Greek model of the globe portrayed an

---

48 See J.B. Harley and David Woodward, eds., *The History of Cartography*, vol. 1, *Cartography in Prehistoric, Ancient, and Medieval Europe and the Mediterranean* (Chicago: University of Chicago Press, 1987), 171–175.

49 See Duane W. Roller, *Through the Pillars of Herakles: Greco-Roman Exploration of the Atlantic* (London: Routledge, 2006), 44, n. 7, as well as 50.

50 Note the entry under "Atlantis" in William Smith, ed., *A Dictionary of Greek and Roman Geography*, 2 vols. (London: John Murray, 1878), 1.314.

51 See, for example, p. 144 of Papamarinopoulos, "Atlantis in Spain V", *Bulletin of the Geological Society of Greece* 43.1 (2010): 138–146, as well as Mark Adams, *Meet Me in Atlantis: My Obsessive Quest to Find the Sunken City* (New York: Dutton, 2015), 198–202, and Robert Bittlestone, *Odysseus Unbound: The Search for Homer's Ithaca* (Cambridge: Cambridge University Press, 2005), 140.

52 Clay and Purvis, 73, n. 46.

oikoumene — a landmass comprising Asia, Africa, and Europe — on a watery sphere of the Earth.[53]

Further, the *Timaeus* account tells us that the island was "in front of the strait that you people say you call the 'Pillars of Heracles.'" The "Pillars of Heracles" or "Pillar of Hercules" is a term used by a number of ancient writers to refer to a location equivalent to the modern Strait of Gibraltar, which leads from the Mediterranean to the Atlantic Ocean.[54] Plato refers to these "pillars" in the *Critias* — again in the context of the Atlantis story — and in the *Phaedo*, with the original Greek term indeed being Ἡράκλειοι στῆλαι. Herodotus, in his *Histories*, labels the sea to the west as the "sea outside the Pillars of Heracles". In fact, the Pillars of Heracles served as a reference point in ancient geographical discussions in distinguishing the Mediterranean and the Atlantic — the sea inside the Pillars versus that outside.[55] In Aristotle and Plutarch, for example, the Mediterranean thus appears as "the sea "inside the Pillars of Hercules" (ἡ ἐντὸς Ἡρακλείων στηλῶν θάλασσα).[56] Again, the geographical terminology that the interlocutors use in Plato's dialogue is clear, with the descriptions genuinely rooted in the traditions of Greek geography.[57]

The *Timaeus* text then adds a further — rather peculiar — geographical detail: that Atlantis "provided passage to the other islands for people who traveled in those days", adding that from "those islands one could then travel to the entire continent on the other side, which surrounds that real sea beyond". The text attempts to clarify that statement by adding the following: "Everything here inside the strait we're talking about seems nothing but a harbor with a narrow entrance, whereas that really is an ocean out there and the land that embraces it all the way around truly deserves to be called a continent." As pointed out earlier, the traditional Greek worldview posited a three-part

---

53 See, for example, the description in Strabo, *Geography* 17.3.1–2, in *The Geography of Strabo*, 8 vols., trans. Horace Leonard Jones (Cambridge: Heinemann, 1917), 8.155–8.156.

54 Clay and Purvis, 66, n. 33. Also see the entry under "Herculis Columnae" in Smith, *A Dictionary of Greek and Roman Geography*, 1.1054–1.1055. The Greek term for the divine hero and son of Zeus is Heracles (Ἡραχλῆς), and the Latin term is Hercules.

55 Herodotus, *Histories* 1.203, trans. Grene, 125.

56 See Aristotle, *Meteorologica* 2.1 / 354a10–15, in Aristotle, *Meteorologica*, trans. H.D.P. Lee. (Cambridge, MA: Harvard University Press, 1952), 128–129, and Plutarch, *Pompey*, 25.1–3, in Perrin, 5.177–5.179.

57 A French geologist has written a sober, if difficult to substantiate, argument that Plato was referring to a very specific location near the Strait of Gibralter; see Jacques Collina-Girard, "La crise finiglaciaire à Gibraltar et l'Atlantide: tradition et géologie", *Préhistoire anthropologie méditerranéennes* 10–11 (2001–2002): 53–60, as well as his subsequent book-length treatment of the subject, *L'Atlantide retrouvée? Enquête scientifique autour d'un mythe* (Paris: Belin, 2009).

landmass, with Asia, Africa, and Europe considered together as kind of a huge island landmass, called the *oikoumene*, residing on an otherwise ocean-filled globe. We find this traditional description in a number of ancient writers on geography; for example, Strabo speaks of "one confluent and continuous sea" surrounding the *oikoumene*, and Agathemerus describes a vast ocean encompassing all the land.[58]

In the *Timaeus*, however, the geographical description presents a different picture. In that configuration, not only is there the island continent of Atlantis, but also another continent beyond the oceans. How does this geographical artifact fit with the Atlantis story? It does *not*, and perhaps that is what is interesting about it. The inclusion of this encircling continent in the overall geographical preface to the story adds nothing. So, again, the Egyptian priest presents a series of descriptions that do not have the makings of a purely imaginary construct. These are simply details that are presented to the reader as part of a general geographical description.

Furthermore, a particular component of the description in the *Timaeus* — a "continent on the other side, which surrounds that real sea beyond" — has echoes in other texts. We find it most clearly in the sixth-century CE *Christian Topography* of Cosmas Indicopleustes, which presents a geographical picture of the world based on a very literal interpretation of the Scriptures. Cosmas was a monk, and author of several works, all now lost; he apparently travelled widely.[59] Nonetheless, his geographical descriptions of the earth and cosmos as a whole are quite abstract. His *oikoumene* is surrounded by ocean, but beyond lays another continent. This is home to Paradise, and the abode of man. At the time of the Flood, Noah was carried over on the waves to our *oikoumene*:

> We have said that the figure of the earth is lengthwise from east to west, and breadthwise from north to south, and that it is divided into two parts; this part which we, the men of the present day, inhabit, and which is all around encircled by the intermedial sea, called the Ocean by the pagans, and that part which encircles the Ocean … which men at one time inhabited to eastward, before the flood in the days of Noah occurred, and in which Paradise was also situated. Men, strange to say, having crossed the

58 Strabo, *Geography* 1.1.8, trans. Jones, 1.17; also see Smith, *A Dictionary of Greek and Roman Geography*, 1.313, for various quotes from ancient authors on the idea of an all-encircling sea.

59 For details on the life and works of Cosmas Indicopleustes, see J.W. McCrindle, *The Christian Topography of Cosmas, An Egyptian Monk* (London: Hakluyt Society, 1897).

> Ocean in the Ark at the time of the Deluge, reached our part of the earth and settled in Persian territory.[60]

Of course, Cosmas was writing long after Plato, and may have derived at least part of this depiction directly from Plato. But there are earlier echoes. This encircling continent has a peculiar history; it does not seem to stem from any ancient Greek depictions of the world, and indeed Plato seems to be the only Greek source that speaks of it explicitly. Furthermore, the concept of this distant circular continent does not appear in any medieval sources, most of which work from Strabo and other writers who articulate only a single landmass, the *oikoumene*. Thus the idea of an encircling continent seems to follow a separate tradition; the Italian scholar Carlo Alfonso Nallino believed that the notion stemmed from Iranian cosmography:

> The origin of this idea of an inaccessible land surrounding the ocean seems to me to come from Iranian cosmography. According to Avesta, on the first day of rain, the earth was divided by the waters into seven parts (*karsvare*); man can inhabit only the *karsvare* called *kwaniratha*, around which, separated by insurmountable abysses and seas, are set the six other [*karsvare*s] concentrically. In short, it is a conception analogous to that of the seven Indian *dvîpa*. I think it likely that this Iranian idea was passed on, like so many other [ideas] of a religious nature, into Christianity.[61]

A twelfth-century CE Syrian map presents another example of this world picture. There, the *oikoumene* is surrounded by an impenetrable ocean. This ocean, in turn, is surrounded by another ring of land, where, the author of the map says, the descendants of Adam lived up until the time of the Deluge. Beyond this, there is a final ring of land, the "Land of Paradise".[62]

Again, this is a later depiction, but the lineage is clear here; Nallino had indicated a source in Iran, and there are further traces in East Asia. A series of Korean world maps clearly display a cartographic system like that found in the

60 Cosmas Indicopleustes, *Christian Topography* 2.131, in McCrindle, 33.

61 See p. 523 of Carlo Alfonso Nallino, "Al-Khuwarizmi e il suo rifacimento della Geografia di Tolomeo", in Maria Nallino, ed., *Raccolta di Scritti Editi Inediti*, vol. 5, *Astrologia–Astronomia–Geografia* (Rome: Istituto per l'Oriente, 1944), 458–532.

62 See pp. 100–101 and fig. A of J.B. Chabot, "Notice sur une Mappemonde Syrienne du XIIIe siècle", *Bulletin de géographie historique et descriptive* (1897): 98–112.

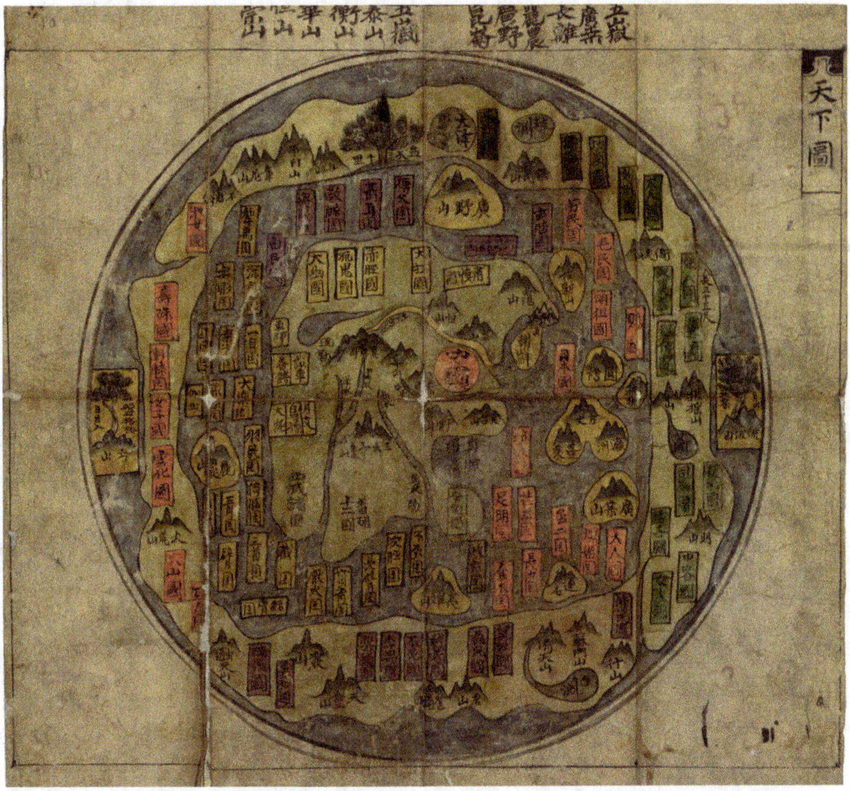

FIGURE 7 An early Korean world map

Timaeus.[63] One such map is entitled *Cheonhado* (天下圖 "Map of [the World] Beneath the Heavens") and synthesizes some geographical knowledge from the West, but working within traditional Asian cartographic styles (fig. 7).

China occupies the main area of this central continent, with Korea, Japan, India, and Southeast Asia nearby. The Western countries are crowded to one side.[64] The central continent is surrounded by an ocean, which contains many islands. Beyond this ocean, there is another continent; unlike in Cosmas's work, here this second continent is surrounded, in turn, by water. In this final outer sea there are islands to the west and east. Though Korean, the maps are derived from Chinese sources, and the various place names on them come from the

63 See Hiroshi Nakamura, "Old Chinese World Maps Preserved by the Koreans", *Imago Mundi* 4 (1947): 3–22; also note Shannon McCune, "World Maps by Korean Cartographers", *Journal of Social Sciences and Humanities* 45 (June 1977): 1–8.

64 Nakamura, 3.

Chinese classics, such as the *Shan Hai Jing* (山海經 "Classic of the Mountains and Seas"), a book on geography, fantastic and otherwise.[65] It is interesting to note that this work dates from the Zhou or Han periods (sixth century BCE to first century CE), while Plato was writing in the middle of the fourth century BCE. Plato, then, was dealing with a geographical depiction which existed in Asia at the time, so perhaps there was a common source. The *Shan Hai Jing* includes an account, too, of a great flood.[66]

There are, however, a few echoes of Plato's "outer continent" in ancient sources in the Western world, as well; in the pseudo-Aristotleian *On the Cosmos*, we read the following:

> The *inhabited world* is divided by the usual account into islands and continents, since it is not recognized that the whole of it is really one island [i.e., the *oikoumene*], surrounded by the sea which is called *Atlantic*. Far away from this one, on the opposite side of the intervening seas, there are probably many other inhabited worlds, some greater than this, some smaller.[67]

Pliny, in his *Natural History*, says:

> [T]here are seas encircling the globe on every side, and dividing it in two, so robbing us of half the world, since there is no region affording us a passage from there to here or from here to there.[68]

Other than these echoes, the geographical picture set out in the *Timaeus* Atlantis story is unusual in terms of geographical discussions in ancient sources. Plato's inclusion of "other islands", along with an "entire continent on the other side, which surrounds that real sea beyond" is a strange technical detail, which does not seem to serve the Atlantis account in any particular way. Is this detail a piece of context which an ancient reader might have understood, but is now lost to the modern reader?

---

65 Ibid., 10–12; on the *Shan Hai Jing*, see Loewe, cited earlier, 359–60.

66 Anne Birrell, "The Four Flood Myth Traditions of Classical China", *T'oung Pao* 83.4–5 (1997): 213–259.

67 Pseudo-Aristotle, *On the Cosmos* 3 / 392b20–30, in Pseudo-Aristotle, *On the Cosmos*, trans. D.J. Furley (Cambridge, MA: Harvard University Press, 1978), 355.

68 Pliny the Elder, *Natural History* 2.170, trans. Eicholz, Jones, and Rackham, 1.304–305.

5 Fiction, Myth, and History

In the *Timaeus*, the Egyptian priest provides both an historical apparatus for the account and a geographical setting. There has been debate as to the specificity of this geography — as pointed out earlier, some have even argued that Plato was actually talking about places *inside* the Mediterranean — but further details in the text actually reinforce the clarity of the setting. The priest goes on to tell Solon:

> Now on this Isle of Atlantis a great and marvelous royal power established itself, and ruled not only the whole island, but many of the other islands and parts of the continent as well. What's more, their rule extended even inside the strait, over Libya as far as Egypt, and over Europe as far as Tyrrhenia.[69]

Again the details are specific. Could the story still be a fabrication? Certainly, but the point here is that the setting of the story is anything but mythical, especially in the sense that it is recounted as history in terms of its specificity in time and space.

The Atlantis story in the *Timaeus* concludes with a series of events: the invasion of the lands inside the straits by the Atlanteans and their subsequent defeat by the Hellenes. Following this is the famous destruction of Atlantis, which, were are told, "sank below the sea and disappeared." But there is some ambiguity in this section of the story, for the text in full here reads:

> Now one day this power [i.e., Atlantis] gathered all of itself together, and set out to enslave all of the territory inside the strait, including your region and ours, in one fell swoop. Then it was, Solon, that your city's might shone bright with excellence and strength, for all humankind to see. Preeminent among all others in the nobility of her spirit and in her use of all the arts of war, she first rose to the leadership of the Greek cause. Later, forced to stand alone, deserted by her allies, she reached a point of extreme peril. Nevertheless she overcame the invaders and erected her monument of victory. She prevented the enslavement of those not yet enslaved, and generously freed all the rest of us who lived within the boundaries of Heracles. Some time later excessively violent earthquakes and floods occurred, and after the onset of an unbearable day and a night,

69 Plato, *Timaeus* 25a–25b, trans. Zeyl, in Cooper and Hutchinson, 1232; Tyrrhenia (Τυρρηνία) was the ancient term for part of central Italy.

> your entire warrior force sank below the earth all at once, and the Isle of Atlantis likewise sank below the sea and disappeared. That is how the ocean in that region has come to be even now unnavigable and unexplorable, obstructed as it is by a layer of mud at a shallow depth, the residue of the island as it settled.[70]

The text says that even the "entire warrior force" fighting against the Atlanteans "sank below the earth all at once", and that Atlantis disappeared "likewise sank below the sea and disappeared". Does this mean that there was some kind of cataclysm both inside *and* outside the Pillars of Hercules? If the "warriors" had repelled the Atlanteans and driven them back to their homeland of Atlantis, and then (we presume), returned home, why did they disappear? The text does not explain this strange turn of events, or perhaps the warriors were still stuck in Atlantis when it was struck by the cataclysm.

Returning to the question of fabrication versus history, it is worth noting what the Egyptian priest says right after he discusses the sinking of Atlantis. He states: "That is how the ocean in that region has come to be even now unnavigable and unexplorable, obstructed as it is by a layer of mud at a shallow depth, the residue of the island as it settled." Again, this is not a mythical element, but simply a topographical observation. Moreover, it is an observation of the kind that is also found in much more mundane early works concerning geography. Avienus, in his *Ora Maritima* from the fourth century CE, talks about the encounter of Himilco, a Carthaginian navigator, with shallows in the Atlantic.[71] Aristotle comments that, "the water outside the Pillars of Hercules is shallow because of the mud".[72] Similarly, in the pseudo-Aristotelian *On Marvelous Things Heard*, there is a passage that notes that on sailing outside the Pillars of Hercules for four days with an eastern wind, the Phoenicians encountered partially submerged islands.[73]

The fact that the Atlantis account in the *Timaeus* contains sober geographical discussions does not necessarily mean that the story is an actual historical account of a particular place and time. In early sources, authors indeed used real geographical settings for their stories. Antonius Diogenes, writing in the second or third century CE, cleverly combines geography with poetic imagery. Although Diogenes wrote what is clearly an imaginary account in his *Wonders*

70 Plato, *Timaeus* 25b–25d, trans. Zeyl, in Cooper and Hutchinson, 1232–1233.

71 See the discussion in Roller, 27.

72 Aristotle, *Meteorologica* 2.1 / 354a20–25, trans. Lee, 128–129.

73 Pseudo-Aristotle, *On Marvelous Things Heard*, 844a, trans. Hett, 308–309; also see Clay and Purvis, 73, n. 47.

Beyond Thule, the ninth-century CE commentator Photius points out that at the beginning of each section of the tale, the writer gave his sources for the subsequent geographical flights of fancy.[74] Concerning the Atlantis tale, the classicist Diskin Clay is clear in his assessment:

> The diagnostic sign that we are dealing with a Platonic fiction is Critias' anxious claim that the strange or unbelievable tale he has to tell is not a fabricated story but *true* history ... Such apologies draw the reader's attention to the possibility that the narrator might be accusing himself by the very act of excusing himself.[75]

The subsequent details provided in the telling of the Atlantis tale, in Clay's opinion, merely reinforce the idea that it is all invention.

Plato himself does not comment on the matter of whether or not the Atlantis story is fabrication, but other authors more clearly expound on this question. For example, Lucian tells fanciful tales similar to those of Antonius Diogenes. Writing in the second century CE, he goes so far as to parody encyclopedic geographical works such as those of Pliny: the "Isles of the Blest" in Lucian's *A True Story* (sometimes known as *A True History*) are carefully noted as having "365 springs of water, 365 springs of honey, 500 springs of myrrh ... seven rivers of milk, and eight of wine".[76] To parody such descriptions, Lucian obviously had to be familiar with more serious works; Lucian himself tells us that readers will find his work

> enticing not only for the novelty of its subject, for the humour of its plan, and because I tell all kinds of lies in a plausible and specious way, but also

74 For Photius' summary of the *Wonders Beyond Thule* of Antonius Diogenes, see Susan A. Stephens and John J. Winkler, eds., *Ancient Greek Novels: The Fragments* (Princeton: Princeton University Press, 1995), 158–172, and "Wonders Beyond Thule", trans. Gerald N. Sandy, in B.P. Reardon, ed., *Collected Ancient Greek Novels* (Berkeley: University of California Press, 1989), 775–782. Also see William Hansen, "Strategies of Authentication in Ancient Popular Fiction", in Stelios Panayotakis, Maaike Zimmerman, and Wytse Keulen, eds., *The Ancient Novel and Beyond* (Leiden: Brill, 2003), 301–314, and Romm, *The Edges of the Earth in Ancient Thought: Geography, Exploration, and Fiction* (Princeton: Princeton University Press, 1992).

75 See p. 9 of Clay, "Plato's Atlantis: The Anatomy of a Fiction", in John J. Cleary and Gary M. Gurtler, eds., *Proceedings of the Boston Area Colloquium in Ancient Philosophy, 1999* (Leiden: Brill, 2000), 1–22.

76 Christopher Robinson, *Lucian and his Influence in Europe* (Chapel Hill, North Carolina: University of North Carolina Press, 1979), 24.

> because everything in my story is more or less comical parody of one or another of the poets, historians, and philosophers of old.[77]

Lucian even cites names of the other authors from whom he is deriving his material, but with humorously subtle provisos:

> One of them is Ctesias, son of Ctesiochus, of Cnido, who wrote a great deal about India and its characteristics that he had never seen himself nor heard from anyone else with a reputation for truthfulness. Iambulus, who also wrote much that was strange about the countries in the great sea: he made up a falsehood that is patent to everybody, but wrote a story that is not uninteresting for all that.[78]

Ancient authors such as Lucian, in other words, seem to have been aware of the difference between history, fiction, parody, and even — as we have seen — the mythologizing of knowledge.

Just as Lucian used his readings of geography to launch his parodies, Plutarch used geography to explore philosophy. His famous *Concerning the Face which Appears in the Orb of the Moon* presents a series of interlocutors — not unlike the setup in Plato's dialogues — speaking on various subjects; the text itself is quite complex in terms of themes, details, and references.[79] One particular part of it, however, draws from Homer's *Odyssey*: "An isle Ogygia lies far out at Sea".[80] Plutarch then goes on to describe a voyage there, and includes a number of geographical details. Again, it is worth noting that Plutarch has a sound grasp of astronomy and geography, reflected both in this text and in other works. Plutarch's interlocutors, like Plato's, speak in specifics: Sulla, the narrator, tells us that this island of Ogygia is "a run of five days off from Britain as you sail westward".[81]

In the case of Lucian, the writer himself makes a clear differentiation between his sober geographical settings and subsequent parody. Plutarch starts out with an explicit reference to the tale of Homer. In the *Timaeus*, the material concerning the placing of Atlantis and the historical framework is done in straightforward manner, and in fact, does not always relate to the story that

---

77 Lucian, "A True Story" 1.2, trans. Harmon, Kilburn, and Macleod, 1.248–1.249.

78 Lucian, "A True Story" 1.3, trans. Harmon, Kilburn, and Macleod, 1.250–1.251.

79 See Plutarch, "Concerning the Face which Appears in the Orb of the Moon" 941a–942c, trans. Cherniss, et al., 12.180–12.193.

80 Cherniss, et al., 12.181, note *d*.

81 Plutarch, "Concerning the Face Which Appears in the Orb of the Moon" 941a, trans. Cherniss, et al., 12.180–12.181.

follows. This, again, leads us away from seeing the narrative as purely fictional invention. Perhaps the text, again, needs to be read in a different way.

Concerning the question of fiction, myth, and history, then, the modern reader is left with a sense of ambiguity, and perhaps that is purposeful. One study notes that in the *Phaedrus* of Plato, discussed further below, there is exactly such an ambiguity, since Socrates argues that there is "the need to believe things based not on their authoritative source but their truth."[82] For a modern reader, by contrast, the truth of a text is in many respects *based* on the nature of the source — the two aspects of a narrative, its veracity and its source, are considered inseparable. Part of the problem may be that Plato was working during a period that "marks a key moment of transition in the cultural history of Athens, as the long-engrained habits of an oral culture were gradually being displaced by the new technology of writing", even while "the increased use of writing did not constitute a straightforward or unidirectional process, nor did it occur without resistance and tension."[83]

The scholar Christopher Gill also discusses this ambiguity concerning fiction, myth, and history in Plato's text, in the context of a discussion of the *Statesman*:

> In this story, which anticipates the Atlantis story at a number of points, Plato disposes us to expect a fiction, and then, as it were, plays with the reader, offering an account which might seem authentically historical (or pre-historical), but which is gradually revealed as a functional fable. The game with the reader (played out much more fully in the Atlantis story) is a minor feature in the *Statesman*. But in both cases it is as though Plato, having explained the distinction between surface fiction and deeper-level truth, deliberately blurs the distinction.[84]

For Gill, then, Plato is penning a fiction, one that seems "authentically historical". A reader, says Gill, will eventually be pushed into "recognizing it for himself" — that is, recognizing that they are looking at a fictional narrative.[85] The problem with this, however, is that in Plato's works and in other early Classical texts, it is clear that "even fictionality and mythic composition or

---

82 John Conor David Barry, "The Seal of the Author: Paradigm, Logos and Myth in Plato's Sophist and Statesman" (PhD diss., University of Ottawa, 2014), 226.

83 Daniel S. Werner, *Myth and Philosophy in Plato's Phaedrus* (Cambridge: Cambridge University Press, 2012), 183.

84 See p. 71 of Gill, "Plato's Atlantis Story and the Birth of Fiction", *Philosophy and Literature* 3.1 (1979): 64–78.

85 Ibid., 71.

creation can bear truth".[86] That is, myths and even folktales may contain historical information, or even technical knowledge. Indeed, as Tarrant points out in the case of Proclus, a fifth-century CE commentator on Plato, the term "*historia* [was] used for an ordinary narrative exposition, while *mythos* implies the presence of a hidden meaning within the narrative".[87] Therefore *mythos* did not necessarily mean "fictional" in the modern sense of the word, and so a text with a mythical structure indeed might contain, again, historical or technical knowledge.

6 Transmission, Memory, and Text

The account concerning Atlantis in the *Timaeus* ends with the speakers turning to other subjects. But before they do, Critias returns once more to the issue of information preservation and transmission. He addresses Socrates:

> Because it had been so long ago, I didn't remember Solon's story very well. So I realized that I would first have to recover the whole story for myself well enough, and then to tell it that way. That's why I was so quick to agree to your assignment yesterday.[88]

Again, there is the idea that the "story" is *recovered* — not invented. Critias goes on:

> And that's how — as Hermocrates has already said — the moment I left here yesterday, I began to repeat the story to him and to Timaeus as it came back to me. After I left them I concentrated on it during the night and recovered just about the whole thing. They say that the lessons of childhood have a marvelous way of being retained. How true that is! In my case, I don't know if I'd be able to recall everything I heard yesterday, but I'd be extremely surprised if any part of this story has gotten away from me, even though it's been a very long time since I heard it. What I heard then gave me so much childlike pleasure — the old man was so eager to teach me because I kept on asking one question after

86 Barry, 226.

87 Harold Tarrant, ed. and trans., *Proclus: Commentary on Plato's Timaeus, Volume 1, Book I: Proclus on the Socratic State and Atlantis* (Cambridge: Cambridge University Press, 2007), 75.

88 Plato, *Timaeus* 26a, trans. Zeyl, in Cooper and Hutchinson, 1233.

> another — that the story has stayed with me like the indelible markings of a picture with the colors burnt in.[89]

Critias is quite humble in his explanations of how he came to know the story, and the nature of memory. In a way that a modern reader would well understand, Critias points out how a memory of long ago actually might be more distinct in one's mind than something "heard yesterday". Also, Plato sets up an interesting intersection here: he is presenting the reader with a *written* text, but in this passage in particular he is talking about the process of *oral* transmission of information.

In fact, none of this "postscript" to the story fits with a simple fiction, insofar as the narrator is quite honest about the fate of information over time. If the story were pure invention, there would be no need for any of this, nor would it add any particular tone of verisimilitude. Rather, the passage here describes how human societies and their cultural memories actually work — the process of an oral tradition and all its vagaries.

More particularly, Plato's works seem to address the question of the transmission of knowledge both directly, as in the text here, and indirectly. In terms of an indirect approach, Plato presents what one specialist in rhetoric calls "productive ambiguities".[90] One of those ambiguities, this commentator notes, is that Plato is,

> the writer who writes that nothing of importance can be conveyed through writing; the word-smith who argues that words are but imitations of imitations at the same time that he insists on precise definitions, divisions, and usages of words.[91]

Indeed, one of the challenges of the Atlantis account in Plato stems from the general problem concerning interpretation of Plato's works — a "complication in reading Plato is how frequently the dialogues contradict specific pronouncements in them."[92]

From an even broader perspective, the *Phaedrus* dialogue has one of Plato's own characters — Socrates — criticize the idea that writing can somehow

---

89 Plato, *Timaeus* 26a–26c, trans. Zeyl, in Cooper and Hutchinson, 1233.

90 Jean Nienkamp, ed., *Plato on Rhetoric and Language: Four Key Dialogues* (Mahwah, NJ: Lawrence Erlbaum, 1999), 2.

91 Ibid., 1.

92 Ibid., 4.

embody "clear knowledge of lasting importance".[93] Earlier in that dialogue, Plato has Socrates speaking as follows:

> Well, then, those who think they can leave written instructions for an art, as well as those who accept them, thinking that writing can yield results that are clear or certain must be quite naive and truly ignorant of Ammon's prophetic judgment: otherwise, how could they possibly think that words that have been written down can do more than remind those who already know what the writing is about?[94]

Here, the character of Socrates chooses an interesting example: "written instructions for an art". The implication here, then, is that certain knowledge cannot be written, but must be *embodied*, just as an art form is learned by observing it, engaging in it, and practicing it. The word here for "art" is *technê* (τέχνη), a term meaning art, skill, or craft; such things for millennia have been learned and passed down through system of active learning and embodiment, in the form of apprenticeships, guilds, and so on. This passage in the *Phaedrus* refers to technical knowledge in particular, then, once more indicating Plato's interest in its fate over time.[95] Socrates makes clear that one cannot learn an art, skill, or craft simply through written instructions — there must be a parallel or background process that involves both an oral tradition and "hands-on" learning.

Moreover, Socrates here also refers to a story that he has just told in this same dialogue: it is a story about the Egyptians — one that describes how Thamus or Ammon, the king of the gods, has a negative view of the god Theuth's gift of writing to mankind. Thamus says to Theuth that writing will "introduce forgetfulness", because human beings "will put their trust in writing which is external and depends on signs that belong to others."[96] It is interesting, then, to compare this to how knowledge about Atlantis is held by the Egyptians, according to Plato's account in the *Timaeus*. There, the Egyptian priest refers to "records inscribed in our sacred documents", and later notes that, "the records speak of

93 Plato, *Phaedrus* 277d, trans. Alexander Nehamas and Paul Woodruff, in Cooper and Hutchinson, 554; also see the discussion in Werner, 181–235.

94 Plato, *Phaedrus* 275c–275d, trans. Nehamas and Woodruff, in Cooper and Hutchinson, 552.

95 For a study of *technê* in Plato's works generally, and a discussion of Plato's perspective on the relationship between *technê* and other kinds of knowledge, see David Roochnik, *Of Art and Wisdom: Plato's Understanding of Techne* (University Park, PA: Pennsylvania State University Press, 1996).

96 Plato, *Phaedrus* 275a, trans. Nehamas and Woodruff, in Cooper and Hutchinson, 551–552.

a vast power". So, if writing is unreliable, why does Plato have the Egyptians use written records as their source for the Atlantis tale? Is the implication that the Egyptians have been duped?

Perhaps, but that is contradicted by the care with which the priest notes to the Greek visitors as to how much more continuous the Egyptian cultural memory is — the priest states clearly to them, "Your souls are devoid of beliefs about antiquity handed down by ancient tradition. Your souls lack any learning made hoary by time." He also adds that, "the antiquities preserved here [that is, Egypt] are said to be the most ancient." The priest's argument is that the Egyptians have continuity in terms of their transmission of knowledge, because of the lack of cataclysms and the stability of their natural environment — "here, in this place, water does not flow from on high onto our fields." But the priest does not necessarily say that the resulting continuity of transmission is due to a reliable *oral* tradition. Rather, he speaks as follows:

> Now of all the events reported to us, no matter where they've occurred — in your parts or in ours — if there are any that are noble or great or distinguished in some other way, they've all been inscribed here in our temples and preserved from antiquity on.[97]

So, despite the apparent ambiguous perspective that Plato has concerning writing, the Egyptians are described as a culture whose deep knowledge and strong retention of history is due to a *written* record.

Yet it may be that what Plato really is arguing here is something that, in fact, is akin to the discussion in the *Phaedrus*. For even as the character of Socrates criticizes writing, the reader realizes that his words are there written on a page. But because Plato uses the form of a dialogue, the writing does not necessarily take the form of something that "continues to signify just that very same thing forever".[98] Indeed, the "dialogues are not depictions of strictly logical dialectic, but rather conversations among complex characters drawn in a full dramatic context."[99] So, Plato can be seen here as using Egyptian culture as a model for knowledge transmission: for those priests, the "records inscribed in … sacred documents" are no more than a tool: the priesthood itself is the dynamic preservation of those documents, to keep their meaning and interpretation fresh and clear — even over a period of millennia. The Egyptian priesthood is depicted in Plato as doing exactly what Socrates notes: to "remind those

97 Plato, *Timaeus* 23a, trans. Zeyl, in Cooper and Hutchinson, 1230.

98 Plato, *Phaedrus* 275d–275e, trans. Nehamas and Woodruff, in Cooper and Hutchinson, 552.

99 Nienkamp, 2–3.

who already know what the writing is about". The sacred documents are the reminder, then, but it is really the priesthood that retains and transmits the Atlantis story. Therefore, both the Platonic format itself of the text-as-dialogue, and the portrayal of the Egyptian use of written records supported by a priesthood, illustrate a clear awareness of the synthesis of textual and oral forms required in the encoding and transmitting of knowledge.

7 Atlantis in Plato's *Critias*

The *Timaeus* is Plato's introduction, in a sense, to the Atlantis story. Plato's comments concerning Atlantis in the *Critias*, in turn, provide various details concerning this mysterious land. The *Timaeus* and the *Critias* both address the issue of the "ideal state" of Socrates discussed elsewhere in the Platonic dialogues. In the *Timaeus*, Socrates requests an example of an ideal state taking action of some kind, such as waging a war. The *Timaeus* provides this, but the description — the Atlantis story reproduced above — sits in the midst of other discussions about unrelated topics, such as cosmology. In the *Critias*, the reader can return to the account of Atlantis, although the dialogue as it stands is incomplete. In the *Timaeus*, Atlantis is depicted as an enemy, whose forces fail in their attack on prehistoric Athens. The account of Atlantis in the *Critias* focuses on its history and structure, but then leaves the reader with questions about how that fits with the description in the *Timaeus*.[100]

In the *Critias* text, Critias begins by making an invocation, to help him recount the story. Addressing Hermocrates, he invokes Mnemosyne, the goddess of memory:

> CRITIAS: ... But I must pay attention to your exhortation and encouragement, and, in addition to the gods you just named, invoke the other gods and make a special prayer to Mnemosyne. The success or failure of just about everything that is most important in our speech lies in the lap of this goddess. For, if we can sufficiently recall and relate what was said long ago by the priests and brought here to Athens by Solon, you the audience in our theater will find, I am confident, that we have put on a worthy

100 For some interesting conjectures as to the connection between the two texts, see Thomas G. Rosenmeyer, "Plato's Atlantis Myth: 'Timaeus' or 'Critias'?" *Phoenix* 10.4 (Winter 1956): 163–172.

> performance and acquitted ourselves of our task. So much said. Now we must act. Let us delay no more.[101]

Notice again the introduction of a key theme in the Atlantis story here — the question of memory, the ability to "recall and relate" information accurately, and the transmission of knowledge. Critias continues:

> We should recall at the very beginning that, in very rough terms, it was some nine thousand years since the time when a war is recorded as having broken out between the peoples dwelling outside the pillars of Heracles and all those dwelling within. This war I must now describe. Now they said that this city of Athens was the ruler of the [Mediterranean] peoples and fought for the duration of the entire war. They said, too, that the kings of the island of Atlantis were the rulers of the other peoples. This island, as we were saying, was at one time greater than both Libya and Asia combined. But now because of earthquakes it has subsided into the great Ocean and has produced a vast sea of mud that blocks the passage of mariners who would sail into the great Ocean from Greek waters and for this reason it is no longer navigable.
>
> In its progress, our tale will describe, as if it were unrolled, the many barbarian nations and all the different Greek peoples of that time, encountering them as they emerge from place to place. It is first necessary at the beginning of this tale to describe the condition of the Athenians of that age and the adversaries with whom they waged war: their respective power and their respective constitutions. But of these themes, pride and place must go to the condition of Athens before this war.[102]

Here, Critias has reiterated the historical and geographical setting. The description is detailed, noting that not only has Atlantis "subsided into the great Ocean", but that this has resulted in a "vast sea of mud". The reader is given detail that seems almost excessive, if this is only intended to be a fiction. Critias' comments then turns to what strikes the modern reader as a strange combination of religious discussion and pragmatic analysis:

> At one time, the gods received their due portions over the entire earth region by region — and without strife. To claim that gods did not recognize what was proper to each would not be fitting, nor would it be right to

101 Plato, *Critias* 108c–108d, trans. Clay, in Cooper and Hutchinson, 1294.

102 Plato, *Critias* 108e–109a, trans. Clay, in Cooper and Hutchinson, 1294–1295.

> say that, although they recognized what belonged by just title to others, some would attempt to take possession of this for themselves — in open strife. But, as they received what was naturally theirs in the allotment of justice, they began to settle their lands. Once they had settled them, they began to raise us as their own chattel and livestock, as do shepherds their sheep. But they did not compel us by exerting bodily force on our bodies, as do shepherds who drive their flocks to pasture by blows, but rather, by what makes a creature turn course most easily; as they pursued their own plans, they directed us from the stern, as if they were applying to the soul the rudder of Persuasion. And in this manner they directed everything mortal as do helmsmen their ships.
>
> Now, as the gods received their various regions lot by lot, they began to improve their possessions. But, in the case of Hephaestus and Athena, since they possessed a common nature, both because she was his sister of the same father and because they had entered the same pursuits in their love of wisdom and the arts, they both received this land as their portion in a single lot, because it was congenial to their character and was naturally suited to them in its excellence and intelligence. And they fashioned in it good men sprung from the land itself and gave them a conception of how to govern their society. The names of these first inhabitants have been preserved, but their deeds have perished on account of the catastrophes that befell those who succeeded them and the long passage of time intervening.[103]

In the passage above, Critias is providing an origin, or an origin myth, for a human civilization. Again, it is not clear how literally or how metaphorically an ancient reader would have taken these comments. But note in the last line how Critias' focus shifts to the pragmatic: once more, the discussion turns to the transmission of knowledge, particularly in the event of a catastrophe. Critias goes on to elaborate that point:

> Those of their race who survived these successive destructions were, as I said before, left as an illiterate mountain people who had only heard the tradition of the names of the rulers of their country and beyond these only little of their deeds. Now, they were pleased to give their descendants the names of these rulers, even though they were unaware of their ancestors' virtues and institutions — except for some dim legends concerning each of them. Then, for many generations, these survivors and their children lived in distress for their survival and gave thought to

103 Plato, *Critias* 109b–109d, trans. Clay, in Cooper and Hutchinson, 1295.

> their needs; they spoke only of supplying these needs, and had no interest in the events of the distant past. For it is in the train of Leisure that Mythology and Inquiry into the Past arrive in cities, once they have observed that in the case of some peoples the necessities of life have been secured, but not before.[104]

So, here the text presents a very objective analysis of what happens to knowledge in human civilizations, with several clear points. First, that when there are "successive destructions", it may be the case that only "illiterate ... people" with fragmentary knowledge will survive. Critias states that people will be given names of distinguished rulers, even while the actual "virtues and institutions" will have been forgotten other than "some dim legends". Finally, Critias makes the quite realistic point that cultures that live in "in distress for their survival" don't have a lot of time for knowledge transmission. Only cultures with leisure time can engage in "Mythology and Inquiry into the Past". Critias goes on:

> This is why the names of the ancients have been preserved but not their deeds. I make this claim and cite as my evidence the statement of Solon, who said that, in their account of the war of that time, the Egyptian priests gave for the most part names such as Cecrops and Erechtheus, and Erichthonius, and Erysichthon, and the names of most of the others which have come down in tradition before the generation of Theseus.[105] And the same is true of the names of the women. Consider too the attributes of the goddess Athena and her statue. At that time the military training of women and men was common. For this reason the people of that time fashioned the statue of the goddess as armed to reflect that ancient custom — an indication that all the female and male creatures that live together in a flock can very well pursue in common, as much as is possible, the special talents that are suited to each species.[106]

Again, Critias is providing pragmatic analysis: he is looking at how names are evidence of incomplete knowledge transmission, for example, and using artistic evidence — the accoutrements of representations of Athena — to try to reach some conclusions about how men and women lived together in the past. Critias then turns to further societal, geographical, and agricultural details:

104 Plato, *Critias* 109d–110a, trans. Clay, in Cooper and Hutchinson, 1296.

105 Cecrops, Erechtheus, and Erichthonius were legendary rulers of Athens; Erysichthon was the son of Cecrops.

106 Plato, *Critias* 110a–110c, trans. Clay, in Cooper and Hutchinson, 1296.

Now, at that time, the other classes of citizens who dwelt in our city were engaged in manufacture and producing food from the earth, but the warrior class that had originally been separated from them by god-like men lived apart. They had all that was appropriate to their training and education. None of them had any private possession, but they thought of all their possessions as the common property of all, and they asked to receive nothing from the other citizens beyond what they needed to live. Their activities were all of the activities that were spoken of yesterday, when the guardians proposed by our theory were discussed.

The report of the Egyptian priests concerning our territory was plausible and true. First of all, at that time its boundaries extended to the Isthmus of Corinth, and, on the mainland to the north, they extended to the summits of Cithaeron and Parnes. And, descending to the east, the boundaries extended down to the region of Oropus to the north and they were defined by the Asopus river down to the sea. In its great fertility our land far surpassed every other, for it was then capable of supporting a great army of men who did not work the land. There is impressive evidence for this excellence. What has now survived of this land can rival any other land in the variety and quality of its crops and the pasture it offers all species of animals. But, at that time, our land produced all this not only of high quality but in great abundance. You might ask how this is credible and how our present land could possibly be called a vestige of our earlier land.

From the interior this entire land extends a great distance into the sea, as if it jutted out as a promontory. It so happens that the entire basin of the sea that surrounds falls off precipitously. Many and great were the floods that occurred in the space of nine thousand years — for this is the number of years between that time and the present — and during this succession of natural disasters the soil was washed down from the high places. It did not form any considerable alluvial deposits, as in other regions, but it disappeared into the deep, as in flood after flood it was continuously washed into the sea from all sides. What actually remains is like our small and barren islands, and, compared to the land it once was, Attica of today is like the skeleton revealed by a wasting disease, once all the rich topsoil has been eroded and only the thin body of the land remains. But in that age our land was undiminished and had high hills with soil upon them; what we now call the Rocky Barrens were covered with deep rich soil. And in the mountains there were dense forests of which there still

> survives clear evidence. Some of our mountains can now grow just barely enough for bees, but it was not so long ago that [lofty trees grew there]. There can still be found intact rafters cut from trees that were felled and brought down to be used for the greatest building projects. And there were many trees that were cultivated for their fruit and they provided limitless fodder for flocks of sheep and goats.
>
> Every year there was a harvest of Zeus-sent rain. It was not lost, as it is now, as it flows off the hard surface of the ground into the sea, but the deep soil absorbed the rain and it stored it away as it created a reservoir with a covering of clay soil above it; and, as it distributed the water it had absorbed from the high places into its hollows, it produced an abundant flow of water to feed springs and rivers throughout every region of the country. There are even today some sacred monuments at these ancient springs that are evidence of the truth of what we are now saying about our country.
>
> This was the nature of the countryside. The land was cultivated with great skill, as we can reasonably conjecture, by farmers who were farmers in the true sense of the word and who devoted themselves to this single occupation — but farmers who had an eye for beauty and were of a truly noble nature, and who in addition possessed a most fertile land and water in abundance, and above this land a climate and seasons that were most temperate.[107]

Again, consider this text from the perspective of both the writer and the reader. Why does Plato have Critias expound on these details? What is Plato's point here? If this Atlantis story is a fiction, it could be that the details — even the rather mundane ones about landforms, erosion, and so on — are there to induce the reader to take the story as something true or almost true. From the perspective of the reader, it is important to ask whether an ancient reader would have been convinced by such a technique.

Moreover, at least from a modern perspective, the detail is not particularly interesting. The technical comments that from "the interior this entire land extends a great distance into the sea", and that "the entire basin of the sea that surrounds falls off precipitously" are only intriguing because of their odd placement in a story — the Atlantis story — that is usually considered a fiction. The mention of "considerable alluvial deposits" also seems to a modern

107 Plato, *Critias* 110c–111e, trans. Clay, in Cooper and Hutchinson, 1296–1297.

reader to be an oddly sober geological reference coming from Plato, a writer best known for philosophy. The only thing that then seems to pull the reader back to the peculiarity that is the hallmark of the Atlantis narrative is the note of "floods that occurred in the space of nine thousand years", a span of time that rings as pure fiction to a modern reader.

Critias then begins a detailed description of Athens:

> As for the city itself, it was laid out at that time in a plan that I will now describe. First of all, the acropolis was very different then than it is now. A single night of torrential rain stripped the acropolis of its soil and reduced it to bare limestone in a storm that was accompanied by earthquakes. Before the destructive flood of Deucalion, this was the third such cataclysmic storm. In the past, the acropolis extended to the Eridanus and Ilisus and held within its circuit the Pnyx and Mt. Lycabettus that faces the Pnyx. It was entirely covered by soil and, except for some small outcroppings, level on top. Outside the acropolis and under its slopes there lived the class of artisans and those of the farmers who worked the neighboring land. But on the heights the class of warriors lived in isolation, as if they belonged to a single household, around the sanctuary of Athena and Hephaestos, which they had enclosed by a single garden wall. On the far northern edge of the acropolis they inhabited common dwellings and ate together in common messes in buildings they had constructed for their winter quarters. And they had a supply of all that was needed for their communal institutions — both in buildings for themselves and for the priests. They made no use of gold or silver — possessions they never had any need of. But, in pursuing a mean between ostentation and servility, they built for themselves tasteful houses and they grew old in them in the company of their grandchildren; and for generation after generation they passed these dwellings down to descendants who were like themselves. As for the south of the acropolis, when they left their orchards, gymnasia, and common messes, as they would for the summer season, they converted it to these uses.
>
> There was a single spring in the location of the present acropolis, but it has been choked by the debris of the earthquakes [of that night], and its waters now flow only in a trickle about the circuit wall. But it provided the men of that age with an abundant supply of water, since it was situated in a location that made it neither too cold in the winter nor too hot in the summer.

> This was the manner of their life: they were the guardians of their own citizens and the leaders of the rest of the Greek world, which followed them willingly. And they kept their population stable as far as they could — both of men and women — for generation after generation, maintaining the population of those who had reached military age or were still of military age at close to twenty thousand at most.
>
> Such, to conclude, was the character of this people and such was their life generation after generation as they directed the life of their city and of Greece with justice. Their fame for the beauty of their bodies and for the variety and range of their mental and spiritual qualities spread through all of Asia and all of Europe. And the consideration in which they were held and their renown was the greatest of all the nations of that age.
>
> As for the state of those who went to war against them and the origins of that state, we will now openly reveal its history to you our friends, as the common property of friends, if we have not lost the memory of what we heard when we were still boys. I must explain one small point before I enter into my history so that you will not be astonished as you hear Greek names frequently used for people who are not Greek. You will now learn the origins of these names. Solon, when he was contemplating his own poetic version of this legend and was inquiring into the meaning of these names, discovered that his Egyptian sources had been the first to record them, once they had translated their meaning into their own language. He, in his turn, recovered the meaning of each of these names and recorded it as he translated them into Greek. These very manuscripts were in the possession of my grandfather and they now remain in my possession. When I was a boy, I studied them carefully. Consequently, do not be astonished if you hear names that sound like Greek names; you now know their explanation.[108]

Again, the way the story runs is peculiar. There is the extensive description of the Athenians and their society, followed by a note that the discussion will now turn to "the state of those who went to war against them". There is also the somewhat convoluted treatment of the issue of Greek names; whether or not this is supposed to convince the reader of the historicity of this narrative is not clear. Rather, one might think that an ancient reader would have liked to have

108 Plato, *Critias* 111e–113b, trans. Clay, in Cooper and Hutchinson, 1297–1299.

heard the names in their original form, or at least how the "Egyptian sources" had recorded them — rather than Solon's translation. Note, too, that the theme of the transmission of knowledge is alluded to again with Critias' comment about "manuscripts [that] were in the possession of my grandfather".

After the passage above, Critias finally turns to a long description of Atlantis:

> What follows, approximately, was the introduction to the long account I heard then. As I said before concerning the distribution of lands among the gods, in some regions they divided the entire earth into greater apportionments and in others into lesser apportionments, as they established sanctuaries and sacrifices for themselves. So it was that Posidon received as one of his domains the island of Atlantis and he established dwelling places for the children he had fathered of a mortal woman in a certain place on the island that I shall describe.
>
> Now seaward, but running along the middle of the entire island, was a plain which is said to have been the loveliest of all plains and quite fertile. Near this plain in the middle of the island and at about fifty stades' distance was a uniformly low and flat hill. Now, there lived on this hill one of the people of this island who had originally sprung up from the earth. His name was Evenor and he dwelt there with his wife Leucippe. They had an only child, a daughter by the name of Clito. When this girl grew to marriageable age, both her mother and father died. It was then that Posidon conceived a desire for her and slept with her. To make the hill on which she lived a strong enclosure he broke it to form a circle and he created alternating rings of sea and land around it. Some he made wider and some he made more narrow. He made two rings of land and three of sea as round as if he had laid them out with compass and lathe.
>
> They were perfectly equidistant from one another. And so the hill became inaccessible to humans. For at that time ships and the art of navigation had not yet come into existence.
>
> And the god himself greatly beautified the island he had created in the middle to make it a dwelling suitable for a god. Because he was a god, he did this with little effort. He drew up two subterranean streams into springs. One gushed out in a warm fountain and the other in a cold fountain. And from the earth he produced all varieties of crops that were sufficient to his island. He sired five pairs of twin sons and he raised them to manhood. He divided the entire island of Atlantis into ten districts: to the

first born of the first set of twins he gave as his portion the dwelling of his mother and the circular island, since it was the largest and the best. And he made him king over the others. The other sons he made governors and to each of these he gave the rule over many men and a great extent of land.

And he gave each of his sons names. To the son who was oldest and king he gave the name from which the entire island and its surrounding sea derive their names, because he was the first of the kings of that time. His name was Atlas; the island is called Atlantis and the sea Atlantic after him. To the twin born after him, who had received as his portion the cape of the island facing the pillars of Heracles opposite what is now called the territory of Gadira after this region, he gave the name that translates into Greek as Eumelos, but in the language of Atlantis, it is Gadirus. It would seem that he gave his name to the region of Cadiz. The two brothers of the second set of twins he called Ampheres and the Euaemon. To the third set he gave the name Mneseas to the first-born and Autochthon to the second-born. Of the fourth set Elasippus was the first-born, Mestor the second. For the fifth set he gave the name Azaes to the first-born and the name Diaprepes to the second. Now all of these sons inhabited the island, as did their sons and descendants over many generations. They were the rulers of many other islands in the Atlantic and, as I have said, they even extended their rule into the Mediterranean as near to us as Etruria and Egypt.

The race of Atlas increased greatly and became greatly honored. And they maintained their kingdom through many generations, as the oldest king would hand his kingship on to his oldest son. They amassed more wealth than had ever been amassed before in the rule of any previous kings or could easily be amassed after them. And they provided for everything that was needed, both in the city and in the rest of the island. For their empire brought them many imports from outside, and the island itself provided most of what was needed for their livelihood. First, there were the mines that produced both hard and fusible ore. And in many regions of the island they exploited that metal which is now only a name to us, but which was then more than a name — *oreichalkos* [literally, "mountain copper", i.e., copper ore]. In that age it was valued only less than gold. And the island provided all trees to be hewn and worked by builders and this in great abundance. It also produced abundant animal life, both domestic and wild. In addition to these there was a great population of elephants. There was pasture land for the other animals who graze in marshlands and along lakes and rivers and on mountainsides and plains, and there was plenty for them and for this the greatest of animals, which consumes the most fodder. The island produced in addition all the aromatic plants the earth produces now — sweet smelling roots and greens, herbs, trees, and gums from flowers and fruits as well, and they flourished there. The island also produced the domesticated crop of grains on which we live and all the other crops on which we depend for our food. It also produced the kinds of crops we call "pulse" and the trees that give us our drink, food, and oils — and the crop that sprung up for the sake of our entertainment and pleasure, is hard to preserve, and comes from tree tops; it produced the side dishes we offer the weary guest as a relief after he has eaten his fill and that refresh him after dinner. All of these did that sacred island once bear in that age under a fostering sun — products lovely, marvelous, and of abundant bounty. And they took all these products from the earth and from their proceeds they constructed their sanctuaries and their palaces, their harbors and their ship-sheds, and they improved the rest of their land according to the plan I will now describe.

First, they constructed bridges joining the rings of sea, which surrounded the ancient metropolis, making a road out from the palace and in to the palace. Their first project was to build a palace in the dwelling of the god and of their ancestors. One king inherited the project from his predecessor, and, as he improved on the beauty of what had already been improved, he would surpass to the extent of his resources what his predecessor had been able to achieve. They continued this progress until they had created for themselves a dwelling astonishing in its size and in its manifold beauty. And starting at the sea they excavated a canal three plethra in width, one hundred feet in depth, and fifty stades in length up to the outermost sea ring. They then made passage from the sea into the interior possible by opening a channel into the sea ring that was wide enough for the largest ships to sail into it as if it were a harbor. And, as for the land rings that separated the rings of sea, they pierced them at the point of the bridges, and thus joined them by water. The resulting canal was wide enough for a single trireme to sail through as it passed into a ring of water. They constructed a roof over the channel to protect the passage of ships, for the walls of the canal through the land rings were high enough from the sea to the bridge above to allow ships to pass under. The largest of the water rings into which the passage from the sea had been excavated was three stades in width and the next land ring was equal to it. Of the next rings of water and land, the ring of water was two stades wide and, as in the first case, the land ring was equal to it as well. And, finally, the ring of water running around the island in the middle was a stade wide.

The island where the palace was located had a diameter of five stades. They threw up an unbroken stone circuit wall around this island, and they also walled the land rings, and the bridge, which was a plethron wide. They built towers and gates at the point where the bridges crossed over the rings of water. They quarried stone from under the circular island that formed the center ring and from the inner and outer land rings as well. There were three colors of stone: white, black, and red. As they quarried this stone, they fashioned ship sheds for two ships in the rock roofed by the stone of the quarry itself.

Some of their buildings they constructed of stones of uniform color. But to delight themselves they made of others a tapestry of stones of different colors, variegating the colors to bring out their natural charm. And they invested the entire circuit wall of the outermost land ring with bronze, as if the bronze revetment were a bright dye. The interior of the land wall they invested with tin. And the wall surrounding the acropolis itself they invested with *oreichalkos*, which glittered like darting fire.

I will now describe the palace buildings erected within the acropolis. At its center was the shrine of Clito and Posidon. It was kept consecrated and no one was permitted to enter it. It was surrounded by a wall of gold. It was here that Posidon and Clito first begot and produced the race of the ten kings. It was to this shrine that each of the ten divisions came to offer their first fruits to each of these original kings in a yearly festival. The temple of Posidon was in this area. It was one stade long, three plethra wide, and of a height that appeared to be proportional to its length and width, but it had something barbaric about its appearance. They invested the entire exterior of the temple with silver, except for the acroteria, which they gilded with gold. The interior presented a roof of solid ivory inlaid with gold, silver, and *oreichalkos*; and they plated all the other areas of the temple with this same metal — the cella walls, the interior columns, and the floors. They placed gold statues within the temple. There was a statue of Posidon standing in a chariot with a team of six winged horses. This statue was so tall that his head touched the rafter of the temple roof; there were a hundred Nereids riding dolphins and arranged in a circle about him, for men of that age thought that the Nereids were a hundred in number; and there were many other statues inside which were the offerings of private individuals.

Outside and surrounding the temple there stood gold statues of all the descendants of the ten kings and their wives and many other dedications of great size made by the kings and private individuals who came from the city of Atlantis itself and from the subject peoples elsewhere. There was an altar on the same scale as the temple and its workmanship was equally lavish. The palace was magnificent in its monumental architecture and it was worthy of the greatness of their empire and the adornment of the temple and shrines.

They drew their water from two springs — a spring of cold water and a spring of hot water. Both had an abundant flow and in the amazing natural freshness and quality of its waters each had its own use. They built fountain houses around them and plantations of trees suitable to the temperature of the waters. And they also built reservoirs around the springs. Some they left open, but to the north they covered the reservoirs to convert them to warm baths. The reservoirs of the kings were separate from those of the rest of the population. Some reservoirs were reserved for the use of women, others for watering horses and other draft animals, and each they fashioned appropriately to its use. The overflow they channeled into the grove of Posidon, where, thanks to the fertility of the soil, there grew all varieties of trees of extraordinary beauty and height. They also irrigated the outer land rings by means of canals that crossed over along the bridges joining them.

Here there were constructed numerous shrines to numerous gods and the land was laid out for many orchards and gymnasia. There were gymnasia for men on each of the two ring islands and tracks for horses were set apart as well. And, remarkably, through the middle of the greatest of the islands they laid out a separate race course for horses, one stade wide, and it extended in a circle around the entire island. Located on each side of the central race course were quarters for the palace guard.

The garrison of the most reliable soldiers was established on the smaller of the ring islands, the island situated nearest to the acropolis. And quarters were built on the acropolis for the most reliable soldiers of all, surrounding the palaces of the kings themselves. The ship-sheds were filled with triremes and all the fittings needed for triremes, and all were in good working order. Such, then, were the buildings they constructed around the [dwellings of the] kings themselves.

Now, once you had crossed over the three rings of water, you would come to a circuit wall that began at the sea and surrounded the greatest of the land rings on all sides at a uniform distance of fifty stades from the greatest land ring and its harbor. It began at the point where the channel had been dug through to the sea. The entire area within was settled by a dense population whose houses were crowded close together. The water-way into the interior and the greatest harbor was teeming with ships and crowds of merchants who had arrived from all over the world and whose voices and bustle produced a commotion and hubbub that could be heard day and night.

I have recalled this description of the capital and the ancient dwelling of the kings pretty much as it was told [to Solon] at that time. But now I must attempt to recall the nature of the rest of the country and the manner in which it was improved. To begin with, the priests said that the entire country was very high and that it rose sheer from the sea. The entire plain that surrounded the capital was itself surrounded by a ring of mountains that sloped down as far as the sea. The plain was smooth and level and entirely rectangular. On its long sides it extended for three thousand stades and, as measured from the sea, it was over two thousand stades across. The slope of the island was to the south and it was protected from the northerly winds. The mountains surrounding the plain were legendary for their number and size and beauty. None of the mountain ranges that exist today can compare with them. They contained on their slopes and in their valleys many populous and wealthy villages. And they contained rivers and lakes and meadows that supplied enough to feed all the animals there, both domesticated and wild. In their abundance and variety, the shrubs and trees were plentiful for all kinds of constructions and uses.

I will now relate how this plain had been developed by nature, and by many kings and over a long period of time. For the most part, the plain was naturally rectangular, regular, and oblong. Where it was not perfectly straight and even they evened it out by excavating a Great Canal around it. As described, its depth and width and length provoke disbelief, since it was the work of human hands and so vast when compared to the other building projects. Nevertheless, I must repeat precisely what we heard then. The Great Canal was excavated to the depth of a plethron, it measured a stade wide along its entire length, and as it framed the entire plain it came to a total length of ten thousand stades. As it received the flow of water that came off the mountains, and as this water circulated and reached the city on two sides, the trench allowed the water to flow out to the sea. Towards the interior, canals were cut in straight lines from the city over the plain a hundred feet broad at most and these emptied their waters into the Great Canal facing the sea. These were spaced at an interval of a hundred stades. They also cut horizontal connecting channels linking one canal with another and with the city, and it is by these canals that they transported timber and the other products of the land on barges from the mountains to the city. They harvested their crops twice a year. In the winter season they relied on the water of Zeus-sent rains, and in the summer season they used the waters stored in the earth drawing it into their canal system to irrigate the crops.

Now, as for the numbers of the men of the plain who were fit to serve in the army: each military district was assigned to contribute one commander. The area of each district was as much as a hundred stades. The total of these districts came to sixty thousand. And as far as the population of the mountainous regions and the rest of the country goes, it was said to be too large to calculate. But, counted by regions and villages, all men fit for military service were assigned to one of the sixty thousand military districts and they served under the commander of each district. In times of war each commander was assigned to have in readiness a sixth part of the complement of a war chariot as a contribution to a force of ten thousand chariots; and in addition, two horses and two riders, a pair of horses without a chariot, with its complement of two riders, a runner, a rider who could fight on foot armed with a small shield, and serving as a charioteer a rider who could mount either horse, two hoplites, two archers, and two sling men; three light armed soldiers with stones and three with javelins. He also had to contribute four sailors to the crews manning twelve thousand ships. These were the principles for raising an army in the royal city. The formulas varied in the nine other cities, and it would take a long time to describe them.

The original ordering of powers and honors in Atlantis was as follows. Within his own patrimony and in his own city, each of the ten kings held power over the inhabitants and over most of the laws, and he could punish or put to death whomever he wished. But, as for their common empire and federation, the kings were regulated by the laws of Posidon as these had been passed down by tradition and according to an inscription which the first kings had cut on a stele of *oreichalkos*. This inscription was placed in the middle of the island in the sanctuary of Posidon. Here in every fifth or sixth year, and in alternating sequence, it was their custom to gather. To both the even and to the odd they accorded an equal share. Once they had assembled, they deliberated on matters of common concern and held an assize to determine if anyone of them had broken the law, and they gave judgment. Whenever they were about to declare judgment, they first offered one another pledges in this manner: as all ten kings were alone in the sanctuary of Posidon, where bulls had been allowed to run free, they joined in prayer to ask the god to be allowed to capture the bull which would be the most acceptable offering to him. They pursued the bulls with staffs and nooses — but with no iron weapon, and they led the bull they had captured to the stele. There they slaughtered it on the crest of the stele and let its blood spill down over the inscription. In addition to the laws written on the stele there was an oath inscribed calling terrible curses down upon those who broke them. And, when they had then sacrificed the bull following this ritual, they would burn all the limbs of the bull and, mixing his blood in a mixing-bowl, they would pour a clot of his blood over the head of each of them, and, once they had scrubbed the stele clean, they would bring the remaining blood over to the fire.

After this, they would draw the blood from the mixing-bowl into gold pouring vessels. Pouring the blood over the fire they would take an oath to render justice according to the laws inscribed on the stele and to punish anyone who had violated these laws since last they met. They swore that in the future they would not willingly violate any of the provisions of the inscription and that they would neither rule nor obey a ruler if either they or he did not issue commands that were in conformity with the laws of their father. When each of the kings had made this oath and engaged both himself and his descendants, they drank and dedicated their pouring-vessels in the sanctuary of the god. And, once they had finished with their dinner and everything else they had to do and night had fallen and the fire about the sacrificial offerings had subsided, they all put on a deep blue robe of the most splendid appearance and, sitting on the ground next to the embers of the sacrificial victim, at night, they put

> out the fire still flickering in the sanctuary and judged anyone accused of violating any of their laws and were judged themselves. Once they had passed judgment, when day dawned, they recorded their judgments on a gold tablet which they dedicated as a memorial offering along with their robes.
>
> There were many other particular laws concerning the prerogatives of each of the kings, but the most important of these were those forbidding them to bear arms against one another and commanding them to help one another should anyone in any of their cities make an attempt to overturn the divine family; that they should deliberate together, as had their ancestors before them, over their decisions concerning war and their other actions, but that they should cede leadership to the royal family of Atlantis; and, finally, that the king should have power to put none of his kinsmen to death, if he could not obtain the approval of the majority of the ten kings.[109]

The passage presents a combination of precise technical detail in terms of architecture and geography along with expositions on history, ritual, law, and society. Critias then provides a comment on Atlantis as a whole:

> Now, this was the power, so great and so extraordinary, that existed in that distant region at that time. This was the power the god mustered and brought against these [Mediterranean] lands. It was said that his pretense was something like what I shall describe. For many generations and as long as enough of their divine nature survived, they were obedient unto their laws and they were well disposed to the divinity they were kin to. They possessed conceptions that were true and entirely lofty. And in their attitude to the disasters and chance events that constantly befall men and in their relations with one another they exhibited a combination of mildness and prudence, because, except for virtue, they held all else in disdain and thought of their present good fortune of no consequence. They bore their vast wealth of gold and other possessions without difficulty, treating them as if they were a burden. They did not become intoxicated with the luxury of the life their wealth made possible; they did not lose their self-control and slip into decline, but in their sober judgment they could see distinctly that even their very wealth increased with their amity and its companion, virtue. But they saw that

109 Plato, *Critias* 113b–120d, trans. Clay, in Cooper and Hutchinson, 1299–1306.

> both wealth and concord decline as possessions become pursued and honored. And virtue perishes with them as well.
>
> Now, because these were their thoughts and because of the divine nature that survived in them, they prospered greatly as we have already related. But when the divine portion in them began to grow faint as it was often blended with great quantities of mortality and as their human nature gradually gained ascendancy, at that moment, in their inability to bear their great good fortune, they became disordered. To whoever had eyes to see they appeared hideous, since they were losing the finest of what were once their most treasured possessions. But to those who were blind to the true way of life oriented to happiness it was at this time that they gave the semblance of being supremely beauteous and blessed. Yet inwardly they were filled with an unjust lust for possessions and power. But as Zeus, god of the gods, reigning as king according to law, could clearly see this state of affairs, he observed this noble race lying in this abject state and resolved to punish them and to make them more careful and harmonious as a result of their chastisement. To this end he called all the gods to their most honored abode, which stands at the middle of the universe and looks down upon all that has a share in generation. And when he had gathered them together, he said …[110]

The *Critias* ends there, leaving the reader with a story that is clearly incomplete. Some modern commentators believe that Plato abandoned the work.[111] This leaves us with the fact that the *Critias* is taken up almost entirely with the description of Atlantis. This final passage seems designed to provide a social critique of this advanced society, and to link its inhabitants' "inability to bear their great good fortune" and the society's becoming "disordered" to their ultimate fate. Although indirect, there is once again the theme of how technologically advanced societies fall: initially, although the Atlanteans have sophisticated architecture and a high standard of living, their focus is solely on virtue. However, as time passes, "possessions become pursued and honored" and "virtue perishes".

---

110 Plato, *Critias* 120d–121c, trans. Clay, in Cooper and Hutchinson, 1306.

111 Gill, *Plato: The Atlantis Story*, 72; also see 298–304 and 304, n. 74 of idem, "The Genre of the Atlantis Story", *Classical Philology* 72.4 (October 1977): 287–304. Also note 51 of Clay, "The Plan of Plato's *Critias*", in Tomás Calvo and Luc Brisson, eds., *Interpreting the Timaeus–Critias: Proceedings of the IV Symposium Platonicum — Selected Papers* (Sankt Augustin, Germany: Academia Verlag, 1997), 49–54.

8 Atlas, Atlantis, and a Question of Interpretation

The *Critias* passage provides a note as to the origin of the name "Atlantis"; one of Poseidon's offspring, "who was oldest and king" is given the name Atlas, and so "the island is called Atlantis and the sea Atlantic after him."[112] Herodotus labeled the sea beyond Europe as the "the entire sea outside the Pillars of Heracles, which is called the Atlantic".[113] In Greek, this is *thalassa Atlantis* (θάλασσα Ἀτλαντὶς), with *Atlantis* (Ἀτλαντὶς) being the possessive form of *Atlas*; the literal meaning, then, is "Sea of Atlas". This "Atlas" refers both the mythical character and to the Atlas mountains in western Africa that marked the western limit of the continental *oikoumene*.

As a strictly geographical term, it seems that Atlas first appears in Herodotus, as a single mountain, with the surrounding area inhabited by the *Atlantes* (Ἄτλαντες).[114] Euripides, in his play *Hippolytus*, makes a number of references to Atlas, using phrases that mix mythology and geography: "Of all who dwell between the Euxine Sea and the Pillars of Atlas"; "To the apple-bearing shore of the melodious Hesperides would I go my way, there were the lord of the sea forbids sailors further passage in the deep blue mere, fixing the sacred boundary of the skies, the pillar held up by Atlas"; and "Yes, beyond the Euxine Sea and the Pillars of Atlas".[115] In the *Critias*, the Atlas connection comes as part of the origin story of Atlantis: "To the son who was oldest and king he gave the name from which the entire island and its surrounding sea derive their names ... His name was Atlas; the island is called Atlantis and the sea Atlantic after him."

There may have been other associations for Plato between Atlas and the narrative in the *Critias*.[116] In Greek mythology, Atlas is a Giant, one of the four sons of a Titan, Iapetus, and Clymene, the daughter of Oceanus. Iapetus, in turn, was the son of the Heavens and Earth — Uranus and Gaea. The Titans were the primeval deities in the Greek worldview. Led by Cronos, they ruled the Earth until they were overthrown by the Olympians, under Zeus. After this struggle between the Giants and the Olympians, Zeus punished Atlas, condemning

112 Plato, *Critias* 114a–114b, trans. Clay, in Cooper and Hutchinson, 1300.

113 Herodotus, *Histories* 1.203, trans. Grene, 125.

114 From Herodotus, *Histories* 4.184, trans. Grene, 349; for an overview of references to "Atlas" in classical sources, see the entry in Charlton T. Lewis and Charles Short, *Harper's Latin Dictionary: A New Latin Dictionary Founded on the Translation of Freund's Latin-German Lexicon* (New York: Harper & Brothers, 1907), 189.

115 See Euripides, *Hippolytus* 3, 742–747, and 1053, in *Euripides: Children of Heracles, Hippolytus, Andromache, Hecuba*, trans. David Kovacs (Cambridge, MA: Harvard University Press, 1995), 125, 197, and 225.

116 Gill, "The Genre of the Atlantis Story", 295, n. 41.

him to bear the sky on his shoulders. Here mythology and geography connect, as Atlas was said to live in the far west, in the country of the Hesperides, his daughters by Hesperis. The Hesperides came to be the name of islands said to be in the western reaches of the ocean. In his *Histories*, Herodotus speaks of Mount Atlas, and notes that nearby residents — the Atlantes — believed it to be the point on which the heavens rested. Herodotus goes on to say that these Atlantes are the last people he knows of in the region:

> As far as the Atlantes I can tell the names of those peoples who live along the ridge, but from there on I no longer can. Yet the ridge goes right through to the Pillars of Heracles and beyond.[117]

Herodotus was the first to employ the term "Atlantic" for the sea beyond the Pillars of Hercules; before that, it had the more generic title of "ocean."[118] In Herodotus' work there is again the integration of myth and geography. He does not discuss Atlas the mythical figure explicitly, but he does make an oblique reference in his recounting of how the locals believe that the heavens rested on the mountain of that name. Atlas appears in Plato's work in the *Phaedo*, in a reference to the myth of his supporting the heavens.

As in the *Timaeus*, Plato's basic structure in the *Critias* Plato is historical, but with geographical settings, and even geological observations. There is the integration of mythical references, as we see in the very term "Atlantis" and its connection with Atlas, but even this has a geographical foundation, as outlined in Herodotus. The idea of the Atlantis story as a purely "mythical" fabrication, in the modern sense of a pure fiction, is difficult to see: there are too many additional strands of history, geography, topography, and geology, along with notes on the transmission of knowledge. So, what *is* going on here?

In the Chinese texts concerning flight, there is a reasonably clear differentiation between two types of story or account. One type of story is fanciful in the sense that it deals with things that normally are not encountered by an ancient or modern reader: flying immortals or spirits. Such images may have been metaphorical, or pure invention. The other type of story talks about machines — mechanical devices for flying. The latter type implies to the reader some lost technological knowledge. In Plato, there is a more complex mixture of story types — so that it is unclear what, exactly, the reader is encountering. Numerous elements in the account given in the *Timaeus* and the *Critias* involve historical structuring of the story and geographical placement

117 Herodotus, *Histories* 4.185, trans. Grene, 349.
118 Herodotus, *Histories* 1.203, trans. Grene, 125.

of the setting. But in the *Critias* account in particular, there are many additional elements of the story — such as the long and detailed description of Atlantis — which are not so simple to understand in terms of such "framing". Those additional aspects of the story do not present themselves as simple fictional elements, either. The *Critias* begins with a reiteration of the figure of nine thousand years. Critias then says that is he going to describe the war between Athens and Atlantis, something that is treated only very briefly in the *Timaeus*.

In fact, the *Critias* does not get to the war at all, and is completely consumed with a description of Athens and Atlantis. Moreover, a major emphasis in the *Critias* text is on history and knowledge. As in the *Timaeus*, the *Critias* gives a kind of historiographic preamble to the Atlantis story that is to come. But the preamble also *uses* history to talk about how cultures, in turn, *engage with* history; as one study notes, "Critias uses Solon's story as evidence of his theory about the gradual development of historical inquiry in civilizations."[119] Again, later in the text Critias puts together two ideas about history when he makes the point that once a civilization reaches a certain level of sophistication, it can engage in "Mythology and Inquiry into the Past". Note there is not really a distinction made between the two activities in the sentence. The first is the Greek word *mythología* (μυθολογία) and the second is *anazétesis* (ἀναζήτησις); they have different meanings, with the latter defined as "investigation" or "examination". But the fact that they are placed together here implies to the reader that the two processes work together to preserve and transmit knowledge about the past.[120] Plato is arguing, in a sense, that *mythología* has a role in these processes, and so to the modern reader the message might be that "myth" here cannot necessarily be equated with "fiction".

9 Ancient Views of the Remote Past

The French historian Paul Veyne, in his book *Did the Greeks Believe in Their Myths? An Essay on the Constitutive Imagination*, examines complex question

119 Gill, *Plato: The Atlantis Story*, 55.

120 Note the comments in Brisson, *Plato the Myth Maker*, ed. and trans. Gerard Naddaf (Chicago: University of Chicago Press, 1998), 149 et ff.; also see idem, *How Philosophers Saved Myths: Allegorical Interpretation and Classical Mythology*, trans. Catherine Tihanyi (Chicago: University of Chicago Press, 2004). In this regard, another commentator outlines the changing nature and definition of myth in ancient Greece; see Marcel Detienne, *The Writing of Orpheus: Greek Myth in Cultural Context*, trans. Janet Lloyd (Baltimore: The Johns Hopkins University Press, 2002), 140 et ff., and especially 149–151.

of the nature of history, the concept of "truth", and the quality of myth in the eyes of ancient writers. At one point, he states that the Greeks

> never asked themselves why or how the traditions were handed down. They were simply there, and that was enough for the Greeks. They were not for a moment surprised that reflection of the past were among them. They gathered myths everywhere ... They do not think about it; perceiving only the message, they do not see the medium. Nor are they surprised that the past has left a memory. It is self-evident [to them] that everything has its reflection, just as bodies have shadows. The explanation of myth is the historical reality that it reflects ... They do not wonder how the reflections could traverse so many centuries, or by what means or for what purpose.[121]

But as can be seen in the passage from the *Critias* cited above, and from the introduction to the treatment of Atlantis in the *Timaeus*, this is not quite the case. In the latter, in fact, the Atlantis narrative is prefaced by a discussion of history — Plato's interlocutors do indeed ask "themselves why or how the traditions were handed down". In the *Timaeus*, there is the geographical placement of the story; in the *Critias*, the next element is a brief treatment of geological and other issues. In other words, history, information, and knowledge *are* examined, and questions indeed *are* posed as to how "reflections could traverse so many centuries". While Veyne may be right that in a general sense the Greeks unconsciously accepted that the past "left a memory" in a way that is unfamiliar in our modern society, these Platonic texts display something more subtle. There is both the natural acceptance of a kind of "living history" as Veyne implies, and, as well, a distinctly self-conscious process going on in these texts — a conscious awareness of the passage of time, history, and change.

There are further examples of this awareness when one looks at how ancient writers addressed what has been termed "deep time" and the issue of geological change over time. Plato, as the passages cited above show, was certainly aware of how landmasses evolve over time, with erosion, subsidence, and so on. In the current age, even educated people still sometimes look at our Earth as a relatively static place, focusing on the "here and now". Other than the weather, the Earth seems to us to be relatively unchanging, and unless we are geologists or paleontologists, we tend to have a very short-term understanding of our planet. But ancient thinkers pondered the long-term history of

121 Paul Veyne, *Did the Greeks Believe in Their Myths? An Essay on the Constitutive Imagination*, trans. Paula Wissing (Chicago: University of Chicago Press, 1988), 68.

the Earth, and how the surface of the planet had changed through time. These thinkers were keen observers, gathering evidence to explain their theories, and trying to piece together an understanding of geological and topographic changes from their observations.

Three texts illustrate this; the first is a brief passage describing the ideas of Xenophanes of Colophon (ca. 580–478 BCE). The passage comes from the early Christian author Hippolytus, in his *Refutation of All Heresies*:

> Xenophanes thinks that a mixture of the earth with the sea is taking place, and that the earth in the course of time is being dissolved by the moisture; he states that he the following proofs: in the interior parts of the land and in mountains are found shells, and he says that in the stone-quarries at Syracuse were found imprints of a fish and of sea-weed, and in Paros the imprint of an anchovy in the depth of the stone, and in Melite … flat outlines of all and sundry sea-creatures. And he says that this came about when everything was "mudded" long ago, and that the imprint was fixed dry in the mud; all mankind is destroyed whenever the earth is brought down into the sea and has become mud, and then begins the process of birth all over again.[122]

The second passage is from the *Geography* of Strabo. Strabo quotes Eratosthenes — a Greek thinker who flourished some two hundred years prior to Strabo himself — as well as some other earlier Greek writers:

> Eratosthenes says further that this question in particular has presented a problem: how does it come about that large quantities of mussel-shells, oyster-shells, scallop-shells, and also salt-marshes are found in many places in the interior at a distance of two thousand or three thousand *stadia* from the sea — for instance (to quote Eratosthenes) in the neighborhood of the temple of Ammon and along the road, three thousand *stadia* in length, that leads to it? At that place, he says, there is a large deposit of oyster-shells, and many beds of salt are still to be found there … [H]e praises the opinion of Xanthus, who says that … he himself had often seen, in many places, stones in the shape of a bivalve, shells of the pecten order, impressions of scallop-shells, and a salt-marsh, and therefore was persuaded that these plains were once sea … Now one may admit that a great part of the continents was once covered by water for certain

122 George Kish, ed., *A Source Book in Geography* (Cambridge, MA: Harvard University Press, 1978), 19.

> periods and was then left bare again; and in the same way one may admit also that the whole surface of the earth now submerged is uneven, at the bottom of the sea, just as we might admit, of course, that the part of the earth above water, on which we live, is subject to all the changes mentioned by Eratosthenes himself.[123]

The third passage is from the *Histories* of the Herodotus:

> There is in Arabia, not far from Egypt, a gulf of the sea entering in from the sea called Red; its length and narrowness are as I shall show. For length, if one begins a voyage from its inner end, to sail right through into the broad sea is a matter of forty days for a boat that is rowed. In breadth, at its broadest, the gulf is only a half day's voyage. It has floodtide and ebb every day. I think that once on a time Egypt was just such another gulf; there was one gulf running from the northern sea toward Ethiopia, the other, the Arabian, of which I shall speak, bearing from the south toward Syria; their ends bored into the land near to one another but left a small strip of ground in between. If the Nile should now turn its stream into the Arabian Gulf, what would hinder it from being silted up inside of twenty thousand years? For myself, I could well believe that it would do so within ten thousand. How, then, in the huge lapse of time before my birth, would a gulf not be silted up — a gulf even much larger than this one — when the river concerned was so vast and so hard-working?
>
> So I believe those who say these things about Egypt and am myself convinced that it is so; for I have seen that Egypt projects into the sea beyond the neighboring land, and that seashells show up on the mountains, and that brine-salt comes to the surface, so that even the pyramids are corroded with it, and the only sand mountain in Egypt is above Memphis; and besides, in soil Egypt is not at all like those countries that march on its borders, neither like Arabia nor like Libya nor yet Syria (for Syrians live along the seacoast of Arabia); Egypt is a land of black soil, crumbly, so that it must be swampland and silt from Ethiopia, carried down here by the river. We know that Libya is a redder earth, and a trifle sandy, and Arabia and Syria lighter clay, with stones underneath.[124]

Note that Herodotus is not afraid to consider times spans of ten thousand and twenty thousand years in his discussion of the history of geological change.

---

123 Ibid., 89–91.

124 Herodotus, *Histories* 2.11–12, trans. Grene, 135–136.

Also note that Plato displays a similar understanding of geological change in the *Critias*. His narrator discusses regional soil erosion in relation to the nine thousand year time scale the story presents:

> Many and great were the floods that occurred in the space of nine thousand years — for this is the number of years between that time and the present — and during this succession of natural disasters the soil was washed down from the high places. It did not form any considerable alluvial deposits, as in other regions, but it disappeared into the deep, as in flood after flood it was continuously washed into the sea from all sides.[125]

The narrator in the *Critias* also makes observations as to ecological change over time:

> What actually remains is like our small and barren islands, and, compared to the land it once was, Attica of today is like the skeleton revealed by a wasting disease, once all the rich topsoil has been eroded and only the thin body of the land remains. But in that age our land was undiminished and had high hills with soil upon them; what we now call the Rocky Barrens were covered with deep rich soil. And in the mountains there were dense forests of which there still survives clear evidence. Some of our mountains can now grow just barely enough for bees, but it was not so long ago that [lofty trees grew there]. There can still be found intact rafters cut from trees that were felled and brought down to be used for the greatest building projects. And there were many trees that were cultivated for their fruit and they provided limitless fodder for flocks of sheep and goats.[126]

Gill has commented that Plato's ecological aside indicates his "concern with providing proof for his theories", and that this concern "is a sign of his growing interest in the techniques of pre-historical inquiry". However, Gill adds that these ecological "phenomena" that Plato includes are "also determined by the requirements of his politico-philosophical fable".[127] In other words, Gill believes that these details do necessarily imply that Plato's tale has any veracity.

125 Plato, *Critias* 111a–111b, trans. Clay, in Cooper and Hutchinson, 1297.

126 Plato, *Critias* 111b–111c, trans. Clay, in Cooper and Hutchinson, 1297.

127 See Gill, *Plato: The Atlantis Story*, 56–57.

Similarly, the classicist Diskin Clay, cited earlier, also sees the details that Plato provides as further evidence that the Atlantis account is a "Platonic fiction". Clay argues that the

> diagnostic test in this anatomy of Platonic fiction presents itself the elaborate guarantees and pedigrees Critias gives to justify his claim that he is telling the truth. These guarantees are offered within the complex narrative structure of the *Timaeus* Prologue ... and they give a complex stemma for how an event which happened some 9,000 years before Solon's visit to the temple of Neith in Egypt reached the younger Critias and his small audience — and ultimately how it reaches Plato's readers. Let us call this feature of Platonic invention "The Genealogy of Knowledge."[128]

However, there is no *necessity* in adding such technical details to a mere fable, and one could even argue that they would have been a distraction to an ancient reader. Moreover, it is not clear that such a reader of this tale would be convinced of its truthfulness by the inclusion of these details. The problem here is more of a functional one — the details and the historical framework in both the *Timaeus* and the *Critias* may serve some other purpose. Indeed the Atlantis story as a whole — given that there are the "guarantees and pedigrees" surrounding the account — perhaps can be taken in a way that is not pure fable at all. This does not mean that Plato is writing pure historical fact, but rather that the tale encodes something of substance in what comes across to the modern reader as a peculiar synthesis of history, geological and topographic detail, and mythical structure. This idea is examined in the conclusion to this chapter, with a note that Gill himself provides some advocacy for a new reading of the Atlantis story.

Following the discussion of prehistoric Athens in the *Critias*, the narrator turns to a description of Atlantis. However, as noted earlier, before that description there is a brief transition, where Critias makes a comment about names:

> I must explain one small point before I enter into my history so that you will not be astonished as you hear Greek names frequently used for people who are not Greek. You will now learn the origins of these names. Solon, when he was contemplating his own poetic version of this legend and was inquiring into the meaning of these names, discovered that his Egyptian sources had been the first to record them, once they had translated their meaning into their own language. He, in his turn, recovered

128 Clay, "Plato's Atlantis: The Anatomy of a Fiction", 10.

> the meaning of each of these names and recorded it as he translated them into Greek. These very manuscripts were in the possession of my grandfather and they now remain in my possession. When I was a boy, I studied them carefully. Consequently, do not be astonished if you hear names that sound like Greek names; you now know their explanation.[129]

One interesting aspect of this passage is the explicit reference to written records. In the *Timaeus*, particularly at the beginning of the story, there is considerable emphasis on the transmission of the account orally. Plato's purpose in including this reference to written records may be to explain the transmission of names from such remote antiquity. Gill comments that this discussion of a written record "seems to be introduced simply to explain how Critias can present, accurately and in Greek form, so many names from barbarous pre-history."[130] However, why would the story here require such a rationalization? Again, would the original readers of such a "fable" require what a modern a reader perceives as a rather awkward interjection?

The central story in the *Critias* begins at this point, although it is not really a story in the sense of a narrative, but rather a description of Atlantis itself. The description is sufficiently detailed that one is able to draw a diagrammatic representation. Critias actually begins with a discussion of the divine origins of Atlantis, with Poseidon and a human companion. The description of the land in many ways sounds idyllic, with Poseidon having "produced all varieties of crops".[131] The actual configuration is slightly ambiguous, but again it is sufficiently articulated to allow a modern reconstruction of the basic design. The account at this point also begins to include a puzzling series of numerological elements, which are discussed further below. Among these elements is the comment that Poseidon is said to have had five pairs of twin males, thus dividing up Atlantis into ten portions.

The circular design of the capital of Atlantis may have some significance, but Plato does not provide any clear explanation. Gill conjectures as follows concerning its relative complexity:

> In primaeval Athens, which is characterised by unity and stability, the city-plan is adapted to the natural resources available … But Poseidon and the Atlantean kings superimpose on the natural landscape an

129 Plato, *Critias* 113a–113b, trans. Clay, in Cooper and Hutchinson, 1298–1299.

130 Gill, *Plato: The Atlantis Story*, 59.

131 Plato, *Critias* 113e, trans. Clay, in Cooper and Hutchinson, 1299.

> architectural (and political) superstructure of lavish variety and mathematical complexity ... And this construction eventually collapses (in the sense that its moral and political cohesion falls apart), unable to bear the "weight" of the materialist greed it has encouraged in the Atlanteans.[132]

However, the story in the *Critias* does not draw any explicit connection between the architectural configuration of the city as a whole and its demise. Moreover, again there is the question of specificity: if the "moral" here is that over-building and excessive design leads to moral degradation and collapse, why spend so much time elaborating details? Why not just recount that the capital city of Atlantis was excessive and lavish in its design? If Gill is correct, it may be that the ancient reader required such detail, and that the circular design of Atlantis would have had a symbolism and resonance in terms of meaning that is lost to the modern reader.

The description in the *Critias* also connects back to the geographical details given in the *Timaeus*, with another reference concerning the geographical relationship between Atlantis and the Greek world. Critias notes of the Atlanteans that they

> were the rulers of many other islands in the Atlantic and, as I have said, they even extended their rule into the Mediterranean as near to us as Etruria and Egypt.[133]

This statement neatly ties together the account of Atlantis and its very ancient history to the known lands of Mediterranean world. However, once more it is uncertain whether or not such a statement is simply a ploy by Plato to add a degree of verisimilitude.

10 Numbers and Technical Detail in the Story of Atlantis

Further detail in the Atlantis narrative is evidenced by the fact that the *Critias* includes a detailed discussion concerning the design of Atlantis. In this discussion, the text includes an odd specificity about numbers:

---

132 Gill, *Plato: The Atlantis Story*, 60.

133 Plato, *Critias* 114c, trans. Clay, in Cooper and Hutchinson, 1300; also see *Timaeus* 25a–25b, trans. Zeyl, in ibid., 1232.

- "[Poseidon] sired five pairs of twin sons and he raised them to manhood."[134]
- "[Poseidon] divided the entire island of Atlantis into ten districts."[135]
- "And starting at the sea they excavated a canal three plethra in width, one hundred feet in depth, and fifty stades in length up to the outermost sea ring."[136]
- "The largest of the water rings into which the passage from the sea had been excavated was three stades in width and the next land ring was equal to it. Of the next rings of water and land, the ring of water was two stades wide and, as in the first case, the land ring was equal to it as well. And, finally, the ring of water running around the island in the middle was a stade wide."[137]
- "The island where the palace was located had a diameter of five stades … They threw up an unbroken stone circuit wall around this island, and they also walled the land rings, and the bridge, which was a plethron wide."[138]
- "The temple of Posidon was in this area. It was one stade long, three plethra wide, and of a height that appeared to be proportional to its length and width, but it had something barbaric about its appearance."[139]
- "They placed gold statues within the temple. There was a statue of Posidon standing in a chariot with a team of six winged horses. This statue was so tall that his head touched the rafter of the temple roof; there were a hundred Nereids riding dolphins and arranged in a circle about him, for men of that age thought that the Nereids were a hundred in number; and there were many other statues inside which were the offerings of private individuals."[140]
- "Now, once you had crossed over the three rings of water, you would come to a circuit wall that began at the sea and surrounded the greatest of the land rings on all sides at a uniform distance of fifty stades from the greatest land ring and its harbor."[141]

These specific measurements continue as the narrator moves from his description of the city and the palace to the rest of Atlantis:

- "The entire plain that surrounded the capital was itself surrounded by a ring of mountains that sloped down as far as the sea. The plain was smooth and level and entirely rectangular. On its long sides it extended for three

134 Plato, *Critias* 113e, trans. Clay, in Cooper and Hutchinson, 1299.
135 Plato, *Critias* 113e, trans. Clay, in Cooper and Hutchinson, 1299.
136 Plato, *Critias* 115d, trans. Clay, in Cooper and Hutchinson, 1301.
137 Plato, *Critias* 115e, trans. Clay, in Cooper and Hutchinson, 1301.
138 Plato, *Critias* 116a, trans. Clay, in Cooper and Hutchinson, 1301.
139 Plato, *Critias* 116c–116d, trans. Clay, in Cooper and Hutchinson, 1302.
140 Plato, *Critias* 116d–116e, trans. Clay, in Cooper and Hutchinson, 1302.
141 Plato, *Critias* 117d–117e, trans. Clay, in Cooper and Hutchinson, 1303.

thousand stades and, as measured from the sea, it was over two thousand stades across."[142]

- "The Great Canal was excavated to the depth of a plethron, it measured a stade wide along its entire length, and as it framed the entire plain it came to a total length of ten thousand stades."[143]
- "Towards the interior, canals were cut in straight lines from the city over the plain a hundred feet broad at most and these emptied their waters into the Great Canal facing the sea. They were spaced at an interval of a hundred stades."[144]

The text also has numerical details concerning military organization:

> Now, as for the numbers of the men of the plain who were fit to serve in the army: each military district was assigned to contribute one commander. The area of each district was as much as a hundred stades. The total of these districts came to sixty thousand ... [A]ll men fit for military service were assigned to one of the sixty thousand military districts and they served under the commander of each district. In times of war each commander was assigned to have in readiness a sixth part of the complement of a war chariot as a contribution to a force of ten thousand chariots; and in addition, two horses and two riders, a pair of horses without a chariot, with its complement of two riders, a runner, a rider who could fight on foot armed with a small shield, and serving as a charioteer a rider who could mount either horse, two hoplites, two archers, and two sling men; three light armed soldiers with stones and three with javelins. He also had to contribute four sailors to the crews manning twelve thousand ships. These were the principles for raising an army in the royal city. The formulas varied in the nine other cities, and it would take a long time to describe them.[145]

The final use of number in the text is the one that makes the most explicit reference to the idea of number itself:

> The original ordering of powers and honors in Atlantis was as follows. Within his own patrimony and in his own city, each of the ten kings held power over the inhabitants and over most of the laws, and he could

142 Plato, *Critias* 118a–118b, trans. Clay, in Cooper and Hutchinson, 1303.
143 Plato, *Critias* 118c–118d, trans. Clay, in Cooper and Hutchinson, 1303–1304.
144 Plato, *Critias* 118d, trans. Clay, in Cooper and Hutchinson, 1304.
145 Plato, *Critias* 119a–119b, trans. Clay, in Cooper and Hutchinson, 1304.

> punish or put to death whomever he wished. But, as for their common empire and federation, the kings were regulated by the laws of Posidon as these had been passed down by tradition and according to an inscription which the first kings had cut on a stele of oreichalkos. This inscription was placed in the middle of the island in the sanctuary of Posidon. Here in every fifth or sixth year, and in alternating sequence, it was their custom to gather. To both the even and to the odd they accorded an equal share.[146]

Given these frequent mentions of numbers in the text, it is worth asking why the Plato includes such numerical detail. In fact, Plato employs numbers in several of his works.[147] The analysis at the beginning of this chapter suggested that Plato's Atlantis story in the *Timaeus* and the *Critias* does not follow what we might consider pure fabrication in the modern of sense of a simple, fanciful tale. The historical introduction, the details about geography, and the explicit references to the transmission of knowledge, all give the Atlantis account a distinct timbre. This has led to great interest in the narrative, but also to to the extensive speculation that the story has suffered. The numbers here, in turn, provide what might be termed "technical verisimilitude". However, numbers — particularly numbers such as three — play a role in many folktales and legends, where the purpose is symbolic rather than factual, even if the precise symbolism is no longer apparent to the modern reader. In some cases, the numbers given may be there simply to communicate a sense of grandness; the historians Pierre Lévêque and Pierre Vidal-Naquet comment that the figure of "sixty thousand" in the description above was "dû sans doute au désir de montrer l'extraordinaire richesse de l'île".[148]

But does Plato have any other purpose in including numbers in this discussion of Atlantis? At one point in the text, the numbers seem to be used

146 Plato, *Critias* 119c–119d, trans. Clay, in Cooper and Hutchinson, 1304–1305.

147 There are several investigations into Plato's use of number in his writings; see, for example, Robert S. Brumbaugh, *Plato's Mathematical Imagination: The Mathematical Passages in the Dialogues and their Interpretation* (Bloomington: Indiana University Press, 1954), and Ernest G. McClain, "A New Look at Plato's Timaeus", *Journal of Musicological Research* 1.4 (October–December 1975): 341–360. Also see the broader study in Irving Elgar Miller, *The Significance of the Mathematical Element in the Philosophy of Plato* (Chicago, University of Chicago Press, 1904), and in Paul W. Pritchard, *Plato's Philosophy of Mathematics* (Sankt Augustin: Academia Verlag, 1995).

148 Pierre Lévêque and Pierre Vidal-Naquet, *Clisthène l'Athénien: essai sur la représentation de l'espace et du temps dans la pensée politique grecque, de la fin du VIe siècle à la mort de Platon* (Paris: Macula, 1964), 135.

to support the realism of the account, in the same way as the historical, geographical, and other elements were utilized:

> As described, its depth and width and length provoke disbelief, since it was the work of human hands and so vast when compared to the other building projects. Nevertheless, I must repeat precisely what we heard then. The Great Canal was excavated to the depth of a plethron, it measured a stade wide along its entire length, and as it framed the entire plain it came to a total length of ten thousand stades.[149]

"Nevertheless, I must repeat precisely what we heard" — this aside by the narrator highlights the nature of the story as being one that is being reported or transmitted. This may be a kind of "false framing", but even then, its full purpose is unclear. The mere use of numbers here indicates that Plato was trying to do something — again, perhaps that was to make this passage either resonate with a certain verisimilitude or serve as a historical reference of some kind.

Commentators have pointed out that in the discussion of prehistoric Athens in the *Critias* there is only one numerical detail, while "figures abound in the description" of Atlantis.[150] In providing measurements for this "Great Canal", for example, Plato may have been trying to provide his readers as vivid and precise a picture as possible of the supposed technological capacity of the Atlanteans. That impression of such technical prowess is reinforced by the mention of the walls, towers, bridges, and ring structures of Atlantis, as well as the use of bronze and tin, along with *oreichalkos*, a metal whose identification is uncertain. Plato's text here also mentions that for certain portions of Atlantlean architecture, there were three colors of stone used: white, black, and red. It is interesting that a number of ancient Greek sources speak of four elemental colors, not three: white, black, red, and yellow.[151] Perhaps this technical detail — that Atlantis apparently used three basic colors rather than four — is something that would have held some symbolic meaning to an ancient reader but that is lost to us today. Precision and design are the hallmarks

149 Plato, *Critias* 118c–118d, trans. Clay, in Cooper and Hutchinson, 1304.

150 See p. 40 of Brumbaugh, "Note on the Numbers in Plato's *Critias*", *Classical Philology* 43.1 (January 1948): 40–42, and p. 119 of Thomas G. Rosenmeyer, "The Numbers in Plato's *Critias*: A Reply", *Classical Philology* 44.2 (April 1949): 117–120. Also see the brief discussion on pp. 225–229 of Yves Charbit, "The Platonic City: History and Utopia", trans. Arundhati Virmani, *Population* (*English Edition*) 57.2 (March–April 2002): 207–235.

151 See the discussion in, for example, John Gage, *Color and Culture: Practice and Meaning from Antiquity to Abstraction* (Berkeley: University of California Press, 1999), 29–30.

of a technologically advanced civilization, and that may be Plato's overall point here, a point that he communicates through the use of numbers and other specific descriptors in the passage. Of course, in Plato's own writings there is also the idea that one can tell falsehoods even about something about as precise as numerical detail — as Socrates says in a different context in the *Lesser Hippias*, "there is such a person as a liar about calculation and number".[152]

Scholarly interpretations of the use of numbers in the Atlantis story vary widely. Some studies have put forward the idea that the specific choice of figures was intended to make a statement about the institutions and philosophical mindset of the Atlanteans. The philosopher Robert S. Brumbaugh, in his "Note on the Numbers in Plato's *Critias*", cited above, takes this approach, and concludes that Plato's particular choice of numbers add to his "eloquent disapproval and condemnation of Atlantis as a whole".[153] In particular, Brumbaugh's analysis suggests that the use of five and six together — for example, "Here in every fifth or sixth year, and in alternating sequence, it was their custom to gather" — is a sign of "a lack of rational statesmanship" on the part of the Atlanteans.[154] Brumbaugh also interprets the other numbers in the *Critias* all as multiples of five and six, or related through various ratios to five and six.[155] The historian Vidal-Naquet, cited earlier, limits himself in drawing any deep conclusion about this five/six ratio by stating the following:

> Je renonce à interpreter ici ces faits en detail, me bornant à remarquer que Platon souligne lui-même que l'opposition entre cinque et six est une forme de l'opposition entre l'impair et le pair, c'est-à-dire, selon la συστοιχία pythagoricienne, entre le bien et le mal.[156]

This συστοιχία is what is known in English as the "Table of Opposites".[157] Aristotle attributes this list — including ten pairs of opposites such as odd/even, right/left, straight, curved, and so on — to the Pythagoreans. The conclusion of Vidal-Naquet, then, seems to be that Plato is simply implying alternating

152 Plato, *Lesser Hippias* 367a, trans. Nicholas D. Smith, in Cooper and Hutchinson, 926.

153 Brumbaugh, "Note on the Numbers in Plato's *Critias*", 42.

154 Ibid., 40–41.

155 Ibid., 41.

156 See p. 438, n. 89 of Vidal-Naquet, "Athènes et l'Atlantide. Structure et signification d'un mythe platonicien", *Revue des Études Grecques* 77.366–368 (July–December 1964): 420–444.

157 See Philip Sidney Horky, *Plato and Pythagoreanism* (Oxford: Oxford University Press, 2013), 26 et ff.; also note Aristotle, *Metaphysics: Book Iota*, trans. Laura M. Castelli (Oxford: Clarendon Press, 2018), 118–119, and Aristotle, *Metaphysics*, trans. C.D.C Reeve (Cambridge: Hackett, 2016), 281.

periods of instability (five, an odd number) and stability (six, an even number), by using this Pythagorean metaphysical way of looking at the world.

Recall that in the *Critias*, the narrative notes that for the Atlanteans "in every fifth or sixth year, and in alternating sequence, it was their custom to gather", with the further note that to "both the even and to the odd they accorded an equal share." This would imply a somewhat deviant behavior, given that elsewhere in Plato's writings — for example, the *Philebus* — there are arguments that imply that those things that contain "equal and double, and whatever else puts an end to the conflicts among opposites" are best for creating and maintaining harmony.[158] Nowhere is it argued that odd numbers or unequal amounts — as are apparently found in Atlantis — are good for a society.

A reading of the text as a whole may bear out Brumbaugh's assertion that Plato's particular choice of these numbers, that is, five and six, highlight his "disapproval and condemnation of Atlantis as a whole" — after all, Atlantis does indeed perish. However, there are two problems. First of all, Brumbaugh's analysis has been critiqued by another commentator, the classicist Thomas G. Rosenmeyer, who states that the numerical content of the *Critias* can be interpreted in an entirely different way. In particular, in this alternative interpretation, the use of fives and sixes

> serve[s] the same purposes as the roundness and the immense size exhibited by the figures of the armed strength of Atlantis. They combine to raise the island realm to the status of a utopia.[159]

But this assumption, too, is based on a perhaps too simple reading of the *Critias* as "a nostalgic trifle in which Plato tries to recapture the intellectual atmosphere of his childhood".[160] For such a reductionist conclusion, there is simply no evidence.

However, Rosenmeyer makes a critical and potentially more fruitful observation, which is that five and six may be seen as representatives of two different systems of counting: the decimal and duodecimal. No further discussion of this important insight is made in his analysis, except that

> by the peculiarly nonchalant way in which Plato intersperses his description of Atlantis with multiples and quotients of 10 and 12 he means to

158 Plato, *Philebus* 25d–25e, trans. Dorothea Frede, in Cooper and Hutchinson, 413.

159 Rosenmeyer, 120.

160 Ibid., 120.

> carry us back to an older age, when utopian constructs, political or astronomical, were liberally spiced with ideal figures.[161]

But are these simply "ideal figures", or instead specific choices of five and six, which, indeed, reflect two very basic systems of human counting? A counting system based on fives clearly relates to the digits of the human body, while a scheme based on six or twelve may relate to the duodecimal system found in a number of cultures — including ours — for measuring time. One culture that preceded the Greek culture is that of the Babylonians, which had a sexagesimal system, the root of our sixty-minute clock today, while the Egyptians used a decimal system. Five and six provide a crude way to "translate" between these two systems, since $60 = 12 \times 5$ or 6×10. Five and six also represent odd and even numbers, and the Pythagoreans apparently understood these as reflecting male and female.[162] Perhaps, then, the use of numbers in the text reflects Plato's wish to describe Atlantean society through an allusion to other ancient civilizations and cultures, or through a Pythagorean perspective. But these are speculations; for now, one can only conjecture that Plato put these numerological items in the *Critias* for a reason — a reason perhaps lost to modern readers, whose view of mathematics does not include the symbolic or metaphorical uses familiar to the ancient Greeks.

In a discussion of Plato's system for organizing a society as recounted in the *Laws*, the political scientist Ernest Barker pointed out that

> The State of the *Laws* is ... self-contained ... In the size of its population it will be a mean between Athens and Sparta; and Plato fixes the number of its citizens at 5040. The number is in no sense arbitrarily chosen. Plato had always believed in the significance of number; and in the last period of his life the Pythagorean element grew even stronger and Platonic philosophy became still more a philosophy of number. The number is primarily selected because it is capable of being divided by a number of different divisors into a variety of divisions. It will thus be useful in war, as a basis of military divisions: it will be equally useful in peace for the assessment of taxes and the distribution of land or other public property among the citizens. The main divisor which Plato apparently suggests

161 Ibid., 120.

162 See Macrobius, *Commentary on the Dream of Scipio*, trans. William Harris Stahl (New York: Columbia University Press, 1990), 99, n. 1, as well as the comments in Eleanor Webster Bulatkin, *Structural Arithmetic Metaphor in the Oxford "Roland"* (Columbus, OH: Ohio State University Press, 1972), 15.

> is the number 12; and the system which he advocates is thus a duodecimal system.[163]

This preference on the part of Plato for duodecimal system would imply that he views the Atlanteans, perhaps, as somehow incompletely evolved, as they still hold on to a system that uses not just sixes, but also fives.

But what might have been Plato's primary point here? For one, the fact that Atlantis is portrayed as employing mathematical ratios and set numerical quantities at all may echo Plato's belief that "by mathematics men may transcend sense, and rise into the region of pure thought" and that a "State rightly divided on true mathematical principles will accord with the world and its structure, which is a structure based on number."[164] As Plato states in the *Laws*:

> Our grand total [of 5040] is divisible by twelve, and so is the number of persons in a tribe (420) and in each case this subdivision must be regarded as holy, a gift of God, corresponding to the months of the year and the revolution of the universe. This is exactly why every state is guided by innate intuition to give these fractions the sanction of religion.[165]

In the *Laws*, then, the correct form of society is outlined — a form set by number. In the *Timaeus*, in a discussion of the gift of sight, a further explicit connection is made between number, the order of the universe, and order in society:

> [O]ur ability to see the periods of day-and-night, of months and of years, of equinoxes and solstices, has led to the invention of number and has given us the idea of time and opened the path to inquiry into the nature of the universe. These pursuits have given us philosophy, a gift from the gods to the mortal race whose value neither has been nor ever will be surpassed. I'm quite prepared to declare this to be the supreme our eyesight offers us. Why then should we exalt all the lesser good things, which a

163 Ernest Barker, *Greek Political Theory: Plato and His Predecessors* (London: Methuen & Co., 1951), 317. Also note the analysis in Charbit, cited earlier. For Plato's original discussion concerning the number 5040, see Plato, *Laws* 738a–738b, 746d–746e, and 771a–771d, trans. Saunders, in Cooper and Hutchinson, 1419 and 1444–1445.

164 Barker, 318.

165 Plato, *Laws* 771b, trans. Saunders, in Cooper and Hutchinson, 1444. Similarly, in *Epinomis* (considered by scholars to be a work not by Plato, but still comprising a further discussion of the *Laws*), there is the idea that "the greatest thing [is] for a person to receive from ... [God" the gift of numbers", and "if the human race were deprived of number, we would never to come to be intelligent in anything" — see *Epinomis* 977b and 977c, trans. Richard D. McKirahan, in Cooper and Hutchinson, 1621.

> non-philosopher struck blind would "lament and bewail in vain"? Let us rather declare that the cause and purpose of this supreme good is this: the god invented sight and gave it to us so that we might observe the orbits of intelligence in the heavens and apply them to the revolutions of our own understanding. For there is a kinship between them, even though our revolutions are disturbed, whereas the universal orbits are undisturbed. So once we have come to know them and to share in the ability to make correct calculations according to nature, we should stabilize the straying revolutions within ourselves by imitating the completely unstraying revolutions of the god.[166]

The essential point here is that human society can become stable and harmonious if it follows the essential nature of the universe, and that essential nature is numerical. One commentator notes:

> What is distinctive about Plato is his systematic exploitation of the fact that Greek value-concepts like concord, proportion, and order are also central to contemporary mathematics. The fundamental concepts of ethics and aesthetics as well, so that to study mathematics is simultaneously to study, at a very abstract level, the principles of value.[167]

Plato's writings even include the idea that human reproduction might be numerically "ordered", with the consequence of ignoring such an order being no less than civil war.[168]

The Atlanteans certainly seem to have followed the principle of using numbers to order their society, as Plato would recommend. For example, as quoted above, the *Critias* account also includes a description involving numerical configurations of the Atlantean military forces, with "men fit for military service … assigned to one of the sixty thousand military districts", with each commanding officer "assigned to have in readiness a sixth part of the complement of a war chariot as a contribution to a force of ten thousand chariots", and so on.[169] In the *Republic*, there is the argument that even warfare must employ "num-

166 Plato, *Timaeus* 47a–47c, trans. Zeyl, in Cooper and Hutchinson, 1249–1250.

167 See p. 76 of M.F. Burnyeat, "Plato on Why Mathematics is Good for the Soul", *Proceedings of the British Academy* 103 (2000): 1–81.

168 See Plato, *Republic* 546a–547a, trans. Grube, rev. Reeve, in Cooper and Hutchinson, 1158. On the idea of a decaying State as expressed in Plato's *Republic*, and the use of Atlantis to illustrate this in the *Timaeus*, also note pp. 29–30 of Slobodan Dušanić, "Plato's Atlantis", *L'Antiquité Classique* 51 (1982): 25–52.

169 Plato, *Critias* 119a–119b, trans. Clay, in Cooper and Hutchinson, 1304.

ber and calculation" and that counting is a subject that is "compulsory for a warrior" if "he's to understand anything about setting his troops in order".[170] This all in turn may imply that while inhabitants of Atlantis were sophisticated enough to have built their society on number — the right thing to do, as the *Laws* notes — they were still, as noted above, trying to use both an old and new numerical system at the same time.

In his use of historical frameworks, geography, geology, and ecology in the *Critias* to tell the Atlantis story, Plato is doing something more than penning pure fiction. It is possible that the use of number here is another way of talking about history, particularly the prehistory that plays such a significant role in the *Timaeus* and the *Critias*. Recall that the Atlantis story is being recounted to Solon by the Egyptians: Egyptian culture is portrayed as much older — and much more conscious of history and "deep time" — than Greek culture. Plato's idea here might be that the history being presented in the Atlantis narrative is so far in the past that even the numerical systems were different. The references to numerical systems of five and six also may reflect a point noted earlier — the intersection of an oral culture and a written culture. The Greeks employed a decimal system, so Plato may be indicating the system of six was the one used by an older, prehistoric culture. Atlantis, again, then would be seen as a nexus of the old and new, the oral and the written, and the prehistoric and the historic.

11 Atlantis: in Search of an Interpretation

The analysis in this chapter attempts to suggest an underlying complexity in both the *Timaeus* and *Critias* texts in their discussion of Atlantis. The point of all this is to cast doubt not on the narrative itself, but rather on the modern tendency to feel a need to judge the texts categorically as fiction or nonfiction. As noted in the discussions concerning ancient tales of flying machines and "magic mirrors", early texts of the type cited here at times were syncretic in terms of combining the historical, the technical, and the fantastical. The Atlantis narrative may be categorized in this general way, as well.

Generally, writers on Atlantis tend to fall into two camps: those who feel that Plato was putting together a story simply to illustrate philosophical concepts, and those who feel that we are being presented with nothing less than a road map to sunken ruins somewhere in the Atlantic Ocean. An example of the first style of analysis:

170 Plato, *Republic*, 522c–522e, trans. Grube, rev. Reeve, in Cooper and Hutchinson, 1139.

> The lost continent is little more than a plot device. The story is about an ostensible ancient *Athens*. Athens is the protagonist, the hero, and the focus of Plato's tale. Atlantis is the antagonist, the empire gone bad in whose military defeat by Athens the functioning of a perfect society as defined by Socrates can be exemplified.[171]

Another writer notes:

> After all ... the story of Atlantis, despite its mythological framework and despite Plato's indications that it is only a creation of his mind, so fires the imagination of its readers that they are endlessly searching and "discovering" the lost island.[172]

While skepticism is healthy, this is simply a kind of reductionism. If the "lost continent" is simply a plot device, why did Plato choose such a complex one? Indeed, it is so complex that in the *Critias*, it seems to have been abandoned.[173]

The problem with this first style of interpretation, too, is that it rides roughshod over what the texts themselves actually say. There *is* an abundance of detail, a marked degree of geographical specificity, and a heavily emphasized historical framework. This is not simply a casually told tale. Some argue that the detail found in the accounts in both the *Timaeus* and the *Critias* is there simply to flesh out the narrative in some way — to make the "fiction" more engaging. And, as this chapter has pointed out, other writers in antiquity such as Plutarch, Antonius Diogenes, and Lucian, used genuine geographical knowledge to inform their storytelling. But again, that does not mean that because those writers used real geography as their starting point for speculative writing, Plato worked in the same way.

It has been argued that the actual phenomenon of "freely invented narrative fiction" came long after Plato in the Greek world.[174] However, Gill, cited earlier, argues the opposite:

> There is a sense in which Plato's Atlantis story is the earliest example of narrative fiction in Greek literature; which is also to say it is the earliest

171 Feder, *Frauds, Myths, and Mysteries: Science and Pseudoscience in Archaeology*, 2nd ed., (Mountain View, CA: Mayfield Publishing Company, 1995), 145.

172 See p. 340 of Leonardo Tarán, "The Creation Myth in Plato's *Timaeus*", in idem, *Collected Papers: (1962–1999)* (Leiden: Brill, 2001), 303–340.

173 Gill, "The Genre of the Atlantis Story", 304.

174 Ibid., 291, n. 13, citing Ben Edwin Perry, *The Ancient Romances: A Literary-historical Account of Their Origins* (Berkeley: University of California Press, 1967), 72 et ff.

> example in Western literature. This may seem a surprising claim. Plato's story is introduced in the Timaeus as the record of a factual event and as one which is "absolutely true." If the story is conceded, nonetheless, to be an invention, one might suppose that earlier works of literature, such as Homer's epics, have an equal claim to be considered fictional.[175]

Gill gives a very precise definition of fiction, but notes the inherent vagaries of the term: "Fiction, one may say, is a kind of game, which both participants share in a willed pretense, treating what is unreal as real, and what is invented as actual."[176] Referring specifically to the Atlantis material, Gill adds that it:

> can be regarded as a pastiche of history used as a means of self-conscious experiment in fiction. In fact, I think Plato deliberately frames his story in such a way as to invite his readers to play the (still unfamiliar) game of fiction, to share in the willing and conscious acceptance of the false as true.[177]

This analysis indeed may be accurate, but it would be interesting if one could go back in time and do a survey of early readers of Plato's text and see how they perceived the Atlantis narrative.[178] That conjecture in and of itself leads to the question of what such a survey would ask, given that the Greek readers of Plato's period might not have had clear definitions of ideas such as "pastiche", "fiction", or "conscious acceptance". The classicist Alan Cameron, however, has given this question a succinct answer, stating that "it is only in modern times that people have taken the Atlantis story seriously; no one did so in antiquity."[179] But there survive no detailed contemporary comments on Plato's Atlantis narrative, so it is impossible to tell what readers of the time thought. Moreover, readers in antiquity might not have "taken the Atlantis story seriously" in the sense of not accepting it as historical record, but they may have taken it quite seriously in other senses: as a reference to or reflection of various historical events, or as a narrative encoding various philosophical and historical ideas through the symbolism of geography, architecture, number, and so on.

Plato's playing with history versus fabrication may reflect the Greek world's own understanding that in some sense the past — their past — could be

175 Gill, "Plato's Atlantis Story and the Birth of Fiction", 64.
176 Ibid., 65.
177 Ibid., 65.
178 Gill makes some concluding remarks concerning audiences — see ibid., 77.
179 Cameron, *Greek Mythology in the Roman World*, 124, n. 5.

understood only as an essentially ambiguous record. Even with modern scholarship, and with much more powerful tools to investigate history, academics of the present day acknowledge that much of what we take as historical fact is a combination of invention, exclusion, and reconstruction. Moreover, it seems clear that Plato was also using the Atlantis text, in part, to discuss politics — one scholar notes that as for the *Critias*, its "polemical objective is part and parcel of the political struggle that opposed oligarchy and democracy".[180] What the Atlantis text does to modern readers — who are habituated to separate categories of fiction, history, political science, and so on — is to make them want to set it into only one such category, period. There currently exist categories such as "historical fiction", of course, but that is not quite what Plato seems to be up to. Perhaps Greek readers understood the Atlantis story for the genre it is — a genre for which we do not even have a label. Indeed, Plato's "system of references is alien to us".[181] Could we call the genre "packaged myth", perhaps? That would be myth which packages encoded information about history, lessons concerning political systems, commentary on overly sophisticated civilizations, and even references to numerical symbolism.

In contrast with Plato, Lucian and Plutarch in their writing make clear between the difference between their geographical setting or background information and the subsequent parody or morality tale. Plato does this neither in the *Timaeus* nor in the *Critias*. Rather, the narrator places Atlantis in a historical framework with no provisos. The geographical setting of Atlantis is stated in the *Timaeus* in a very matter-of-fact manner. It is then left uncommented upon; in no way is the reader explicitly led to believe that the setting is included as part of some fantastical structure.

Those who believe that Plato's narrative provides readers with sufficient information to pack some pith helmets — or underwater gear — and begin looking for sunken ruins are equally misguided. The text simply does not provide specific enough information for such a venture. However, this caveat applies equally to those who claim that Plato is referring in his account to the Minoan civilization on Crete. These interpreters are simply ignoring what the text says, both in terms of location — the text states clearly that Atlantis is outside the Pillars of Hercules — and in terms of its scale and configuration.

The question, then, is how the Atlantis account "fits" in Plato's thinking. If it is not just a plot device, what is it? If it is not a textual map to ancient ruins, then how should it be read? In one analysis, the narrative concerning

180 Charbit, 227.

181 Ibid., 233.

Atlantis is interpreted as "a distant echo of the old flood myth."[182] According to that model, what is new in Plato's rendering is the focus on the destruction of a civilization by a marked catastrophe — with catastrophes, in turn, being "illustrations of a universal pattern of cyclical time".[183] The result of this model is quite interesting:

> Plato's account of Atlantis cannot be separated from his quasi-Euhemeristic theory of myth which it documents, and, finally, Plato's self-conscious orientation toward the Golden Age reinforces the conclusion that the Atlantis tale is based upon earlier Greek myth. In the face of these determining factors, all modern scholarship which believes Plato is describing a real prehistoric event has fallen into the logical fallacy of recapitulating a false ancient theory of myth.[184]

In other words, modern commentators who believe that Atlantis — or the ruins of Atlantis — will be found through archaeological exploration are misunderstanding how Plato viewed and used myth.

It has been pointed out that the claim in *Timaeus* of veracity for what appears as a mythical story is not that unusual in Plato. Plato employs a kind of affirmation of this kind in the *Georgias*, in the *Meno*, the *Laws*, and the *Republic*.[185] But the fact that Plato prefaces his passages with such claims does not mean that he is simply trying to write some kind of "true-to-life" fiction. Rather, this means that one needs to look at what kind of narrative process he actually is engaging in.[186] One commentator has even mused on the possibility — rather remote, but interesting — that the Atlantis tale as it

182 Fredericks, 98.

183 Ibid., 98.

184 Ibid., 98–99.

185 Feder, *Frauds, Myths, and Mysteries*, 146, citing William Stiebing, *Ancient Astronauts, Cosmic Collisions and Other Popular Theories About Man's Past* (Buffalo, NY: Prometheus Press, 1984), 51–52.

186 For some perspectives on this subject, see Johansen, "Truth, Lies and History in Plato's *Timaeus-Critias*", *Histos* 2 (1998): 192–215, reproduced as Chapter 2, "The Status of the Atlantis Story" in idem, *Plato's Natural Philosophy*, cited earlier. Also see Sarah Broadie, "Theodicy and Pseudo-History in the *Timaeus*", *Oxford Studies in Ancient Philosophy* 21 (2001): 1–28; Kathryn A. Morgan, "Designer History: Plato's Atlantis Story and Fourth-Century Ideology", *The Journal of Hellenic Studies* 118 (1998): 101–118; and Christopher Rowe, "On Plato, Homer, and Archaeology", *Arion*, 3rd series, 5.3 (Winter 1998): 134–144.

appears in Plato is a kind of remnant of story that was passed on through a series of "garbled oral retellings".[187]

As with the Chinese and other sources examined earlier in this book, what is presented here in Plato is a type of textual construction concerning knowledge that does not quite exist in our contemporary paradigm. The Chinese passages about flying machines, for example, come across as textual records and accounts that strike the modern reader as anomalous. Those records and accounts take the form of technical descriptions, or semi-technical descriptions, in texts that are not in themselves technical. In Plato, there are philosophical arguments — but ones that contain elements of history, geography, mathematics, and elements of natural science. Plato's works are no more of a specific genre than the Chinese ones.

It is also fairly clear that the account of Atlantis no more fits into our current conception of human history than do the Chinese "flying vehicle" stories. The nine-thousand year time scale, along with the account of the lost civilization and its disappearance, do not match any kind of historical records that we possess. The few other sources that mention Atlantis all come after Plato, and therefore it must be assumed that they are working directly from his *Timaeus* and *Critias*. But again, this does not mean that Plato, in writing about Atlantis, is merely dealing in invention.

Nor is Plato simply using the Atlantis "story" as a vehicle for talking about philosophy. He had no need to fabricate a complex story — and the Atlantis story is a complex one — to put forward his ideas about an ideal state. These ideas had already been presented elsewhere in Plato's works, without the intricate narrative of an ancient island continent and its structure. Moreover, if Plato had wanted to set up a contrast, a starting point to talk about the Athens of his day, he could have simply employed a description of ancient or prehistoric Athens — there was no need to introduce an entirely new civilization. Finally, if Plato is talking about the concept of an ideal state, then it is far from clear how the Atlantis account really elucidates this concept. There are pieces here and there in the narrative: a highly structured state with a rigid architecture for its capital, an ancient war with Athens, and so on. But the story does not really come together as a coherent piece of philosophical metaphor. Finally, the Atlantis material in the *Timaeus* is presented with an extensive historical preamble, which must serve some purpose other than simple "window dressing" for a piece of fiction. It may have been that for the ancient reader, the story was not intended to convey a single message — it may be that the

187 See p. 15 of Oliver D. Smith, "The Atlantis Story: An Authentic Oral Tradition?", *Shima: The International Journal of Research into Island Cultures* 10.2 (2016), 10–17.

meaning was to be found in all the details and references to history, geography, and even number, which are seen by modern readers almost as distractions. The preamble in particular may have served a function which is obscure to today's readers.

Some early commentators had their own views on the Atlantis narrative.[188] Crantor, a philosopher whose works are now lost, was a student of Xenocrates, around 300 BCE. He was the earliest writer of commentaries on the works of Plato. Crantor's take on the Atlantis account is given by Proclus: "Some say that all this tale about the Atlantines is straightforward narrative, like the first of Plato's interpreters, Crantor."[189] An expert on Plato, the scholar Harold Tarrant, believes that Crantor's actual take on the Platonic account is more ambiguous than it seems.[190] Tarrant also notes that, "we may be confident that Crantor was not necessarily taking the story as history in our sense of the word."[191] Another writer comments that Proclus had a "tendency to accept as literally true stories like the myth of Er in the *Republic*", so that "Crantor's literal interpretation of the Atlantis myth may be nothing but a mistaken inference on Proclus' part."[192] Indeed, Tarrant also argues that, "Proclus actually knew of no Platonist authorities who maintained that the account of Atlantis was historically true."[193]

Strabo also mentions the Atlantis account, saying that the Greek polymath Posidonius (ca. 135–ca. 51 BCE).

> correctly sets down in his work the fact that the earth sometimes rises and undergoes settling processes, and undergoes changes that result

188 See the discussion in Tarrant, *Proclus: Commentary on Plato's Timaeus, Volume 1, Book I*, 60–84. On Proclus's reading of Plato, note Tim Buckley, "A Historical Cycle of Hermeneutics in Proclus' *Platonic Theology*", in Harold Tarrant and Dirk Baltzly, eds., *Reading Plato in Antiquity* (London: Duckworth, 2006), 125–134. Also see John J. Cleary, "Proclus as a Reader of Plato's *Timaeus*", in Tarant and Baltzly, 135–150, and Marije Martijn, "The *eikôs mythos* in Proclus' Commentary on the *Timaeus*", in Tarant and Baltzly, 151–167.

189 Tarrant, *Proclus: Commentary on Plato's Timaeus, Volume 1, Book I*, 168; in this same work, see Tarrant's comments on pp. 63 et ff. concerning Crantor. An extensive analysis of the passage in Proclus cited above is found in Cameron, "Crantor and Posidonius on Atlantis", *The Classical Quarterly*, new series, 33.1 (1983): 81–91; also note Emilie Kutash, *Ten Gifts of the Demiurge: Proclus on Plato's Timaeus* (London: Bloomsbury Academic, 2011), 55.

190 See Tarrant, "Atlantis: Myths Ancient and Modern", *The European Legacy* 12.2 (April 2007): 159–172.

191 Tarrant, *Proclus: Commentary on Plato's Timaeus, Volume 1, Book I*, 75.

192 See p. 618 of Tarán, "Proclus on the Old Academy", in idem, *Collected Papers: (1962–1999)*, 564–622.

193 Tarrant, *Proclus: Commentary on Plato's Timaeus, Volume 1, Book I*, 23; also see the discussion concerning Crantor in ibid., 27–28.

> from earthquakes and the other similar agencies, all of which I too have enumerated above. And on this point he does well to cite the statement of Plato that it is possible that the story about the island of Atlantis is not a fiction. Concerning Atlantis Plato relates that Solon, after having made inquiry of the Egyptian priests, reported that Atlantis did once exist, but disappeared — an island no smaller in size than a continent; and Poseidonius thinks that it is better to put the matter in that way than to say of Atlantis: "Its inventor caused it to disappear, just as did the Poet the wall of the Achaeans." And Poseidonius also conjectures that the migration of the Cimbrians and their kinsfolk from their native country occurred as the result of an inundation of the sea that came on all of a sudden.[194]

Note Strabo's comment: "And on this point [Poseidonius] does well to cite the statement of Plato that it is possible that the story about the island of Atlantis is not a fiction." Strabo here may be referring to a particular passage in the *Timaeus*; in that dialogue, this issue of the apparent truthful or false nature of the story is raised when Socrates addresses Critias, saying:

> Well, Critias, what other speech [besides that concerning Atlantis] could we possibility prefer to this one? We're in the midst of celebrating the festival of the goddess, and this speech really fits the occasion. So it couldn't be more appropriate. And of course the fact that it's no made-up story but a true account is no small matter. How and where shall we find others to celebrate if we let these men go? We've no choice. Go on with your speech, then, and good luck! It's my turn now to sit back and listen to your speeches that pay back mine of yesterday.[195]

Also looking at Strabo's commentary, note his inclusion of this line: "Its inventor caused it to disappear, just as did the Poet the wall of the Achaeans." With this citation, Strabo is implying that the storyteller made things easier for himself by sinking Atlantis, in the same way that Homer did when the latter, in *The Iliad*, has "Poseidon and Apollo take counsel to sweep away the wall, bringing against it the might of all the rivers that flow forth from the mountains

194 Strabo, *Geography* 2.3.6, trans. Jones, 1.391–393.

195 Plato, *Timaeus* 26e–27a, trans. Zeyl, in Cooper and Hutchinson, 1233–1234. For an interesting commentary on this passage, see E. Douka Kabitoglou, *Plato and the English Romantics: Dialogoi* (London: Routledge, 1990), 215–217.

of Ida to the sea".[196] Overall, Strabo's commentary does not seem to suggest that Posidonius believed Plato's account to be a genuine historical record. But Strabo also does not suggest that Posidonius took it as pure fiction either, since the mention of Atlantis is put in the context of Posidonius' conjectures on the history of the Cimbrians, a northern European people, and his comments on earthquakes — that is, less speculative subjects.[197]

Proclus, in a discussion of the soul, talks about the use of mathematical forms, such as the line, to represent abstract ideas. At the conclusion of that discussion, he states that

> doubtless Plato secretly used mathematical terminology like a disguise for the truth of things, just as the [Orphic] Theologians used myths and the Pythagoreans used symbols.[198]

In turn, when Proclus examines the Atlantis story, he does so by "regarding its account as a symbolic presentation of the truth of things, on the part of the Egyptian priests, much as Pythagorean philosophy interprets nature through number and shapes".[199] Proclus himself writes that "the myth of Atlantis" is an example "of the study of the cosmos through images."[200]

Indeed, the use of numbers in the Platonic description of Atlantis may have reinforced for Proclus this idea of the tale as a set of symbols or images. It is also clear that Proclus is interested in an important idea in Plato's account, "the preservation of what is permanent within the compass of priests, temples and the recorded history connected with them."[201] "What is permanent" refers to "permanent knowledge that survives the periods of destruction", and one modern study notes that Proclus must have understood his own place and time in Hellenic history as part of this continuum, and one that similarly was under threat of destruction.[202]

---

196 See Homer, *The Iliad*, 12.1–33, in *The Iliad*, 2 vols., trans. A.T. Murray (London: Heineman, 1965), 1.545–1.547.

197 On this issue, see Cameron, "Crantor and Posidonius on Atlantis", 89–90; his comments imply that Plato himself actually modeled his own story — which Cameron believes is fictional — on prior historical records of floods and inundations of coastal cities.

198 Baltzly, trans., *Proclus: Commentary on Plato's Timaeus, Volume 4, Book 3, Part II: Proclus on the World Soul* (Cambridge: Cambridge University Press, 2009), 231; also see Marije Martijin, *Proclus on Nature: Philosophy of Nature and Its Methods in Proclus' Commentary on Plato's* Timaeus (Leiden: Brill, 2010), 197.

199 Kutash, 58.

200 Tarrant, *Proclus: Commentary on Plato's Timaeus, Volume 1, Book I*, 95.

201 Kutash, 58.

202 Ibid., 58.

What *was* Plato writing about when he put together the Atlantis account? As Gill puts it:

> The general reader usually assumes it is the true story it is said to be, while most classical scholars, as readily, take it as the invented myth it is explicitly denied to be. There have been some attempts to describe its character as intermediate between these two alternatives ([i.e.,] a likely story), or as a synthesis of them (a myth written in the form of history).[203]

Saying that Plato was simply carrying out some combination of genres is convenient, but it sidesteps the question of what Plato himself felt he was writing. It is clear that Plato was interested in communicating knowledge — the dialogues set forth all kinds of ideas about politics, metaphysics, and ethics. Indeed, the dialogue format allowed him to do that in a very precise yet comprehensive way, as Plato could choose what a particular dialogue or section of dialogue might contain. Different perspectives — even contradictory perspectives — could be presented readily through the format of different speakers.

In the beginning of the account in the *Timaeus*, Plato outlines what he is doing fairly specifically. Socrates has asked for a story about the ideal state; Critias responds with alacrity, and it appears to the reader that the account Critias gives is not something he is inventing on the spot. He is dealing with *alétheia* (ἀλήθεια) — truth.[204] In this introductory part of the Atlantis account in the *Timaeus*, however, there is also ambiguity. But that is not surprising, in that one should not believe that Plato can be read like a history textbook — textbooks, of course, having their own ambiguities. The ambiguity seems intentional, almost frustrating the reader into wanting to ask Plato to just *tell* us whether he is talking about history or invention.

But if the ambiguity is intentional, then perhaps it is Plato's way of telling us that this is the nature of history, in that it takes — as he says elsewhere — the form of myth.[205] In taking that form, history is ambiguous, but nonetheless contains truth. As a modern commentator has noted, Critias, before he launches into the Atlantis account in the *Critias*, invokes not the Muses — who inspire invention — but the *mother* of the Muses, *Mnemosyne*, or Memory:

---

203 Gill, "The Genre of the Atlantis Story", 287.

204 Ibid., 288.

205 For a different view — that in Plato myth is distinguishable from "true discourse about the past" — see Brisson, 22–23; also see Naddaf, "The Atlantis Myth: An Introduction to Plato's Later Philosophy of History", *Phoenix* 48.3 (Autumn 1994): 189–209.

> I would specially invoke Mnemosyne; for all the important part of my discourse is dependent on her favour, and if I can recollect and recite enough of what was said by the priests and brought hither by Solon, I doubt not that I shall satisfy the requirements of this theatre. And now, making no more excuses, I will proceed.

In other words, Critias assumes "the persona of the historian to begin the actual story".[206] Indeed, a study on the *Critias* has noted:

> The function of the Muses and their mother, Mnemosyne, was to inspire humans with a knowledge of the past that would be unavailable to them otherwise.[207]

So, there seems to be history here in terms of a record of events, while there is also the ambiguous nature of the Atlantis account that transcends the task of the "persona of the historian". The text itself allows for multiple readings: "factual report", "philosophical myth-making", and "free invention".[208] These readings can be syncretized in a way that nonetheless gives a harmonious understanding, although one alien to modern readers.

As we have shown above, the Atlantis narrative provides sufficient detail to be read in some way as a "factual report". Despite the one commentator's note, outlined earlier, that Plato uses claims of truth in a number of his works, another points out that outside of this Atlantis discussion, "Plato introduces no other illustrative narrative into his dialogues with such affirmations of its factual truth".[209] In addition, the historical, geographical, and other details make the story resonate with other early historical works, such as those of Herodotus and Strabo.[210]

In allowing multiple readings, Plato builds a model of how information transmission itself works. For Plato, then, the way information is passed on — the chain of recounting from the Egyptian priests to Solon to Dropides, the inherent ambiguity of the account, the invocation to Memory — *all* comprises history. One scholar has talked about this integration of fact, philosophy, and invention in terms of what he calls "coherence". Commentators on

206 Ibid., 289.

207 Diskin and Purvis, 72, n. 44.

208 Gill, "The Genre of the Atlantis Story", 291.

209 See Feder, *Frauds, Myths, and Mysteries*, 146, cited above, and the contrasting opinion in Gill, "The Genre of the Atlantis Story", 292 and n. 15.

210 Gill, "The Genre of the Atlantis Story", 29 and n. 21.

the Atlantis account in the *Timaeus* have labeled the passages as unconnected or even irrelevant to the rest of the text. But in fact the pieces fit together.[211]

How they fit together presents us with some questions. If one takes the view that the Atlantis account is a "politico-philosophical myth", there is the question of the apparently extraneous detail, and the questions as to why the *Critias* account is unfinished and why Plato chose such a complex structure to make a political point.[212] Indeed, as Gill notes:

> The political significance of the picture is by no means on the surface; and, while I have suggested reasons why this restraint, in itself, served those political purposes, it is still remarkable — in a Platonic text — just how unsymbolic the picture superficially seems. There is an uncomfortable haste about the way Plato introduces a moral dimension into the story in the final paragraph of the *Critias*.[213]

Perhaps, then, the wealth of detail in the *Critias* account is not serving a strictly politico-philosophical purpose, but has some other goal. In fact, if we compare the *Timaeus* account with the *Critias*, on the one hand the latter seems to continue the former, but on the other hand, they seem to be two different kinds of writing. The *Timaeus* account of Atlantis is recounted almost as straight historical narrative, while the account in the *Critias* seems to be a kind of "package", with its detail — architectural and mathematical — waiting to be opened and examined. Concerning the historical nature of the *Timaeus* Atlantis account, Gill notes that Plato seems genuinely interested in history. Noting that the interpretation of the account simply as "politico-philosophical myth" leaves us with unanswered questions, he comments:

> There is another element in the story which this interpretation does not quite explain ... This is the stress on the factual truth of the story, the presentation of the account as if it really were a historical narrative, the

211 See Francisco Rodriquez Adrados, "Coherencia e Incoherencia en la Forma y Contenido del *Timeo*", in Calvo and Brisson, *Interpreting the Timaeus–Critias*, 37–47. At the beginning of his essay (p. 37), Adrados points out the problem of how the Atlantis account "fits" with the rest of the *Timaeus*: "El mito de la Atlántida es también habitualmente considerado irrelevante dentro del diálogo." He goes on to explain how the pieces of the Atlantis account in the *Timaeus* and the *Critias* are part of a larger whole in Plato's philosophical discourse, adding that myth and the descriptive writing which characterizes other parts of these works actually do have what he calls a "coherence" (see ibid., 41).

212 Gill, "The Genre of the Atlantis Story", 299; for other comments on why the *Critias* is unfinished, see Clay and Purvis, 61, n. 21.

213 Gill, "The Genre of the Atlantis Story", 299.

> evidence of a genuine interest in historic and prehistoric investigation. It is true that Critias' claim to present a history (and he makes allusions to one history, at least, that of Herodotus) helps to draw attention to the historical implications of the story.[214]

One can also conjecture about Plato's ultimate plan for the seemingly abandoned *Critias*, and Gill believes that:

> Plato intended eventually to make plain that his story was "true" in a sense ironically divergent from Critias': his story would be an invention, but one which would reshape the history of Athens in such a way as to demonstrate the moral truth which underlay it.[215]

But that is only conjecture, and as Gill himself then notes:

> But this does not quite explain Plato's seemingly authentic interest in prehistory (his account of soil erosion in Attica, *Criti.* 110E ff., does not seem to be motivated only by his political theme) and in the development and discontinuity in civilization (*Ti.* 22B–23C, *Criti.* passim). Nor does it explain why he chooses to make his philosophical myth so full of historical implications.[216]

The accounts taken together, then, provide us with a kind of historical narrative and something else — the "package" mentioned above, where a myth or tale actually contains a broad range of information including history, geography, and so on. To the passage above, Gill adds that the historical "elements seem to point to the growth of an interest in prehistory and history" that is somewhat unique to this part of Plato's works.[217] Plato is working, then, at what seems to be a synthesis of history as a record of events, narratives concerning political philosophy, and an analysis of "the development and discontinuity in civilization". But does that mean that the account of Atlantis is simply *not* history?

That is too reductive a conjecture; rather, Plato is dealing in a different kind of history. Looking at a different section of the *Timaeus*, where cosmology is discussed, the Italian historian of philosophy Giovanni Reale noted the following:

---

214 Ibid., 299.
215 Ibid., 299.
216 Ibid., 299.
217 Ibid., 299.

> If we are to read and understand the *Timaeus*' doctrine of the creation of the world, we must free ourselves also of the other misleading prejudices. In particular, we must rid ourselves of reductive and unhistorical readings of the Platonic *myths*, given that this text presents itself as mythic.[218]

So, Plato's myth is something more: as suggested, it is some kind of history combined with transmitted knowledge or a message about transmitted knowledge. But this presents complexities, as Reale points out:

> Despite recent interest in the issue, this is not the place to take up the related theoretical debate about the value of myth in knowledge-gathering. Nor should we get side-tracked into the complexities of the various uses of the terms "myth" in Plato, especially given that he appears to think of his writings, including even the *Republic*, as myths, in the sense of narratives, as we find him saying explicitly and incontrovertibly not only at the end of the *Phaedrus* but also in the *Republic* itself.[219]

But in the *Timaeus*, the "home" of the Atlantis account, there may be some clarity:

> For the purposes at hand, it is, however, worth stressing that, in the *Timaeus*, "myth" has it own *precisely defined role within the structure of theory*, one that Plato is deeply committed to. Moreover, we should bear in mind that the type of myth that makes up the bulk of the *Timaeus* is what Plato frequently calls in that very dialogue εικως λογος, a probable or plausible or conjectural story.[220]

This is nearer, then, to understanding what Plato is doing, and what kind of text he has put together. The art historian, Ananda Coomaraswamy, writing on his own conception of myth, perhaps brings us closest to what might have been Plato's view:

> [M]yth is a symbol, a representation ('as in a glass darkly') of the reality that underlies all fact but never itself becomes fact ... I can and do believe in the myth far more profoundly than in any historical event which may

218 See p. 151 of Giovanni Reale, "Plato's Doctrine of the Origin of the World, with Special Reference to the *Timaeus*", in Calvo and Brisson, *Interpreting the Timaeus–Critias*, 149–164.

219 Ibid., 151.

220 Ibid., 151–152.

> or may not have taken place ... 'History' is the least convincing level of truth, the myth and the (genuine) fairy-tale the most convincing."[221]

For Coomaraswamy, myth is "a metaphysical statement of a truth to be understood ... the myth was always true, proving of course that the myth was genuine and not the creation of 'fallen' man."[222] Certainly, this echoes Plato: first of all, myth is taken as vitally important, as a bearer of universal truths that underpin reality. Consider Plato's "Allegory of the Cave", for example, where the highest truths are represented by the metaphysical "Forms". These "Forms" are — like myth — far removed from the material and the sensory, but for Plato they are for that very reason also far more representative of the truth behind the world and human existence in that world.

Moreover, historical events, particularly in Plato's times, were not easily verifiable in terms of written records or systematic archaeological investigations (as one would have today), and thus his skepticism concerning both the material and the sensory makes sense from that more pragmatic perspective, as well. Finally, for Plato both myth and apparently simple tales — such as the "Ring of Gyges", examined in the next chapter — held a culture's knowledge, since it was clear that all cultures had these modes of transmission. The myths and tales in various cultures of Plato's time already had long lineages and universal themes, and that thematic universality itself likely indicated to Plato that such narratives embodied profound knowledge.

As moderns, we sometimes view history simply as a list of past events, supported by textual and archaeological evidence. But even our history serves as — and is often taught as — allegory, and the very decision to preserve and transmit historical knowledge is a choice replete with philosophical implications. Moreover, there are pieces of our past that we do not, and perhaps cannot, know. But that does not mean that we discard them when talking about history. The historian of philosophy Giovanni Reale argues that Plato, in the cosmological part of the *Timaeus*, realizes this:

> The cosmos is a perceptible thing, a *being-in-becoming* and, so, there can be no absolute knowledge of it, as there can be of eternal beings. But, in virtue of its very nature, there can be only a sort of *probable* or *conjectural*

221 Ananda K. Coomaraswamy and Rama P. Coomaraswamy, *The Door in the Sky: Coomaraswamy on Myth and Meaning* (Princeton: Princeton University Press, 1997), xiii.

222 Ibid., xiii.

> acquaintance with it, which is expressible by means of myth, by means of a sort of narrative and not through dialectic.[223]

Plato also makes it clear that as far as prehistory is concerned, "there can be no absolute knowledge of it" either. As another commentator puts it, for Plato, "Traditional myth and genealogy will then be a reminder of the inaccessibility of the past", thus serving as an historical tool.[224] But to address this inaccessibility, Plato indeed does construct a "*probable* or *conjectural* acquaintance" with that remote period, using his Atlantis account, with its careful historical preamble, the articulated connection to events nine millennia in the past. Plato's account of Atlantis is emblematic of this "narrative and not ... dialectic" definition of history. The historical preamble, the geographical description, the technical details of the architecture, the geological asides, and the puzzling numerological components together present a profound lesson about advanced civilizations and their fate. It is a lesson, too, about history and the transmission of knowledge — although it is also a lesson that ironically has reached the present day in a form that is no longer completely intelligible.

---

223 Reale, 152.

224 Kathryn A. Morgan, *Myth and Philosophy from the Presocratics to Plato* (Cambridge: Cambridge University Press, 2000), 270.

CHAPTER 5

Rings and Dangerous Powers

1 The Nature of a Folktale

An ancient myth — really a folktale — found in Plato's works suggests the idea of lost technical knowledge and catastrophe in a subtle way. The writer William Irwin Thompson has argued that folktales can serve the function of preserving information.[1] In fact, even folk narratives that appear to be no more than simple morality tales may contain elements that transcend the basic theme of a moral lesson for the story's hearers. The tale of the ring of Gyges has several such elements: it speaks of a particular technology that is magical — a fantastic ring — but the idea is set in the wider context of various technological and cultural themes, including the potential of technical abilities to corrupt a civilization.

The story of Gyges and his magic ring is best known as a morality tale. Indeed, Plato's version of the story, found in the *Republic*, is framed by a discussion of morality and ethical behavior.[2] Gyges obtains a ring that bestows upon him the power of invisibility. He then proceeds to behave in a deceitful and selfish manner. In the Platonic dialogue, Glaucon uses Gyges' behavior as a launching point for an argument concerning the fragility of human character.

But other elements of the story provide intriguing hints as to further possible meanings of the tale. This chapter offers a tentative investigation of those elements and provides a conjectural interpretation. The primary argument here is that Plato has adopted the tale of Gyges to talk about something more than just the question of human moral choices; Plato includes the tale to provide a commentary — if a rather subtle one — on the idea of technological power and its dangers. This is not unlike the Atlantis narrative, in that the setup concerns a society that is technologically sophisticated; in the Atlantis story, such sophistication is represented in terms of architecture and in terms of military organization. In the Gyges story, other elements are used to place the tale in a similar, if implied, context of an advanced civilization.

---

1 William Irwin Thompson, *Imaginary Landscape: Making Worlds of Myth and Science* (New York: St. Martin's Press, 1989), 33.

2 The story is in Plato, *Republic*, 358e–360d, trans. Grube, rev. Reeve, in Cooper and Hutchinson, 999–1001.

 | DOI:10.1163/9789004352728_006

The term "folktale" as used here in this chapter can be defined approximately as an "anonymous narrative transmitted in a popular milieu": that is, the idea here is that Plato picked up a tale that was already in circulation in the wider culture.[3] If a stricter definition is required, then one can say that Plato is using a story that has Märchen-type motifs. The end result is the same, as Plato is presenting a universal theme: a magical ability conferred by a device, one that has serious moral consequences. In turn, this compels the reader to think about the conduct of society as a whole.

The interpretation of the Gyges tale offered here also operates from the idea that almost nothing in a folktale is wasted. In other words, all elements in a folktale have some kind of symbolic meaning or serve some kind of instructive function. The various elements in the tale are not merely descriptive, but hold significance related to the larger theme. In some cases, the meaning or functionality of these elements may become lost over time, but the elements themselves do not disappear. Many decades ago, the anthropologist and folklorist J.L. Fischer noted insightfully that in folktales it "is notable that often the events which appear *least* plausible from a scientific viewpoint are alleged by the local people to be *most* certainly true and *most* important."[4] Given that, in the case of the story of Gyges, this chapter examines not necessarily the "least plausible" events, but those that seem to the modern reader to be the least relevant. Again, it is argued that those minor events and contextual elements convey one of the story's most important messages, and it is argued that a reader at the time of Plato would have understood that.

Folktales express the popular beliefs, customs, and ethics of a particular culture. The culture that devised the Gyges folk tale — and the tale's origins are, in fact, unclear — had a certain worldview.[5] Parts of that worldview can be gleaned from examining the setting and plot of this tale. Indeed, while much of the modern commentary has focused on the actions of Gyges in the story, there has been less examination of the setting and related elements of the story. The analysis will examine the peculiar, cave-like setting of the tale as it appears in Plato's rendering, the body on which Gyges finds the ring, the giant bronze horse, and the nature of the ring itself. It is important to point out that

3 Graham Anderson, *Greek and Roman Folklore: A Handbook* (Westport, CT: Greenword, 2006), 64.

4 See p. 244 of J.L. Fischer, "The Sociopsychological Analysis of Folktales", *Current Anthropology* 4.3 (June 1963): 235–295.

5 There is also some uncertainty and speculation about the name as well; for example, concerning the possible connection between the name "Gyges" and "Gog", see Sverre Bøe, *Gog and Magog: Ezekiel 38–39 as Pre-text for Revelation 19, 17–21 and 20, 7–10* (Tuebingen: Mohr Siebeck, 2001), 91–93.

although the story of Gyges appears in a political context in Plato and an historical context in Herodotus, it "combines and incorporates a number of narrative motifs that have parallels in Greek myth and other folk tale traditions."[6] In short, this tale is reaching back into the deep past, carrying on a folkloric tradition and conveying a cultural lesson to its audience.

2 The Tale of Gyges in Plato's *Republic*

According to historical tradition, Gyges was the founder of the Mermnad dynasty in Lydia, an ancient kingdom in western Anatolia (modern Turkey). He reigned some time around the seventh century BCE.[7] To secure his kingship over the reluctant Lydians, he gave numerous and extravagant gifts to the temple; his authority as monarch was further established by the oracle at Delphi.[8] Later, he waged war with various cities in Asia Minor, including Smyrna, Miletus, and Colophon. However Herodotus notes that, "he did nothing else great in his reign of thirty-eight years".[9] All else aside, the riches of Gyges were said to have been remarkable.[10] Moreover, this historical figure — despite Herodotus' claim of his uneventful reign — soon became part of a folkloric tradition. As early as the period of the poet Archilochus of Paros, cited by Herodotus, there is an account of Gyges' wealth.[11] The particularly interesting

6 Note p. 66 of Ivan M. Cohen, "Herodotus and the Story of Gyges: Traditional Motifs in Historical Narrative", *Fabula* 45.1–2 (March 2004): 55–68; also see the discussion in Stanley Isser, *The Sword of Goliath: David in Heroic Literature* (Leiden: Brill, 2003), 90 et ff.

7 Concerning dates for Gyges, see Anthony J. Spalinger, "The Date of the Death of Gyges and Its Historical Implications", *Journal of the American Oriental Society* 98.4 (October–December 1978): 400–409.

8 Herodotus, *Histories* 1.13–1.14, trans. Grene, 38–39.

9 Herodotus, *Histories* 1.15, trans. Grene, 39.

10 Note Aristotle, *Art of Rhetoric* 3.17, in Aristotle, *Art of Rhetoric*, trans. J.H. Freese (Cambridge, MA: Harvard University Press, 1926), 460–461, and see *The Art of Rhetoric*, trans. Hugh Lawson-Tancred (London: Penguin Books, 1991), 258. Also note Lucian, "The Parasite", 58, in *The Works of Lucian of Samosata*, 4 vols., trans. H.W. Fowler and F.G. Fowler (Oxford: The Clarendon Press, 1905), 3.189.

11 Archilochus, fr. 19 in Douglas E. Gerber, ed, and trans., *Greek Iambic Poetry: From the Seventh to the Fifth Centuries BC* (Cambridge, MA: Harvard University Press, 1999), 92–93; see pp. 344–349 of B.M. Lavelle, "The Apollodoran Date for Archilochus", *Classical Philology* 97.4 (October 2002): 344–351, and Andrea Rotstein, *The Idea of Iambos* (Oxford, Oxford University Press, 2010), 200. Also note p. 6 of A.J. Podlecki, "Archilochus and Apollo", *Phoenix* 28.1, *Studies Presented to Mary E. White on the Occasion of Her Sixty-Fifth Birthday* (Spring 1974): 1–17; Podlecki comments: "If we may trust a comment by a late grammarian, there was considerably more of the Gyges story from Archilochus' pen, for we are told that the *Gygae fabulam optime complexus est*."

story that arose, however, concerns the path of Gyges' ascent to power as the king of Lydia.

Several discussions of Gyges survive, either in the form of longer tales or in passing references. The longer accounts appear in the works of Nicolaus of Damascus, Herodotus, and Plato. The shorter references to Gyges are found in a variety of sources, including Plutarch and Cicero.[12] Plato has one of the longer and more engaging discussions.[13] In a debate in the *Republic*, Plato has Glaucon tell a story about Gyges. In the following passage, Glaucon presents the Sophist view concerning justice — namely that only the weak desire others to behave in a just manner. Moreover, he says, anyone who might be able to behave in an unjust way while still appearing to be just would gladly do so. Glaucon tells the tale of Gyges and his magic ring to illustrate this point. In the tale, the shepherd Gyges obtains the power of invisibility — and so, too, the ability to act unjustly without anyone finding out:

> They say that to do injustice is naturally good and to suffer injustice bad, but that the badness of suffering it so far exceeds the goodness of doing it that those who have done and suffered injustice and tasted both, but who lack the power to do it and avoid suffering it, decide that it is profitable to come to an agreement with each other neither to do injustice nor to suffer it. As a result, they begin to make laws and covenants, and what the law commands they call lawful and just. This, they say, is the origin and essence of justice. It is intermediate between the best and the worst. The best is to do injustice without paying the penalty; the worst is to suffer it without being able to take revenge. Justice is a mean between these two extremes. People value it not as a good but because they are too weak to do injustice with impunity. Someone who has the power to do this, however, and is a true man wouldn't make an agreement with anyone not to

12 For a survey of the various mentions of Gyges, see Kirby Flower Smith, "The Literary Tradition of Gyges and Candaules", *The American Journal of Philology* 41.1 (1920): 1–37. Also note Theodor von Liebenau, "Der Ring des Gyges in der Schweiz", *Schweizer Archiv für Volkskunde* 4 (1900): 220–221, and the brief listing in Marc Shell, *The Economy of Literature* (Baltimore: Johns Hopkins University Press, 1978), 14, n. 12. A short overview of the lineage of the story is in pp. 169–172 of Gabriel Danzig, "Rhetoric and the Ring: Herodotus and Plato on the Story of Gyges as a Politically Expedient Tale", *Greece & Rome*, 2nd series, 55.2 (October 2008): 169–192.

13 For a discussion of the relationship between the story of Gyges in Herodotus and in Plato, see Karl Reinhardt, "Gyges und sein Ring", in Carl Becker, ed., *Vermächtnis der Antike: Gesammelte Essays zur Philosophie und Geschichtsschreibung* (Göttingen: Vandenhoeck and Ruprecht, 1960), 175–183, and note 139–143; also see Adi Ophir, *Plato's Invisible Cities: Discourse and Power in the Republic* (London: Routledge, 1991), 15–17.

do injustice in order not to suffer it. For him that would be madness. This is the nature of justice, according to the argument, Socrates, and these are its natural origins.

We can see most clearly that those who practice justice do it unwillingly and because they lack the power to do injustice, if in our thoughts we grant to a just and an unjust person the freedom to do whatever they like. We can then follow both of them and see where their desires would lead. And we'll catch the just person red-handed traveling the same road as the unjust. The reason for this is the desire to outdo others and get more and more. This is what anyone's nature naturally pursues as good, but nature is forced by law into the perversion of treating fairness with respect.

The freedom I mentioned would be most easily realized if both people had the power they say the ancestor of Gyges of Lydia possessed. The story goes that he was a shepherd in the service of the ruler of Lydia. There was a violent thunderstorm, and an earthquake broke open the ground and created a chasm at the place where he was tending his sheep. Seeing this, he was filled with amazement and went down into it. And there, in addition to many other wonders of which we're told, he saw a hollow bronze horse. There were window-like openings in it, and, peeping in, he saw a corpse, which seemed to be of more than human size, wearing nothing but a gold ring on its finger. He took the ring and came out of the chasm. He wore the ring at the usual monthly meeting that reported to the king on the state of the flocks. And as he was sitting among the others, he happened to turn the setting of the ring towards himself to the inside of his hand. When he did this, he became invisible to those sitting near him, and they went on talking as if he had gone. He wondered at this, and fingering the ring, he turned the setting outwards again and became visible. So he experimented with the ring to test whether it indeed had this power — and it did. If he turned the setting inward, he became invisible; if he turned it outward, he became visible again. When he realized this, he at once arranged to become one of the messengers sent to report to the king. And when he arrived there, he seduced the king's wife, attacked the king with her help, killed him, and took over the kingdom.

Let's suppose, then, that there were two such rings, one worn by a just and the other by an unjust person. Now, no one, it seems, would be so incorruptible that he would stay on the path of justice or stay away from other people's property, when he could take whatever he wanted from

> the market place with impunity, go into people's houses and have sex with anyone he wished, kill or release from prison anyone he wished, and do all the other things that would make him like a god among humans. Rather his actions would be in no way different from those of an unjust person, and both would follow the same path. This, some would say, is a great proof that one is never just willingly but only when compelled to be. No one believes justice to be a good when it is kept private, since, wherever either person thinks he can do injustice with impunity, he does it. Indeed, every man believes that injustice is far more profitable to himself than justice. And any exponent of this argument will say he is right, for someone who didn't want to do injustice, given this sort of opportunity, and who didn't touch other people's property would be thought wretched and stupid by everyone aware of this situation, though, of course, they'd praise him in public, deceiving each other for fear of suffering injustice.[14]

An analysis of this story can invoke William Irwin Thompson's citation, drawing from Dante, of the "four levels of meaning" in a tale: the narrative, the allegorical, the moral, and the anagogical.[15] Plato fits the story of Gyges into Glaucon's argument that men are prevented from doing wrong only through a fear of being caught. The just man as well as the unjust man, the reader is told, will behave in precisely the same way given free license. Only the power of law, says Glaucon, keeps us from behaving immorally. The narrative concerning Gyges is an efficient mode of setting up an allegory about operating outside of the law. The moral piece is the lesson of the tale: that given the chance, we will all behave unjustly. Finally, there is the anagogical aspect — the spiritual interpretation, or the more abstract mythological information embedded in the tale, especially in its conclusion. This final aspect of the tale of Gyges is outlined more completely later in this chapter; for now, it is important to note that the anagogical sense of the tale here includes its wider implication that not only the individual but whole civilizations can become corrupt and perish.

As we have seen in the story above, Glaucon presents the "thought experiment" that addresses the moral matter as follows:

> We can see most clearly that those who practice justice do it unwillingly and because they lack the power to do injustice, if in our thoughts we grant to a just and an unjust person the freedom to do whatever they like. We can then follow both of them and see where their desires would lead.

14 Plato, *Republic*, 358e–360d, trans. Grube, rev. Reeve, in Cooper and Hutchinson, 999–1001.
15 Thompson, *Imaginary Landscape*, 10.

> And we'll catch the just person red-handed traveling the same road as the unjust. The reason for this is the desire to outdo others and get more and more. This is what anyone's nature naturally pursues as good, but nature is forced by law into the perversion of treating fairness with respect.[16]

Glaucon concludes pessimistically in regards to human nature:

> Let's suppose, then, that there were two such rings, one worn by a just and the other by an unjust person. Now, no one, it seems, would be so incorruptible that he would stay on the path of justice or stay away from other people's property, when he could take whatever he wanted from the market place with impunity, go into people's houses and have sex with anyone he wished, kill or release from prison anyone he wished, and do all the other things that would make him like a god among humans.[17]

So much for just behavior: Glaucon believes that individual human beings are inherently weak in terms of moral conduct. Plato, through the dialogue between Socrates and Glaucon, obviously is considering human nature as a whole through use of this individual example, since Plato's larger "project" here is an examination of how one can build an ideal state — a state built of many individuals. But Glaucon's particular arguments concerning weakness and the consequences of giving into the temptations of power clearly have implications not only for Greek society of Plato's time but also human civilization generally. Those implications, as subsequent analysis here will show, are not laid out explicitly in the narrative path of the figure of Gyges himself but rather in the background elements that frame the tale.

3 The Background and Setting of the Tale of Gyges

Before looking at that background, it is useful to note other accounts of Gyges. Herodotus' version of the tale of Gyges contains no magical or fantastic elements, and differs from that found in Plato in terms of various aspects of structure and plot.[18] First, we are told that Candaules was the last of the descendants

16 Plato, *Republic*, 359b–359c, trans. Grube, rev. Reeve, in Cooper and Hutchinson, 1000.

17 Plato, *Republic*, 360b–360c, trans. Grube, rev. Reeve, in Cooper and Hutchinson, 1001.

18 Note pp. 382–383 of Smith, "The Tale of Gyges and the King of Lydia [part 2]", *American Journal of Philology* 23.4 (1902): 361–387; in this essay, Smith examines the various extant forms of the tale, and conjectures as to an original form that the story may have possessed.

of Hercules to rule Sardis.[19] Herodotus then tells us that Candaules believed that his wife was the fairest in the world. In the story, Candaules speaks of her beauty to his bodyguard, Gyges, and to convince the reluctant Gyges of this, arranges for him to see her undressing. Gyges responds to Candaules that he believes this to be very inappropriate, indeed "lawless". Candaules is unconvinced. In the end, his wife sees Gyges spying on her, and figures out what has happened.[20] The next day, the queen — who is unnamed in the story — summons Gyges. She convinces him to murder her husband, marry her, and seize the Lydian throne.

The magic ring has been left out entirely. Perhaps Herodotus sought to make a narrative appear more historically sound, and thus deleted the fantastic elements of the tale. However, is possible to reconcile the two versions of the tale.[21] Plato's account of Gyges' discovery of the ring and his subsequent entry into the king's court — as a messenger rather than the bodyguard of Plato's version — supplies a kind of preface to the text in Herodotus. The story that Glaucon tells in the *Republic* then could be seen as explaining how Gyges, a simple shepherd, first becomes acquainted with the king. After that, the version of Herodotus picks up the story line there, with Gyges now as a bodyguard rather than as a messenger. Recall that Plato gives us just one sentence covering the seduction, murder, and taking of the throne by Gyges:

> And when he arrived there, he seduced the king's wife, attacked the king with her help, killed him, and took over the kingdom.

Herodotus, by contrast, gives us all the intermediate details concerning the queen's ultimatum. Thus the two versions complement each other; combining the two one could perhaps reconstruct the folktale from which Plato and Herodotus may have been working.[22]

The deeper origins of the tale are very unclear. One scholar, Andrew Laird, suggests that Plato himself might have constructed the tale, especially some of

---

19 Sardis was the capital of the ancient kingdom of Lydia, and is now a city in Turkey.

20 The English artist William Etty beautifully — and rather explicitly — captured this scene in his 1830 painting entitled "Candaules, King of Lydia, Shews his Wife by Stealth to Gyges, One of his Ministers, as She Goes to Bed".

21 For a discussion of the differences between Herodotus' account of Gyges and Plato's account, and possible reasons for such differences, see Danzig, cited earlier, as well as James Adam. *The Republic of Plato: Volume 1, Books I–V* (Cambridge: Cambridge University Press, 1905), 126–127.

22 Smith, 383–385.

the more fantastic elements, stating that, "the Herodotean text prompts Plato's fictive invention of the ring."[23] Laird goes on to say:

> Glaucon's narrative also contains a great deal of descriptive detail, apparently superfluous for the ends to which the story is told. Such detail increases in richness and specificity as the story goes on … At the very least, this suggests that the story is very much subject to Glaucon's manipulation and narrative design.[24]

However, if the tale told by Glaucon is really Plato engaging in a fictional construct, there remain a number of questions. Were these other details — the cave setting, the bronze horse, the giant-like stature of the dead body — really included just to produce a rhetorical effect? If so, why those particular details and not others? In fact, one commentator has noted that "Glaucon describes the circumstances in which his hero finds his magic ring, but the significance of many of the details is not obvious." It still seems quite plausible that Plato is drawing from an existing folktale, which may well have had *more* detail in its earlier form, rather than less.[25] One scholar even speaks of "the old popular tale of Gyges which appears to have been current in the times of Herodotus and Plato."[26] The argument here is that the existence of this "descriptive detail" supports the idea that this was an existing folktale, since many folktales include symbolism in the form of such details as the location, objects encountered by the protagonist, and so on. The more fundamental argument, then, is that these background elements found in the version that Plato cites are *not* "superfluous for the ends to which the story is told", but in fact reinforce and expand the moral message of the tale. To understand how this is the case, it is necessary to look at the setting as it is presented in Plato's text, and, again, to accept that each detail is there for a reason.

Three motifs in the story mark it as a fanciful tale. First, there is the fantastic earthquake and the marvels revealed to Gyges within the gaping hole in the earth. Next, there is the magic ring itself, granting the power of invisibility. Finally, there is Gyges' rapid ascent from poverty and obscurity to a position of wealth and power, following his taking of the ring. Magic rings appear quite

23 See pp. 17–19 of Andrew Laird, "Ringing the Changes on Gyges: Philosophy and the Formation of Fiction in Plato's *Republic*", *The Journal of Hellenic Studies* 121 (2001): 12–29.

24 Ibid., 19.

25 See p. 35 of Mary P. Nichols, "Glaucon's Adaptation of the Story of Gyges & Its Implications for Plato's Political Teaching", *Polity* 17.1 (Autumn 1984): 30–39.

26 See p. 263 of Smith, "The Tale of Gyges and the King of Lydia [part 1]", *American Journal of Philology* 23.3 (1902): 261–282.

frequently in stories from antiquity, although a ring rendering its wearer unseen — the case in the tale of Gyges — is less common.[27] Despite the presence of these fantastic elements, it is clear that Plato has placed the story so as to fit the pragmatic philosophical argument that Glaucon articulates. The primary concerns in this passage for Glaucon are with the hypothetical power of the ring and the simple fact that Gyges uses it for immoral purposes. All the evil deeds of Gyges are condensed into one final sentence in the Platonic rendition of the tale.

This Platonic rendition states that as the shepherd entered the chasm in the earth, "in addition to many other wonders … he saw a hollow bronze horse". The "other wonders" are never described. Perhaps an earlier recounting of the tale had more details, and perhaps these also would have explained the significance of the bronze horse. In Plato's text, the primary focus appears to be on the usefulness of the story as a commentary about the morality of human beings. But the other elements of the tale can still be examined — and these elements all play a part in the moral content of the story.

4 The Elements of the Tale: the Cave

Moreover, again, the individual elements in the setting of this tale have implications beyond Glaucon's argument concerning the lack of morals of individual human beings. The first clue is the earthquake that reveals an opening in the earth, an opening into which Gyges descends as if entering into a cave. Not only is there this earthquake in the Platonic rendering of the tale, in fact,

---

27 Concerning the magical power of invisibility, also note p. 13 of James G. Frazer, "Some Folk-Lore of Ferns", *American Fern Journal* 24.1 (January–March, 1934): 10–15; see, too, Arthur Stanley Pease, "Some Aspects of Invisibility", *Harvard Studies in Classical Philology* 53 (1942): 1–36. For a discussion of the many appearances of magic rings in literature, see p. 268, n. 2 of Smith, "The Tale of Gyges and the King of Lydia [part 1]", cited above. Also see the study of magic rings in the chapter entitled "Das märchen der zauberring" in Antti Aarne, *Vergleichende Märchenforschungen* (Helsingfors: Druckerei der Finnischen Literaturgesellschaft, 1908), 3–82. For a more general listing of magic ring tales, note story types D865.1, D1076, D1310.4.1, D1316.4, D1316.4.1, D1317.5, D1317.5.1, D1317.5.2, D1335.5, D1335.5.1, D1342.1, D1344.1, D1361.17, D1368.2, D1372.1, D1376.1.2, D1380.23, D1381.7, D1382.11, D1383.3, D1384.1, D1384.7, D1385.3.1, D1388.0.1, D1389.9.2, D1395.5, D1400.1.2, D1402.0.2.3, D1402.12, D1405.1, D1407.1, D1421.1.6, D1421.1.13, D1426.2, D1431.3, D1456.2, D1456.2.1, D1477.2, D1487.1, D1507.1, D1561.1.2, D1592.1, D1602.17.1, D1612.2.1, D1662.1, and D1662.1.1 in Thompson, *Motif-Index of Folk Literature.*

but also it is prefaced by a storm. Both, naturally, are quite negative omens.[28] An earthquake appears elsewhere in Plato, of course — in the Atlantis tale. In the treatment of Atlantis in the *Timaeus*, there are dreadful earthquakes and floods ending in the disastrous destruction of the Atlantean civilization.[29] In the Gyges story, the earthquake serves as a portent of evil in two ways: as an ominous introduction to the evil acts in which Gyges is going to engage, and more simply as revealing a cave and tomb which would have been better off remaining hidden. In Plato, an earthquake can either cause destruction directly or reveal the necessary ingredients for one's own destruction. The setting of the cave is certainly resonant in its symbolism, both of Plato's other famous allegorical cave, as well as of the underworld.[30] A political scientist writing on the *Republic* has pointed out that Plato uses the "symbolism of depth and descent" in several places in this work.[31] Gyges descent foretells his future immoral actions, since even this initial act is immoral; he "descends into the earth and robs a corpse without apparent fear of divine retribution."[32]

But in the tale of Gyges, the cave is also very much of this Earth. Just as we dig into the ground to find the remains of past civilizations, ancient peoples understood that time and geology had buried the remains of long-dead peoples and their artifacts. A folklorist specializing in the classical period even has argued that some Greek and Roman writers were familiar with fossils and their meaning.[33] Thucydides tells us that the Greeks in the fifth century BCE dug up (although not systematically) material from a sanctuary on the island of

28 For earthquakes as portents of trouble in Greek and Roman sources, see the brief mentions in p. 13 of R.F. Newbold, "Pliny HN 2.199", *Classical Philology* 68.3 (July 1973): 211–213, and p. 862 of Clifford Orwin, "Review Essay on Thucydides", *Political Theory* 28.6 (December 2000): 861–869. Also see Richard Lasch, "Die Ursache und Bedeutung der Erdbeben im Volksglauben und Volksbrauch", *Archiv für Religionswissenschaft* (1902): 236–257 and 369–383; that study provides a survey of folk beliefs about both the cause of earthquakes and their meaning as portents.

29 For some brief comments on these interesting parallel uses of earthquakes in Plato, see Ophir, 21–22.

30 Danzig (pp. 188–189) also sees a number of parallels between the cavern entered by Gyges and the cave of Plato's famous allegory.

31 See p. 231 and p. 250, n. 15 of John Evan Seery, "Politics as Ironic Community: On the Themes of Descent and Return in Plato's Republic", *Political Theory* 16.2 (May 1988): 229–256.

32 Note p. 188 of Jacob Howland, "Plato's Reply to Lysias: *Republic* 1 and 2 and *Against Erastosthenes*", *American Journal of Philology* 125.2 (Summer 2004): 179–208.

33 See Adrienne Mayor, *The First Fossil Hunters: Paleontology in Greek and Roman Times* (Princeton: Princeton University Press, 2000).

Delos.[34] More specifically, a scholar of classical archaeology has noted a number of examples of "arguments from archaeological evidence" in Greek texts.[35] These arguments were not always historically sound, but they do reveal that the inhabitants of the Graeco-Roman world had a sense of something more ancient, a sense of the cultures and peoples that had preceded them.[36]

The Gyges tale suggests that the earthquake has revealed remnants of a civilization that was lost and buried. The tale recounts that there are artifacts — "wonders" — including a horse of bronze, and, of course, a ring. The body that Gyges finds is inside this metallic horse, furthering the image that the reader gets of an archaeological site; it is like a tomb find unearthed by some intrepid nineteenth-century British explorer.[37] Gyges the shepherd has come across the remains of a lost world. It is here that a careful reader can begin to see that the tale of Gyges might be about not just an individual's morals, but about the knowledge, actions, and morality of a civilization as a whole.

5 The Elements of the Tale: the Horse

The story's mention of a bronze horse with doors is very peculiar. At first it does not seem to add much to the tale, other than emphasizing that Gyges has come across a tomb. The purpose of this bronze horse in the Gyges tale is indeed obscure, although the famed archaeologist and art historian George M.A. Hanfmann saw it as a reference to a kind of ancient religious ritual involving a mechanical facsimile of a horse:

> The bronze horse, presumably a man-made construction, is puzzling. A possible explanation may lie in the existence of a ritual of a god, in which

34 James Whitley, *The Archaeology of Ancient Greece* (Cambridge: Cambridge University Press, 2001), 70.

35 Anthony M. Snodgrass, *The Dark Age of Greece: An Archaeological Survey of the Eleventh to the Eighth Centuries BC* (Edinburgh: Edinburgh University Press, 1971), 17.

36 Ibid., 16–18. For a brief, if somewhat parsimonious, view of classical attitudes to archaeology, see Bruce G. Trigger, *A History of Archaeological Thought* (Cambridge: Cambridge University Press, 1989), 27–31.

37 Concerning the bronze horse as tomb, see p. 79 of George M.A. Hanfmann, "Lydiaka", *Harvard Studies in Classical Philology* 63 (1958): 65–88. Also note the mention of a bronze horse in Xenophon's *On the Art of Horsemanship* 1.1, in Xenophon, *Hiero. Agesilaus. Constitution of the Lacedaemonians. Ways and Means. Cavalry Commander. Art of Horsemanship. On Hunting. Constitution of the Athenians*, trans. E.C. Marchant and G.W. Bowersock (Cambridge, MA: Harvard University Press, 1925), 296–297.

> an artificial horse played a part. Such a ritual did exist in Anatolia; and a bronze horse played a part in it.[38]

Hanfmann goes on to point out possible religious and historical connections between such artificial horses, divinities, and the famous Trojan horse. Herodotus has a story, too, that combines the use of a tomb in the form of an animal — a gilded wooden cow — with religious symbolism.[39]

For the purposes of the discussion here, the most revealing fact is that the Greeks may have received a garbled tradition concerning horses from the Hittites:

> In the Hittite representations the god stands on the horse or is seated on a throne which is placed on the back of the horse. If the statue of the god was taken down to be carried into a sanctuary, some confusion concerning the whereabouts of the god might easily result ... [A] Hittite explaining the procession in which a horse (statue) was carried by itself might try to explain that "the divine power of the god was within the horse".[40]

Pausanias, too, speaks of a bronze horse with a peculiar magical power, a story also found in Pliny and Aelian.[41] In the version of the story in Plato, Gyges does not find a divinity inside the bronze horse, but he does find a body of very large stature possessing a ring that indeed has a divine power.

The Gyges story also clearly denotes this horse as artificial, mechanical — which means that it has been made by a civilization. So, Gyges has stumbled upon the remains of a race of powerful people who were artificers, and no mean artificers at that. By comparison, Gyges, a simple shepherd, is living in an age that is relatively unsophisticated. Elsewhere, too, Plato muses more explicitly upon the theme of more sophisticated and noble cultures that had existed in the long-distant past. In the famous Atlantis tale in the *Timaeus*, Critias recounts something he has heard:

---

38 Hanfmann, 76.

39 See the story concerning the daughter of Mycerinus in Herodotus, *Histories* 2.129–2.133, trans. Grene, 187–188.

40 Hanfmann, 77.

41 See p. 15 of Jaś Elsner "Structuring 'Greece': Pausanias' *Periegesis* as a Literary Construct", in Susan E. Alcock, John F. Cherry, and Jaś Elsner, eds., *Pausanias: Travel and Memory in Roman Greece: Travel and Memory in Roman Greece* (Oxford: Oxford University Press, 2001), 3–20.

> The story is that our city had performed great and marvelous deeds in ancient times, which, owing to the passage of time and to the destruction of human life, have vanished.[42]

Later in the same dialogue, Plato's text has an Egyptian priest remark:

> Ah, Solon, Solon, you Greeks are ever children. There isn't an old man among you ... You are young ... young in soul, every one of you. Your souls are devoid of beliefs about antiquity handed down by ancient tradition. Your souls lack any learning made hoary by time. The reason for that is this: There have been, and there will continue to be, numerous disasters that have destroyed human life in many kinds of ways. The most serious of these involve fire and water, while the lesser ones have numerous other causes.[43]

In the tale of Gyges, then, a reader could perceive a similar scenario: this underground tomb is a surviving remnant of one of the "many destructions of mankind". The image is of a technological culture that has vanished, save for the remnants in this tomb.

6 The Elements of the Tale: the Body

Inside the bronze horse, Gyges finds "a corpse ... of more than human size". Why so? The story could have functioned with a regular body, of average size. Why the implication of this being the body of some kind of giant? Immediately, the reader may be reminded of giants in other folktales and myths, as well as classical Biblical traditions. In Jewish accounts, there were the *rephaim*, giants who were "dwelling in the Land of Canaan prior to the Conquest".[44] Again there is the connection between this race of large stature and a fall from grace, as an early "Hebrew myth propounds that the original antediluvian race of giants, called at different times *nephilim* or *rephaim*, was destroyed by God and banished to the underworld."[45]

---

42 Plato, *Timaeus* 20e–21a, trans. Zeyl, in Cooper and Hutchinson, 1228.

43 Plato, *Timaeus* 22b–22c, trans. Zeyl, in Cooper and Hutchinson, 1230.

44 Simcha Paull Raphael, *Jewish Views of the Afterlife* (Lanham, MD: Rowman & Littlefield, 2009), 55.

45 Ibid., 55.

In Greek tradition, there was a race of giants who battled with the Olympian gods and were defeated with the help of Heracles.[46] In an interesting connection between the buried giants and earthquakes — as we see in the Gyges story — the Greeks believed that tremors near Mt. Etna were caused by the movements of the wounded giant Enceladus buried below.[47] Similarly, old "myths and folktales that are indigenous to the Caucasus region explain earthquakes as being caused by the efforts of a giant to free himself from a divinity's punishment."[48]

In Plato's version, the reader is told nothing about the identity of the giant corpse that is wearing the ring. However, Pliny's *Natural History* has the following: "As to the ring of Midas, which, upon being turned around, no one could see the wearer — who would not admit this story to be even more fabulous?"[49] Perhaps the corpse here was that of Midas, but Pliny's identification as such appears unique.[50] Regardless, a way of interpreting this element of the story is that Gyges has found the remains of a race of giants, powerful beings more advanced than the simple humans of his own time. This is not at all an unusual idea; again, folktales from a number of regions of the world speak of races of giants as living before mankind. A reader of Plato's time might well have read the setting of the tomb in this way, especially if that reader already possessed a shared cultural memory of similar folktales about giants, earthquakes, and antediluvian civilizations.

In several classical texts, too, there are discussions of finds of giant bones, and the peculiar identification of them as belonging to the bodies of heroes.[51] For example, a character in the *Heroikos* (*On the Heroes*), ascribed to

46 See pp. 209–212 of Marie Delcourt and Robert L. Rankin, "The Last Giants", *History of Religions* 4.2 (Winter 1965): 209–242. Chinese lore also includes giants; in one of their creation myths, there is the giant Pangu (盤古), who from chaos separates *yin* and *yang*, creates Earth and sky, and all the other parts of the world. On this story, see Paul R. Goldin, "The Myth that China Has No Creation Myth", *Monumenta Serica* 56 (2008): 1–22, and Marc Kalinowski, "Mythe, cosmogénèse et théogonie dans la Chine ancienne", *L'Homme* 36.137 (January–March 1996): 41–60.

47 Note p. 473 of H.N. Hutchinson, "Earthquakes", *Folklore* 19.4 (30 December 1908): 472–473.

48 Donna Rosenberg, *Folklore, Myths, and Legends: A World Perspective* (Lincolnwood, IL: LTC Publishing Group, 1997), 164.

49 "Midae quidem anulum, quo circumacto habentem nemo cerneret, quis non etiam fabulosiorem fateatur?" See Pliny the Elder, *Natural History*, 33.3, trans. Eichholz, Jones, and Rackham, 9.6–9.9.

50 On this line in Pliny the Elder, see pp. 273–274 of Smith, "The Tale of Gyges and the King of Lydia [part 1]" and p. 2 of Smith's article "The Literary Tradition of Gyges and Candaules", cited earlier. It could be that Pliny is confusing Midas with Gyges here.

51 For an extensive discussion of this topic, see pp. 148–157 of Jeffrey Rusten, "Living in the Past: Allusive Narratives and Elusive Authorities in the World of the *Heroikes*", in Ellen

Philostratus (ca. 172–ca. 250 CE), gives a litany on the remains of giants — and sure enough, the body found by Gyges is included in this classification:

> If I were fond of telling stories I would have told you about Orestes' body, which the Spartans discovered in Nemea — it was seven cubits tall ... or about the body in the Lydian bronze horse, which had been buried in Lydia still before Gyges' time, and miraculously appeared after an earthquake to some shepherds in Lydia, one of whom was Gyges. The horse contained windows on both sides, and in its hollow inside had been stuffed a body, which was greater than a man could imagine. Not long ago an excavation on the banks of the river Orontes brought to light Aryadês — thirty cubits tall — who had been buried in Assyria.[52]

This is evidence that an early reader — in this case Philostratus — indeed would interpret the "corpse ... of more than human size" in the Gyges tale in the wider context of cultural beliefs about giants.

There are various notices of giants in early texts; several examples can be found in the curious work by Phlegon of Tralles, entitled *Book of Marvels*, composed in the second century CE.[53] Naturally, Phlegon's work is a miscellany, mixing all kinds of sources, including reports of what might have been actual paleontological finds.[54] Ancient Greek sources speak of giants in a number of contexts, with sources dating back at least to Homer.[55] More specifically, there are stories of bones of giants found in coffins, such as a tale recounted by Herodotus, where Lichas, a Spartan, discovers the bones of Orestes. Similarly,

---

Bradshaw Aitken and Jennifer K. Berenson Maclean, eds., *Philostratus's Heroikos: Religion and Cultural Identity in the Third Century CE.* (Leiden: Brill, 2004), 143–158.

52 Rusten, 149. Concerning this text, see Christopher P. Jones, "Philostratus' *Heroikos* and Its Setting in Reality", *The Journal of Hellenic Studies* 121 (2001): 141–149. As to the remains of the giant found on the banks of the Orontes, and the connection between the *Heroikos* and some passages in Pausanias about giants, see idem, "The Emperor and the Giant", *Classical Philology* 95.4 (October 2000): 476–481.

53 See William Hansen, *Phlegon of Tralles' Book of Marvels* (Exeter: University of Exeter Press, 1996), 43–45 and 137–148.

54 For a brief discussion of this paleontological possibility, see Dóra Pataricza, "Phlegon and the Giants", in György Németh, ed., *Kalendae: Studia sollemnia in memoriam Johannis Sarkady* (Budapest-Debrecen: University of Debrecen, 2008), 271–282. Also note Mayor, *The First Fossil Hunters*, cited earlier.

55 Pataricza, 274.

Plutarch tells a tale of how the giant body of Theseus was found in a coffin, with a bronze spear-head and a sword.[56]

Phleglon gives no more than brief passages on alleged finds of giants. However, some of these provide peculiar echoes of the Gyges story. In one passage, entitled "Idas", we read the following:

> In Messene not many years ago, as Apollonios says, it happened that a storage jar made of stone broke apart in a powerful storm when it was pounded by much water, and there came out of it a triple head of a human body. It had two sets of teeth. They sought to discover whose head it was, and the inscription explained it: "Idas" was inscribed thereon. So the Messenians prepared another storage jar at public expense, placed the hero in it, and tended him more carefully, since they perceived that he was the man about whom Homer says
>
> And of Idas, who of men on earth at that time
> Was the strongest. He drew his bow against lord Phoibos
> Apollo for the sake of his lovely-ankled bride.[57]

As a commentator notes, "In this anecdote, there are many clichés or elements of folktale — the storm that reveals an ancient artifact, a monstrous relic, an inscription, the identification of a Homeric hero's remains, and the creation of a hero shrine."[58]

The particular similarities with the Gyges story are the giant, of course, but also the storm both as an omen (not stated, but implied) and as a plot device to reveal the tomb or receptacle of the body. In Pausanias, there is also an interesting connection between Lydia, where Gyges reigned, and giants:

> There is a small city of upper Lydia called "The Doors of Temenus". There a crest broke away in a storm, and there appeared bones the shape of which led one to suppose that they were human, but from their size one would never have thought it.[59]

Note that the storm motif is also repeated here.

---

56 Ibid., 275–277; the author there provides a long list of ancient sources that discuss the finding of the bones of giants.

57 Hansen, *Phlegon of Tralles' Book of Marvels*, 43.

58 See p. 240 of Carolyn Higbie, "Hellenistic Mythographers", in Roger D. Woodward, ed., *The Cambridge Companion to Greek Mythology* (Cambridge: Cambridge University Press, 2007), 237–254.

59 Pausanias, *Description of Greece* 1.35.7, trans. Jones, 1.191.

Another tale in Phlegon's text talks about bones of giant bodies on display in a town in Egypt. Phlegon comments:

> One should not disbelieve in these bones either, considering that in the beginning when nature was in her prime she reared everything near to gods, but just as time is running down, so the sizes of creatures have been shrinking.[60]

This echoes what is argued here to be a key aspect of the Gyges story: the use of giants as a way of referencing a past — and more advanced — age. A modern commentator points out that it "was a common notion among the Greeks that mortals of earlier times were of greater stature than the folk of the present day, and in this trait resembled the gods", and that "the generally accepted interpretation of large, humanoid bones was they were the remains of earlier mortal inhabitants of the earth."[61] There was also the belief that giants had built megalithic structures, and one could see this belief, too, as aligning with the idea that this race had its own civilization.[62] Pausanias comments on a statue by the Greek sculptor Lysippus saying that, "it represents the tallest of all men except those called heroes and any other mortal race that may have existed before the heroes."[63]

Similarly, Pliny implies a connection between a past race of taller humans and the "conflagration" of civilizations, stating that

> it is almost a matter of observation that with the entire human race the stature on the whole is becoming smaller daily, and that few men are taller than their fathers, as the conflagration that is the crisis towards which the age is now verging is exhausting the fertility of the semen.[64]

Pliny here also alludes to an ancient idea, that of cyclic destructions of the world alternately by flood and fire.[65] So, again, there is the intersection of the

60 Hansen, *Phlegon of Tralles' Book of Marvels*, 44.

61 Ibid., 137 and 139; also see the detailed listing and discussion of ancient accounts of giants in Rusten, cited earlier.

62 Ibid., 138.

63 Pausanias, *Description of Greece* 6.5.1, trans. Jones, 3.30–3.31.

64 Pliny the Elder, *Natural History* 7.73, trans. Eicholz, Jones, and Rackham, 2.552–2.553; also note Hansen, *Phlegon of Tralles' Book of Marvels*, 143–145.

65 See p. 97 of Ernesto Paparazzo, "Philosophy and Science in the Elder Pliny's *Naturalis Historia*", in Roy Gibson and Ruth Morello, eds., *Pliny the Elder: Themes and Contexts* (Leiden: Brill, 2011), 89–112; also note pp. 72–73 of James I. Porter, "Ideals and Ruins: Pausanias, Longinus, and the Second Sophistic", in Alcock, et al., 63–92. The concept of

idea of fallen giants and fallen civilizations. Pliny treats the subjects of giants in another passage that appears in his *Natural History*, stating that according to Aristotle, the Cyclopes — a primordial race of giants — created stone towers.[66] Pausanias also states that parts of the city wall of Mycenae were said to be the work of the Cyclopes.[67] These structures are still referred to, poetically, as "Cyclopean masonry".

What is interesting is that Pliny places his recounting of this not uncommon attribution in a long list of technologies. Like other early writers, he places this technical knowledge in the past. This idea is discussed further in the conclusion of this book, but for now it is important to illustrate that writers such as Pliny at the same time could speak of the tales of the Cyclopes together with what a modern reader would consider more serious topics. To cite just part of Pliny's listing of technological achievements:

> Brick-kilns and houses were first introduced by the brothers Euryalus and Hyperbius at Athens; previously caves had served for dwellings. Gellius accepts Toxius, son of Uranus, as the inventor of building with clay, the example having been taken from swallows' nests. Cecrops named after himself the first town, Cecropia, which is now the Acropolis at Athens; though some hold that Argos had been founded before by King Phoroneus, and certain authorities say Sicyon also, but the Egyptians hold that Diospolis was founded in their country long before. Tiles were invented by Cinyra, son of Agriopa, as well as mining for copper, both in the island of Cyprus, and also the tongs, hammer, crowbar and anvil; wells by Danaus who came from Egypt to Greece to the region that used to be called Dry Argos; stone quarrying by Cadmus at Thebes, or according to Theophrastus, in Phoenicia; walls were introduced by Thrason, towers by the Cyclopes according to Aristotle but according to Theophrastus by the Tirynthians; woven fabrics by the Egyptians, dyeing woollen stuffs by the Lydians at Sardis, the use of the spindle in the manufacture of woollen by Closter son of Arachne, linen and nets by Arachne, the fuller's craft by

---

these cyclic patterns in both Pliny and in biblical texts is detailed further in Eric C. Steward, *Gathered Around Jesus: An Alternative Spatial Practice in the Gospel of Mark* (Cambridge: James Clarke, 2009), 123–135. St. Augustine, working from Virgil and other sources, expresses a similar idea about figures of the past being greater in physical stature; see Augustine, *The City of God*, trans. Marcus Dods (Peabody, MA: Hendrickson, 2009), 441–442, and Hansen, *Phlegon of Tralles' Book of Marvels*, 144–145.

66 Pliny the Elder, *Natural History* 7.195, trans. Eicholz, Jones, and Rackham, 2.638–2.639.

67 See Pausanias, *Description of Greece* 2.16.5 and 2.25.8, trans. Jones, 1.330–1.331 and 1.382–1.383.

> Nicias of Megara, the shoemaker's by Tychius of Boeotia; medicine according to the Egyptians was discovered among themselves, but according to others through the agency of Arabus, son of Babylon and Apollo; and the science of herbs and drugs was discovered by Chiron, the son of Saturn and Philyra. Aristotle thinks that Lydus the Scythian showed how to melt and work copper, but Theophrastus holds that it was the Phrygian Delas; manufactures of bronze some ascribe to the Chalybes and others to the Cyclopes; the forging of iron Hesiod ascribes to the people called the Dactyli of Ida in Crete. Erichthonius of Athens, or according to others Aeacus, discovered silver; mining and smelting gold was invented by Cadmus the Phoenician at Mount Pangaeus, or according to others by Thoas or Aeacus in Panchaia, or by the Sun, son of Oceanus, to whom Gellius also assigns the discovery of medicine derived from minerals. Tin was first imported by Midacritus from the island of Cassiteris. Working in iron was invented by the Cyclopes, potteries by Coroebus of Athens, the potter's wheel by the Scythian Anacharsis, or according to others by Hyperbius of Corinth. Carpentry was invented by Daedalus, and with it the saw, axe, plumb-line, gimlet, glue, [and] isinglass; but the square, the plummet, the lathe and the lever by Theodorus of Samos, measures and weights by Phidon of Argos, or, as Gellius preferred, Palamedes; fire from flint by Pyrodes, son of Cilix, the storing of fire in a fennel-stalk by Prometheus; [and] a vehicle with four wheels by the Phrygians.[68]

For the modern reader, the list seems rather unsystematic, but the point here is that Pliny was interested in these technologies, and interested in coming up with at least conjectural attributions. Also notable is that there is an acceptance of the idea that advanced technologies, such as medicine, could be invented by a culture — such as the Egyptians or the Phoenicians — in the distant past that had then fallen into decline. In terms of the discussion about the Gyges story here, it is also interesting that Pliny states that, "manufactures of bronze some ascribe to the Chalybes and others to the Cyclopes", thus drawing a connection between the tales of a race of giants and the more pragmatic practice of the working of bronze.

In Plato's recounting of the story, Gyges finds that perhaps at least one of these giants of the past possessed the ability to forge special rings. A commentator points out the following:

---

68 Pliny the Elder, *Natural History* 7.194–7.199, trans. Eicholz, Jones, and Rackham, 2.636–2.641; a gimlet is a small hand drill, and isinglass is a kind of gelatin obtained from the swim bladders of fish and used for clarifying beer as well as for making glue.

> The ring is found on the hand of a large corpse, evidently its previous owner. The large size of the corpse serves as an excellent indication of the material advantages that accrued to him as a result of the use of rhetoric. The fact that he is now dead, however, serves to show that such advantages do not necessarily serve their owners' best interests. *Republic* as a whole illustrates the damage that is caused to the soul by acting without knowledge of good and bad, and in particular the damage incurred by the tyrannical individual, by means of his unregulated power to obtain the objects of his desire.[69]

Note that "such advantages" — and here one could read "*technê*" — as the ring provides "do not necessarily serve their owners' best interests". Technology and its capabilities, along with materialism (those "objects of desire") have consequences, and the corpse is a symbol of this for the reader. Gyges, moreover, then commits a damaging transgression — stealing from the dead — and steals the ring that the giant had on his finger.

In the story the giant is naked except for the ring.[70] Few funerary traditions seem to dictate burial of a body naked — usually, people are buried in their finery or at least a simple gown. One scholar has conjectured that this nakedness is an adaptation put into the story for moral purposes, so that Glaucon can make Gyges out to be truly a reprobate:

> When he describes the corpse as naked, Glaucon carries Herodotus' theme of nakedness into his own story. But it is a corpse rather than a beautiful woman who is naked. Glaucon's hero ignores what is ugly, whereas Herodotus' Gyges recognizes the queen's beauty.[71]

This supposes that the nakedness of the corpse is something that Plato invents, and did not exist in some earlier version of the tale. Whether this is accurate or not is unclear; regardless, nakedness would seem to imply some kind of violation. At the death of the philosopher Polemarchus, the Thirty Tyrants prevented his family carrying out funerary rights — and so "violated funerary custom, according to which one's corpse is washed, dressed, and laid out for viewing in one's house".[72] Antigone wailed when she found the corpse of

69 Danzig, 189.

70 Danzig (p. 189) notes that in the Gyges story, "Nakedness may serve ... as a symbol of the natural reality that lies under the conventions of human culture".

71 Nichols, 35.

72 Howland, 188.

Polynices — "Seeing that the corpse is bare, she straightway reacts", notes one commentator.[73] The contrast is clear: a "corpse left naked of caretakers and thus abandoned to savagery as opposed to the civilized realm of burial rites."[74] An ancient reader of this Gyges tale would understand that this nakedness is a symbol on a large scale of violation, pertaining to a whole society. Moreover, Gyges brings that sense of societal failure to the present, as his reaction to the naked corpse is simply to commit a theft. In short, the reader of the Gyges story in Plato would see this giant body and its nakedness as significant, as elements that gave further meaning to the tale as a whole. The "background" details of this folktale are all connected and all highly relevant: the giants, their apparent technological abilities, and their moral decline. A reader of Plato's time would have received from the text a subtle message that those with advanced powers, including technological ones, can have their civilization disappear — just as in the Atlantis tale.

7 The Elements of the Tale: the Ring

The next focus of the story as it appears in Plato is the ring. A ring that can render their wearer invisible, while not common, is not unique to the tale of Gyges — in Sanskrit tradition, for example, there is a story of a magic ring that gives a hero invisibility and allows him to visit his lover.[75] There is also a Jewish tradition of a ring that will render one invisible to one's enemies.[76] A ring with such a power is even found in a Welsh legend, and there has been conjecture of a connection between the story of the ring of Gyges and a tale of an Indian magic stone of invisibility.[77] A Hermetic compilation of Greek magical texts

---

73 See p. 213 of Richard M. Rothaus, "The Single Burial of Polyneices", *The Classical Journal* 85.3 (February–March 1990): 209–217; also note William Blake Tyrrell and Larry J. Bennett, *Recapturing Sophocles' Antigone* (Lanham, Maryland: Rowman and Littlefield, 1998), 65. The relevant passage can be found in *Antigone*, 422–427; the term used in *Antigone* for "bare" here is ψιλὸν; this word does not mean "naked", *per se*, but rather "uncovered" — that is, not yet properly buried. The sense of impropriety, however, is the same.

74 Tyrrell and Bennett, 65.

75 See p. 110 of Virginia Saunders, "Magic in the Sanskrit Drama", *Journal of the American Oriental Society* 45 (1925): 110–114.

76 See p. 21 of Dennis C. Duling, "The Eleazar Miracle and Solomon's Magical Wisdom in Flavius Josephus's 'Antiquitates Judaicae' 8.42–49", *The Harvard Theological Review* 78.1–2 (January–April 1985): 1–25.

77 On the Welsh story, see George F. Kunz, *Rings for the Finger, from the Earliest Times to the Present* (Philadelphia and London: J.B. Lippincott, 1917), 307; on the India connection, see ibid., 291. A ring with a magic stone from India with the power of invisibility is also

from late antiquity, known as the *Cyranides*, recounts a method for the construction of a ring granting the power of invisibility.[78]

In the Greek tradition, there were other tales, too, of rings that either protected the wearer or gave them special abilities.[79] Lucian, in his comedic tale entitled "The Ship: Or, the Wishes", has his character Timolaus say:

> [M]y wish is that Hermes should appear and present me with certain rings, possessed of certain powers. One should ensure its wearer continual health and strength, invulnerability, insensibility to pain. Another, like that of Gyges, should make me invisible.[80]

Lucian makes reference to the ring of Gyges in another tale, as well, repeating Glaucon's cynical take on human nature:

> We could scarcely ask for a better judge of the matter than Dionysius himself. He was as familiar with the Stoic doctrines as any man, and held at one time that virtue was the only Good: but he presently discovered that toil was an evil: he then chose what seemed to him the better course.

---

found in an Arthurian legend; see p. 402 of Edward Thorstenberg, "'Duke Frederick of Normandy,' an Arthurian Romance", *Modern Philology* 7.3 (January 1910): 395–409. For a brief discussion of magic rings and invisibility, see Mario N. Pavia, *Drama of the Siglo de Oro: A Study of Magic, Witchcraft and Other Occult Beliefs* (New York: Hispanic Institute in the United States, 1959), 61 et ff.

78 Note p. 39 of Campbell Bonner, "The Technique of Exorcism", *The Harvard Theological Review* 36.1 (January 1943): 39–49; also see Louis Delatte, *Textes Latins et Vieux Francais Relatifs aux Cyranides* (Paris: Droz, 1942), as well as p. 64 of Christopher A. Faraone, "From Magic Ritual to Semiotic Game: The Transformation of Neo-Assyrian Love Spells in Classical and Hellenistic Greece", in Antonio Panaino and Giovanni Pettinato, eds., *Ideologies as Intercultural Phenomena: Proceedings of the Third Annual Symposium of the Assyrian and Babylonian Intellectual Heritage Project* (Milan: Mimesis Edizioni, 2002), 61–74, and Daniel Ogden, *Magic, Witchcraft, and Ghosts in the Greek and Roman Worlds* (Oxford: Oxford University Press, 2002), 273.

79 See Katherine Morris Lester, Bess Viola, and Helen Westermann Oerke, *Accessories of Dress: An Illustrated Encyclopedia* (Mineola, NY: Dover, 2004), 327. There are also many other tales in various cultures of magic rings with various powers; see Smith, "The Tale of Gyges and the King of Lydia," 268, n. 2, cited earlier, as well as the many mentions in Lyons, *The Arabian Epic: Heroic and Oral Story-Telling*, 3 vols. (Cambridge: Cambridge University Press, 1995). Magic rings appear, too, in the folktales of Indonesia; see p. 105 and p. 109, n. 14, of C.M. Pleyte, "An Unpublished Batak Creation Legend", *The Journal of the Anthropological Institute of Great Britain and Ireland* 26 (1897): 103–109.

80 See Lucian, "The Ship: Or, the Wishes" 42–45, trans. Fowler and Fowler, 4.49. Also note Kunz, 295, as well as Hansen, *Ariadne's Thread: A Guide to International Tales Found in Classical Literature* (Ithaca: Cornell University Press, 2002), 13.

> He would no doubt observe that those philosophers who had so much to say on the subject of patience and endurance under toil were secretly the servants of Pleasure, carefully abiding by her laws in their own homes, though they made so free with her name in their discourses. They cannot bear to be detected in any relaxation, or any departure from their principles: but, poor men, they lead a Tantalus life of it in consequence, and when they do get a chance of sinning without being found out, they drink down pleasure by the bucketful. Depend on it, if some one would make them a present of Gyges's ring of invisibility, or Hades's cap, they would cut the acquaintance of toil without further ceremony, and elbow their way into the presence of Pleasure; they would all be Dionysiuses then.[81]

Lucian again presents a ring connected to invisibility in another work.[82] His story *The Liar* presents Eucrates and a magic ring, which — when turned — reveals the underworld, in the form of a chasm in the ground. This is an interesting echo of the Gyges story.[83] Note here how the ring is explicitly connected with the underworld, much as in the tale of Gyges, where there is the clear link between Gyges' descent and his finding of the magic ring. Plato's use of the ring, then, resonates through other early texts. The ring means invisibility, authority (as Gyges becomes king), and power.[84] Stories of magical rings also were associated with King Solomon and with Moses.[85]

But recall that the ring in the Gyges story was on the body of a giant. Again, it is argued here that in the Platonic context, the ring and its magical aspect imply that the ancient race of giants had powers that no longer existed. This magic ring might serve as a metaphor for superior civilizations, in addition to

---

81 Lucian, "The Double Indictment" 21, trans. Fowler and Fowler, 4.158–4.159.

82 Note the brief mention in William Reginald Halliday, *Greek Divination: A Study of Its Methods and Principles* (London: Macmillan, 1913), 221.

83 Lucian, "The Liar" 24, trans. Fowler and Fowler, 3.243. Also see Ogden, *Magic, Witchcraft, and Ghosts in the Greek and Roman Worlds*, 273, as well as idem, *In Search of the Sorcerer's Apprentice: The Traditional Tales of Lucian's Lover of Lies* (Swansea: The Classical Press of Wales, 2007), 161–170.

84 One commentator has interpreted the story of Gyges and the ring as symbolic of the development of *economic* power in particular; see Richard Seaford, *Reciprocity and Ritual: Homer and Tragedy in the Developing City-State* (Oxford: Oxford University Press, 1994), 224–226, as well as Percy Neville Ure, *The Origin of Tyranny* (Cambridge: Cambridge University Press, 1922), 146–155, drawing from and commenting on Georges Albert Radet, *La Lydie et le monde grec au temps des Mermnades (687–546)* (Paris: Thorin et fils, 1893), 151–186.

85 See p. 196 et ff. of DiMarco, "The Squire's Tale", in Correale and Hamel, 1.169–1.210.

its role as motivator for Gyges' subsequent actions.[86] As the famous modern science fiction writer, Arthur C. Clarke put it: "Any sufficiently advanced civilization is indistinguishable from magic."[87] A further clue to this kind of deeper interpretation of the tale comes from Glaucon's statement that the possessor of a magic ring of this type would "in all respects be like a God among men".[88] For indeed, again, where does Gyges find the ring? On "a corpse ... of more than human size" — that is, the giant.

There is an interesting connection between giants and invisibility found in a medieval source, the history of Denmark by the thirteenth-century writer Saxo Grammaticus. A passage in his preface articulates a connection between giants and past civilizations, in an odd echo of Plato:

> That the country of Denmark was once cultivated and worked by giants, is attested by the enormous stones attached to the barrows and caves of the ancients. Should any man question that this is accomplished by superhuman force, let him look up at the tops of certain mountains and say, if he knows how, what man hath carried such immense boulders up to their crests. For anyone considering this marvel will mark that it is inconceivable how a mass, hardly at all or but with difficulty movable upon a level, could have been raised to so mighty a peak of so lofty a mountain by mere human effort, or by the ordinary exertion of human strength. But as to whether, after the Deluge went forth, there existed giants who could do such deeds, or men endowed beyond others with bodily force, there is scant tradition to tell us.[89]

The author then talks about the power of those mysterious peoples:

> But, as our countrymen aver, those who even to-day are said to dwell in that rugged and inaccessible desert aforesaid, are, by the mutable nature of their bodies, vouchsafed the power of being now near, now far, and of appearing and vanishing in turn. The approach to this desert is beset with

86 The mere fact of a ring in the Gyges story indicates a folkloric origin, as such an object "fulfills a definite structure function" in a folktale; see p. 45 of Udo Strutynski, "The Survival of Indo-European Mythology in Germanic Legendry: Toward an Interdisciplinary Nexus", *The Journal of American Folklore* 97.383 (January–March 1984): 43–56.

87 Arthur C. Clarke, *Profiles of the Future* (New York: Harper and Row, 1973), 21.

88 Pease's essay "Some Aspects of Invisibility", cited earlier, points out that in historical terms invisibility was a quality attributed to the divine (see Pease, 3–6).

89 *The Nine Books of the Danish History of Saxo Grammaticus*, 2 vols., trans. Oliver Elton (London: Norrœna Society, 1905), 1.89.

> perils of a fearful kind, and has seldom granted to those who attempted it an unscathed return.[90]

So, these giants — or their descendants — have the power "of appearing and vanishing in turn", a power that Gyges gains when he robs the giant's corpse.[91] The connection between giants and special powers is not unique to the tale of Gyges.[92]

However, although the ring has this magical power, the ring itself is *not* obtained magically, that is, through the workings of a mischievous god or demon spirit. The implication is that the ring has been forged by mortal beings of some kind, giant beings of a culture in the distant past. The ring, then, represents magic in a way, but not divine magic. Rather, it represents the "magic" of a people with an advanced ability — again, *technê* — the artifice or skill in making, much like the Atlanteans with their fantastic architecture and ability to work with the metal *oreichalkos*. This reinforces the idea that the tale of Gyges reflects a commentary not so much on the divine or magical, but on human civilization, its inventive capacity, and its patterns of rise and fall. This connection of a fall is echoed in other Greek sources: the fifth-century BC Greek poet Bacchylides writes that *Hybris* (Ὕβρις) — the goddess or personification of excessive pride — "destroyed the arrogant sons of Gaia, the Giants".[93]

Plato uses rings to suggest the idea of *technê* in another text as well. In *Lesser Hippias*, the sophist Hippias boasts of his polymathic abilities. He appears at Olympia garbed completely in his own creations, including a ring. In fact, the ring is the first thing mentioned in the list given by Socrates: "You said that you had once gone to Olympia with everything you had on your body the product of your own work. First, the ring you are wearing — you began with that — was your own work."[94] Plato's point here is that Hippias may have *technical* knowledge — symbolized by the ring — but that it comes without any kind

---

90 Ibid., 90.

91 See Shell, 28, n. 39.

92 Note pp. 137, 144, and 173 et ff. of Justin Glenn, "The Polyphemus Folktale and Homer's Kyklôpeia", *Transactions and Proceedings of the American Philological Association* 102 (1971): 133–181, and Ian Morris and Barry B. Powell, *A New Companion to Homer* (Leiden: Brill, 1997), 450 et ff.

93 See David Castriota, *Myth, Ethos, and Actuality: Official Art in Fifth-century BC Athens* (Madison, WI: University of Wisconsin Press, 1992), 234, and the discussion on p. 139; also note pp. 342–344 of Claude Calame, "The Dithyramb, a Dionysiac Poetic Form: Genre Rules and Cultic Contexts", in Barbara Kowalzig and Peter Wilson, eds., *Dithyramb in Context* (Oxford: Oxford University Press, 2013), 332–352.

94 Plato, *Lesser Hippias* 368b–368c, trans. Smith, in Cooper and Hutchinson, 928.

of more profound understanding.[95] In fact, rings generally may have been perceived by Plato to be a sign of decadence: Aelian wrote that Plato criticized Aristotle's elaborate dress and his wearing of many rings.[96]

8 Conclusions: Technology and the Fate of a Civilization

Overtly, the story of Gyges in Plato's texts functions as a morality tale. This is the context in which Glaucon presents the story. The background elements of the story also concern morality, but again with the implication of a larger context. As one commentator notes:

> For the purpose of Plato's argument it is the ring which makes [one] invisible that counts; for it enables a man to commit unjust deeds without fear of detection and presents, so to speak, an unlimited opportunity for temptation. This context argues that the unusual and curious features of the tale are not invented *ad hoc*; for they have no value for the argument.[97]

Again, we return to the idea that these additional elements serve some important purpose in the tale, and that they existed, perhaps, in a common version circulating at or before Plato's time. As Fischer notes, the message in a folktale can be "given added impact by being embodied in concrete images".[98] Those images in the Gyges story — the almost "archaeological" setting, the peculiar bronze horse and "other wonders", the superhuman dead body, its nakedness, and the power of the ring — give a particular impact, one of a great loss long ago.

Indeed, the underground setting, the bronze horse and "other wonders", the giant body, and the peculiar and dangerous ring — all serve as the evidentiary trappings of a civilization that has destroyed itself. None of these symbols in the story are without purpose:

---

95 For a rather humorous take on the boastful style of Hippias, see pp. 1–2 of Paul Shorey, "Hippias Paidagogos", *The School Review* 17.1 (January 1909): 1–9.

96 Aelian, *Historical Miscellany* 3.19, in Aelian, *Historical Miscellany*, trans. Nigel G. Wilson (Cambridge, Massachusetts: Harvard University Press, 1997), 148–149. Also note Charles William King, *Antique Gems: Their Origin, Uses, And Value As Interpreters of Ancient History; And As Illustrative Of Ancient Art: With Hints To Gem Collectors* (London: John Murray, 1860), 281.

97 Hanfmann, 76.

98 Fischer, 245.

> We should recall, moreover, that folktales ... are social products and that in the long run only what is significant to a sizable segment of the society will be retained in a tale. This eliminates the idiosyncratic "accidental" symbolism from folktales.[99]

Later in the tale, we see how Gyges uses the ring to carry out immoral acts, overthrow the king, and seize power. The folktale implies this: "Doesn't it seem likely, then, that this lost race of giants carried out even greater acts of immorality? This lost civilization, one that could create such rings, must have let their powers of *technê* get the better of them."

The tale then can be read, too, as a morality lesson not just for individuals but also for entire civilizations.[100] It is true that the in the Gyges story the tomb contains only one giant, and so one cannot assume with certainty that a race of giants is implied. But the combination of elements here — the body of large stature, the metal horse, the ring, and so on — suggest a larger whole: a culture that built that tomb, a culture that indeed was sophisticated and possessed certain technical powers. Certainly, this kind of broader interpretation is what Glaucon intends anyway, as he is talking about the corruptibility of human beings *generally*, not just Gyges. Glaucon only cites the Gyges story specifically as a way of presenting a clear "case study" of an everyman. But Glaucon is not talking only about the idea that his contemporary fellow men behave in a just manner "not willingly ... but of necessity"; the background elements of the Gyges tale as he recounts it show, in a subtle way, what happened to men (or supermen) in a culture as a whole who indeed lived with the power to "safely be unjust". Not only did they behave unjustly, but they also destroyed their civilization in doing so. As noted earlier, this theme of arrogance was a common one in early literature.[101] The theme of a destruction of civilization, in turn, is found in Plato.

Moreover, the implication of Glaucon's statements is that not only would the ring-wearer behave unjustly, but that there would be anarchy: "No man would keep his hands off what was not his own when he could safely take what

99 Ibid., 248.

100 Ovid and Hesiod, of course, both outlined the various — declining — ages of human civilizations; in the *Metamorphoses*, moreover, Ovid explicitly connects this idea with the giants' impious actions; see Stephen M. Wheeler, *Narrative Dynamics in Ovid's Metamorphoses* (Tübingen: Gunter Narr Verlag, 2000), 23–26.

101 See the discussion, for example, in Eugene Dwyer, "Excess", in Helene E. Roberts, ed., *Encyclopedia of Comparative Iconography: Themes Depicted in Works of Art*, 2 vols. (Chicago: Fitzroy Dearborn, 1998), 1.293–1.297, and the treatment in Steele Commager, *The Odes of Horace: A Critical Study* (New Haven: Yale University Press, 1962), 118–120.

he liked out of the market, or go into houses and lie with any one at his pleasure, or kill or release from prison whom he would." The tale itself, particularly in the version found in Herodotus, seems to revolve around the fate of individuals who violate the laws of society and moral guidelines. Candaules perishes because of his excessive pride in the beauty of his wife. In the case of the tomb here, there may be a question about this anarchy: If the civilization of the race of giants degenerated into anarchy, how did it happen that they managed to bury the huge body carefully in a tomb, inside a bronze horse that had to be manufactured? How could a civilization in anarchy do this? The implication here, rather, is that the civilization was in decline, rather than already completely collapsed — indeed, the story is all the more disturbing for that fact. The tomb seems quite orderly, and quite sophisticated, except for the nakedness of the body. It is that single slightly "off-key" aspect that makes the story troubling. Indeed, the setup is quite similar to the "City of Brass" in *The Arabian Nights* tale, discussed below; in that city, everything is perfect, with everyone in place — except that everyone is also completely dead.

Again, if the morality tale recounted by Glaucon is taken in combination with the elements of the giant, the horse, and the ring, the reader can perceive a larger theme of the fall of civilizations as a whole. One commentator has noted that there is "no earlier version of the Gyges story in which a magic ring figures." He adds that "while Plato may well be drawing on a general acquaintance with folk stories, or another version of the Gyges story, it is quite possible that he is himself responsible for some of the unique features of his version."[102] In many case, Plato surely was working with a trope that would be familiar to his readers — magic powers in the form of a manufactured object, their corrupting influence, and a civilization's decline. Indeed, the latter part of that trope certainly appears, as noted, elsewhere in the *Republic*.[103]

The ring in the story, moreover, is placed in a kind of technological context. A study of Chaucer looks at how stories of magical objects in his stories may have roots in Tartar tales, and in a discussion of magic rings in both Chaucer and the Gyges tale, the author of the study notes the following:

> This story, no less than that of the Colchian ram with the golden fleece … suggests that at a very early date romances of metallurgy wandered into

---

102 Danzig, 185.

103 Danzig (p. 186) notes: "It is precisely the folk-story-like quality of these details that led scholars such as Smith to assume that Plato was basing himself on a pre-existing story. If that is wrong, and Plato is responsible for them, he must have had reasons for including them."

> Greece from what was afterward Tartar territory. Before leaving Greek mythology it is worth mentioning that Hephaestus is attended by feminine figures of gold and endowed with life and motion (*Iliad* 18.417–421); that Daedelus flies with artificial wings; that the Colchian Medea rides in a chariot through the air (Euripides *Media* vv. 1321–1322); and that Bellerophon rides the winged horse Pegasus.[104]

These "romances of metallurgy" again show how non-technological sources, such as fabulous tales and romances, can include technological themes. Such themes might stem from a "prehistoric tradition, originating, no doubt, when the craft of the smith, metallurgy, was invested with superstitious awe".[105] Certainly, the magic ring has a long pedigree up to Chaucer, with rings with special powers being attributed even back to Moses.[106]

Fischer, cited earlier, comments that folktale elements or images "can be interpreted as symbols referring simultaneously to several 'orders' of reality", by which he means that a given folktale can talk about personality conflicts, relationships in a family, community issues, and so on.[107] Lévi-Strauss, notes Fischer, would add other "orders", including a "technical-economic" one, and that seems to be one that is relevant to the Gyges story here and its peculiar background elements.[108]

Most important, however, is Fischer's comment that the fact "that a folktale is meaningful when interpreted on one of these levels is no evidence that it is meaningless on any other level". In this case, the basic level on which the Gyges story has been interpreted on is one of personal immorality. But there is meaning on another level — one about civilization as a whole. Even in the Herodotean version of the tale of Gyges, there is the idea of the fall of civilizations, as the oracle predicts that in the fifth generation after the reign of Gyges, the kingdom would be lost. The prophecy is indeed fulfilled, when Croesus is defeated by the Persian armies of Cyrus the Great. The fall is not of a person, but of a kingdom, for "scores are balanced by the gods, not in the span of one's life but rather in the span of one's dynasty."[109] Marc Shell, a scholar of comparative literature, provides an extensive and interesting interpretation of both Herodotus' and Plato's version of the tale of Gyges in terms of

104 Henry Barrett Hinckley, *Notes on Chaucer: A Commentary on the Prolog and Six Canterbury Tales* (Northampton, MA: Nonotuck Press, 1907), 210.

105 Ibid., 211.

106 See Correale and Hamel, 1.195–1.200.

107 Fischer, 263.

108 Ibid., 263.

109 Ophir, 16.

monetary history and economics, foundations of modern — and perhaps decadent — civilizations.[110] As one reviewer of Shell's analysis notes:

> Tracing the myth of Gyges' Ring, Shell recounts how the ring, when the collet is turned palm-side, makes Gyges invisible and eventually turns him into a murderer and then a tyrant. Similarly, the Sophists, of whom Thrasymachus is regarded by Plato as the prototype of the tyrant, begin by teaching for money, allege that justice is power, and end as tyrants made invisible by the bureaucracies they create.[111]

Perhaps, too, this was the fate of the race of giants whose remains Gyges finds.

Moreover, in the tale in Plato, Gyges is a shepherd, the "archetype of the ruler as well as the ruled", and his decadent behavior then is a model of the decline of both leadership and the populace.[112] It is interesting to compare the tale of Gyges with the story of Polycrates and his ring. In that narrative — which, like a version of the Gyges story, is also recounted by Herodotus — the pharaoh Amasis urges Polycrates to throw away that which he valued most in order to avoid tragedy. As with the Gyges story, the ring not only controls the fate of an individual, but the fate of a kingdom.[113]

The tale itself does not spell out how the civilization to which the man "of more than human size" belonged finally fell into oblivion, but clearly that civilization had fallen, and was already buried deep in the ground at the time of Gyges. Gyges only comes across the remains when an earthquake accidentally opens up the ground to reveal them. Just as Gyges comes upon a race of powerful giants that has fallen, his own immoral acts through the power of the giant's ring threatens civilized society: "The myth of Gyges problematizes the demarcation of Greek man and the way he belongs to civic space by depicting a man who resembles a god."[114]

Plato dealt with the idea of lost civilizations elsewhere in his work, again notably in his presentation of the Atlantis myth. There, too, a powerful civilization disappears, a civilization that is articulated as one more advanced than that of Greece in Plato's own time. In Plato's Atlantis tale, the fall of the earlier

110 The first chapter of Shell's book, cited earlier, looks at the tale of Gyges; see Shell, 11–62.

111 See p. 141 of Bertram Morris, review of *The Economy of Literature* by Marc Shell, *Technology and Culture* 21.1 (January 1980): 140–142.

112 Shell, 27.

113 See Leslie Kurke, *Coins, Bodies, Games, and Gold: The Politics of Meaning in Archaic Greece* (Princeton: Princeton University Press, 1999), 107 et ff.; like Shell, Kurke sees economic and monetary symbolism in these stories.

114 Shell, 102.

civilization is brought on by decadence in terms of power, and unbridled ambition — exactly what Gyges draws up in the form of a magic ring from the strange tomb of the giant's lost civilization.[115] Again, folktales can retain such "reminiscences" of the distant past:

> Among the many narratives that have come down to us from antiquity, several closely resemble the complex *magic tales* found in modern oral tradition ... [M]otifs found scattered in different authors have struck scholars as possible reminiscences of ancient wonder, such as the magic table that provides a banquet whenever its possessor utters the command "Table, set yourself!" mentioned in a fragment of a lost Greek comedy. So also the magic ring of Gyges, which, when Gyges turns it a certain way, makes him invisible, and the imaginary magic rings with various powers wished for by a character in a dialogue of Lucian.[116]

Even Herodotus' account of Gyges has the nature of a folktale, as noted earlier — a folktale that recounts the fall of a kingdom.[117] Do the peculiar ring and its power of invisibility make us doubt that the tale of Gyges might be speaking of a genuine fall of a civilization in the distant past? In discussing ancient stories of the powers of invisibility, the classicist Arthur Stanley Pease noted:

> [A]s with other forms of the miraculous, the superficial ascription of more than human powers is no sufficient evidence against an underlying historicity. For example, no one doubts the historic character of Alexander the Great or of the Roman Emperors, though we may fairly doubt particular incidents about them which ancient writers, relying upon popular tradition, may have reported.[118]

The tale of Gyges and its "miraculous" aspects may, too, retain traces of historicity. While such historical details may have been lost, the arc of the tale nonetheless reflects historical cycles — rise and fall — writ large.

Moreover, this idea of a popular folk tale dealing the concept of lost technologies and vanished civilizations is not that unusual. In the beginning of "The Story of the City of Brass" in *The Arabian Nights*, there is a strange description

115 See Bury, 88.

116 Hansen, *Ariadne's Thread*, 13.

117 For a look at the folktale elements of the Herodotean narrative concerning Gyges, see ibid., 319–320.

118 Pease, 35.

that provides an interesting echo of the Gyges story.[119] A group of kings and sultans are talking about "traditions of former nations":

> They called to mind the stories of our lord Solomon the son of David … and the dominion and authority which God (whose name be exalted!) had bestowed upon him over mankind, and the Genii and the bird and the wild beast and other things; and they said, We have heard from those who were before us, that God … bestowed not upon any one the like of that which He bestowed upon our lord Solomon, and that he attained to that to which none other attained, so that he used to imprison the Genii and Marids and the Devils in bottles of brass, and pour molten lead over them, and seal this over them with his signet.[120]

A trip is then described — really an expedition — to a land in the distant west where these bottles might still be found. As with Gyges, the expeditionary group encounters objects that are "ominous", and it is noteworthy that these objects are technological in nature.[121] The first such marvel they encounter is a robotic "horseman of brass" that indicates the direction of the "City of Brass". They similarly encounter automata in the form of enticing maidens, a special locking device at the gates, and two statues guarding a body — one of those statues slays a member of the group who attempts to steal jewels from the corpse.[122] There are echoes of the Gyges story in that detail, but also in the more general concept of a civilization that had great technological abilities, yet perished. Indeed, inside the "City of Brass", everyone is dead, the bodies frozen where they where at the moment of death. In the bazaar, the group sees "merchants dead in their shops", victims of some sudden catastrophe. One commentator notes:

---

119 On the "City of Brass" tale, see the extensive discussion in David Pinault, *Story-Telling Techniques in the Arabian Nights* (Leiden: Brill, 1992), 148–239, and in Mia I. Gerhardt, *The Art of Story-telling: A Literary Study of the Thousand and One Nights* (Leiden: Brill, 1963), 195–235. Also note the shorter overview in Marzolph and van Leeuwen, 1.146–150.

120 *The Arabian Nights Entertainments; or, The Thousand and One Nights*, trans. Lane (New York: George Routledge & Sons, 1886), 521; also see the summary of the "City of Brass" story in Bruce Fudge, "Signs of Scripture in the 'City of Brass'", *Journal of Qur'anic Studies* 8.1 (2006): 88–118, as well as Warner, 55–58.

121 See p. 11 of Andras Hamori, "An Allegory from the Arabian Nights: The City of Brass", *Bulletin of the School of Oriental and African Studies, University of London* 34.1 (1971): 9–19.

122 Lane, 530–531 and 535–536; also note Hamori, 12–13. For a detailed study of early views of automata, see E.R. Truitt, *Medieval Robots: Mechanism, Magic, Nature, and Art* (Philadelphia: University of Pennsylvania Press, 2015).

> On their way to the palace, they [i.e., the explorers] must pass through the bazaar where the merchants' dead bodies are still sitting behind their wares, looking literally as if merely asleep. It is a hideous kind of limbo. Dying is here made perpetual, although death in action is nothing but apparitions and appearances. The inhabitants, who created the former, are now themselves reduced to the latter.[123]

Another commentator on this story points out that this "theme, of a people whose wealth and power is long vanished, appears throughout various genres of Islamic writings".[124] Yet another adds that the tale "stresses at every turn the fall of the most ambitious and grandest of civilisations".[125]

Once more this is a civilization that had technological abilities — one that could devise "apparitions and appearances" — but could not save itself.[126] This "City of Brass"

> itself is a work of craftsmanship and technology, raised and adorned by human skills and trades, overflowing with luxury and plenty — once upon a time; a specifically brazen city perhaps because brass, unlike gold or silver, is a metal alloy made by human craft.[127]

The technological aspect here, then, is clear. This is human artifice that has destroyed itself, and its precise history has been lost. Interestingly, another story in *The Arabian Nights*, entitled "Bulūqiyā", and its framing story, entitled "Ḥāsib Karīm al-Dīn", together contain the ideas of both the "preservation of knowledge" and hidden knowledge, and even include the theft of a ring from a corpse, with negative consequences.[128]

This theme is global: a similar story pattern of too much power incurring an unpleasant fate appears in the *Popol Vuh*, a work written in the middle of the

123 Hamori, 13.

124 Fudge, 97.

125 Warner, 61.

126 For an interesting look at how this idea of civilizational loss articulated by the "City of Brass" story persists in the modern era in the form of "obsessive nostalgias", see Katarzyna Pieprzak, "Ruins, Rumors and Traces of the City of Brass: Moroccan Modernity and Memories of the Arab Global City", *Research in African Literatures* 38.4 (Winter 2007): 187–203.

127 Warner, 62.

128 Fudge, 110–111.

sixteenth century in Guatemala.[129] The book comprises a series of myths and narratives about the indigenous peoples there. The *Popol Vuh* also presents a mythological version of the creation of the world, and that section of the work recounts how the gods initially created a culture rather more powerful than what came later:

> They were excellent and chosen people. Their faces were manly in appearance ... Perfect was their sight, and perfect was their knowledge of everything beneath the sky. If they gazed about them, looking intently, they beheld that which was in the sky and that which was upon the earth. Instantly they were able to behold everything. They did not have to walk to see all that existed beneath the sky. They merely saw it from where they were. Thus their knowledge became full. Their vision passed beyond the trees and the rocks, beyond the lakes and the seas, beyond the mountains and the valleys.[130]

But such a power cannot last. These powerful people express their thanks to the deities who created them:

> We are able to ponder and to move about. We know much, for we have learned that which is far and near. We have seen the great and the small, all that exists in the sky and on the earth. We thank you, therefore, that we were created, that we were given frame and shape.[131]

But the creators grow uneasy, and say amongst themselves: "It is not good what they have said, they that we have framed and shaped. They said, 'We have learned everything, great and small.'"[132] So, there are the consequences:

> Thus their knowledge was taken back by She Who Has Borne Children and He Who Has Begotten Sons: "What now can be done to them so that their vision reaches only nearby, so that only a little of the face of the earth can be seen by them? For it is not good what they say ... It is a mistake that they have become like gods. "But if they do not multiply or are

129 See Allen J. Christenson, *Popol Vuh: The Sacred Book of the Maya: The Great Classic of Central American Spirituality, Translated from the Original Maya Text* (Norman, OK: University of Oklahoma Press, 2007).

130 Ibid., 197–198.

131 Ibid., 199.

132 Ibid., 199.

> increased, when will the first planting be? When will it dawn? If they do not increase, when will it be so? Therefore we will merely undo them a little now … Thus they remade the essence of that which they had framed and shaped … They were blinded like breath upon the face of a mirror … They could see only nearby; things were clear to them only where they were. Thus their knowledge was lost.[133]

Again we see a magical age that disappears, and god-like abilities that vanish.

Returning to the tale of Gyges and its appearance in Plato's text, it should be recalled that the Greeks were quite aware of themselves as a *civilization*, as the apogee of distinct developments in politics, philosophy, art, and culture. As such, they were conscious of *prior civilizations* — real and mythical — and their fragile nature. Excessive powers, be they political or "magical", could bring a civilization down. Now the question is whether this tale of Gyges had Greek origins. Certainly, the story demonstrates an awareness of the pattern of the rise and fall of civilized societies that appears elsewhere in Plato, and in other Greek sources. But it may be that the Gyges story was indeed a folktale in circulation before being picked up by the Greeks, since it concerns a non-Greek figure.[134] That folktale would have been recounted and passed on by ordinary peoples, but harking back to some deep memory — real or imagined — of a powerful civilization that had brought about its own tragic end.

Earlier, the discussion here noted the concept of a tale having narrative, allegorical, moral, and anagogical elements.[135] The anagogical aspect of the tale of Gyges remains to be examined. "Anagogical" has as its root *anagoge*

133 Ibid., 200. Concerning these passages, note Stephen D. Houston, David Stuart, and Karl A. Taube, *The Memory of Bones: Body, Being, and Experience among the Classic Maya* (Austin: University of Texas Press, 2006), 167.

134 Concerning the question of the versions of the Gyges story in Herodotus and in Plato, and their relationship to folktale traditions, see Hansen, 327, n. 28. While Smith argues for an "old popular tale of Gyges which appears to have been current in the times of Herodotus and Plato" (see Smith, "The Tale of Gyges and the King of Lydia [part 1]", 263), and which may have been known to Pliny the Elder (Smith, 273–274), Laird argues against the existence of a "lost 'original' to inspire Plato's story", and suggests the "the possibility of Plato drawing, perhaps quite directly, from Herodotus"; see Laird, "Ringing the Changes on Gyges", 12. Ivan M. Cohen, cited earlier, supports Smith and says that the Herodotean version of the tale of Gyges fits in with a "number of stories [in Herodotus] that exhibit themes or motifs which are characteristic of folk narrative"; see Cohen, "Herodotus and the Story of Gyges", 55. For another brief discussion of folk themes involving the ring tradition that entered the classical tradition, see Xavier François Marie Gérard Wolters, *Notes on Antique Folklore on the Basis of Pliny's Natural History Bk. XXVIII 22–29*, (Amsterdam: H.J. Paris, 1935), 61–67.

135 Thompson, *Imaginary Landscape*, 10.

(ἀναγωγή), a word referring to an ascent, or "leading up". In terms of a tale, this can be taken as the part of the story that refers to the final circumstances, the conclusion, and, as noted, the more spiritual or mythological aspect. It is also, then, the part of the folktale that contains the deep meaning, the information beyond the narrative itself and the moral values. The anagogical elements of the story draw together the ring itself, the circumstances of its discovery, its power, and how that power is used or abused.

In Plato's recounting, the shepherd enters a chasm in the earth; there, the reader is told, "in addition to many other wonders ... he saw a hollow bronze horse" — the other "wonders" (θαυμάσαντα) are never described. Perhaps an earlier version of the Gyges tale had more details, which would have also explained the significance of the bronze horse. This horse may remind the reader of the Trojan horse, especially insofar as bad things — enemy soldiers — came from the inside of that artificial equine.

But the fact that a person "of more than human size" wore the ring is a key anagogical element here. The ring can be seen as a metaphor for technical knowledge, the ability to manipulate forces with something forged, something crafted. A reader of this story would have understood that the shepherd should have taken note of the scene: a dead giant, wearing a spectacular ring, and surrounded by "marvels". The anagogical implication is clear: the ring, the marvels, and the technological and intellectual sophistication suggested by the setting in the story all point to a fatal conclusion. Although the tale ends with Gyges getting away with his abuse of power — he indeed becomes king — the long view implies that a society as a whole cannot continue in this immoral way. The fact that in the tale those "of more than human size" and who had such power now exist as no more than naked corpses in underground tombs, suggests that such magical-technological ability inevitably will end up corrupting any civilization that possesses it, including that of Gyges — and that of the reader.

CHAPTER 6

The Nature, Encoding, and Transmission of Knowledge

1 Storing Knowledge

The knowledge of early civilizations included many of the same fields found in modern societies: history, law, geography, astronomy, agriculture, architecture, metallurgy, and so on. Each of these areas of knowledge involves a great degree of complexity in terms of information, context, and skills, and so transmission is often uneven and incomplete; in fact, knowledge in a particular field may be lost entirely over time. In the process of "encoding" — putting the knowledge in the form of a formal text, a folktale, an epic poem, a song, a diagram, or a ritual for transmission — it is simplified, and that can be the beginning of its journey to obscurity. Simplification can mean oblivion. The "vehicle" for the information — the text, the tale, the poem, and so on — comes to be seen for no more than its apparent face value as a piece of literature, a fanciful story, or an oratorical performance. Over time, eventually no single person, nor the collective memory of society, perceives or remembers that there is information within this "vehicle". Knowledge, especially complex knowledge, when "encoded" in the form of a story, song, or ritual, for example, runs the danger of being remembered solely as the story, song, or ritual in and of itself.

Similarly, the key to decipher the encoded knowledge may disappear over the millennia or slip from the shared memory of a culture or civilization. As one study says in a discussion of myths as forms of knowledge transmission:

> First, myths, as they are passed down through the ages, are inevitably distorted by peculiarities of our cognition until finally, if there is nothing nearby to remind people of the original story — a volcano, or stars, for example — they typically become unintelligible. Second, in certain cases, we can reconstruct some of the original information by paying close attention to these cognitive peculiarities — analyzing backwards from them, in effect. But we still have little hope of finding independent verification of our reconstruction unless we too can look at the object

 | DOI:10.1163/9789004352728_007

> that kept the original story more or less intact — that is, study, say, the actual volcano.[1]

It is indeed fortunate that any knowledge at all reaches us across the ages. The systems that transmit information, particularly information encoded in myth, are so long-lived and complex that one researcher has examined them using phylogenetic networks.[2]

The thoughts of Marcus Aurelius, an example of profound and complex knowledge, speak clearly to a modern reader as that reader leafs through the pages of a printed volume of the *Meditations*. It is a small, compact book, with clear, crisp type. A reader can peruse it at their leisure, and tuck it into their pocket to peruse during their daily commute. But this simple act of knowledge transmission — one's ability to read and understand the words of Marcus Aurelius that have been transmitted across many centuries — relies on any number of textual, historical, and linguistic conditions. If any of these conditions had not been met, the knowledge — the text with all its knowledge, insight, and wisdom — would have been lost forever. The thoughts of Marcus Aurelius, the epigrams penned in Greek by this Roman emperor who is now mere dust, reach the reader, centuries later and thousands of kilometers away: this is no small feat. First, there is the critical issue of language: fortunately for readers, Greek and Latin were preserved beyond the end of the Classical period, well into the Renaissance. Even as these languages' places were taken by the other tongues or evolved into our present languages, the grammar and vocabulary of the Classical forms were kept alive. Moreover, some Greek and Latin texts came to be translated into Arabic, and into European languages through that route. The result is that now a reader can "access the information" of Aurelius's text by way of one of these translations, or one can learn Greek — thanks to the grammars which also have been preserved — and read the text in its original form. Moreover, there is the preservation of context:

1 Barber and Barber, 163.

2 See Julien d'Huy, "A Cosmic Hunt in the Berber Sky: A Phylogenetic Reconstruction of a Palaeolithic Mythology", *Les Cahiers de l'AARS* [*Association des Amis de l'Art Rupestre Saharien*] 16 (2013): 93–106. Also see idem, "A Phylogenetic Approach of Mythology and its Archaeological Consequences", *Rock Art Research* 30.1 (May 2013): 115–118, and idem, "Polyphemus (*Aa. Th. 1137*): A Phylogenetic Reconstruction of a Prehistoric Tale", *Nouvelle Mythologie Comparée* 1 (2013): 3–18. In addition, note the same author's study, "Peut-on tester scientifiquement la phylogénétique des mythes? A propos de la chasse cosmique", *Mythologie française* 263 (June 2016): 20–24. A collection of these types of analyses is also found in Ralph Kenna, Máirín MacCarron, and Pádraig MacCarron, eds., *Maths Meets Myths: Quantitative Approaches to Ancient Narratives* (Switzerland: Springer, 2016). I wish to thank Don Bohn for bringing this material to my attention.

other documents, also preserved and translated, tell us who Marcus Aurelius was, the years he lived, his society and its achievements, and so on. In short, there are many tools at a reader's disposal with which they can obtain an accurate and meaningful interpretation of the information in the text in front of them.

Finally, there is the question of conservation, in the literal, physical sense. The "data", the text of Marcus Aurelius' *Meditations*, has to have been physically conserved from the time it was composed until now. Of course, the original material on which he wrote has not survived, but copies — reasonably accurate ones, it can be assumed — were made on a consistent basis. After the development of printing, a much greater number of copies were made of this text, thus increasing the chance of survival for the text, and decreasing the chance of any textual corruption. Through a combination of diligence by a whole series of anonymous scribes and then by the grace of printing technology, the *Meditations* have survived the ages, and can be understood and enjoyed by readers today.

2 Transmission through Time

Despite wars, social upheaval, natural disasters, and other disruptions in the history of human civilization, a surprising amount of knowledge has been preserved and remains intelligible to this day. However, two critical issues still stand. The first concerns how can we preserve our knowledge from the present day so that it is both available and intelligible to our descendants after many epochs. Jeff Rothenberg, a computer scientist, examined this problem in some depth in a discussion concerning digital documents.[3] He noted that to preserve knowledge in the form of intelligible information for the distant future, not only would a society have to develop media storage systems which would be both durable and free from obsolescence, it would also have to preserve the "interpretive" systems — for example, software as well as hardware — that could decode the media. As Rothenberg pointed out, there are organizations today that maintain just such older and obsolete software and hardware so that firms that have information stored in earlier computer formats can retrieve and convert important data. Despite the convenience that these services provide, there is the considerable cost of maintaining the antiquated software

3 Jeff Rothenberg, "Ensuring the Longevity of Digital Documents", *Scientific American* 272.1 (January 1995): 42–47.

and hardware. The context required for the preservation of even simple information at some point becomes too cumbersome.

The problem of information transmission, however, also has more philosophical roots. First, there is the question of how knowledge can be encoded in a way that requires the minimal possible contextual support — that is, one could assume that some cataclysm might do away with all of the "context", the hardware, software, dictionaries, and so on — with only the media and the knowledge embedded in that media surviving. Next, there is the problem of how knowledge can be preserved and transmitted to future generations. The issue with all storage systems is that the storage formats, parameters, and paradigms change. New standards, as Rothenberg noted in his story, are constantly emerging. Technology in the area of hardware and software-based informational retrieval systems will develop in directions still unknown to us. He added that, "it is naive to think that the encoding of any document — however natural it seems to us — will remain readable by future software for very long".[4]

As far as the textual sources examined earlier in this book are concerned, modern readers are fortunate that the "media" of ancient Chinese and Greek are still readable without any supporting apparatus or software. A page in the *Shan Hai Jing* can be read as directly and simply now on the page as it was by an ancient reader. Given that, one anachronistic solution to making our present knowledge readable in the future has been to go back to plain text: engraving or micro-etching information at an atomic scale on disks made of nickel. All that is needed to read the etched information is a powerful microscope — no proprietary software is required. A company in Los Alamos, New Mexico has designed such a system, knowing that to preserve information, one needs "a medium that is compact, durable and immune to the fashions of information technology".[5] As the author of a review of this "retro" technology has pointed out, "digital media, regardless of their physical durability, rely on the correct hardware and software being available to transcribe and decode them — a problem that is already making some early archives inaccessible". But will a future reader, even armed with a powerful magnifying device, be able to understand the carefully engraved words? As readers today, we are able to read the tales in the *Shan Hai Jing* — but that does not mean that we have retained the context to interpret them properly, and future generations may have the same challenge with the knowledge that we are conserving for them.

Transforming and encoding the information that carries our knowledge is not problematic: all words and images can be reduced to a simple code.

---

4 Ibid., 44.

5 "Storing Information: Come Up and See my Etchings", *The Economist* (30 May 1998): 77.

The modern approach to the problem of information preservation has been to develop the simplest possible encoding system simply utilizing zero and one — but this is a system that typically requires one of the most complex possible retrieval systems, namely, computers. The binary code, simple yet powerful, can retain a remarkable array of information, but still relies on context. Is the encryption five-bit? Seven-bit? Is it text that is encrypted, or an image? We will always need the "context", the "program", to answer such questions and allow us to retrieve the information in its original form.

Ancient societies developed extensive systems for the preservation of information: complex epics, elaborate rituals and traditions, and huge monuments laden with meaning.[6] These systems required largely non-technological retrieval methods. Most societies had a person such as a wise man or shaman, or a group of persons, perhaps forming a brotherhood, priesthood, or secret society within the society as a whole. These individuals or groups were the caretakers of the passed-on knowledge of the epics, possessing an understanding of how to "read" the ancient monuments, along with the knowledge of how to carry on the society's traditions. They preserved the knowledge, the rituals, and other forms into which that knowledge had been encoded, as well as the overall context that would allow the "unpacking" of that knowledge when needed.

These societies attempted to make their systems of knowledge encoding and preservation as efficient as possible — epics were composed in patterns that were easy to remember, traditions were connected with certain songs, foods, and festival days to fix the knowledge in societal memory, and in some cultures monuments were engraved with information to avoid the need for further memorization. But memories and monuments fade with time. The contextual world of shamans, high priests, and teachers, too, changes and sometimes disappears entirely. In the science fiction novel, *A Canticle for Leibowitz*, a discarded shopping list becomes a sacred document after a nuclear holocaust, since the entire context for the list has been lost. The future monks who find the list do not know how to "read" the text, since they have no cultural memory of the society that created it.[7]

---

6 For a discussion concerning the use of architecture to encode and preserve information through time — as a "comprehensive mnemonic sign", with "buildings as tectonic repositories of meaning" — see pp. 290–291 of Brian Lukacher, "Joseph Gandy and the Mythography of Architecture", *The Journal of the Society of Architectural Historians* 53.3 (September 1994): 280–299.

7 Walter M. Miller, Jr., *A Canticle for Leibowitz* (Philadelphia: Lippincott, 1959); on a similar theme, see David Macaulay's *Motel of the Mysteries* (Boston: Houghton Mifflin, 1979). For a look at how scientific knowledge can be lost over time due to cultural decline, see Russo, 388–398.

Even systems based on apparently "eternal" entities are fallible. The mythic figures by which the ancients encoded the constellations still rely on an astrological context for their meaning to be preserved. Moreover, even the stars themselves continue their gradual meandering across the heavens, causing the constellations to lose their shapes. Even if the appellation, "Orion the Hunter" is remembered by humanity in some distant future age, these descendants will not find him in their changed night sky.

Before one can seek to build an information storage and retrieval system that will survive the ages, then, there must a simple encoding scheme, like the binary code — but with a decoding key that is both simple and easy to preserve intact. This key should be fixed to some aspect of existence that is apparently indestructible and immutable (Benjamin Franklin might say that it would be "death and taxes"). A civilization must create an "epochal code", one that can survive for many millennia. Let us say the choice is the atomic weight of the hydrogen atom, or even the speed of light, *c*, as our decoding key, since both are universal constants. This is an insightful idea — except for the fact that future societies may use different units for these measures, thus yielding different numerical values and corrupting the key.

One possible long-term key could be the human body; of course, it too has changed and evolved, and will continue to do so, but its form has remained relatively stable for many millennia. Languages, systems of measurement, and monuments — all have come and gone, while the basic configuration of the body has been stable: two arms, two legs, five digits each on the hands and feet; two eyes, two ears, one nose, one mouth. Barring any radical natural mutation or genetic engineering, the "human measure", or "man as measure of all things" might be a good standard on which to base a decoding scheme. The sociologist Paul Connerton, in his work on societal memory, speaks of memory literally incorporated in the body, through posture, gesture, and other configurations of the human form.[8]

In fact, a system of encoding information in the body may have been used in the past. In religious writings, there is often the idea of the "macrocosm in the microcosm", and especially the concept that human beings, as products of a divine creator, must themselves contain the pattern of all creation. The Jewish mystical tradition of Kabbalah, for example, takes the human body as being made in the image of God, and further reasons that therefore the body also must contain, or rather "embody", the knowledge and meaning of the universe:

8 Paul Connerton, *How Societies Remember* (Cambridge: Cambridge University Press, 1989), 72.

> [T]he different parts of the body conform to the secrets of the supreme wisdom. The skin represents the firmament, which extends everywhere and covers everything, like a cloak ... The bones and veins represent the celestial chariot, the forces that exist within, the servants of God ... [A]s different figures formed by the stars and planes in the firmament that envelops us betoken hidden matters and profound mysteries so do the figures and lines on the skin which encompasses the human body and are the body's stars and planets. All these signs have a hidden meaning and are the objects of attention of wise men who know how to read the face of man.[9]

This passage clearly reflects the Kabbalistic concept of "as above, so below". But what is most interesting here is the last line, explicitly stating that the body itself can contain information.

The Chinese, too, consider the human body to be a microcosm of the universal creation. This is reflected in their concern with physiognomy and the belief in the existence of meridians in both the Earth and the body. In the Chinese practice of face-reading or physiognomy, it is said that one can determine a person's personality, health, longevity, and even fate by a careful study of the facial features.[10] Again, information, be it mystical or practical, is seen as "encoded" in the body.

The problem of encoding knowledge, however, is found not only philosophical and mystical speculation but also in pragmatic concerns. In a project dating back several decades, at Yucca Mountain in Nevada, spent nuclear fuel was to be stored in vast underground chambers carved into a ridge of volcanic rock. This kind of radioactive waste must be kept protected and isolated in the site for thousands of years. Geologists and other scientists studied the Yucca Mountain site to try to foresee the possibility of earthquakes, seepage, the rise of the water table, and other natural activity. But as one engineer pointed out, human activity in the distant future around the site is much less predictable: "The speculation about whether and how inadvertent human intrusion might occur is much like the attempts to determine the type of society that might occupy Yucca Mountain: interesting to think about but unknowable."[11]

---

9 Adolphe Franck, *The Kabbalah: The Religious Philosophy of the Hebrews* (Secacus, NJ: Citadel Press, 1967), 123.

10 See Richard J. Smith, *Fortune-Tellers & Philosophers: Divination in Traditional Chinese Society* (Boulder: Westview Press, 1991): 187–195.

11 See p. 77 of Chris G. Whipple, "Can Nuclear Waste Be Stored Safely at Yucca Mountain?", *Scientific American* 274.6 (June 1996): 72–79. I wish to give a special thanks to Chris G.

Radioactive waste that would have been stored in the site will remain dangerous for millennia, so somehow future generations would have to be warned not to dig in the area or intrude into the site in any way. In the early 1990's, Sandia National Laboratories put together a team comprising an anthropologist, an astronomer, an archaeologist, an environmental designer, a linguist, and a materials scientist to design a 10,000-year marking system for the Yucca Mountain site. The criteria for the design were several, but the primary purpose of the marking system was to inform any future intruders onto the site that the area is highly dangerous.[12]

The sheer scope of the project is mind-numbing: to think of humanity's descendants (if there are any) 10,000 years or more into the future is to reach into the realm of pure speculation. What will be their technology? Fantastically advanced? Or will they, perhaps, have returned to the ways of our hunter-gatherer ancestors? Will they have any record or understanding of any of our current languages and symbols?

In a 1994 article, Kai T. Erikson, a sociologist interested in communities, conflict, and catastrophe, commented on the profound problem of communicating about this dangerous site to distant future generations. In his essay, he asks: "But what monuments can be counted on to last? What warnings can be counted on to be both decipherable and impressive several thousand years from now?" Aside from the problem of the erosion of stone, metal, or other materials used for the monuments, there is again the problem of what markings should be put on them anyway. Erikson says of the Yucca signage plan:

> And what symbols or words shall we call on to make our point? Languages have half-lives far shorter than nuclear wastes, or so our experience to date would suggest, and even if we suppose English survives, the 20th-century expressions we select to convey the awe in which we hold those wastes are almost sure to have a different meaning and a different emphasis in the distant future.[13]

Whipple for providing me with a number of insights and sources concerning the Yucca Mountain project.

12 See Kathleen M. Trauth, Stephen C. Hora, and Robert C. Guzowski, *Expert Judgement on Markers to Deter Inadvertent Human Intrusion into the Waste Isolation Pilot Plant — Sandia National Laboratories Report SAND92-1382 / UC-721* (Albuquerque, NM: Sandia National Laboratories, 1993).

13 See p. 50 of Kai T. Erikson, "Out of Sight, Out of Our Minds", *The New York Times Magazine* (6 March 1994): 36–41; 50; 61.

Any encoding scheme involving language suffers from the mutability of the written and spoken word. A system involving symbols and markings is flawed in that these may be misread or misinterpreted. At the most fundamental level, a system of markings intended as a warning instead may be read by members of a future society as an invitation. As this book has shown, even when an early text is readable, it is unclear whether what one is reading is simple historical record, pure invention, a myth encoding some historical fact, or some other form of knowledge for which we do not even have the terminology. Future readers may have as much trouble deciphering present texts as we have had with texts from the past.

The designers of the Yucca site markings indeed wanted the signs there to be understood by future generations even without any context preserved. But it is unlikely that the markings' meaning could be understood for millennia without a contextual structure. A 1990 book entitled *Warnungen an die ferne Zukunft: Atommüll als Kommunikationsproblem* ("Warnings to the Far Future: Nuclear Waste As Communication Problem") advocated the establishment of a system to preserve such context. The work promoted the idea that "long-term institutions must be established to take responsibility for maintaining information about nuclear waste over millennia". It is interesting that this same book suggests that this plan "may require radical transformation of the procedures of communication".[14] The idea of "long-term institutions" reminds us of Plato's tale of Egyptian priests and their mission to preserve knowledge of the distant past over the millennia.

A conference entitled "Transmittal of Information Over Extremely Long Periods of Time", held in Norway in 1991, put forward the idea of preserving nuclear waste warning information through the medium of religion. One participant in the conference stated: "The danger symbols must be included in the set of holy symbols of each religion".[15] A consultant with Bechtel proposed an "atomic priesthood", suggesting that, "information be launched and artificially passed on into the short-term and long-term future with the supplementary aid of folkloristic devices, in particular a combination of an artificially created

14 Wendy Oser, "So That Posterity May Know", review of *Warnungen an die ferne Zukunft: Atommüll als Kommunikationsproblem* ["Warnings to the Far Future: Nuclear Waste As Communication Problem"], ed. Roland Posner, *Nuclear Guardianship Forum* 3 (Spring 1994): 3.

15 Susan Garfield, "Oslo Conference Suggests that World Religions Carry Nuclear Waste Danger Warnings into the Far Future", *The Nuclear Guardianship Library* (n.d.), http://www.nonukes.org/w28relig.htm.

and nurtured ritual-and-legend".[16] It seems we have returned to the archaic world of the Gyges tale and other ancient stories recounted in this book.

When we look at ancient symbols we often know or can guess the meaning because enough context has been preserved. But sometimes we are faced with symbols where we can perceive no meaning, and cannot even tell whether or not there is encoded information present at all. In such cases, the entire context has been lost, and the symbols, or text, do not themselves seem to embody any discernible decoding device or key. There is also the danger of seeing hidden meaning where there perhaps is none. This may have been the road taken by the otherwise highly distinguished linguist Ferdinand de Saussure, who extended his theory of hidden "anagrams" to apply to a vast corpus of early poetic texts.[17]

William Irwin Thompson, cited earlier, devotes a chapter in his book *Imaginary Landscape: Making Worlds of Myth and Science* to the Rapunzel fairy tale. There he muses whether this tale, through its series of scenes, images, characters, and plots, embodies encoded ancient knowledge of a technical nature concerning cosmology and botany. After noting Dante's narrative, allegorical, moral, and analogical levels of a tale, Thompson puts forward four new "levels of reading" that can be found in tales such as the Rapunzel story: the literal, the structural, the anthropological, and the cosmological.[18]

For our purposes, it is anthropological level that is most important; as Thompson notes, this is "level in which we recognize that the narration is telling more than one story".[19] Thompson later adds: "Stories ... are themselves a form of cultural storage, for knowledge will survive the transmission through

16 Idem, "Proposed 'Atomic Priesthood' is NOT Nuclear Guardianship! A Critique of Thomas Sebeok's Vision of the Future", *Nuclear Guardianship Forum* 3 (Spring 1994): n.p.

17 See Peter Wunderli, "Saussure's Anagrams and the Analysis of Literary Texts", in Carol Sanders, ed., *The Cambridge Companion to Saussure* (Cambridge: Cambridge University Press, 2004), 174–185 and Demelza Marlin, "Saussure and the Elusive Question of the Origin", *Semiotica* 2008.172 (January 2008): 185–200; also note Samuel Kinser, "Saussure's Anagrams: Ideological Work", *Modern Language Notes* 94.5 (December 1979): 1105–1138 and Jean Starobinski, *Words Upon Words: The Anagrams of Ferdinand de Saussure* (New Haven: Yale University Press, 1979). On the actual practice in early works of including hidden texts, note William Levitan, "Dancing at the End of the Rope: Optation Porfyry and the Field of Roman Verse", *Transactions of the American Philological Association* 115 (1985): 245–269.

18 Thompson, *Imaginary Landscape*, 10–11. It is interesting to note that a series of early Chinese myths may also "embody" cosmological information; see Deborah Lynn Porter's chapter entitled "See Yü Later: A New Interpretation of Chinese Flood Myths", in her book *From Deluge to Discourse: Myth, History, and the Generation of Chinese Fiction* (Albany: State University of New York Press, 1996).

19 Thompson, *Imaginary Landscape*, 10.

images for many more generations than it ever could through mere conceptual duplication."[20] This book — with its studies of strange texts about flying machines and other fantastical technologies — asks whether there might be other such tales, texts, or symbols for which we have failed to apply the proper decoding keys to understand the knowledge within.

A further question is what key would people from earlier times have used to encode their knowledge: indeed, was the knowledge even encoded consciously, or was it simply incorporated gradually into myths and fairy tales with the passage of time? Thompson comments that one could interpret the presence of the complex cosmological information in the Rapunzel tale as meaning that "the story had to have an author or authoress who could compose the story to sneak lost knowledge into a society." He points out, however, that in modern cultural theory, "we do not need to be that simplistic to insist that an authorial person is required for the existence of such a cultural story", and argues that a form of "autopoesis" took place.[21] The story forms naturally around the key knowledge that must be transmitted through time, preserving that knowledge, just as amber slowly surrounds a piece of life, congealing and hardening, thus leaving a window back to a distant, ancient age.

3 The Concept of "Encoding"

The idea that myths contain some kind of "encoded" information is not an exclusively modern conception. Ancient writers also spoke of the idea of lost knowledge that was framed in mythical or metaphorical knowledge. Recall that Plato, in his *Timaeus*, includes the following passage, with the words articulated by an Egyptian priest to his Greek visitor, Solon:

> Now Solon said that when he arrived there the people began to revere him. Furthermore, he said that when he asked those priests of theirs who were scholars of antiquity about ancient times, he discovered that just about every Greek, including himself, was all but completely ignorant about such matters ...
>
> [O]ne of the priests, a very old man, said, 'Ah, Solon, Solon, you Greeks are ever children. There isn't an old man among you.' On hearing this, Solon said, 'What? What do you mean?' 'You are young,' the old priest replied, 'young in soul, every one of you. Your souls are devoid of beliefs

20 Ibid., 33.
21 Ibid., 41–42.

> about antiquity handed down by ancient tradition. Your souls lack any learning made hoary by time. The reason for that is this: There have been, and there will continue to be, numerous disasters that have destroyed human life in many kinds of ways. The most serious of these involve fire and water, while the lesser ones have numerous other causes. And so also among your people the tale is told that Phaethon, child of the Sun, once harnessed his father's chariot, but was unable to drive it along his father's course. He ended up burning everything on the earth's surface and was destroyed himself when a lightning bolt struck him. *This tale is told as a myth, but the truth behind it is that there is a deviation in the heavenly bodies that travel around the earth*, which causes huge fires that destroy what is on the earth across vast stretches of time.[22]

This passage clearly illustrates two fundamental ideas that have been examined in the preceding chapters of this book. First, there is the idea that myths contain technical information. In addition, taking a step back, it is clear that early writers such as Plato were well aware of history, the passage of time, and the subsequent risk of the loss of knowledge.

They were also aware of the modes and methods of attempting to preserve that knowledge. As noted earlier in this book, in the *Phaedrus*, Plato employs the myth of the Egyptian deity Theuth to illustrate issues concerning writing and memory. Theuth is said to have invented number, calculation, geometry, astronomy, and most significantly, writing. Theuth comes to an Egyptian king, Thamus, and states: "O King, here is something that, once learned, will make the Egyptians wiser and will improve their memory; I have discovered a potion for memory and wisdom".[23] But it is interesting that Thamus objects to this gift of writing, stating:

> O most expert Theuth, one man can give birth to the elements of an art, but only another can judge how they can benefit or harm those who will use them. And now, since you are the father of writing, your affection for it has made you describe its effects as the opposite of what they really are. In fact, it will introduce forgetfulness into the soul of those who learn it: they will not practice using their memory because they will put their trust in writing, which is external and depends on signs that belong to others, instead of trying to remember from the inside, completely on their own. You have not discovered a potion for remembering, but for

22 Plato, *Timaeus* 22a–22e, trans. Zeyl, in Cooper and Hutchinson, 1229–1230.

23 Plato, *Phaedrus* 274e, trans. Nehamas and Woodruff, in Cooper and Hutchinson, 551.

> reminding; you provide your students with the appearance of wisdom, not with its reality. Your invention will enable them to hear many things without being properly taught, and they will imagine that they have come to know much while for the most part they will know nothing. And they will be difficult to get along with, since they will merely appear to be wise instead of really being so.[24]

Thamus believes that writing will make people lose the discipline of memory. But this seems an odd argument, especially from a modern perspective. Memory strikes us as markedly unreliable, whereas writing is fixed and vital for the verification of events. But Thamus' objections to writing are several and subtle. First, he points out that writing will impair the discipline of memory. He goes on say that writing will make knowledge something dependent on "external ... signs that belong to others". In other words, the actual locus of knowledge itself is moved. Finally, Thamus states that it will lead to incomplete or false knowledge — people will "hear many things without being properly taught", and they will think that they have learned when in fact "for the most part they will know nothing". His point here is quite clear: if knowledge is held externally, it is more subject to corruption and loss — and at the same time, a certain arrogance combined with weakness will set in, as people will become "difficult to get along with, since they will merely appear to be wise instead of really being so".

Thamus' statement about people ending up in a situation where they "know nothing" is a reference to the Socratic idea of knowledge being a "recollection of real truths already known and hidden in the soul."[25] But it is also a reference to the idea that the practice of writing can lead to the transmission of knowledge without the necessary context for proper interpretation by later generations. This is shown a bit later in the text, when Socrates states:

> Well, then, those who think they can leave written instructions for an art, as well as those who accept them, thinking that writing can yield results that are clear or certain, must be quite naïve and truly ignorant of Ammon's prophetic judgment: otherwise, how could they possibly think

24 Plato, *Phaedrus* 274e–275b, trans. Nehamas and Woodruff, in Cooper and Hutchinson, 551–552.

25 Mott T. Greene, *Natural Knowledge in Preclassical Antiquity* (Baltimore and London: The Johns Hopkins University Press, 1992), 159.

> that words that have been written down can do more than remind those who already know what the writing is about?[26]

In other words, a written text serves as no more than a reminder of the existence of a piece of knowledge held within a society. To truly access that knowledge, context is essential, as is the embodiment or continued practice of that knowledge by a certain sector of the society. The *Phaedrus* brings this point home when Socrates presents an additional argument about the weakness of writing as a method of knowledge preservation and transmission:

> You know, Phaedrus, writing shares a strange feature with painting. The offsprings of painting stand there as if they are alive, but if anyone asks them anything, they remain most solemnly silent. The same is true of written words. You'd think they were speaking as if they had some understanding, but if you question anything that has been said because you want to learn more, it continues to signify just that very same thing forever. When it has once been written down, every discourse rolls about everywhere, reaching indiscriminately those with understanding no less than those who have no business with it, and it doesn't know to whom it should speak and to whom it should not. And when it is faulted and attacked unfairly, it always needs its father's support; alone, it can neither defend itself nor come to its own support.[27]

Here, writing is presented almost as an abstract set of symbols, as open to interpretation — and misinterpretation — as a painting. In this passage, context again is presented as essential for the proper interpretation of knowledge, and in fact, Plato uses the expression "needs its father's support" as a metaphor for proper contextual background. As one scholar has put it, the "task of Plato in the dialogues is to 'expel' the errors occasioned by analogical knowledge" — in other words, the goal is to solve the problem of taking the analogy for the thing itself, a fundamental challenge when knowledge in encoded in the analogical language employed in myths.[28] Text without context yields only "discourse [that] rolls about everywhere".

---

26 Plato, *Phaedrus* 275c–275d, trans. Nehamas and Woodruff, in Cooper and Hutchinson, 552.

27 Plato, *Phaedrus* 275d–275e, trans. Nehamas and Woodruff, in Cooper and Hutchinson, 552.

28 See p. 33 of Massimo Verdicchio, "A Reader Like Phaedrus", *Diacritics* 14.1 (Spring 1984): 24–35.

But Plato's Atlantis tale, as this book has shown, exhibits no neat division between history, myth, and fiction as one might define such categories today, and unfortunately the context for that narrative is missing. Modern readers of the Atlantis story at least can perceive that they dealing with a different kind of text. Again, as the classicist Christopher Gill notes:

> Plato disposes us to expect a fiction, and then, as it were, plays with the reader, offering an account which might seem authentically historical (or pre-historical), but which is gradually revealed as a functional fable.[29]

What this means is that the way earlier cultures framed and transmitted knowledge was in a form quite different from our own. A "functional fable" is not something most modern writers would use when packaging or encoding knowledge, especially technical knowledge. But this practice was the norm of human civilizations for a very long time indeed. One study of Renaissance England — a time far removed from Plato, one might suppose — notes

> the ability of Renaissance minds to half-believe or believe in contradictory things ... [like] the not dissimilar case of the ancient Greeks.[30]

That is, just as the Greek readers of Plato were able to embrace the Atlantis story, perhaps, both as a fabrication and as a narrative containing historical information, so, too were Renaissance thinkers. Again, the myths were not perceived as fabrications, but instead were viewed as possessing the ability "to place the problems of contemporary life in a temporal perspective".[31] Even as the Renaissance saw the beginning of a more modern view towards historical verification, there was the retention of older methods of trying to understand the past. As the famed historian Carlo Ginzburg pointed out, the "image of the Renaissance as an age sharply differentiated from the traditional 'medieval' world' is "overly simplified".[32]

But in a discussion of Renaissance astronomy, one commentator indeed notes a shift:

29 See Gill, "Plato's Atlantis Story and the Birth of Fiction", 71.

30 Arthur B. Ferguson, *Utter Antiquity: Perceptions of Prehistory in Renaissance England* (Durham, NC: Duke University Press, 1993), 2.

31 Ibid., 3.

32 See Carlo Ginzburg, *Myths, Emblems, Clues* (London: Hutchinson Radius, 1990), 63–64.

> The majority of astronomers in the Renaissance, like their contemporaries in other disciplines, believed in the existence of knowledge granted to humans by God at the beginning of history, and held that mankind should seek to return to this ancient knowledge. I call this the "retrograde approach". Later in the seventeenth century, this conception was slowly replaced by the belief that astronomy is gradually improved in the course of history: the "progressivist approach".[33]

This is a subtle change in some ways: from framing knowledge as having pre-existed, in an almost Platonic sense, to a more open-ended, forward-looking idea where the quantity (and perhaps quality) of knowledge was less in the past and increased as one moved forward in history. The Renaissance historians

> were coming to recognize ... that society itself had a history apart from patriarchs and kings, and, with an implicit faith in the validating authority of origins and antiquity per se, they were ready to trace that history beyond recorded memory to a mythic past they were only beginning to question, to a primeval age they could recover only by imaginative projection from the observable fact that society had indeed changed.[34]

Such a societal change means that Renaissance writers were aware that human society had developed in terms of complexity, range of knowledge, and use of technology. They desired to learn how this came about, and used traditional methods to do so: a combination of drawing from earlier writers, examining what empirical evidence there might indeed be, and building conjectural models of their own.

A lesson that be drawn from this is that when one looks at an early text, it may indeed contain the products of "imaginative projection", but should not be deemed pure fiction in the modern sense. Although the writer cited in the passage above deems the "bridge between history proper and the quasi-history of myth and legend" created in the Renaissance as having only "limited success", it is perhaps time to revive consideration of such a bridge, since myth and legends were clearly used with success by human cultures for thousands of years to store and transmit knowledge.[35]

---

33 See p. 337 of Špelda, "The Search for Antediluvian Astronomy: Sixteenth- and Seventeenth-Century Astronomers' Conceptions of the Origins of the Science", *Journal for the History of Astronomy* 44.3 (August 2013): 337–362.

34 Ferguson, 3.

35 Ibid., 5; also see the discussion in Robert Goulding, *Defending Hypatia: Ramus, Savile, and the Renaissance Rediscovery of Mathematical History* (Dordrecht: Springer, 2010),

In terms of the history of technological ideas, of course, there may be merit in the ancient practice of synthesizing myths with history in another sense: such a synthesis has the capacity to preserve warnings about the potential dangers of technological development, as evidenced by the idea of an "atomic priesthood", mentioned earlier, with information packaged in "folkloristic devices", such as "nurtured ritual-and-legend".[36] In a fascinating essay entitled, "Remembering the Ancients: Observations on Technoscience in Čapek's *R.U.R.*", the scholar Christine Cornell notes how the playwright engaged the ideas of early writers such as Plutarch — in fact, Karel Čapek names two of his robotic characters Sulla and Marius.[37] What Čapek is doing in his play is mirroring Plutarch's use of history not just as an account of past events, but also as a mode of embodying information about morality and the "dangers of ambition". But as Cornell points out, perhaps Čapek's point was too subtle. Academic readings of early sources such as Plutarch miss out on references that are trying to convey lessons about the past: "For most moderns, the use of *exempla* as Plutarch and others would use them seems unsystematic — that is unscientific."[38] Čapek himself creates a myth, concerning robots in this case, like those found in the "City of Brass" story. His construction is also like Plato's Atlantis story — in the form of a myth, but one that contains historical references and framing which serve to communicate moral insights about technological power.

Čapek, like ancient writers before him, aims in his tale to present "a mean between the more-than-human aspirations and the less-than-human consequences of technoscientific endeavors".[39] It is a tale reminiscent of the Prometheus story, as Cornell concludes in her study of Čapek's *R.U.R.*:

1–18. Another study of Renaissance and Enlightenment period attitudes towards past knowledge is Stephen Gaukroger, ed., *The Uses of Antiquity: The Scientific Revolution and the Classical Tradition* (Dordrecht: Kluwer, 1991). Also see Jan W. Wojcik, "Pursuing Knowledge: Robert Boyle and Isaac Newton", in Margaret J. Osle, ed., *Rethinking the Scientific Revolution* (Cambridge: Cambridge University Press, 2000), 183–200, and Richard H. Popkin, "Newton and Spinoza and the Bible Scholarship of the Day", in ibid., 297–314.

36 See Garfield, "Proposed 'Atomic Priesthood' is NOT Nuclear Guardianship!", cited earlier.

37 Christine Cornell, "Remembering the Ancients: Observations on Technoscience in Čapek's *R.U.R.*", *Bulletin of Science, Technology & Society* 31.2 (April 2011): 103–112; Lucius Cornelius Sulla and Gaius Marius were Roman statesmen who are among the subjects of Plutarch's *Parallel Lives*.

38 Ibid., 104.

39 Ibid., 110.

> In Hesiod's telling of the Promethean myth, man may retain the fire stolen by Prometheus, but not without cost. Zeus gloats, "things will go hard for you and for humans after this. / I'm going to give them Evil in exchange for fire, / Their very own Evil to love and embrace". That Evil is Pandora with her jar of miseries. As Roger Shattuck ... has observed, our tellings of this myth routinely leave out Pandora, and "thus they avoid dealing with the full consequences to humankind of the knowledge Prometheus brings".[40] The original myth presents knowledge and its consequences together: we tell ourselves stories of knowledge without consequences ... Čapek reminds us of the ancients and of ways of thinking and asking questions about what we create without abandoning the accomplishments of modernity or succumbing entirely to them.[41]

So, encoding the traces of history and ideas about technology continues throughs work such as Čapek's *R.U.R.* Čapek, like Plutarch, and Plato before that, is part of a tradition of writers preserving knowledge and encoding — in Čapek's case, through theater — reflections on technology. But even here there is the familiar loss of knowledge over time, for "even as he presents his warning, Čapek shows us a lesson completely lost on his characters and likely on a considerable portion of his audience."[42] Čapek wrote *R.U.R.* in 1920, and a hundred years later, his warnings about technology and its consequences go unheeded, perhaps obscured by the drama in which those warnings are embedded.

4 Knowledge and Loss

How is knowledge best preserved? How can we ensure that our descendants will find and understand the knowledge that we have left for them? Should we even make a conscious effort to preserve our knowledge, or will such preservation take place naturally, as William Irwin Thompson suggests? These questions still linger, as our society seeks encryption, storage, and retrieval schemes that have the capacity survive into the distant future, despite all the vagaries of our mortal existence.

---

40 Cornell here is citing Roger Shattuck, *Forbidden Knowledge: From Prometheus to Pornography* (New York: St. Martin's Press, 1996), 15.

41 Cornell, 110.

42 Ibid., 104.

The examples presented in this book show that the ancients, too, struggled with the question of how knowledge might be preserved — in physical media, or through ritual and tradition, or some combination. In a study of *The Arabian Nights* cited earlier in this book, one writer notes the following:

> In the *Nights*, knowledge of a story and the ability to tell it may assure the survival of an individual. Analogously it may be that in real life too knowledge of stories assists the survival of communities.[43]

Indeed, the very structure of the *Nights* tales — Scheherazade recounting story after story to preserve her life — serves as an analogy for the process of an entire civilization. Maintaining a series of linked tales, and forming a fuller narrative, preserves the cultural memory of the civilization and all of its embedded knowledge: moral, religious, and even technological. Certainly, in the Islamic world there was an understanding of knowledge lost over the vast reaches of time. In his fourteenth-century *Muqaddimah* — a work dealing with the nature of history, theology, politics, and science — Ibn Khaldūn notes

> There are many sciences. There have been numerous sages among the nations of mankind. The knowledge that has not come down to us is larger than the knowledge that has. Where are the sciences of the Persians that 'Umar ordered to be wiped out at the time of the conquest? Where are the sciences of the Chaladaeans, the Syrians and the Babylonians, and the scholarly products and results that were theirs? Where are the sciences of the Copts, their predecessors?[44]

Again note the twin themes of knowledge becoming lost and former knowledge having been greater that what has survived. The perspective expressed by both "City of Brass" story and Ibn Khaldūn's comment, in short, is this: "the disappearance … of the learning of olden times".[45]

---

43 Irwin, 236; also see Irwin's interesting discussion of the analogy between biological adaptation and survival and the similar process in stories and tales over the ages — see ibid., 234–237. In addition, note one of the original formulations of this concept in the profoundly titled last chapter "Le royaume et les ténèbres" of Jacques Monod, *Le Hasard et la Nécessité: Essai sur la philosophie naturelle de la biologie moderne* (Paris: Éditions du Seuil, 1970).

44 Ibn Khaldūn, *The Muqaddimah: An Introduction to History*, trans. Franz Rosenthal, ed. N.J. Dawood (Princeton: Princeton University Press, 1967), 39.

45 Gerhardt, 209.

There are several ancient accounts of knowledge being recorded for posterity on physical media. Certainly, these accounts have a fanciful quality, but the very existence of such narratives points to a fundamental concern in a society about passing on knowledge to future generations. Perhaps the best known such account is in the *Jewish Antiquities* of Flavius Josephus; speaking of the "descendants of Seth", Josephus states that they

> discovered the science of the heavenly bodies and their orderly array. Moreover, to prevent their discoveries from being lost to mankind and perishing before they become known — Adam having predicted a destruction of the universe, at one time by a violent fire and at another by a mighty deluge of water — they erected two pillars, one of brick and the other of stone, and inscribed these discoveries on both; so that, if the pillar of brick disappeared in the deluge, that of stone would remain to teach men what was graven thereon and to inform them that they had also erected one of brick.[46]

This narrative, though fanciful, reflects precisely two elements discussed above: the concern for preserving knowledge, and the fact that the knowledge is technical in nature. Of course, too, there is the idea that the knowledge is being guarded to survive a cataclysmic event, something presented in the Atlantis narrative, as well. Indeed, there exists a separate tradition of knowledge about being Atlantis being preserved in the "hard media" of stone pillars or stelae.[47]

Josephus' story is just one of several versions of the "two pillars" tradition. In a surprising source — one of Galen's medical works, *Against the Opinions of Julian Concerning Hippocrates' Aphorisms* — there is the following passage:

> It was a law in old Egypt that all inventions in handicrafts had to be judged by an assembly of educated men and be written on pillars in a sacred place. Likewise, we should have an assembly of just and equally well-educated men. They should scrutinize all that has been written, and

46 Flavius Josephus, *Jewish Antiquities*, 1.70–71, in Flavius Josephus, *Jewish Antiquities*, 9 vols., trans. Louis H. Feldman, et al. (Cambridge, MA: Harvard University Press, 1930–1965), 4.33. Also see Preston Cory, *Cory's Ancient Fragments of the Phoenician, Carthaginian, Babylonian, Egyptian and Other Writers* (London: Reeves & Turner, 1876), 151–152.

47 For an analysis of pillars in relation to Atlantis, see Cameron, "Crantor and Posidonius on Atlantis", 81–84; also note the discussion in Adler, *Time Immemorial*, 196–198, concerning the concept of knowledge being storied in stelae.

> deposit in a public place only what appears worthwhile, but destroy what is worthless.[48]

Again one sees a writer in antiquity quite conscious of history, and the key issues of the storage and transmission of knowledge.

Josephus is very explicit in terms of talking about a sense of history, and about the desire to gain accurate knowledge of the past. In his work *Against Apion*, he talks at length about the preservation of historical records:

> Of the care bestowed by the Egyptians and Babylonians on their chronicles from the remotest ages, and how the charge and exposition of these was entrusted, in the former country to the priests, in the latter to the Chaldaeans; and how, among the nations in touch with the Greeks, it was the Phoenicians who made the largest use of writing, both for the ordinary affairs of life and for the commemoration of public events; of all this I think I need say nothing, as the facts are universally admitted. But that our forefathers took no less, not to say even greater, care than the nations I have mentioned in the keeping of their records — a task which they assigned to their chief priests and prophets — and that down to our own times these records have been, and if I may venture to say so, will continue to be, preserved with scrupulous accuracy, I will now endeavour briefly to demonstrate.[49]

The veracity of his claims in this case are less important than the clear concern that Josephus has with the *idea* of history, and that a civilization's identity is dependent on the preservation of knowledge. Indeed, in *Against Apion*, Josephus delivers a lengthy critique about the relatively limited historical sense of the Greeks:

> My first thought is one of intense astonishment at the current opinion that, in the study of primeval history, the Greeks alone deserve serious attention, that the truth should be sought from them, and that neither we nor any others in the world are to be trusted. In my view the very reverse

48 This translation is from Rudolph E. Siegel, *Galen's System of Physiology and Medicine: An Analysis of his Doctrines and Observations on Bloodflow, Respiration, Tumors and Internal Diseases* (Basel: S. Karger, 1968), 11; also see the brief comments on this passage in Plinio Prioreschi, *A History of Medicine*, vol. 3, *Roman Medicine* (Omaha, NE: Horatius Press, 1998), 334.

49 Flavius Josephus, *Against Apion*, 1.28–1.29, in Flavius Josephus, *Against Apion*, trans. H. St. J. Thackeray (Cambridge, MA: Harvard University Press, 1926), 174–175.

> of this is the case, if, that is to say, we are not to take idle prejudices as our guide, but to extract the truth from the facts themselves. For in the Greek world everything will be found to be modern, and dating, so to speak, from yesterday or the day before: I refer to the foundation of their cities, the invention of the arts, and the compilation of a code of laws; but the most recent, or nearly the most recent, of all their attainments is care in historical composition. On the contrary, as is admitted even by themselves, the Egyptians, the Chaldaeans, and the Phoenicians — for the moment I omit to add our nation [i.e., the Jews] to the list — possess a very ancient and permanent record of the past. For all these nations inhabit countries which are least exposed to the ravages of the atmosphere, and they have been very careful to let none of the events in their history be forgotten, but always to have them enshrined in official records written by their greatest sages. The land of Greece, on the contrary, has experienced countless catastrophes which have obliterated the memory of the past; and as one civilization succeeded another the men of each epoch believed that the world began with them. They were late in learning the alphabet and found the lesson difficult; for those who would assign the earliest date to its use pride themselves on having learnt it from the Phoenicians and Cadmus. Even of that date no record, preserved either in temples or on public monuments, could now be produced; seeing that it is a highly controversial and disputed question whether even those who took part in the Trojan campaign so many years later made use of letters, and the true and prevalent view is rather that they were ignorant of the present-day mode of writing. Throughout the whole range of Greek literature no undisputed work is found more ancient than the poetry of Homer.[50]

Here, there is a clear echo of the comments in Plato's Atlantis tale wherein the Egyptian priests present their view that the "Greeks are ever children" due to their short historical memory. Josephus also discusses here the idea that catastrophic events can disrupt the transmission of knowledge.[51] Philo of

50 Josephus, *Against Apion* 1.6–1.12, trans. Thackeray, 164–167; concerning this passage, also see Minna Skafte Jensen, *The Homeric Question and the Oral-formulaic Theory* (Copenhagen: Museum Tusculanum Press, 1980), 154–156.

51 On this critique, see p. 422 of Adler, "Berossus, Manetho, and '1 Enoch' in the World Chronicle of Panodorus", *The Harvard Theological Review* 76.4 (October 1983): 419–442; in this same essay, Adler discusses at some length the issue of the writing of history as practiced by Josephus and other early authors. Also see p. 103 of Mark Chavalas, "Genealogical History as 'Charter': A Study of Old Babylonian Period Historiography and

Alexandria is another early writer who comments on how knowledge of certain rituals becomes lost over time, due to catastrophes.[52] Again, for a modern reader, it is important to see that these kinds of quite contemporary questions about the preservation of knowledge, and even about historical accuracy, were being discussed by writers centuries ago.

The story in Josephus' *Jewish Antiquities*, cited earlier, not only deals with the issue of the preservation of knowledge in the face of societal destruction, but also reflects the idea that such preservation took place through *writing*. For ancient societies, of course, many other modes of information transmission were available: poetry, song, dance, and so on. Josephus' story, however, places prime importance on written media.[53] Moreover, the comprehensive nature of the knowledge preserved in the story in *Jewish Antiquities* is also found in related traditions, such as that of Enoch, where "astronomical, meteorological, calendarical, and eschatological knowledge" is included in the knowledge being articulated and preserved.[54]

Another version of the "two stelae" story makes a further point, noted earlier: the possible corruption of information over time. Again, it is intriguing that an otherwise fanciful tale should include what is, upon reflection, a rather sophisticated understanding of the issues of retention and transmission of knowledge over time. In an Armenian text known as *The History of the Forefathers*, there is a discussion of Enosh, son of Seth:

> [H]e made writings and wrote on stela(e) of baked brick and bronze, and he prophesied that the earth will pass through water and fire on account of the sins of humans. And he cast the baked brick into the water and the bronze into the fire, in order to test (them), if the fire was to come first, the bronze would melt, and if the water was to come first, the brick would be destroyed ... And the writings on the two stelae told the names of all things, for he knew that by lispers, stutterers and stammerers the language was destined to be corrupted ... And they confused and changed the names of the objects that had come into being, which Adam had

---

the Old Testament," in A.R. Millard, J.K. Hoffmeier, and D.W. Baker, eds., *Faith, Tradition, and History: Old Testament Historiography in its Near Eastern Context* (Winona Lake, IN: Eisenbrauns, 1994), 103–128.

52 See Herold Weiss, *A Day of Gladness: The Sabbath among Jews and Christians in Antiquity* (Columbia, SC: University of South Carolina Press, 2003), 46, as well as 192, n. 36.

53 See p. 141 of Andrei A. Orlov, "Overshadowed by Enoch's Greatness: 'Two Tablets' Traditions from the *Book of Giants* to *Palea Historica*", *Journal for the Study of Judaism* 32.2 (2001): 137–158.

54 Ibid., 143.

> named and fixed. On this account he wrote (them) on the two stelae and left them, so that if the water came first and destroyed the pillar of baked brick, the bronze writing and names of things would remain, so that after the flood and the passing of times it might come to use.[55]

One modern commentator notes that the "motif of stelae left for postdiluvian generations was a device commonly employed by oriental historians to explain the survival of pre-flood wisdom after the flood".[56] However, this "device" also reflects a genuine understanding concerning the transmission of knowledge through time.

The third-century BCE writer Berosus provides a story similar to that of Enosh, one connected with an account of the Great Flood and a king named Xisouthros:[57]

> Kronos appeared to Xisouthros in a dream and revealed that on the fifteenth of the month Daisios mankind would be destroyed by a great flood. He then ordered him to bury together all the tablets, the first, the middle, and the last, and hide them in Sippar, the city of the sun.[58]

These tablets contained all of the knowledge of human civilization. The high priest and writer Manetho claimed to have copied his accounts of Egypt from engraved columns that had been set up before the Great Flood.[59] There

---

55 Ibid., 146; also see the discussion of *The History of the Forefathers* text in Michael E. Stone, *Selected Studies in Pseudepigrapha and Apocrypha: With Special Reference to the Armenian Tradition* (Leiden: Brill, 1991), 28–40, and Adler, *Time Immemorial*, 75, as well as Pieter W. van der Horst, *Japheth in the Tents of Shem: Studies on Jewish Hellenism in Antiquity* (Leuven: Peeters, 2002), 152.

56 Adler, *Time Immemorial*, 59.

57 "Xisouthros" is a Greek form of Ziusudra, the Sumerian hero in the flood myth found in the famous "Epic of Gilgamesh"; see Michael Rice, *The Archaeology of the Arabian Gulf, c. 5000–323 BC* (London: Routledge, 1994), 138–139, as well as Ewa Wasilewska, *Creation Stories of the Middle East* (London: Jessica Kingsley Publishers, 2000), 175.

58 Gerald P. Verbrugghe and John M. Wickersham, *Berossos and Manetho, Introduced and Translated: Native Traditions in Ancient Mesopotamia and Egypt* (Ann Arbor: University of Michigan Press, 1996), 49.

59 Cory, 109; also note Adler, *Time Immemorial*, 59, and van der Horst, 144–145. For some interesting early nineteenth-century speculations on these stories concerning antediluvian knowledge that comes to be preserved, see pp. 277–278 of William Drummond, "Memoir on the Antiquity of the Zodiacs of Esneh and Dendera — Part II", *The Classical Journal* 24 (September and December 1821), 251–282, as well as Thomas Maurice, *The History of Hindostan and its Arts, and its Sciences, as Connected with the History of the Other Great Empires of Asia, During the Most Ancient Periods of the World. With Numerous*

also is a related story in the historical work of the Roman writer Ammianus Marcellinus:

> There are also subterranean fissures and winding passages … which, it is said, those acquainted with the ancient rites, since they had foreknowledge that a deluge was coming, and feared that the memory of the ceremonies might be destroyed, dug in the earth in many places with great labour; and on the walls of these caverns they carved many kinds of birds and beasts, and those countless forms of animals which they called hierographic writing.[60]

Again, the accuracy of this account is less important than the cultural idea that is being conveyed: that a sophisticated civilization will wish to preserve its learning and its cultural memory.

In the late nineteenth century, Bernard Carra de Vaux, a French orientalist, translated an Arabic text entitled *Kitāb akhbār al-zamān* — roughly, "Book of the History of the Past". This work, from the tenth century CE, contains a mixture of geography, amateur ethnography, and speculative history; it also recounts another rather fantastic story concerning the preservation of knowledge. The text discusses a figure named Sūrīd, who has had a vision concerning pending destruction of civilization.[61] He then

> shared it with Philemon, chief of the priests, and ordered the priests to consult the stars, to see which events threatened the greatest part of the world. They immediately began observations; they studied the sky with

Illustrative Engravings, 2 vols. (London: W. Bulmer and W. Nicol, 1820), 1.243–1.248 and 1.401–1.406. For further comments concerning these passages and related traditions, note John Garnier, *The Worship of the Dead or the Origin and Nature of Pagan Idolatry and its Bearing upon the Early History of Egypt and Babylonia* (London: Chapman & Hall, 1904), as well as Henry Christmas, *The Cradle of the Twin Giants, Science and History*, 2 vols. (London: Richard Bentley, 1849), 1.35–1.36.

60 Ammianus Marcellinus, *History* 22.15.30, in Ammianus Marcellinus, *History*, 3 vols., trans. J.C. Rolfe (Cambridge, MA: Harvard University Press, 1950–1952), 2.294–2.295.

61 Concerning Sūrīd, see Sándor Fodor, "The Origins of the Arabic Sūrīd Legend", *Zeitschrift für Ägyptische Sprache und Altertumskunde* 96 (December 1970): 103–109, and idem, "The Origins of the Arabic Legends of the Pyramids", *Acta Orientalia Academiae Scientiarum Hungaricae* 23.3 (September 1970): 335–363. Also note the discussion in Michael Cook, "Pharaonic History in Medieval Egypt", *Studia Islamica* 57 (1983): 67–103, and in Joost Kramer, "Wrestling with the Past: The Origins of Pharaonic History in the Islamic Middle Ages", in Carl Graves, et al., eds., *Current Research in Egyptology: Proceedings of the Thirteenth Annual Symposium University of Birmingham, 2012* (Oxford: Oxbow Books, 2013), 94–102.

> great care, and they found that a marvel descending from heaven and rising from the Earth would bring about the swallowing of almost all mankind. This marvel was to be a great deluge, after which nothing would survive. The king asked whether this event would be repeated several times, or whether it would be fulfilled once and for all. They made further observations and replied that the human race with its empires would reappear as before, like all things on the Earth. So the king had temples and grand monuments built for himself and his family, in order to preserve their bodies and all the wealth that they would deposit there. He had all the secrets of the sciences in which the Egyptians excelled more than any other nation inscribed on the roofs, on the walls, and on the columns, and he had painted [there] a figure of the greater and lesser stars, with signs that enabled them to be recognized. Also inscribed there were the names of the plants and their properties, the method for constructing talismans, their description, and the rules of arithmetic and of geometry. All those who know the books and the language of Egypt are able to make use of these images and inscriptions.[62]

The tale clearly outlines a certain perception of history. First of all, history is understood as cyclical. In addition, technical knowledge — such as that concerning astronomy — has tremendous value. Even if the story, then, is completely fictional, it reflects the value system of the author, and the genuine concerns of an ancient culture. A modern reader should not be dismissive of this fact, since our own preservation of knowledge might strike a future civilization as rather random, with everything from profound scientific discoveries to television shows apparently preserved with equal care, at least in terms of ranking by societal value.

Moreover, the author of the passage above again puts forward the idea that there was advanced knowledge in a prior time. This may also strike the modern reader as peculiar; however, it can be understood as a process whereby early writers placed themselves in a historical continuum concerning knowledge. Whereas our present culture values novelty and uniqueness, some other cultures reflected more upon their place in the long timeline of history. As one scholar puts it concerning the early Islamic world:

62 See Bernard Carra de Vaux, *L'abrege des merveilles, traduit de l'arabe d'apres les manuscrits de la Bibliotheque Nationale de Paris par le Baron Carra de Vaux* (Paris: Klincksieck, 1898), 248.

> The transmission and spread of knowledge about the past had stimulating effects on Muslim scholarly consciousness and encouraged Muslim elites to clarify and emphasize the position of their community among past and contemporary cultures. It is no wonder that the rise of universal histories in Muslim tradition also dates to the ninth century onward. It may be useful to draw parallels here between the ideologies of conquest in early Islam and the translation movement as an act not only to retrieve ancient wisdom from its historical, linguistic, and cultural repositories but also to reclaim, repossess, and reuse it in a new context.[63]

That is, the Islamic community emphasized a sense of place both synchronically and diachronically. The pursuit of knowledge — and the retrieval of ancient knowledge — was key in this process, even if it led to fabulous speculation. In this Islamic period

> the pyramids were already imagined to have been built by antediluvian or prophetic figures shortly after the Flood to preserve knowledge both within the walls of the pyramids and in the symbolic language of the hieroglyphic script. As a number of scholars have shown, legends of this sort were quite widespread at the time.[64]

From a modern perspective, one might even use the term "appropriation" for such a belief system: another commentator on this type of tale points out that for the Arabs "it tied some of the most stupendous monuments in human history [that is, the pyramids] to their own ancestral past and attributed them to a figure around whom fantastic stories were circulating."[65] However, one must also remember that the tellers of these stories about the preservation of knowledge and catastrophic deluges did not necessarily see themselves as writers of fiction; rather, the "stories surrounding the pyramids ... can be understood not merely as fanciful compositions, but as tools for giving acceptable meaning to these structures".[66]

---

63 See pp. 524–525 of Hayrettin Yücesoy, "Translation as Self-Consciousness: Ancient Sciences, Antediluvian Wisdom, and the 'Abbāsid Translation Movement", *Journal of World History* 20.4 (December 2009): 523–557.

64 Ibid., 554.

65 See p. 12 of Martyn Smith, "Pyramids in the Medieval Islamic Landscape: Perceptions and Narratives", *Journal of the American Research Center in Egypt* 43 (2007): 1–14.

66 Ibid., 14; for a related discussion of how knowledge concerning such subjects as astrology and alchemy were given an antediluvian framework in early Jewish culture, see David Frankfurter, "Apocalypses Real and Alleged in the Mani Codex", *Numen* 44.1 (January 1997):

In the fifth century CE, the Christian writer John Cassian, in a discussion of Genesis, talked about Ham, son of Noah. Cassian described how Ham had extensive knowledge of the arts and sciences from the period before the Flood. Noah did not permit him to bring the books of this knowledge on board the ark, and so Ham inscribed metal plates with this material and stored them in the ground. He even, it is alleged, was able to retrieve them when the waters receded.[67] Cassian gives the following description:

> And so, as ancient traditions tell us, Ham the son of Noah, who had been taught these superstitions and wicked and profane arts, as he knew that he could not possibly bring any handbook on these subjects into the ark, into which he was to enter with his good father and holy brothers, inscribed these nefarious arts and profane devices on plates of various metals which could not be destroyed by the flood of waters, and on hard rocks, and when the flood was over he hunted for them with the same inquisitiveness with which he had concealed them, and so transmitted to his descendants a seed-bed of profanity and perpetual sin.[68]

Clearly, as in other early sources, there is the idea here that the "lost knowledge" has a corrupting quality, one that was at least partially responsible for a civilizational cataclysm.[69] Just as the *loss* of knowledge was lamented by early writers, so, too, was there concern about mankind possessing the *wrong kind* of knowledge — a theme seen in everything from Plato to the *The Arabian Nights*.

The tradition of knowledge preserved in these kinds of media is found in a number of connected sources. These "traditions [are] ... primarily preoccupied with the idea of copying [information], where the tablets are portrayed as the specific means for the preservation of knowledge in the impending catastrophe."[70] The nature of that knowledge included, as we have pointed out,

---

60–73, noting especially pp. 65–66. Also see the note on p. 179 of David R. Blumenthal, ed. and trans., *The Commentary of R. Ḥōṭer ben Shelōmō to the Thirteen Principles of Maimonides* (Leiden: Brill, 1974).

67 See Needham, 5.4, 346.

68 John Cassian, *Conferences* 21, in Edgar C.S. Gibson, trans., *A Select Library of Nicene and Post-Nicene Fathers of the Christian Church*, 2nd series, vol. 11 (New York: The Christian Literature Company, 1894), 384.

69 For an engaging late nineteenth-century perspective concerning these early stories of lost or concealed knowledge, see Ernest C. Richardson, "Antediluvian Libraries", *Papers and Proceedings of the Eleventh General Meeting of the American Library Association* (Boston: Library Bureau, 1890): 40–44, and note a later discussion of similar topics in idem, *The Beginnings of Libraries* (Princeton: Princeton University Press, 1914).

70 Orlov, 157.

astronomical and calendrical material, and there is an accompanying story of Seth having devised a system of writing itself, namely the Hebrew alphabet.[71] As we have seen, the folktale concerning Gyges also encodes the interconnected ideas of lost knowledge, dangerous knowledge, and catastrophe.[72]

Once again, a modern reader should place less emphasis on the veracity of these accounts of preserved knowledge, and instead note the peculiar framing of the knowledge itself. Such framing shows, as one commentator puts it, that in

> this primordial time, so it was thought, mankind certainly possessed precious knowledge now lost, great wisdom now only attainable to those who are fortunate enough to lay hold of documents that survived.[73]

This idea — that "precious knowledge" was lost — is peculiar in our present age of novelty and innovation, where the present takes a clear priority over the past, and history tends to be seen as a linear path of ever-increasing progress, especially in the technological sense. By contrast, in these earlier times, there was a great deal of speculation about "the origins of the arts and sciences" as having arisen at some point in the *past*.[74] As this same commentator points out, early cultures such as the Babylonians and Egyptians actually competed over whose knowledge of astronomy, for example, was most deeply rooted in the past.

At the same time, there was an ambiguity about this technical knowledge and the antediluvian cultures that possessed it. In the Jewish tradition, the opinions of early authors were "equivocal about the ethical quality of the

71 Louis Ginzberg, *Legends of the Jews*, 2 vols. (Philadelphia: Jewish Publication Society, 2003), 1.116, n. 53.

72 For a modern look at the problem of retaining knowledge and passing it down despite catastrophes and other interruptions, see Gregory Benford, *Deep Time: How Humanity Communicates Across Millennia* (New York: Bard, 1999). A famous work in this area is Giorgio de Santillana and Hertha von Dechend, *Hamlet's Mill: An Essay on Myth and the Frame of Time* (Boston: Gambit, 1969); however, the arguments presented in that study are explicated in a very dense and convoluted fashion — see the critique by the famous historian of technology Lynn White, Jr. in *Isis* 61.4 (Winter 1970): 540–541, and the detailed review by the classicist Harald A.T. Reiche in *The Classical Journal* 69.1 (October–November 1973): 81–83. An approach similar to that found in *Hamlet's Mill* is Reiche's "The Language of Archaic Astronomy: A Clue to the Atlantis Myth?", in Kenneth Brecher and Michael Feirtag, eds., *Astronomy of the Ancients* (Cambridge, MA: MIT Press, 1979), 155–189; in that essay, Reiche suggests that Plato understood the Atlantis tale as a mode of "encoding" astronomical information.

73 van der Horst, 139.

74 Ibid., 139.

generations immediately preceding the flood."[75] The stories of antediluvian knowledge have a possible origin in Mesopotamia, and took root in cultures in Egypt, Greece, and elsewhere, and continued into the time of early Christian writers such as Tertullian.[76] That such a concept was so pervasive and long-lasting indicates a belief not only that knowledge can disappear, but also that it can then be re-discovered. In all these stories, it is intriguing how little the writers talk about humans as innovators or inventors — even technical knowledge comes not from the human mind, but from such entities as "fallen angels."[77]

In the modern period, the idea of advanced, ancient knowledge appears primarily in highly speculative works and in fiction of varying types. In the 1920's, the pulp fiction author Talbot Mundy (born William Lancaster Gribbon) wrote a story entitled *The Nine Unknown*, a work that outlined how the Indian emperor Ashoka entrusted a group of nine mysterious individuals to guard a set of books of secret knowledge.[78] Perhaps not surprisingly, Mundy was a Theosophist, and the concept of advanced technological knowledge from the past was part of Theosophist beliefs, as outlined in the discussion of Frederick Soddy in an earlier chapter here. Interestingly, Mundy's story — while entirely fictional — was recounted as fact in the 1960 occult classic, *Le matin des magiciens*.[79] A similar "synthesis" of scholarship and fantasy can be found in some of the work of the nineteenth-century French traveler and writer Louis Jacolliot — another figure influential for the Theosophists.[80] The process of fantastic tales becoming interpolated into historical narrative continues, perhaps, into the present, even if it is happening only on the margins of our culture.

75 Ibid., 155.

76 Ibid., 156–157.

77 Ibid., 157.

78 Talbot Mundy, *The Nine Unknown* (Indianapolis: Bobbs-Merrill, 1924); see the discussion in Brian Taves, *Talbot Mundy, Philosopher of Adventure: A Critical Biography* (Jefferson, NC: McFarland & Company, 2005), 94 et ff.

79 Louis Pauwels and Jacques Bergier, *Le matin des magiciens: Introduction au réalisme fantastique* (Paris: Gallimard 1960).

80 See the brief critique in René Guénon, *The King of the World*, trans. Henry D. Fohr, ed. Samuel D. Fohr (Hillsdale, NY: Sophia Perennis, 2004), 1, as well as Daniel Caracostea, "Louis-Francois Jacolliot (1837–1890): A Biographical Essay", *Theosophical History* 9.1 (January 2003): 12–39.

5 Knowledge and Myth, Knowledge *in* Myth

Again, our inclination nowadays is to regard the myths of ancient societies or even those of peoples in the contemporary "developing" world simply as fictional tales. The myths are regarded by modern societies as absurd inventions of superstitious cultures. In a book on Stonehenge, the historian and philosopher of science John North provides a careful analysis of the role and definition of myth:

> There is no sharp division between primitive myth and primitive science ... Myths are not usually seen as fictions, parables, or allegories by those who tell them. They are considered as in some sense true, but even so, their meanings are usually held to be somehow "below the surface".[81]

North goes on to look at several types of myths, and notes that they

> often contain abstract impersonations of rivers and trees, the sea and earth and sky, fire and earthquake — their behaviour being explained in terms that every person can comprehend.[82]

He acknowledges that myths deal with real information, such as seasonal changes, lunar phases, and so on. However, North is somewhat skeptical of the idea of the early third century BC Greek writer on myth and its relationship to history, Euhemerus, who felt that myths were an "echo" of actual historical events.

Nonetheless, North also says that it is "absurd" to maintain that, "primitive people invariably explain natural phenomena in terms of mystical and generally unseen powers, and that their mentality is 'prelogical'". The polymath Athanasius Kircher actually put forward a quite similar interpretation centuries before. He noted that, "although some myths are nothing but allegories and moral tales, the real reason for mythology is that precious things

---

81 John North, *Stonehenge: A New Interpretation of Prehistoric Man and the Cosmos* (New York: The Free Press, 1996), 531. A brief look at the intersection of myth and scientific knowledge in early cultures is also found in the opening chapter, "Voices from the Sands", of David Park's book *The Grand Contraption: The World as Myth, Number and Chance* (Princeton: Princeton University Press, 2005).

82 North, 531. For an interesting view of how earthquakes are framed in terms of myth in early cultures, see Chapter 3, "History, Myth, and the Reliability of the Written Record", in Amos Nur, *Apocalypse: Earthquakes, Archaeology, and the Wrath of God* (Princeton: Princeton University Press, 2008).

are always hidden, in order to avoid misunderstanding."[83] Kircher went on to make the point that "few people in the world ... are truly religious", his meaning being that religious language — that is, mythical language — actually should be understood as encoding genuine, practical knowledge.[84] Of course, Kircher at times was *too* inclusive in his sources: for his study of the history of ancient Latium, for example, he used the bogus work of Giovanni Nanni (Annius of Viterbo).[85] But such an error came from his sincere concern to obtain as much information as he could on a particular subject, and his understanding that information could come from many and varied sources.

Similarly, in speaking about what he calls "aboriginal ontology" in Australia, the anthropologist W.E.H. Stanner states the following about deductive reasoning and insight in a non-Western framework:

> The truth of it seems to be that man, society, and nature and past, present, and future are at one together within a unitary system of such a kind that its ontology cannot illuminate minds too much under the influence of humanism, rationalism, and science. One cannot easily, in the mobility of modern life and thought, grasp the vast intuitions of stability and permanence, and of life and man, at the heart of aboriginal ontology.
>
> It is fatally easy for Europeans, encountering such things for the first time, to go on to suppose that "mysticism" of this kind rules all aboriginal thought. It is not so. "Logical" thought and "rational" conduct are about as widely present in aboriginal life as they are on the simpler levels of European life. Once one understands three things — the primary intuitions which the aborigine has formed about the nature of the universe and man, those things in both which he thinks interesting and significant, and the conceptual system from within which he reasons about them — then the suppositions about prelogicality, illogicality, and nonrationality can be seen to be merely absurd. And if one wishes to see a really brilliant demonstration of deductive thought, one has only to see an aborigine tracking a wounded kangaroo, and persuade him to say why he interprets given signs in a certain way.[86]

83 Joscelyn Godwin, *Athanasius Kircher: A Renaissance Man and the Quest for Lost Knowledge* (London: Thames and Hudson, 1979), 22.

84 Ibid., 22.

85 See the discussion on pp. 33–34 of Harry B. Evans, *Exploring the Kingdom of Saturn: Kircher's* Latium *and its Legacy* (Ann Arbor: University of Michigan Press, 2012).

86 See p. 60 of W.E.H. Stanner, "The Dreaming", in T.A.G. Hungerford, ed., *Australian Signpost: An Anthology* (Melbourne: F.W. Cheshire, 1956), 51–65.

Stanner is making several points here. The first is that non-Western cultures have a completely different ontology, which one can define here as a way-of-being in the world. Moreover, Stanner is saying that "translating" that ontology into Western terms is difficult, if not impossible. But it is the final point that is most relevant here: that this other ontology provides its own, perfectly successful way for the aborigine to navigate in the world. Just because a Westerner does not *understand* that way, does not mean that it will not work in terms of gaining and transmitting knowledge — even technical knowledge, such as how to track an animal.

North, in turn, argues that Stonehenge demonstrates that "even a preliterate society can reason in a highly logical way".[87] North's statement has further resonance when one considers a case where a "mythical" belief held by a "primitive" people was realized in a modern discovery in anthropology. Two scientists, Peter Brown and Michael Morwood, reported in 2004 in the journal *Nature* the discovery of the remains of a variant of *Homo erectus*: this variant stood roughly one meter tall, and was quickly nicknamed "Hobbit".[88] The find was made on the Indonesian island of Flores; it was dated to a period 38,000 to 18,000 years ago, and the scientists noted that this *Homo floresiensis* coexisted with modern humans for tens of thousands of years. In a related piece, Henry Gee, an editor at *Nature*, commented:

> The discovery that *Homo floresiensis* survived until so very recently, in geological terms, makes it more likely that stories of other mythical, human-like creatures such as yetis are founded on grains of truth. In the light of the Flores skeleton, a recent initiative to scour central Sumatra for "orang pendek" can be viewed in a more serious light. This small, hairy, manlike creature has hitherto been known only from Malay folklore …

87 North, 533.

88 P. Brown, et al., "A New Small-bodied Hominin from the Late Pleistocene of Flores, Indonesia", *Nature* 431 (2004): 1055–1061; also note M.J. Morwood, et al., "Archaeology and Age of a New Hominin from Flores in Eastern Indonesia", *Nature* 431 (2004): 1087–1091. In addition, see M.J. Morwood and Penny Van Oosterzee, *A New Human: The Startling Discovery and Strange Story of the "Hobbits" of Flores, Indonesia* (New York: Smithsonian Books, 2007). The precise nature of these finds in Flores has been the subject of some debate; for a concise overview, see Kenneth W. Krause, "Pathology or Paradigm Shift? Human Evolution, *Ad Hominem* Science, and the Anomalous Hobbits of Flores", *Skeptical Inquirer* 33.4 (July–August 2009): 31–39. For a recent study, see Gerrit D. van den Bergh, et al., "*Homo floresiensis*-like Fossils from the Early Middle Pleistocene of Flores", *Nature* 534 (2016): 245–248, and Adam Brumm, et al., "Age and Context of the Oldest Known Hominin Fossils from Flores", *Nature* 534 (2016): 249–253.

> Now, cryptozoology, the study of such fabulous creatures, can come in from the cold.[89]

What is critical here is the reference to "Malay folklore" — again, this is the idea that such sources have an empirical value, recording and retaining historical and anthropological information: "orang pendek" literally means, "short people".

This folk tradition of "small, hairy, manlike creatures" is quite old, and exists in Asia, Europe, and the Americas.[90] Nor was this tradition obscure; in fact, such entities — sometimes called "wild men" — are not just mentioned in texts, but even find their way into visual culture. For example, they are depicted on a fifteenth-century Spanish piece known as the "'Admiral' Heraldic Carpet".[91] This carpet includes both European and Islamic motifs, as medieval Spain still included both Christian and Muslim cultures at this time. Along the border here are scenes of the "wild men" fighting bears (fig. 8 and fig. 9).

In the Chitral district of northern Pakistan, such beings are called *jangali mosh*, meaning a "man of the forest" or "wild man".[92] Russian scientists have also investigated this phenomenon.[93] The famed American painter and

89 Henry Gee, "Flores, God and Cryptozoology", *Nature News* (27 October 2004), doi:10.1038/news041025-2.

90 For some examples, see the discussion in Gregory Mobley, "The Wild Man in the Bible and the Ancient Near East", *Journal of Biblical Literature* 116.2 (Summer 1997): 217–233, and the discussion on pp. 200 et ff. of John Block Friedman, *The Monstrous Races in Medieval Art and Thought* (Syracuse: Syracuse University Press, 2000), as well as Gregory Forth, "Images of the Wildman Inside and Outside Europe", *Folklore* 118.3 (December 2007): 261–281. Also see the study, with numerous images, in Timothy Husband and Gloria Gilmore-House, *The Wild Man: Medieval Myth and Symbolism* (New York: The Metropolitan Museum of Art, 1980).

91 This carpet is part of a series of medieval Spanish rugs; they bear the name "admiral" carpets because they include the coat of arms of Fadrique Enríquez de Mendoza (1390–1473), who was second admiral of Castile, and grandfather of Ferdinand II, the first king of a united Spain. The carpet shown here is on display at the Philadelphia Museum of Art.

92 See p. 353 of Jordi Magraner, "Oral Statements Concerning Living Unknown Hominids: Analysis, Criticism, and Implication for Language Origins", in Bernard H. Bichakjian, et al., eds., *Becoming Loquens: More Studies in Language Origins* (Frankfurt am Main: Peter Lang Europäischer Verlag der Wissenschaften 2000), 339–385.

93 There is an extensive literature and much debate concerning Russian investigations of the persistence of non-human hominids. See the discussion about research of "relic hominids" in Magraner, 341. Also see, for example, B.F. Porshnev, Dmitri Bayanov, and Igor Bourtsev, "The Troglodytidae and the Hominidae in the Taxonomy and Evolution of Higher Primates", *Current Anthropology* 15.4 (December 1974): 449–456; Gordon Strasenburgh, "On Paranthropus and 'Relic Hominoids'", *Current Anthropology* 16.3 (September 1975): 486–487; and Bayanov and Bourtsev, "On Neanderthal vs. Paranthropus",

FIGURE 8 "Admiral" Heraldic Carpet; Spanish, possibly woven by Muslim craftsmen, ca. 1429–1473, wool

FIGURE 9 "Admiral" Heraldic Carpet — detail of "wild man"

naturalist Charles Willson Peale (1741–1827) spoke of the *ourang outang*, in that case clearly conflating it with the idea of a "wild man of the woods".[94]

In terms of early sources, there are descriptions of "primitive humans" in writers such as Lucretius; in talking about the remote past, he notes — in poetic but precise form — the following:

> The human beings who lived on earth in those early days were far tougher than we are ... larger and more solid bones formed the inner framework of their bodies, while their flesh was knit with strong sinews, and they were not easily affected by heat or cold or unaccustomed food or any physical malady ... No sturdy farmer guided the curved plow; no one knew how to work the fields with iron implements or plant young saplings in the earth or cut the old boughs from tall trees with pruning hooks ... For the most part they nourished their bodies among the acorn-bearing oaks; and arbute berries ... As yet they had no knowledge of how to utilize fire or clothe their bodies in skins stripped from wild beasts. They lived in woods and mountain caves and forests and, when compelled to escape from the lashing of wind and rain, sheltered their shaggy limbs among

---

Current Anthropology 17.2 (June 1976): 312–318. Also note p. 54 and p. 65, n. 44 of Adrienne Mayor and Michael Heaney, "Griffins and Arimaspeans", *Folklore* 104.1/2 (1993): 40–66; and Michael Heaney, "The Mongolian Almas: A Historical Re-evaluation of the Sighting by Baradin", *Cryptozoology* 2 (1983): 40–52.

94 See David R. Brigham, *Public Culture in the Early Republic: Peale's Museum and Its Audience* (Washington, D.C.: Smithsonian Institution Press, 1995), 130–131, as well as the longer discussion in Brett Mizelle, "'Man Cannot Behold It Without Contemplating Himself': Monkeys, Apes and Human Identity in the Early American Republic", *Pennsylvania History: A Journal of Mid-Atlantic Studies* 66 (1999): 144–173. On the concept of the "wild man" generally, see Richard Bernheimer, *Wild Men in the Middle Ages: A Study in Art, Sentiment, and Demonology* (Cambridge, MA: Harvard University Press, 1952), and Marjorie Halpin and Michael M. Ames, eds., *Manlike Monsters on Trial: Early Records and Modern Evidence* (Vancouver: University of British Columbia Press, 1980). Also see the detailed discussion and numerous examples of "wild men" in medieval and early Renaissance sources in Barbara Baert, "Caliban as a Wild-Man: An Iconographical Approach", in Nadia Lie and Theo D'Haen, eds., *Constellation Caliban: Figurations of a Character* (Amsterdam: Rodopi, 1997), 43–60. Brigham notes: "In Basel at the Historical Museum, I saw at least one tapestry that depicts a "wild man" with hair covering its body. This is apparently a folkoric tradition of medieval Europe, including Switzerland" (David R. Brigham, private communication, 14 March 2001); the museum has, in fact, both a tapestry ("Wirkteppich mit Sinnbildern für hoffnungsvolle und enttäuschte Liebe", 1470, Inv. Nr. 2003.6) and a statue ("Zeichen des ehemaligen Gasthauses 'Zum wilden Mann'", 1600, Inv. Nr. 1870.1202) depicting "wild men". Moreover, in Basel there is still a traditional festival, the Vogel Gryff Volksfest, where a lion, a griffin and a "wild man of the woods" go down the Rhine River on a raft.

> the thickets. They were unable to look to the common interest, and had no knowledge of the mutual benefits of any customs or laws. Individuals instinctively seized whatever prize fortune had offered to them, trained as they were to live and use strength for themselves alone.[95]

Lucretius, in the same passage, goes on to comment that early humans lacked speech, sophisticated tools, clothing, and fire. Lucretius may have been extrapolating from accounts that he had received concerning groups of humans less developed than those in his own society — groups of hunter-gatherers in regions beyond the Roman world — or perhaps he was working from some kind of common cultural memory. Regardless, this is another example of an author who is aware of humanity's remote past, and interested in providing the reader with a rather vivid picture of the life of prehistoric human beings.[96]

In addition to *Homo floresiensis*, there are others hints that various hominids were observed by early humans. In 2012, a scientific paper was published recounting a find in China of fossils that may represent a "late-surviving archaic population". That population has been dated to just 11,500 to 14,500 years ago — quite recent in terms of human evolution.[97] In terms of the theme here — how knowledge is encoded and the reliability of such knowledge, one researcher on this subject of "relic hominids" put it this way:

> There are two ways of interpreting these stories which describe beings that are close enough to Man to be related to Mankind rather than to apes. Either the stories are made up, and totally imagined, as prompted by oral tradition, or they describe a reality that does not belong to the witnesses' make-up. The arguments which enable us to discuss the relative weight of each of these two interpretations concern, on the one hand, the consistency between the various stories and, on the other, the

95 Lucretius, *On the Nature of Things*, 5.925–5.962, trans. Smith, 161–163.

96 For an analysis of the speculative anthropology of Lucretius and other early writers, see Wiktor Stoczkowski, *Explaining Human Origins: Myth, Imagination, and Conjecture* (Cambridge: Cambridge University Press, 2002). For a discussion of the Lucretius passage cited here, as well as a look at medieval sources on early humans, see Bayanov and Bourtsev, "On Neanderthal vs. Paranthropus", cited above. In Barber and Barber (pp. 94–95), there is also a brief discussion of how early foiktales and fairly tales may contain anthropological knowledge.

97 Darren Curnoe, et al., "Human Remains from the Pleistocene-Holocene Transition of Southwest China Suggest a Complex Evolutionary History for East Asians", *PLoS ONE* 7.3 (March 2012): e31918, doi:10.1371/journal.pone.0031918.

> consistency between the environment and the conditions allowing wild human populations to exist.[98]

If a piece of knowledge that "does not belong to the witnesses' make-up" ends up being preserved, then that knowledge certainly may take the form of what strikes us as a peculiar oral tradition, as the storyteller tries to capture and make sense of something that is outside their area of understanding and experience. But that does not mean, then, that the tales are "totally imagined"; as a number of other examples in this book have shown, technical knowledge (and in this case, perhaps, anthropological knowledge) can be preserved in forms that capture the essential idea, such as mechanized flight, and even some mechanical details, such as the need for a switch to activate the mechanism. In the case of the "wild men", it is very possible that human cultures preserved, over the millennia, knowledge concerning encounters with other people — but people not quite like *Homo sapiens*.

6 Changing Knowledge, Changing History

Euhemerus penned a now-lost work that seems to have taken the form of a travel narrative. The journey is one to a mysterious Indian Ocean island, a utopian locale called Panchaia. Through this tale, Euhemerus promotes a conjecture that the gods were actually powerful human rulers who lived in the remote past, and were deified by subsequent generations. Although the tale is clearly fanciful, a serious point about myth is being made. Again, too, as in the story of Seth, there is a reference to material means of preserving knowledge — the narrator boldly claims to have received at least some of his information from inscriptions on gold stelae in the temple of Zeus.[99]

William Irwin Thompson puts the problem of historicity and mythical tradition in another framework in his book *At the Edge of History*. There, in an essay entitled "The Re-Visioning of History", he notes that even as we read the most peculiar fantasies of novels, we are struck by something peculiar:

> But as the fantasy crops up in several clearly separated traditions, it begins to insinuate its way into the reader's imagination until it becomes

98 Magraner, 377.

99 On Euhemerus, see the extensive study by Marek Winiarczyk, *Euhemeros von Messene* (Munich: K.G. Saur, 2002); an English translation of the Panchaia tale is found in Clay and Purvis, *Four Island Utopias*, cited earlier, 98–106.

> slightly unsettling. At the back of one's head a thought begins to push toward the front of one's consciousness: *What if the history of the world is a "myth," but myth is the remains of the real history of the earth?*[100]

Thompson goes on to point out that such an idea can be dangerous, especially in the academic world, where there is what he terms a "university-accepted version of history".[101] While one should certainly disagree with his defense of catastrophists such as Velikovsky and sincere but misguided researchers such as Hapgood, Thompson's critique of the standard approach to history is sound. Of course, the crux of the matter when talking about the relationship between myth and history is *evidence*. Is there any evidence that our standard version of history is incomplete or incorrect? Thompson puts it well when he says:

> Bits and pieces of a new world view seem to lie scattered around us, and as suddenly the imagination perceives them in a new form, the old world views seems all the more incredible. Still, imagination is not knowledge; we do not know — anything yet, but ... I now believe that myth is the detritus of actual history.[102]

In some ways, then, we are at the beginning of such investigations of the "detritus of actual history". Again, Thompson: "At the beginning of a study we do not need to *know* all the answers through *conclusive* proof; we do need to have an imaginatively open mind so that the *incipient* proof is not cast aside."[103]

In his book *Natural Knowledge in Preclassical Antiquity*, the historian of science Mott T. Greene, noted in Chapter 1 here, presents his study as follows:

> This book is a series of meditations on the relationships between mythology and natural knowledge both in antiquity and in the present. I have used the phrase "natural knowledge" rather than "science" in my title because what we generally think of as "science" is only one of many ways of getting, storing, and conveying knowledge about our natural surroundings.[104]

---

100 Thompson, *At the Edge of History* (New York: Harper & Row, 1971), 126.

101 Ibid., 129.

102 Ibid., 132.

103 Ibid., 132.

104 Greene, xi. An engaging series of conjectures on this idea of using myth for "getting, storing, and conveying knowledge about our natural surroundings" can be found in Florence Wood and Kenneth Wood, *Homer's Secret Iliad: The Epic of the Night Skies Decoded* (London: John Murray, 1999).

Even in antiquity, there were arguments about these "relationships between mythology and natural knowledge", focusing on sources such as Homer. The Stoic philosopher, Crates of Mallos, for example,

> asserted that Odysseus had … sailed outside the Mediterranean; that Homer knew this accurately, just as he knew the true structure of the four-part world; and that therefore Books 9–12 of the *Odyssey* should be read quite literally as the record of a voyage into Ocean.[105]

Other writers, such as the historian Polybius,

> adopted a middle ground … agreeing with Crates on the essential historicity of the fabulous wanderings, but seeing them nevertheless as having been "exported" from their original Mediterranean setting.[106]

These debates may illustrate the problems of knowledge transmission in a period where oral and written cultural traditions were intersecting, as discussed earlier in this book.[107]

Another scholar who examines the culture of knowledge notes:

> [*S*]*cience, technology*, and even *history* are present-day terms whose meanings may be inappropriate or misleading for past cultures even when cognate terms exist in the diverse languages of those cultures. For example, *scientia* in Latin means "knowledge" in a broad sense and has none of the methodological or disciplinary meanings that we today associate with *science* and *scientist*. Similarly, the ancient Greek word *techne* refers to material production of all kinds, from making soup to constructing catapults, and the reasoning associated with that production.[108]

Moreover, the categories of knowledge are not fixed — all the more reason that it is necessary to explore myth, poetry, and other forms for historical remnants — Thompson's "detritus" — of what we term today scientific or technical knowledge:

105 Romm, *The Edges of the Earth in Ancient Thought*, 188.
106 Ibid., 189; also see the discussion in ibid., 196 et ff.
107 See Werner, 183.
108 Long, 2.

> Conceptual categories change from one historical period to another. For example, Aristotle delineated three areas of human activity: first, material and technical production (*techne*); second, action (*praxis*), such as political or military action, that requires judgment in contingent or uncertain situations (*phronesis*); and third, theoretical knowledge or knowledge of unchanging things (*episteme*). Aristotle's separation of material production from action and from theoretical knowledge presupposed a hierarchy with techne at the bottom and episteme, or theoretical knowledge, at the top ... Such categories and the relationships between them exert a crucial influence on the construction of knowledge itself.[109]

That "construction of knowledge" can be interpreted several ways. For the discussion here, this term means that certain kinds of knowledge will be packaged in a particular way in mythic or other textual structures. Concerning myth as a retainer of knowledge, Mott says that "[s]ome detailed and accurate knowledge of modern natural science is required to make sense of the mythic material, insofar as this mythic material also is detailed and accurate material about the natural world, albeit in an unfamiliar form."[110] The fourth-century CE philosophical commentator Macrobius felt that "prudent men" discuss nature's "secrets only through the medium of myth".[111]

Mott also chooses a very clear approach to myth, one designed to understand it in a more profound way; his statement on this approach is worth quoting in full:

> I have insisted on challenging the overwhelming diversion of interest in mythology away from its contents towards its forms and transformations. Such an exclusive concentration on forms must been seen to distort not mere details, but the overall function of these entities (myths) and their informative content. While it is possible to learn much about mythology by examining its formal structures, this is not a complete study of mythology. One might say by way of analogy that we may learn much about the principles of mechanics by taking apart a watch to see how it works; and yet we must also have firmly in mind that people make watches for the exclusive purpose of telling time, and not at all to exhibit the principles

109 Ibid., 2.

110 Greene, xii.

111 See p. 9 of David Ashurst, "Journey to the Antipodes. Cosmological and Mythological Themes in *Alexanders Saga*", in Geraldine Barnes and Margaret Clunies Ross, eds., *Proceedings of the Eleventh International Saga Conference* (Sydney: University of Sydney, Centre for Medieval Studies, 2000), 1–13.

> of mechanics. So it is with myths. They were not made to exhibit the structures of human consciousness (though they do); they were made to transmit the information they contain.[112]

Indeed, perhaps because of our excessive focus on forms and structures, the particular uses of myth in the texts of Plato and other early writers — and the meaning and content of those myths — have become unfamiliar to us in our modern, specialized age.[113]

Mott notes, for example, that Plato employs a definition of *mythos* as "a general term for all material not susceptible to dialectical treatment and demonstrative argument by the means of collection and division".[114] Certainly, this definition does *not* denote myth as pure fiction. As Mott concludes, our modern view tends to give the mythical content of "Plato's dialogues a figurative status as literary invention".[115] Plato's definitions are quite different; the idea that Plato was engaged in "literary invention", in fact,

> falsifies the approach taken to rhetoric (and to mythic discourse) by Plato, and clouds the actual distinctions being made between mythic material and other forms of knowing. The distinction Plato makes ... is not between false/figurative myth and true/representative rhetoric. Rather, the distinction is that between classes of things in the world not knowable by dialectic, and things which are knowable by dialectic.[116]

This provides us at least a hint as to a very different view of knowledge than the one employed nowadays. For Plato, there is a tripartite approach: first, knowledge can be gained through dialogue — interlocutors working together, debating towards truth; in addition, knowledge can be accumulated through the gathering of evidence, a method not unfamiliar to our current methods; and finally, knowledge may be gathered in a way that employs neither of these methods, but rather through allegories bearing events, history, and useful insights. The texts that are examined in this book — including texts characterized as "myths" — point to the need for further investigations. The ancient

---

112 Greene, xviii.

113 Note, however, that a number of authors have written on the idea that mythical structures continue to exist in our thinking, in our literature, and so on; see, for example, Eleazar M. Meletinski, *The Poetics of Myth*, trans. Guy Lanoue and Alexandre Sadetsky (New York: Garland, 1998).

114 Greene, 160.

115 Ibid., 162.

116 Ibid., 162.

writers themselves had questions concerning the histories of their own societies and traditions, and questions concerning the origins of ideas, inventions, and even civilization itself; many of these questions remain open. The texts speak to us from ages past *about* ages past, and likely have meanings that are not yet clear to us.

In conclusion here, it is worth noting a philosophical angle to this question of the nature, encoding, and transmission of knowledge. The process of keeping knowledge intact serves a very specific function beyond the preservation of the knowledge itself: that other function is the retention of a culture's, even a civilization's, identity. The use of myths, symbols, and icons play a key role in this process, and the philosophical implications can be profound. The French theorist Jean Baudrillard talks about the problem of the separation of the *representation* — be it a myth or other vehicle — from what is being *represented.* He discusses this in terms of icons and worship of the divine:

> [T]he question returns to religion and the simulacrum of divinity: "I forbade that there be any simulacra in the temples because the divinity that animates nature can never be represented." Indeed it can be. But what becomes of the divinity when it reveals itself in icons, when it is multiplied in simulacra? Does it remain the supreme power that is simply incarnated in images as a visible theology? Or does it volatilize itself in the simulacra that, alone, deploy their power and pomp of fascination — the visible machinery of icons substituted for the pure and intelligible Idea of God? This is precisely what was feared by Iconoclasts, whose millennial quarrel is still with us today. This is precisely because they predicted this omnipotence of simulacra, the faculty simulacra have of effacing God from the conscience of man.[117]

In other words, the representation can become empty, and indeed take the place of the idea being represented. Baudrillard continues:

> One can see that the iconoclasts, whom one accuses of disdaining and negating images, were those who accorded them their true value, in contrast to the iconolaters who only saw reflections in them and were content to venerate a filigree God. On the other hand, one can say that the icon worshipers were the most modern minds, the most adventurous, because, in the guise of having God become apparent in the mirror of

117 Jean Baudrillard, *Simulacra and Simulation*, trans. Sheila Faria Glaser (Ann Arbor: University of Michigan Press, 1994), 4.

> images, they were already enacting his death and his disappearance in the epiphany of his representations (which, perhaps, they already knew no longer represented anything, that they were purely a game, but that it was therein the great game lay — knowing also that it is dangerous to unmask images, since they dissimulate the fact that there is nothing behind them).[118]

In a very similar manner, myths, too, can end up with nothing behind them, representing nothing and devoid of content. A myth has to retain its connection to the knowledge embodied within. If it loses that, it becomes nothing but a simulacrum.

118 Ibid., 5.

CHAPTER 7

Conclusions — What Did They Mean?

1 Technology and the Concept of the "Golden Age"

This book has looked at a sampling of early texts that reflect the idea of "lost knowledge". A number of these texts imply a non-linear history of technological development, the rise and disappearance of an advanced kind of knowledge, skill, or even entire society. But there is no archaeological evidence for such a pattern in history. Atlantis has not been found, and archaeological excavations have unearthed no fragments of early Chinese flying machines, and no one has stumbled upon fragments of ancient X-ray machines.[1]

But it is clear that a number of early stories, folktales, chronicles, and other works were concerned with the idea of lost, advanced knowledge. In addition to the idea of a "golden age", not uncommon in various cultures, there is here the more particular idea of a "golden *technê*" — the ability to devise clever, sophisticated machines. This, in turn, implies that past civilizations were not just a rhetorical tool for early writers, but also a fundamental part of their cultural worldview and their perception of history. Broadly speaking, it was a history that was cyclical, one where mankind's knowledge and skills had flourished and then disappeared some time in the remote past.

In terms of the Chinese stories of flying machines and strange mirrors, one might say that the cyclical pattern of history is being presented on a smaller scale: knowledge that existed just a few generations, say, before the writer's time was now lost, and the technological descriptions came to take on an almost magical quality. There were, interestingly, a few Chinese writers who went against this kind of thinking. The best known such writer is Wang Chong (王充), a Han dynasty scholar who wrote on both philosophy and the sciences in a work entitled *Lunheng* (論衡), a title with the literal meaning of "measured discussion". His work is notable in terms of the analysis here in that he

1 Not surprisingly, there is a large body of "fringe" archaeological evidence for vanished, advanced civilizations. This apparent evidence is known colloquially as "OOParts" — an acronym for "Out Of Place Artifacts"; such objects include the famous "Baghdad battery", whose exact function is subject to speculation, the "Coso artifact", and many others. For a more serious discussion of how the remains of a highly advanced technological society would survive in an archaeological setting, see Frank and Schmidt, "The Silurian Hypothesis", cited earlier.

 | DOI:10.1163/9789004352728_008

"believed in the superiority of his own time by virtue of its moral standards."[2] As outlined below, the Daoist text known as the *Zhuangzi* presented the idea that ancient understanding was *superior* to that of the present age. Although Wang used Daoist perspectives in his writing, in this case he takes an *opposite* approach, arguing that society had advanced rather than regressed in moral terms. Wang presents these ideas at several points in the *Lunheng*; for example, in a section entitled *Xuan Han Pian* (宣漢篇) — roughly meaning "Discussion in Praise of the Han [Dynasty]" — Wang begins:

> The Literati contend that the Five Emperors and the Three Rulers brought the empire the blessings of universal peace, and that since the accession of the Han dynasty there has not yet been general peace.[3]

He then presents an extended counterargument — oddly based on omens — and concludes that, "As a matter of fact, the [present] empire enjoys universal peace."[4] Moreover, Wang, a rationalist, believed that his period of history possessed a "better way of life in material terms".[5] Another section of his book that drives home this point is, in fact, entitled *Qi Shi* (齊世), meaning "The Equality of the Ages" — expressing Wang's belief that the past was not necessarily better. He writes the following in that chapter:

> There is a saying that in ancient times people were tall, good-looking, and strong, and lived to become about a hundred years old, whereas in modern times they are short, ugly, cut off in their prime, and short-lived ... This statement is preposterous. In olden days the rulers were sages, and so they are in modern times. The virtue of the sages then and now does not differ, [and] therefore their government in ancient and modern times cannot be different. The Heaven of antiquity is the Heaven of later ages ... The people of former ages are the same as those of modern times.[6]

Note Wang's connection between moral development and physical development, a connection examined elsewhere in the present study. Thus one might

---

2 See p. 326 of Michael Loewe, "The Cycle of Cathay: Concepts of Time in Han China and Their Problems", in Chun-chieh Huang and Erik Zürcher, eds., *Time & Space in Chinese Culture* (Leiden: Brill, 1995), 305–328.

3 See Chong Wang, *Lunheng*, 2 vols., trans. Alfred Forke (Leipzig: O. Harrassowitz, 1907–1911), 2.192.

4 Ibid., 2.194; also see Forke's comments on Chong's dismissal of a "Golden Age" in ibid., 1.41.

5 Loewe, 326.

6 Forke, 1.471.

say that one ancient view was that as morals decline, so physical development declines, and technology that was once powerful also disappears: such a scenario is evident in the "Ring of Gyges" tale, examined earlier. But of course here, Wang is arguing the opposite — neither moral nor physical qualities have changed. His reasoning is a peculiar combination of the metaphysical and pragmatic: he believed that as all things are imbued with the pervasive essence of *qi* (氣), and that *qi* does not change, then nothing can change. Specifically, Wang writes that *yuanqi chun he gujin buyi* (元氣純和古今不異) — a phrase roughly meaning that the purity of the *qi* of the past is no different to that of the present. Since *qi* does not change, so civilizations do not change.

However, even with exceptions such as that presented by Wang Chong, the idea that there was a "golden age", more sophisticated than the present one, and perhaps more developed technologically and culturally, is found in many cultures, from the Hindu to the Hopi. As one writer notes:

> The belief that the present world and its inhabitants exist in an age of privation and distress is not unique to India, nor is the expectation that this situation is only a temporary one, that it degenerated from earlier better ages.[7]

One might wish to offer some kind of psycho-historical explanation — that is, the universality of the idea suggests a fundamental psychological origin, a kind of universal sense of *temps perdu*, deep in the unconscious of human beings everywhere.[8]

But it seems reasonable to point out that such a feeling would run contrary to our personal development — most of us *gain* knowledge as we mature — and to our evolution as human beings. The ancient Greeks, for example, saw themselves as a sophisticated culture, so why would they believe that humans were *more* developed in the past? An idea of a lost "golden age", of course, could be useful as social criticism of the present, a standard against which the

7 See p. 91 of Ludo Rocher, "Concepts of Time in Classical India", in Ralph M. Rosen, ed., *Time and Temporality in the Ancient World* (Philadelphia: University of Pennsylvania Museum of Archaeology and Anthropology, 2004), 91–110.

8 Geoffrey Ashe, in his book *Dawn Behind the Dawn: A Search for the Earthly Paradise* (New York: Henry Holt, 1992), proposes a possible origin in early Altaic religious and cultural systems for later "golden age" myths among the Greeks and other civilizations; Ashe also has sought historical and archaeological evidence for the legend of King Arthur. Richard Heinberg's *Memories and Visions of Paradise: Exploring the Universal Myth of a Lost Golden Age* (Wheaton, IL: Quest Books, 1995) combines anthropology, archaeology, mythology, and New Age thinking in ruminations on the idea of a "golden age" across cultures.

current, decadent society could be compared. That kind of comparison usually centers on a kind of devolution in morals, not technological knowledge. However, the two are not unconnected: We have seen, in the tale of the "Ring of Gyges", the implied message that the society of giants had superior technological knowledge, and that they likely disappeared due to some resulting moral failure. These "techno-magical" abilities — such as invisibility — inevitably lead to the destruction of a civilization through a degradation of behaviors. Indeed, as soon as Gyges comes into possession of the techno-magical marvel of the ring, he repeats this pattern of bad moral conduct.

It is interesting that Hopi mythology reveals a similar connection, even employing a visual structure similar to the Platonic "cave". Writing about the Hopi "Emergence Myth", one scholars notes that

> Hopi mythology is based *a priori* upon the belief that the present World is one of several (in most cases, it is the Fourth World) all cosmographically situated one upon the other ... Each world begins as a paradise, but is sooner or later disrupted through mankind's evil behaviour. Each world is destroyed by natural catastrophe which forces the faithful few, led by certain clan deities, to climb up to the next world level, the floor of which is the sky of the lower level ... The Emergence Myth continues this line of reasoning by describing the conditions of the Third World prior to the Emergence to the present Fourth World ... Again, the paradisic situation is destroyed through disobedience, evil magic and sexual immorality.[9]

In a famous speech given at the United Nations in 1992, the Hopi representative Thomas Banyacya reiterated this tradition of the rise and fall of civilizations.[10] There is a technological connection here, as well. According to the Hopi, the fallen civilizations had technology more advanced than subsequent civilizations. An example of such a fantastical technology, the *paatuwvota* or "magic flying shield", was discussed earlier in this book. One might label these traditions as fanciful, but they indicate a consciousness on the part of the culture concerning at least the *concept* of advanced knowledge, its dangers, and the cyclical nature of history.

In a similar vein, the Ainu have stories that provide outlines of a more sophisticated past and subsequent decline:

---

9 See pp. 219–220 of Geertz, "A Reed Pierced the Sky: Hopi Indian Cosmography on Third Mesa, Arizona", *Numen* 31.2 (December 1984): 216–241.

10 On Thomas Banyacya, see Robert McGill Thomas, Jr., "Thomas Banyacya, 89, Teller of Hopi Prophecy to World", *The New York Times*, 15 February 1999, sec. B, p. 8, col. 5.

> Those were the days of magic. People were then able to fly six or seven miles, and to light on the trees like birds, when they went out hunting. But now the world is decrepit, and all good things are gone.[11]

An earlier chapter in this book noted how the Ainu, in a way similar to other cultures, have stories of flying machines; again, such technologies are framed as being lost or forgotten in South Asian, Chinese, and other cultures.

In a completely different cultural context, Francis Bacon also expressed a similar uneasiness about technologically advanced societies. In Bacon's *New Atlantis*, there is a

> decidedly mixed Baconian message about the technological future: that the conquest of nature may just as likely be for ill as for good. Bensalem seems haunted by the problems of the soul debased by materialism, bad means used for good ends, and lawless use of technological power. The *New Atlantis* suggests the problem, as much as the promise, of technology and the story raises questions about the ways and means of the scientific project, the ends and the limits of the conquest of nature, and the implications of science and technology for human life and values.[12]

Chaucer also expressed a discomfort about mechanical knowledge. First of all, his framing of the topic sets up an interesting ambiguity:

> That Chaucer was both interested in and knowledgeable about medieval technology is evidenced most clearly by his *Treatise on the Astrolabe*, but also by the occurrence of technological devices and technical descriptions in both the *Canterbury Tales* and the shorter poetry.... When one examines Chaucer's poetic usage of both mechanical and magical devices in the *Canterbury Tales* as whole, however, the line between technology and magic becomes as blurry as that between science and pseudo-science. Technological inventions, particularly mechanical devices, frequently function within the *Tales* as if they were in fact magical, while magic often reveals itself to be mechanically based.[13]

---

11 See pp. 46–47 of Basil Hall Chamberlain, "Aino Folk-Lore", *The Folk-Lore Journal* 6.1 (January–December 1888): 1–51.

12 Weinberger, "On the Miracles in Bacon's *New Atlantis*", 110.

13 See p. 377 of Joyce Tally Lionarons, "Magic, Machines, and Deception: Technology in the 'Canterbury Tales'", *The Chaucer Review* 27.4 (1993): 377–386.

Why does Chaucer draw these two ideas together? It is as a subtle warning:

> Magic and technology are alike in that both are based on knowledge hidden from or at least unavailable to the majority of people; this allows each to function within the *Tales* most often as an aid to fraud and trickery.[14]

As for devices such as the "horse of brass", it is almost magical in its capabilities, but distinctly technological: "What is most interesting about Chaucer's version of the flying horse is its distinctly mechanical nature."[15]

So, Chaucer is speaking about technology *first*, in a sense, and highlighting, then, that it may seem magical in its operation. In a similar way, in Chaucer's tales "magical toys become objects of technological speculation".[16] That is, the technology is at the root, not some fantastical magic. The perspective in Chaucer's stories is that technology is around us, and can be made to deceive. Again there is a warning, because "even if the so-called 'magic' is really technology, it is still suspect, precisely *because* it is based on uncommon knowledge and can therefore be used by the learned to deceive the ignorant" — in fact, Chaucer says as much in the *Squire's Tale*.[17]

What should strike the modern reader as peculiar, however, is the fact that Chaucer — who was, of course, living in a society that was not particularly technologically sophisticated — should have wanted to write warnings about technology. For Chaucer, technology was "dangerous in that it can sometimes deceive the technician into believing that it can provide both god-like control over the natural world and the ultimate answers to philosophical, instead of merely technical, questions."[18] It seems likely that Chaucer formulated this idea from looking *backwards* rather than forwards in time. As he was familiar with history — and perhaps with stories such as the "Ring of Gyges" — he may have developed a strong sense concerning past societies that indeed had developed sophisticated technologies that eventually got the better of them. The idea of civilizations abusing technology, or even destroying themselves with it, appears to be almost universal — with the same universal causes, as well.

With Chaucer's works, and certainly with the much earlier texts, there is the problem of "prime objects", a term used by the art historian George Kubler in reference to his field of study. For Kubler, a "prime object" is either an object

---

14 Ibid., 377.
15 Ibid., 378.
16 Ibid., 379.
17 Ibid., 379.
18 Ibid., 385.

or an idea that is truly original, and that results in replications over time. In terms of history, Kubler notes, "prime objects" are problematic, and "their order in history is enigmatic", an issue which certainly appears in the history of technology.[19] With discussions of technological objects in early texts, the vagueness of many of the descriptions gives one the impression of dealing with what Kubler terms "pale reflections".[20] Kubler notes that in the case of the remnants of works of architecture and art, we only have the "phenotypes" and that "we have to deduce the originating genotypes" from these bits of evidence.[21] The history of technology as reflected in the sources examined in this book can be described in the same way that Kubler describes the history of art:

> The history of art in this sense resembles a broken but much-repaired chain made of string and wire to connect the occasional jeweled links surviving as physical evidences of the invisible original sequence of prime objects.[22]

The accounts from Chaucer, and the narratives from centuries preceding him — those ancient tales of flying machines and other peculiar devices — were clearly part of a long tradition. But the pattern or process by which technological ideas were transmitted through the centuries reveals itself to us only in fragments. Technical details, if there even were any, have become obscured and lost. What survive are intriguing passages concerning flight, the ability to see into the human body, and other technological ideas and speculations. What also survive are the broader comments on technology, its perils, and the loss of knowledge over time. The original sources of all this material and the precise way in which these ideas and conjectures were transmitted over the centuries are obscure.

---

19 George Kubler, *The Shape of Time: Remarks on the History of Things* (New Haven: Yale University Press, 1962), 39.

20 Ibid., 40.

21 Ibid., 41.

22 Ibid., 40.

2 More on the "City of Brass"

Earlier in this book, the theme of technology and its dangers appeared in the context of a discussion of the "City of Brass" in *The Arabian Nights.* One study of those tales notes:

> Predictions of a holocaust and speculations about what post-holocaust society will be like have featured prominently in twentieth-century science fiction. Science fiction's imaginary societies are (usually) set in the future. In the *Nights*, however, they are located in the past, and the authors of its stories speculated about the lost technology of the ancients, vanished civilizations and the catastrophes which had overwhelmed them.[23]

Again, this is a reverse image of history from our own: the past was *more* advanced than the present. Moreover, the past civilizations brought destruction upon themselves:

> The Koran abounds with examples of communities which went astray and which God destroyed, and this same theme is alluded to in the opening exordium of the *Nights* itself, in which praise is offered to God who "destroyed the race of Thamud, Ad and Pharaoh of the vast domain". The *Nights* offers a warning to those who will be instructed. Its audience should consider the hubris of those who sought to dwell in Iram of the Columns. King Shaddad spent 500 years building this city, but when he prepared to enter it, "the Cry of Wrath" of the Angel of Death slew him and all his following. Or they should consider the city of Labtayt, whose king defied the ancient interdiction and opened the tower of treasures, thereby bringing about the sack of his city by the Arabs.[24]

Moreover, in these tales of advanced civilizations and their destruction, technology plays a role — it is the inhabitants of sophisticated, technologically advanced and affluent cities who meet a dreadful fate. Continuing the list above, one also should remember

> the City of Brass, with its petrified inhabitants and their mummified queen (given a ghastly semblance of life by the quicksilver in her

23 Irwin, 212.

24 Ibid., 212–213.

> eye sockets). Or, in the cycle of the "The Porter and the Three Ladies of Baghdad", in the "The Tale of the First Lady", the city of fire-worshippers, who, all save one, failed to heed the mighty and invisible voice, did not turn to God and consequently were turned to stone.[25]

The message here is clear: "Such marvels are also warnings to those who would reflect, and a sense of wonderment is the beginning of philosophy."[26] That is, history is filled with marvels — from Atlantis to the City of Brass — but those marvels should give us pause, encourage us to reflect upon the ending of a "golden age" and why it ended, and compel us to make careful decisions about our present civilization and its technological choices.

The "City of Brass" tale brings together the themes of technology, lost knowledge, and morality in a way that uses truly vivid iconography. There is the "magical engineering" of a metal horseman — "not evil in itself, but ... ominous".[27] There is the clear denotation of an abandoned city, where one "can see the vacant streets and weeds growing in them."[28] Technology is depicted with ominous overtones in other ways, as well, closely connected with issues of morality. Ṭālib, a courtier of the caliph who has sent him on the mission to the City of Brass, decides to despoil the body of the dead queen. The queen's body is lifelike and fully adorned with pearls and jewels.[29] There is a stairway leading to her throne, and on the steps are two "statues" of slaves — actually, automata — both armed. Ṭālib and his party of explorers are given fair warning that this is a dangerous situation, as the queen herself has left an inscription indicating the one should beware the vanities of materialism. Nonetheless, Ṭālib moves forward:

> Then he approached the step and climbed the stair till he went between the columns and arrived between the two figures. Suddenly one of the two figures struck him in the back, and the other struck him with the sword in its hand and flung aside his head; and he fell down dead.[30]

One is reminded of the similar act of Gyges, who steals a ring from a dead body. However, while Gyges goes unpunished, Ṭālib suffers a terrible fate — a fate dealt by the mysterious technology of the City of Brass.

25 Ibid., 213.
26 Ibid., 213.
27 See Hamori, 10.
28 Ibid., 12.
29 Pinault, 204.
30 Ibid., 207.

Even in such an early tale as this, the reader is given specifics about the exact dangers of technology in a way that is quite modern. One commentator notes:

> Seductive girls beckon to those who scale the walls [of the City of Brass], but they are airy marionettes worked by magic, and a bait for the naïve. The men deceived by their charm laugh and clap their hands in delight, then throw themselves down from the battlements and are immediately crushed ... The girls are a product of human dissimulation. They are not there to protect the dead, but rather, left over from the City's lifetime — automata that go on functioning after their engineers have perished.[31]

A modern reader should reflect on the rather sophisticated images here, and should ponder why an early tale of the desert includes robotic beings and a destroyed city.

Another peculiarly modern image in this tale is the street scene that the group exploring the city encounters:

> On their way to the palace, they must pass through the bazaar where the merchants' dead bodies are still sitting behind their wares, looking literally as if merely asleep. It is a hideous limbo.[32]

It is fitting to wonder where the inventor of this tale ever came up with this image. It is a form of death that seems like something from a science fiction movie — the result of some kind of epidemic or mysterious weapon. Certainly, this form of death has a moral element, as proper burial rituals have not been followed, and one could read the death as the result of a "spiritual starvation" rather than a corporeal one.[33] A study of the "City of Brass" tale notes that the image of the rigid bodies here echoes stories where "the inhabitants ... were turned to stone by God for their pride and lack of piety."[34]

In a less vivid but nonetheless profound way, the *Book of Enoch* pairs together an increase in knowledge with a decline in morals:

> And Azazel taught men to make swords, and daggers, and shields, and breastplates. And he showed them the things after these, and the art of making them; bracelets, and ornaments, and the art of making up the

31 Hamori, 12–13.
32 Ibid., 13.
33 Ibid., 14.
34 Pinault, 210.

> eyes, and of beautifying the eyelids, and the most precious stones, and all kinds of coloured dyes. And the world was changed. And there was great impiety, and much fornication, and they went astray, and all their ways became corrupt. Amezarak taught all those who cast spells and cut roots, Armaros the release of spells, and Baraqiel astrologers, and Kokabiel portents, and Tamiel taught astrology, and Asradel taught the path of the Moon. And at the destruction of men they cried out; and their voices reached Heaven. And then Michael, Gabriel, Suriel and Uriel, looked down from Heaven and saw the mass of blood that was being shed on the earth and all the iniquity that was being done on the earth.[35]

A modern commentator puts it succinctly: "In sum: the invention and teaching by the angels of metallurgy, cosmetics, magic, and astrology were the main cause of mankind's going astray."[36] The English epic poem *Beowulf* similarly connects technology and its destructive consequences to an earlier period, with a sword being the work of giants from an ancient time.[37] We see here, then, the theme that has been evidenced in many of the texts presented in this book: that early authors were not only aware of technological ideas, but also their profound effect on society.

3 Knowledge Transmission and Cyclical History

Daoist literature provides another perspective concerning the idea of knowledge transmission, combining it in a subtle way with the idea of a "golden age". In the famous work known as the *Zhuangzi* one finds, as the sinologist Nathan Sivin has noted, an early tradition that "manual skill is not a matter of technical rules".[38] Sivin cites two passages: The first passage talks about a cook who has mastered the art of meat-carving by being "in touch through the daemonic" and so not even using the senses any more.[39] The second passage

---

35 1 *Enoch* 8.1–9.1, in Knibb, 2.79–2.86.

36 van der Horst, 147.

37 See the interesting discussion in Andy Orchard, *Pride and Prodigies: Studies in the Monsters of the* Beowulf-*Manuscript* (Toronto: University of Toronto Press, 2003), 65–66.

38 See p. 186, n. 5 in Nathan Sivin, "On the Limits of Empirical Knowledge in Chinese and Western Science", in J.T. Fraser, N. Lawrence, and F.C. Haber, eds., *Time. Science, and Society in China and the West: The Study of Time V* (Amherst: University of Massachusetts Press, 1986), 151–169.

39 A.C. Graham, trans., *Chuang-tzu: The Seven Inner Chapters and Other Writings from the Book Chuang-tzu* (London: George Allen and Unwin, 1981), 63–64.

gives the story of the wheelwright named Pien, who argues against book learning. He talks of his own knowledge, that of wheel-making:

> Duke Huan was reading a book at the top of the hall, wheelwright Pien was chipping a wheel at the bottom of the hall. He put aside his mallet and chisel and went up to ask Duke Huan
>
> "May I ask what words my lord is reading?"
>
> "The words of a sage."
>
> "Is the sage alive?"
>
> "He is dead."
>
> "In that case what my lord is reading is the dregs of the men of old, isn't it?"
>
> "What business is it of a wheelwright to criticize what I read? If you can explain yourself, well and good; if not, you die."
>
> "Speaking for myself, I see it in terms of my own work. If I chip at a wheel too slowly, the chisel slides and does not grip; if too fast, it jams and catches in the wood. Not too slow, not too fast; I feel it in the hand and respond from the heart, the mouth cannot put it into words, there is a knack in it somewhere which I cannot convey to my son and which my son cannot learn from me. This is how through my seventy years I have grown old chipping at wheels. The men of old and their untransmittable message are dead. Then what my lord is reading is the dregs of men of old, isn't it?"[40]

These passages, certainly, are about a key theme in Daoism — the inability to "name" the *dao* or describe it with words. But at the same time, the metaphors used here — that is, skill-based knowledge in such fundamental professions as meat-carving and wheel-making — tells us something about ancient ideas of how such knowledge is "embodied", practiced, and transmitted.[41] The "knack"

40 Ibid., 139–140.

41 For more on these types of tales (known as "knack stories") in the *Zhuangzi*, see pp. 59 et ff. of Chris Jochim, "Just Say No to 'No Self' in *Zhuangzi*", in Roger T. Ames, ed., *Wandering at Ease in the Zhuangzi* (Albany: State University of New York Press, 1998), 35–74.

mentioned by the wheelwright is that kind of knowledge that can only be gained through engaging in the actual activity itself.

This philosophy also clearly favors such manual, intuitive knowledge, and at other points in the *Zhuangzi*, there are oblique references to a time in the distant past when society had more profound understanding of this kind. One passage in the *Zhuangzi* talks about such understanding, a way both subtle and profound:

> With the ancients, understanding had gotten somewhere. Where was that? Its height, its extreme, that to which no more could be added, was this: some of them thought that there had never begun to be things. The next lot thought that there are things, but that there had never begun to be boundaries among them.[42]

On the one hand, this passage is discussing a classic Daoist idea — that it is we as humans, in our process of naming things, who differentiate and categorize, while nature is always a "whole". But note that the passage also places the human capacity for profound understanding of the natural state of things *in the past*, with "the ancients" described as essentially wiser and closer to a natural state of learning than we are. in the present In essence, this passage suggests that our knowledge and understanding of the nature of the universe — what the *Zhuangzi* calls *wuhua* (物化) or the "transformation of things" — has actually *declined*, with what was intuitive before now consciously having to be relearned.[43]

Regardless of how early writers were using the concept of a lost "golden age" and positing an earlier period of greater understanding, their concern shows an awareness of history, of patterns of change in human society, and of variations in human knowledge. The Renaissance physician Jean Fernel, cited earlier in this book, wrote insightfully concerning his perspective on the history of knowledge:

> To speak frankly for the moment about my own lifetime: branches of study and arts that had been buried for practically twelve hundred years, or, to be more accurate, had perished and been wiped out, are now clearly restored to life, and have resumed their original glory, if not a greater one. So the present age has virtually nothing it need envy in that learned past

---

42 This passage is quoted in David L. Hall and Roger T. Ames, *Thinking from the Han: Self, Truth, And Transcendence in Chinese and Western Culture* (Albany: State University of New York Press, 2003), 248–249.

43 Ibid., 71.

> epoch. Oratory and the heights of eloquence are now flourishing everywhere; all sorts of philosophy are being cultivated; musicians, geometers, craftsmen, painters, architects, sculptors, and other practitioners beyond counting have displayed minds so keen that each group of them has adorned its art with noteworthy and splendid works, which match those of the ancients ... The advance of time has not only contributed luster and additions to what has already been discovered; it has brought forth in addition new arts, to which neither the talent nor the diligence of our predecessors had ever attained.[44]

In short, Fernel looked to the past as a time of great technical and artistic knowledge, and understood that his own age both drew from that, and could formulate new ideas. He understood that the relationship between past and present was subtle and complex. Again, early writers — from ancient China to Renaissance Europe — were not merely isolated figures in history: they were fully aware of their place in it, their serving as a link between past and present, and their role to provide some continuity despite the vagaries of human civilizations.[45]

The idea of lost knowledge is not new, and has been a subject of discussion from at least the time of Plato through the Renaissance. Writers through the centuries have been aware of the rise and fall of a series of human cultures. Francis Bacon, in an essay entitled "Of Vicissitude of Things", wrote:

> Solomon saith, *There is no new thing upon the earth.* So that as Plato had an imagination, *That all knowledge was but remembrance*; so Solomon giveth his sentence, *That all novelty is but oblivion.*[46]

This is rather poetic, but Bacon adds an interesting element about the need for a referent through time to maintain the continuity of human civilization:

> There is an abstruse astrologer that saith, *If it were not for two things that are constant* (*the one is, that the fixed stars ever stand a like distance one from another, and never come nearer together, nor go further asunder; the*

44 Forrester and Henry, 106–107.

45 For another look at classical writers seeking to locate their civilizations in the realm of deep time, see Elias J. Bickerman, "Origines Gentium", *Classical Philology* 47.2. (April 1952): 65–81.

46 Clark Sutherland Northup, ed., *The Essays of Francis Bacon* (Cambridge, MA: Houghton, Mifflin and Company, 1908), 172–173.

> *other, that the diurnal motion perpetually keepeth time), no individual would last one moment.*[47]

Finally, Bacon comments on disappearance and loss:

> Certain it is, that the matter is in a perpetual flux, and never at a stay. The great winding-sheets, that bury all things in oblivion, are two; deluges and earthquakes. As for conflagrations and great droughts, they do not merely dispeople and destroy. Phaeton's car went but a day. And the three years' drought in the time of Elias, was but particular, and left people alive. As for the great burnings by lightnings, which are often in the West Indies, they are but narrow. But in the other two destructions, by deluge and earthquake, it is further to be noted, that the remnant of people which hap to be reserved, are commonly ignorant and mountainous people, that can give no account of the time past; so that the oblivion is all one, as if none had been left.[48]

Clearly, Bacon is drawing from and reinforcing Plato's idea of civilizational collapse, and the resulting lack by the survivors of any ability to retain and transmit forward knowledge of that civilization.

In another essay, entitled "Of the Wisdom of the Ancients", Bacon expresses a similar idea:

> But howsoever the works of wisdom are among human things the most excellent, yet they too have their periods and closes. For so it is that after kingdoms and commonwealths have flourished for a time, there arise perturbations and seditious and wars; amid the uproars of which, first the laws are put to silence, and then men return to the depraved conditions of their nature, and desolation is seen in the fields and cities. And if such troubles last, it is not long before letters also and philosophy are so torn in pieces that no traces of them can be found but a few fragments, scattered here and there like planks from a shipwreck; and then a season of barbarism sets in, the waters of Helicon being sunk under the ground, until, according to the appointed vicissitude of things, they break out and issue forth again, perhaps among other nations, and not in the places where they were before.[49]

---

47 Ibid., 173.

48 Ibid., 173.

49 John Pitcher, ed., *Francis Bacon: The Essays* (London: Penguin, 1985), 268.

Clearly, Bacon believed in a cyclical rather than linear history of human civilization, with "letters … and philosophy" disappearing, save for "a few fragments" that future generations must puzzle over as they rebuild.[50]

A passage in Bacon's famous *New Atlantis* gives a picture of how in the distant past mankind's knowledge of the world was far greater than in later ages:

> You shall understand (that which perhaps you will scarce think credible) that about 3,000 years ago, or somewhat more, the navigation of the world (especially for remote voyages) was greater than at this day.[51]

The text goes on to recount a world of global voyaging on the high seas, with trade between the New World and Old World long before Columbus. Of course, this is a fiction, but Bacon's point is that knowledge — even globalism — can arise in human civilization and then fade over time. This potential discontinuity in human knowledge through time was of particular interest to Bacon, as his *New Atlantis* notes how nature can be explored and understood through analysis and experiment, and mentions the practice of such investigations done both openly and in secret.

4 How Far Back?

An area for further exploration, beyond the scope of this book, concerns how far back cultural memory might extend. This topic has been touched upon already at several points here — could the Egyptians really have had a record of events some *nine thousand years* before them? Could indigenous cultures have retained some kind of collective memory of early hominids living in their area? Just how far back can communal memory reach, and how far can the chain of the transmission of knowledge extend?

There has been idle speculation in this area, notably — perhaps notoriously — by Immanuel Velikovsky, mentioned earlier in this book.[52] In an

50 On this idea in Bacon, see B.H.G. Wormald, *Francis Bacon: History, Politics & Science, 1561–1626* (Cambridge: Cambridge University Press, 1993), 67–69, as well as pp. 40–41 of Paolo Rossi, "Bacon's Idea of Science", in Markku Peltonen, ed., *The Cambridge Companion to Bacon* (Cambridge: Cambridge University Press, 1996), 25–46.

51 Brian Vickers, ed., *Francis Bacon: The Major Works* (Oxford: Oxford University Press, 2002), 467.

52 For a fair and thoughtful critique of Velikovsky's ideas and methods, see Stephen Jay Gould's essay entitled "Velikovsky in Collision" in his book *Ever Since Darwin: Reflections in Natural History* (New York: Norton, 1977), 153–159.

essay entitled "The Earth Without the Moon" in his never-published *In the Beginning*, Velikovsky assembled a series of ancient sources to argue for his eclectic and peculiar catastrophic view of world history.[53] In particular, he quoted ancient authors who wrote of a time before the Earth had a moon. These are unscientific musings, but it is interesting actually to look at the sources Velikovsky cites. Plutarch, in his essay "Roman Questions", makes a brief mention of the Arcadians as a "pre-lunar" people. Hippolytus, in a short discussion of the first peoples of the Earth, speaks of a Greek conception of them as "of greater antiquity than the Moon".[54]

Now, of course, this can simply be read as metaphorical language — with the phrase "older than the Moon" being read as an analogy no more complex than "slower than molasses". But the passages are interesting, nonetheless, as examples of ancient writers pondering an even more ancient past. Lucian, in his work *Astrology*, adds a bit more about the Arcadians; in a passage on divination, he points out that they rejected the practice of astrology, and "in their folly they affirm that they are older than the Moon".[55] It is interesting that Lucian, in this way, highlights an older culture (one perhaps arrogant about its age) as being also a less superstitious one.

Apollonius of Rhodes, in the *Argonautica*, mentions the Arcadians in the context of a discussion on the distant past:

> They eagerly listened when Argus spoke to them:
>
> We were traveling to Orchomenus by the course which that unerring seer directed you to travel, the seer you encountered before meeting us. For there exists another sea route, which the priests of the immortals who were offspring of Triton's daughter Thebe revealed. Not yet did all the stars that revolve in the sky exist, nor yet was there a sacred race of the Danaans for explorers to hear of. Only Apidanian Arcadians existed, Arcadians who are said to have lived in the mountains eating acorns even before the moon existed; nor at that time was the Pelasgian land ruled by the glorious sons of Deucalion, in the days when Egypt, the mother of

53 This somewhat obscure essay and related pieces by Velikovsky can be found in Joseph K. Geiger, *Transition: An Overview of History from a Christian Perspective* (Victoria, BC: Trafford Publishing, 2006), 225–232.

54 Hippolytus, *The Refutation of All Heresies* 5.2; see Alexander Roberts and James Donaldson, eds., *The Ante-Nicene Fathers: Translations of the Writings of the Fathers down to AD 325*, vol. 5, *Hippolytus, Cyprian, Caius, Novatian, Appendix* (Buffalo: The Christian Literature Company, 1886), 48.

55 Lucian, "Astrology" 26, trans. Harmon, Kilburn, and Macleod, 5.366–367.

> men of long ago, was called grain-rich Eërie and the wide-flowing river was called Triton, by which all of Eërie is watered, for Zeus' rain never wets it, but thanks to its streams the fields bear bountiful crops. From here, they say, a man traveled all around Europe and Asia, relying on the strength, might, and courage of his soldiers. He founded countless cities on his way, some of which are perhaps still inhabited, others not, for a great stretch of time has since passed. Aea, at least, has continued to exist to this day, along with the descendants of those men whom that king settled to dwell in Aea. They, in fact, preserve their forefathers' writings, pillars on which are found all the routes and boundaries of the sea and land for those who travel around them.[56]

Note how this passage ties together history, myth, the question of the origins of various peoples, and — with the mention of the pillars, which we have seen earlier — the transmission of knowledge across time.

In Ovid's *Fasti*, a poetic text about the months of the Roman year, there is also mention of these Arcadians:

> The land that rose before the moon (if we may take its word for it) derives its name from the great Arcas.[57]

In another passage, Ovid writes:

> The Arcadians are said to have possessed their land before the birth of Jove, and that folk is older than the moon. Their life was like that of beasts, unprofitably spent; artless as yet and raw was the common herd. Leaves did they use for houses, herbs for corn: water scooped up in two hollows of the hands to them was nectar. No bull panted under the weight of the bent ploughshare: no land was under the dominion of the husbandman: there was as yet no use for horses, every man carried his own weight: the sheep went clothed in its own wool. Under the open sky they lived and went about naked, inured to heavy showers and rainy winds. Even to

56 Apollonius of Rhodes, *Argonautica*, 4.255–275 in Apollonius Rhodius. *Argonautica*, ed. and trans. William H. Race (Cambridge, MA: Harvard University Press, 2009), 348–351.

57 Ovid, *Fasti* 1.469, trans. James G. Frazer and rev. G.P. Goold (Cambridge, MA: Harvard University Press, 1931), 34–35. In Greek mythology, Arcas was the son of Zeus and Callisto, and became king of Arcadia; see Frazer, ed. and trans., *Fastorum libri sex: The Fasti of Ovid*, 2 vols. (Cambridge: Cambridge University Press, 2015), 2.188–2.189 and 2.360.

> this day the unclad ministers recall the memory of the olden custom and attest what comforts the ancients knew.[58]

While the context is poetic, note that the discussion is less so. The phrase "older than the moon", despite Velikovsky's claims, is not to be taken literally. But this passage shows, again, that a writer of antiquity was aware of a more remote time in the past, and able to reflect upon the passage of time and the development of human societies from distant origins.

Censorinus actually provides a critique and interpretation of this question of the Arcadians.[59] In a discussion of the calendrical year, Censorinus notes:

> [I]t is reported that in Egypt, in the most ancient times, the year was composed of two months, then King Isone made it four months, and then Arminon composed it of 13 (lunar) months and 5 days. The same thing in Achaia; the Arcadians commenced by having years of three months which gave to these people the name of Ante-Moonites, not as some have thought, that they existed before the moon was in the heavens, but because they counted by (solar) years (of three months) before the lunar year was established in Greece.[60]

Again the most important thing here is to witness various ancient writers even considering these questions, to see them thinking about deep time and what might be the limits of what one might call human history, and of the memory of such a history. Also of note is how Censorinus deconstructs the myth about the Arcadians having "existed before the moon was in the heavens", by providing a sober, historical interpretation.

Other early writers also took an interest in the content of myths, what they might mean, and their origin. The author Palaephatus penned a work in which he systematically analyzed a series of Greek myths.[61] His approach is a rationalizing, middle path, with the argument that myths were neither completely factual accounts nor pure invention. Rather, Palaephatus says, every myth has

---

58 Ovid, *Fasti* 2.290–2.302, trans. Frazer and rev. Goold, 78–79.

59 For a list of sources on the Arcadians' belief in being "older than the Moon", see Censorinus, *The Birthday Book*, trans. Holt N. Parker (Chicago: University of Chicago Press, 2007), 96, n. 5; also see Emma Stafford and Judith Herrin, eds., *Personification in the Greek World: from Antiquity to Byzantium* (Aldershot: Ashgate, 2005), 120–121.

60 Censorinus, *De die natale* (*"The Natal Day"*), trans. William Maude (New York: Cambridge Encyclopedia Co., 1900), 25–26.

61 The identity and date of this author is uncertain: see Palaephatus, *Περὶ ἀπίστων / On Unbelievable Tales*, trans. Jacob Stern (Wauconda, IL: Bolchazy-Carducci, 1996), 1–3.

its origin in some actual event, but that subsequent tellers of these stories modified them for the purpose of entertaining an audience. As the classicist Jacob Stern points out, the "mythological rationalization" which Palaephatus practiced was of a very particular kind, moreover: he wished to "remove impediments to belief" in ancient tales by pushing away the accretions deposited by the storytellers and by finding the apparent (or supposed) core of the story.[62] In terms of ideas about transmission, Palaephatus also is quite specific: his argument is that "myths ... would not have been transmitted to us if there were not some kernel of historical truth".[63] As he himself writes, "I understand that there would not be stories if there were not really events behind them".[64] Both Stern and another commentator on Palaephatus, George F. Osmun, reject this idea; Stern says that the idea that all myths must contain some historical fact,

> implies ... a fundamental repudiation of the human imagination: its implication is that myth is not the result of creative invention, but that it must originate in actual historical events which at a later time were confused or misunderstood by a naïve or foolish populace.[65]

However, it is not at all clear that the habit of pure, creative invention was one that characterized early storytellers. Moreover, even if there were tales that were purely "made up" — whatever that term may mean — that does not mean that all myths had such an origin. In fact, as another scholar of Greek myth notes, there was a "great eclecticism" in terms of the "ancient approach to myth"; this demonstrates that modern readers should be careful not only before they define what is fact and what is fiction, but also before they claim to know what the ancient beliefs were concerning myth generally.[66]

In the case of both Censorinus and Palaephatus, what is important, first of all, is that these authors are actually "attempting to bridge the gap from known historical events to ... legendary history".[67] History matters to these authors,

---

62 Ibid., 8–9.

63 Ibid., 16.

64 Ibid., 29.

65 Ibid., 17; also see p. 136 of George F. Osmun, "Palaephatus — Pragmatic Mythographer", *Classical Journal* 52.3 (December 1956) 131–137. For more on Stern's take on Palaephatus, see his essay "Rationalizing Myth: Methods and Motives in Palaephatus", in Richard Buxton, ed., *From Myth to Reason? Studies in the Development of Greek Thought* (Oxford: Oxford University Press, 1999): 215–222.

66 See Greta Hawes, *Rationalizing Myth in Antiquity* (Oxford: Oxford University Press, 2014), 11.

67 Palaephatus, *Περὶ ἀπίστων/On Unbelievable Tales*, 10.

as does the complexity of the relationship of myth to history. In addition, if there is some merit to the method of looking to myth for historical knowledge, then this means that history — human history — is long, extending back into a period only recorded in myth. This may also imply that the transmission of knowledge may be able to sustain itself over much longer periods than we can currently imagine. Indeed, as the discussion here has noted, we live in a particularly peculiar age, one where attention spans and memories are unusually short. We cannot apply our modern practice as a model for understanding human transmission of knowledge in the past. Our culture also tends to have a bias wherein human history begins when writing begins, and that anything that preceded that in terms of a narrative tradition is suspect.

But in an interview concerning his 2007 book, *On Deep History and the Brain*, the Harvard historian Daniel Lord Smail argues that a

> student might be told in a Western civilization class that history begins in Sumeria in 4000 BCE. But how can we claim that this is the birthplace for history, when we now have extensive evidence that humans have lived on Earth for tens of thousands of years?[68]

Smail's book presents the question of how we might start constructing a deeper history of human culture, connecting the written records of history with what came before. Moreover, Smail's work addresses the issue of "not only when history began, but how history is both defined and created".[69]

Similarly, in his discussion of megaliths, the architect and scholar Keith Critchlow points out a number of biases that historians have when looking at the knowledge of earlier cultures. He notes, for example, the "inbuilt prejudices against the possibility of a pre-literate and therefore presumed pre-numerate society being capable of ... mathematical activity".[70] He also notes that while we admire the achievements of ancient peoples — for example, their massive stone structures — we "neglect a study of inspiration or *motivation*".[71]

---

68 Amy Lavoie, "Human History Precedes Writing", *Harvard University Gazette Online* (13 December 2007), http://www.news.harvard.edu/gazette/2007/12.13/35-deephistory.html.

69 Ibid. For another type of call to look at deeper human history and to pay greater respect to sources in antiquity that deal with the remote past, see Mary Settegast, *When Zarathustra Spoke: The Reformation of Neolithic Culture and Religion* (Costa Mesa, CA: Mazda Publishers, 2005), and idem, *Plato Prehistorian: 10,000 to 5000 BC in Myth, Religion, Archaeology* (Cambridge, MA: Rotenberg Press, 1987).

70 Keith Critchlow, *Time Stands Still: New Light on Megalithic Science* (New York: St. Martin's Press, 1982), 10.

71 Ibid., 15.

When thinking about the texts examined in this book, then, we need to consider this: What were these ancient writers trying to *do*? The inspirations and motivations of earlier cultures may have been quite different from our own, and therefore what the ancient writers recorded should be understood from *their* perspective, not ours. One can take as an analogy what Critchlow states concerning prehistoric structures: "[O]ur aim is to attempt an understanding of the megalithic builder in his own terms, that is, according to all reasonableness, in *traditional* terms."[72]

In an essay on hermeneutics at the beginning of his book on early Daoist sources, *Tao and Method*, the philosopher Michael LaFargue addresses this general problem of interpretation through the lens of the present. He notes that a

> text becomes ambiguous when removed from its original context. Rather than taking advantage of this ambiguity to read modern meanings into classical texts, historical and cultural study tries systematically to reconnect them with their original contexts, and hence to reconstruct the meaning that the text had for its original authors and audience.[73]

This kind of "historical and cultural study" is what LaFargue calls "historical hermeneutics".[74] But he notes that even with such a method, "we can never achieve an understanding of people culturally distant from ourselves that is completely accurate and completely certain".[75] He advocates a slightly modified approach, where there is an attempt to recover the "original meaning of the text" in question, along with, at the same time, a consideration of "how the text is relevant to us today".[76]

In this book on the idea of lost knowledge, with its examination of strange accounts of flying machines, special mirrors, magic rings, and so on, the emphasis has been on looking at the early texts directly. LaFargue's "original meaning" in the case of the texts presented here can have varied interpretations: in some early texts, technology is described in a way that strikes the modern reader as almost matter-of-fact. At other times, as in the discussion of magic rings and the technological wonders of the "City of Brass", the original meaning — at least as it appears in the accounts that have come down to

72 Ibid., 19.

73 Michael LaFargue, *Tao and Method: A Reasoned Approach to the* Tao Te Ching (Albany: SUNY Press, 1994), 7.

74 Ibid., 7.

75 Ibid., 7–8.

76 Ibid., 9.

us — concerns technological sophistication in relation to moral decline. In terms of "how the text is relevant to us today", the initial "relevance" primarily takes the form of puzzlement for the modern reader: one is struck by the apparent anachronisms, the oddity of a mechanical device appearing in a centuries-old Chinese chronicle or Ainu folktale.

But of course, there is also relevance in the sense of how technology is viewed by some of the ancient sources: as something both magical — which we, even in our modern age still feel — and potentially catastrophic. LaFargue concludes: "[T]he world that ancient writers are generally talking about, the world to which the words of their text refer, is the world as they experience it." This means that the early writers cited in this book indeed were experiencing a range of technologies, if indirectly.[77] That is, they were reading about, hearing about, or otherwise experiencing flying machines and other technologies. This "experiencing" may have comprised or included, of course, speculative thinking about potential technological abilities, based on observation of more mundane devices already existing in that particular period.

5 The Methods for the Transmission of Knowledge

A key factor in interpreting what the texts presented here are talking about is the question of how knowledge is transmitted. This is a theme that has been discussed by a number of authors in a wide range of contexts, but bears repeating here.[78] In our contemporary society, technical knowledge is transmitted in the form of manuals, textbooks, guides, encyclopedias, and so on. There are accepted standards — such as the metric measuring system, longitude and latitude coordinates, and electronic component symbols — that help us exchange and pass down all kinds of information about materials, processes, and techniques through time.

---

77 Ibid., 20.

78 A key work in this field is Barber and Barber's *When They Severed Earth from Sky*, cited earlier. The literature on the transmission of knowledge across time and across cultures is fairly extensive, involving studies of everything from education and medicine to martial arts and the fine arts. See, for example, such titles as Nigel Crook, ed., *The Transmission of Knowledge in South Asia: Essays on Education, Religion, History, and Politics* (New York: Oxford University Press, 1996); Elisabeth Hsu, *The Transmission of Chinese Medicine* (Cambridge: Cambridge University Press, 1999); Michael Rosenbaum, *Kata and the Transmission of Knowledge in Traditional Martial Arts* (Boston: YMAA Publications, 2004); and David Hockney, *Secret Knowledge: Rediscovering the Lost Techniques of the Old Masters* (New York: Viking Studio, 2006).

But these standards and methods are all very recent in historical terms, and it must be understood that there have been — and continue to be — other systems of passing down technological knowledge. Clearly, the aborigines in Australia must have had a system of transmitting knowledge about their environment to their children, so that for some tens of thousands of years they were able to survive in that remote and difficult environment. And that knowledge was not simple: how to hunt an animal and prepare it as food; what plants were edible and which could provide some medicinal properties; how to create and use certain weapons; how to navigate across open terrain; and so on. All this information not only was transmitted without books and manuals, it was passed down over *millennia* without any written language whatsoever.

How was this kind of transmission of knowledge and skills carried out in early societies? Certainly, there was simple instruction, as we still see today in all societies, where children are taught by example and through plain didactic methods. In medieval society, there were apprenticeships and guilds. But a great deal of knowledge was also embedded in the form of stories, poems, songs, dance, and rituals so that it could be more easily remembered and passed on. These formed a complete system of information transmittal that was effective for hundreds and even thousands of years. Sometimes this form of information transmission can be considered superior to one based on texts. The dance critic Martha Ullman West has pointed out how preserving dance routines on film, video, or notations on paper is no substitute for actual apprenticeship — that is, the dance routines being taught live, from dancer to dancer:

> Long after they leave the stage, in their minds and muscles they hold the memory of form, rhythm, mood, and intent, constituting an irreplaceable resource for performers, historians, and frequently the choreographers themselves ... There are other ways of preserving the most ephemeral of the arts: film, video, various forms of notation; the visual record provided by painting, sculpture, and photography; sometimes written accounts. But whether a work's vocabulary is the flexed feet and filigreed hand gestures of South Asian dance, the straight spines and pointed toes of classical ballet, the contracted pelvis and floorbound movement of traditional modern dance, or the arm-swinging walking and running of postmodernism, there is no more viable way to transmit it than dancer to dancer.[79]

79 Martha Ullman West, "Dancers as Living Archives", *The Chronicle Review* 52.31 (7 April 2006), B14.

Dances themselves, of course, are frequently employed as rituals to preserve a society's knowledge, and thus they must be performed precisely. The Sufi mystical order of the Whirling Dervishes, for example, engages in a strict, trance-like dance that represents the cosmos and the connection between mankind and the heavens.[80] Folk dances, too, often encode stories or a common societal narrative.

Elizabeth Wayland Barber and Paul T. Barber, in their book *When They Severed Earth from Sky: How the Human Mind Shapes Myth*, lay out their beliefs about how historical facts and other information can be found in ancient tales.[81] They outline in depth the argument that is sketched out here — that early, preliterate societies have for millennia used myth in a range of forms to transmit knowledge that was important to them. The authors develop a sophisticated series of principles for how mythical structures are able to condense this knowledge, make it memorable, and pass it on. This process of building mythical narratives to capture and transmit knowledge has been used by cultures around the world, and has been employed to retain everything from astronomical data to environmental history.[82]

In the past, even in societies with some degree of literacy, the transmission of knowledge in the form of the written word was relatively rare, due to the slow nature of the process of copying manuscripts by hand and the expense of writing materials. It was only with the advent of the inexpensive printed book that the standard for the transmission of knowledge came to adopt its current — often textual — form. Even in the digital age, where text is supplanted by image and video, there is a reliance on a *medium* (a computer storage device, for example), rather than a ritual or narrative to transmit knowledge.

The important thing to remember here is that there was a need to transmit knowledge even in the earliest periods of human existence on the planet — knowledge about plants, animals, the seasons, and all aspects of the environment that affected human existence. In addition, that knowledge had to be "packaged" or "encoded" in a form that allowed for ready transmission from generation to generation. If the knowledge had been not packaged or encoded

80 See Reza Aslan, *No god but God: The Origins, Evolution, and Future of Islam* (New York: Random House, 2005), 209.

81 For a review of the Barbers' book, with some well-considered critiques as to the problem of recovering information from myths, see Michael Shermer, "Time Capsules", *American Scientist* 93.3 (May–June 2005): 276–277. Another careful examination concerning this subject is found in Jean Markale, *Celtic Civilization* (New York: Gordon & Cremonesi, 1978), 14–20.

82 Note, for example, Patrick D. Nunn, "On the Convergence of Myth and Reality: Examples from the Pacific Islands", *The Geographical Journal* 167.2 (June 2001): 125–138.

in this way, then it would have been lost — and with that vital knowledge lost, the very existence of human beings on the planet would have been at risk.[83] Even the most remote cultures develop what is known as "folk taxonomy", a way of building and transmitting knowledge about flora and fauna that provides the cultures with a way of, for example, organizing certain plants by the medicinal properties that they possess. This taxonomy, while not matching those of Western science in terms of structure, still provides knowledge that is both highly structured and useful.[84] In dealing with knowledge that has been "packaged" or encoded by cultures of the past, or by current indigenous cultures around us, it is also important to remember that such "traditional knowledge is not necessarily accessible by analysis or interrogation".[85]

6 The Idea of "Lost Knowledge" and the Nature of Myth

When information — especially complex information — is "encoded" in traditional formats, there is the danger that the story, poem, ritual, or song will be remembered as an end in and of itself, and not as a vessel for information. The fact that the story, poem, ritual, or song contains encoded or embedded information can be forgotten; similarly, the key to the encoding may be forgotten. In this way, knowledge might become corrupted or lost over time, and these vehicles for information transmission might become empty of meaning.

But stories, poems, rituals, and songs, and the myths they recount, can function as highly efficient transmission modes for information. As "moderns", we tend to view the myths of "primitive" societies as fanciful tales spun by superstitious and ignorant peoples. But the burden is on us, then, to take the time to interpret these tales:

83 For a thoughtful look at how technical knowledge might be encoded and transmitted by indigenous peoples — and how that encoding and transmission might be misinterpreted by modern scholars — see Bradley C. Bennett, "Doctrine of Signatures: An Explanation of Medicinal Plant Discovery or Dissemination of Knowledge?", *Economic Botany* 61.3 (Autumn 2007): 246–255. For an example of the encoding of technical knowledge by indigenous peoples and the interaction of that work with Western knowledge, see Louis De Vorsey, Jr., "The Importance of Native American Maps in the Discovery and Exploration of North America", *Terrae Incognitae* 42.1 (September 2010): 5–18, as well as the analysis in Turnbull, cited earlier.

84 See, for example, John K. Joseph and V.T. Anthony, "Ethnobotanical investigations in the genus *Momordica* L. in the Southern Western Ghats of India", *Genetic Resources and Crop Evolution* 55 (2008): 713–721.

85 I wish to thank Alex Schein for his coining of this apt phrase.

> What, then, of myths that occur over and over, the world around, yet describe something that could not actually happen in the world as we know it? Why, for example, do we find in the native cultures of both hemispheres a story of a catastrophic Flood covering the entire world, like that of Gilgamesh, Genesis, or the Andean peoples? Or of a Conflagration that destroys the whole world, as in Polynesia — sometimes even destroying the gods, as at Norse Ragnarök? A story which always ends with the entire world being renewed pretty much as it was? Geology has so far found no flood — or fire — that covered even half a continent within the existence of Homo sapiens, so a source in real experience seems barred (although stories of actual fires and floods could be *attracted* to this myth, and collisions with comets have such a potential). If not a "real" fire or flood, then what alternatives for interpretation can we deduce?[86]

The myths may well contain information that modern readers simply are misinterpreting. Moreover, the idea that myths contain some kind of "encoded" knowledge, an idea that has gained new credence in recent years, is not just a modern, academic conception. As this book has shown, ancient writers understood that knowledge prior to their own time might have ended up being framed in mythical or metaphorical structures. Recall that Plato in the *Timaeus*, for example, talks about records of historical events that have taken on the "form of a myth".

Diodorus Siculus also discusses the difficulty, even in ancient times, of looking back and talking intelligibly about myth:

> I am not unaware of the fact that those who compile the narratives of ancient mythology labour under many disadvantages in their composition. For, in the first place, the antiquity of the events they have to record, since it makes record difficult, is a cause of much perplexity to those who would compose an account of them; and again, inasmuch as any pronouncement they may make of the dates of events does not admit of the strictest kind of proof or disproof, a feeling of contempt for the narration is aroused in the mind of those who read it; furthermore, the variety and the multitude of the heroes, demi-gods, and men in general whose genealogies must be set down make their recital a difficult thing to achieve; but the greatest and most disconcerting obstacle of all consists in the fact that those who have recorded the deeds and myths of the earliest times are in disagreement among themselves. For these reasons the writers of

86 Barber and Barber, 177.

> greatest reputation among the later historians have stood aloof from the narration of the ancient mythology because of its difficulty, and have undertaken to record only the more recent events. Ephorus of Cymê, for instance, a pupil of Isocrates, when he undertook to write his universal history, passed over the tales of the old mythology and commenced his history with a narration of the events which took place after the Return of the Heracleidae. Likewise Callisthenes and Theopompus, who were contemporaries of Ephorus, held aloof from the old myths.[87]

Diodorus himself, though, then notes:

> We, however, holding the opposite opinion to theirs, have shouldered the labour which such a record involves and have expended all the care within our power upon the ancient legends. For very great and most numerous deeds have been performed by the heroes and demi-gods and by many good men likewise, who, because of the benefits they conferred which have been shared by all men, have been honoured by succeeding generations with sacrifices which in some cases are like those offered to the gods, in other cases like such as are paid to heroes, and of one and all the appropriate praises have been sung by the voice of history for all time.[88]

In other words, Diodorus wants to *include* the mythical narrative, and take it seriously as part of the "voice of history" of human civilization.

In contemporary thinking, we relegate myth to the classification of "fiction", but in other cultures it has played a larger — and more cohesive — role. As the classicist Walter Burkert notes:

> In Greek civilization we see myth dominating art and poetry in a special way, a complex which in turn became the major formative power of cultural progress and set the standard for centuries; we see rational language and thought struggling for emancipation from myth without ever arriving at a radical separation. There are larger and more colorful corpora of mythology in many other civilizations; there are more exotic and more elaborate rituals elsewhere. But the Greeks may well claim to represent

87 Diodorus Siculus, *Library of History* 4.1–4.3, trans. Geer, et al., 2.339–2.341.
88 Diodorus Siculus, *Library of History* 4.4, trans. Geer, et al., 2.341.

> the most advanced among the archaic, and the most archaic among the advanced.[89]

In looking at texts such as Plato's Atlantis tale, then, it may be useful to recontextualize them in this way — not to demand that they "fit" into a modern category. When dealing with history, the relationship between ancient writers and their sources is complex; writing about the Greek world, the historian Andreas Hartmann notes that "we can gain some very revealing, additional insights from literary texts which are not historical works proper, but fictional works pretending in some way to present history".[90]

This does not mean that the original writers intended to deceive, but rather that they were trying to present a genuine idea and had to construct a way to do so: "In ancient historiographical practice, the historian had to present evidence primarily when contradicting traditional accounts or when arguing against a scholarly predecessor. In such contexts, documents — and especially apparent ancient ones — had to play their role."[91] All this points to the conclusion that there was an effort in the ancient world to capture and preserve the past, even if this happened in ways that we might consider contrary to the practice of academic history. That means that modern readers should take ancient tales — from Plato's texts to Ainu folklore — seriously, and understand the context in which they arose. Hartmann says that an understanding of ancient perceptions of the past "should make us aware that there are different ways of conceptualizing history", adding that we must "accept this Otherness in order to overcome dismissive notions of gullibility and naïveté in the discussion of ancient interest in material remains".[92]

William Irwin Thompson, as noted earlier, provides a fresh look at the relationship between myth and history. In addition, however, he examines the deeper question of history itself, and the development of human civilization. In his book *At the Edge of History*, cited earlier, he asks his readers to look beyond the idea of linear history. We are, as a culture, fixated on the idea that in the remote past, there were only "primitive" human cultures, and that we

---

89 Walter Burkert, *Structure and History in Greek Mythology and Ritual* (Berkeley: University of California Press, 1979), xii; on the difficulty of defining myth in modern terms, see the discussion in the first section of Bukert's book, entitled "Tales, Text, and Reference."

90 See p. 39 of Andreas Hartmann, "*Cui vetustas fidem faciat*: Inscriptions and Other Material Relics of the Past in Graeco-Roman Antiquity", in Peter Liddel and Polly Low, eds., *Inscriptions and Their Uses in Greek and Latin Literature* (Oxford: Oxford University Press, 2013), 33–64.

91 Ibid., 39.

92 Ibid., 53.

have evolved in a steady progression since then. According to this conceptual framework, everything from forms of government to technology has progressed forward in a straight line, with virtually no cyclical patterns apart from a few fallen empires.[93] Thompson labels this the "Marxist view of history" and rightly challenges the concept; he states:

> [I]f within the period from the neolithic revolution to the twentieth century, one is interested in the relation of economic change and ideological transformation, then the Marxist formulation has a very high order of elucidation. However, a few small features of the environment go unexplained: namely, the mathematical sophistication of neolithic cultures and the cultural evolution of "the primitive."[94]

Typically, historians might view even impressive neolithic achievements such as Stonehenge with no more than a gentle nod to their anomalous nature. A "primitive" culture, living according to a basic, traditional lifestyle — and yet also building large structures out of tons of stone transported over significant distances? Yes, we can admit that, historians will say. Explanations are typically that the society needed some kind of calendar mechanism to keep track of the seasons for agriculture, and so on. But why build a calendar with massive tons of stone? Other explanations are that these kinds of mammoth building projects derived from the crude theology of these primitive people, some inchoate need to placate their gods with massive engineering projects. Few will admit that these "explanations" are really no more than Whiggish speculations. Indeed, one cannot make more than speculations about these kinds of artifacts, since their context has not survived. But the artifacts indeed do indicate that "primitive" peoples were thinking in a far more profound way than we might give them credit for, and possessed sophistication in certain areas — such as building technology and astronomy — that they might have lacked in others.

Thompson also takes his investigations in another direction, a path that is speculative but relevant to some of the stories examined in this book. That direction leads again to the idea that there were great cultures that rose and fell, leaving behind only traces — traces that were preserved in ritualistic and mythical forms long after the collapse. Thompson poses the question, then, as

---

93 For a detailed study of linear versus cyclical views of history, see Elizabeth L. Eisenstein, "Clio and Chronos: An Essay on the Making and Breaking of History-Book Time", *History and Theory* 5, suppl. 6 (1966): 36–64.

94 Thompson, *At the Edge of History*, 137.

to why "primitive" cultures carried out sophisticated technological tasks, and asks why so many cultures speak — explicitly or implicitly — of complex and powerful civilizations that had come before them:

> To answer these questions, one has to enlarge his conception beyond the Marxist frame of historical reference. One explanation is that the cataclysmic theory of history is not absurd. There are intelligent primitives because they are not the crude beginning of something, but the fragmentary remains of something else: the detritus of a "Lost World", as the ... anthropologist Lévi-Strauss calls it. He does not, of course, invoke the legends of the lost continent of Atlantis (as the first historian of the New World, Oviedo, did in 1550), for Lévi-Strauss is willing to rest with the unknown ... but he does leave things behind for others to pick up.[95]

It is this "picking up" that Thompson implies must be done, a further exploration of the phenomenon of lost knowledge and the passage of information through history and through cultures. As this book has shown, even if there was no advanced civilization in the remote past, a number of early sources from the Chinese to the Ainu, believed in the existence of subsequently vanished technology that might comprise this "detritus of a 'Lost World'" — bits of disappeared knowledge, knowledge that came to be wrapped in layers of archaic and mythical language and poetic images.

7 Looking at the Texts

So, how should a modern reader approach ancient texts that contain strange tales of flying machines and vanished lands? How does one tell which stories are simply pure invention, and which are metaphorical, stylized, or condensed descriptions of genuine technologies or real events? The provisional answer here is as follows: stories that are complex, filled with details, and have a storyline that is peculiarly illogical would seem to indicate some kind of "encoded" information. If a writer is going to spin a tale that is designed to entice a reader, chances are that the author of that text will make it engaging by providing a clear and interesting storyline. But one often finds stories that are quite complex, tales that do not even make sense in terms of the plot and the interconnections between characters. At other times, there is no storyline at all, simply

95 Ibid., 137.

commentary on some event or observation, as is the case with the Chinese accounts of special mirrors.

In discussing ancient Chinese tales of flight and winged beings, the sinologist Berthold Laufer noted:

> My conception of the so-called mythological and legendary period in the history of aviation differs wifely from that of my predecessors. It is a comfortable method and no more than a conventional form of thinking to stamp early traditions as mythological or legendary. This is a scholastic phrase from which little accrues. An inquisitive mind is intent on unravelling the fabric of a myth, on seeking an interpretation of its origins. If myth it is, how did the myth spring into existence? As there is a logic of human reasoning, so there is a logic of human imagination. The imaginative faculty of the human mind cannot conceive things that have absolutely no reality in existence; the product of our imagination is always elicited by something that exists of that we have reason to believe exists.[96]

Laufer clearly sees myth as a fruitful source of research for historical material. At the same time as a reader engages in "unravelling the fabric of a myth", there is also need to look at precisely what the author has written. Frequently, there is a modern tendency to try to "correct" ancient authors, ignoring what the texts actually say.

As noted earlier, a perfect example of this tendency can be found in historians' treatment of the Atlantis narrative. Plato's discussion concerning the location of Atlantis is very clear; in his *Timaeus*, the text points out that this Atlantean civilization

> sprang forth from beyond, from the Atlantic Ocean. For at that time this ocean was passable, since it had an island in it in front of the strait that you people say you call the 'Pillars of Heracles.' This island was larger than Libya and Asia combined, and it provided passage to the other islands for people who traveled in those days. From those islands one could then travel to the entire continent on the other side, which surrounds that real sea beyond.[97]

96 Laufer, *The Prehistory of Aviation*, 8–9.

97 Plato, *Timaeus* 24d–25a, trans. Zeyl, in Cooper and Hutchinson, 1232.

This vast island of Atlantis is described explicitly in the text in terms of both size and location. Yet there are modern writers who insist that Plato was talking about Santorini or other small islands near the Greek mainland. But again, Plato is very clear not only about the location, but also the size of Atlantis: "larger than Libya and Asia put combined". For the ancient Greeks, Asia meant what we might now call the Near East, and Libya was the term for the small part of the African continent of which the Greeks were aware. But even this relatively small area is obviously much too large to correspond to any Mediterranean island; Plato clearly is talking about land out in the Atlantic Ocean, whether imaginary or not. That is, even if Plato is not talk about a real place (in our sense of that term), his geography if quite specific, and that specificity likely has a motive in terms of the tale that he is recounting and the message that he is trying to convey.

One of the reasons that modern readers might feel compelled to "correct" ancient texts is that they are, in special sense, "hard to read". As pointed out earlier, a key reason for this is that many ancient texts do not fall into clear — by modern standards — conceptual categories. As the historian Pamela Long notes in her book *Openness, Secrecy, Authorship: Technical Arts and the Culture of Knowledge from Antiquity to the Renaissance*:

> [S]*cience, technology,* and even *history* are present-day terms whose meanings may be inappropriate or misleading for past cultures even when cognate terms exist in the diverse languages of those cultures. For example, *scientia* in Latin means "knowledge" in a broad sense and has none of the methodological or disciplinary meanings that we today associate with *science* and *scientist.* Similarly, the ancient Greek word *techne* refers to material production of all kinds, from making soup to constructing catapults, and the reasoning associated with that production ... Conceptual categories change from one historical period to another. For example, Aristotle delineated three areas of human activity: first, material and technical production (*technē*); second, action (*praxis*), such as political or military action, that requires judgment in contingent or uncertain situations (*phronesis*); and third, theoretical knowledge or knowledge of unchanging things (*epistemē*). Aristotle's separation of material production from action and from theoretical knowledge presupposed a hierarchy with technē at the bottom and epistemē, or theoretical knowledge, at the top. Such categories and the relationships between them exert a crucial influence on the construction of knowledge itself.[98]

98 Long, 2.

One must read an early text *as it stands*, and again, one must avoid a desire to "correct" ancient authors. If what the text says seems anachronistic or anomalous, then a modern reader must investigate *why* that might be so — not force the text to fit into some kind of already established framework of interpretation based on a linear model of history, or force the text to fit our modern definitions of fiction and non-fiction.

The philosopher Thomas K. Johansen, cited earlier, states that the Atlantis story, for example, "should be seen as 'history' only in a special Platonic sense", adding that it is "a story which is fabricated about the past in order to reflect a general truth" — which is, of course, a very specific and not particularly modern type of "fabrication".[99] Moreover, such a creation does not mean that the description of Atlantis and its fate simply was written *ex nihil*, as some kind of pre-modern novella. Rather, perhaps like the Gyges story, the Atlantis narrative was a pastiche of materials, all serving the larger purpose of communicating historical, cultural, and philosophical content. When looking at early sources such as Plato, too, there is the question of a different value system: the "general truths" communicated through stories such as the Atlantis tale might well have been considered "to be more significant facts that any particular historical facts".[100]

The historian Elizabeth Eisenstein notes that while we might question the follies committed by early chroniclers, we should also note that modern methods of history also have their defects:

> Distracted by the necessity of mastering a large literature devoted to tangential issues, historians have yet to come to terms with the very real problems inherent in the communications systems they use ... Historical perspectives have been set askew by certain distortions that result from the way historical data are at present handled ... [M]ental chronological progression from one hypothetical era to another conveys illusions about the course of history that have to be unlearned with considerable difficulty.[101]

A similar critique is found in Pamela Long's analysis, cited above:

> Autonomous, idealized disciplinary histories are often based on the presupposition that they contain a core of substance that has remained

99 Johansen, *Plato's Natural Philosophy*, 24.
100 Ibid., 29.
101 Eisenstein, 61–62.

> stable for centuries. More recently, historians have approached texts and disciplines in terms of the particular historical cultures in which they developed and have also investigated how disciplinary categories themselves have shifted over time. Central to this latter approach is the notion that a text — or any other historical artifact for that matter — should be understood within its own culture in terms of the categories that belong to historical actors. The meaning of a text, as Quentin Skinner and others have argued, will be misunderstood without careful, detailed attention to the fabric of meaning that exists within the culture to which that text initially belonged."[102]

Plato's work — as we have seen — often operates in the ambiguous area where history meets myth, and the "fabric of meaning" can become difficult to decipher. The philosopher Luc Brisson, in his book *Plato the Myth Maker*, examines the critical role of Plato in discussing myth and its relation to history, and written versus oral cultures:

> [T]his ambivalent feeling with respect to writing — of which Plato is the spokesperson — reflects a precise historical situation in ancient Greece. The oral civilization of which the Homeric epic constitutes the model was not submerged by some civilization of writing ... In fact, an important part of Greek culture remained oral up until — and even beyond — the end of the fifth century. Plato's testimony on myth is thus balanced on a razor's edge. At the turning point between two civilizations, one founded on orality and the other on writing.[103]

Given that the sources in this book often hover at this vital historical nexus, there is a need to re-frame the context of those sources in some sense. The re-framed context is this, then: the cultures behind the sources cited in this book — from Chinese chronicles to Hopi tales — were able to carry on as highly functioning entities even without the rigid categories of myth, fiction, and fact promoted by our current culture. Writers and tellers of stories in those cultures passed on knowledge that served the needs of those cultures, however ambiguous and mysterious that knowledge might appear to us today.

---

102 Long, 3.

103 Brisson, *Plato the Myth Maker*, 38–39. For a related discussion, see pp. 463–472 of Ludwig Edelstein, "The Function of the Myth in Plato's Philosophy", *Journal of the History of Ideas* 10.4 (October 1949): 463–481.

8 Reading Texts

In looking at these narratives and tales, there are pitfalls and problems along the way, of course.[104] There is the pitfall, for example, of looking for patterns where there are none. This is apophenia, a tendency or even compulsion to find connections, patterns, and meaning even where the data is truly random. As noted earlier, the famous linguist Ferdinand de Saussure spent several years researching pre-classical Latin poetry, looking for hidden patterns. He examined phonemes and scrutinized word order, searching for anagrams. These anagrams hid names and phrases, Saussure believed, and he puzzled at how they might have gotten there, or even whether they were placed in the texts purposely by the ancient poets or had appeared by accident.

Mario Livio, an astrophysicist and popular writer in science and mathematics, highlights a similar striving to find hidden patterns in early texts:

> George Eckel Duckworth, a professor of classics at Princeton University, made the most dramatic claim about the appearance of the Golden Ratio in poetry. In his 1962 book, *Structural Patterns and Proportions in Vergil's Aeneid*, Duckworth states that "Vergil composed the *Aeneid* on the basis of mathematical proportion; each book reveals, in small units as well as in the main divisions, the famous numerical ratio known variously as the Golden Section, the Divine Proportion, or the Golden Mean ratio."[105]

The method employed by Duckworth for finding such patterns in the *Aeneid* was complex:

> Duckworth made detailed measurements of the lengths of passages in the *Aeneid* and computed the ratios of these lengths. Specifically, he measured the number of lines in passages characterized as major (and denoted that number by M) and minor (and denoted the number by m), and calculated the ratio of these numbers. The identification of major

104 Concerning the general problem of how we should read ancient historical works such as those of Thucydides, Polybius, et al., see Emilio Gabba, "True History and False History in Classical Antiquity", *The Journal of Roman Studies* 71 (1981): 50–62.

105 Mario Livio, *The Golden Ratio: The Story of Pi, the World's Most Astonishing Number* (New York: Broadway Books, 2002), 198. For a complex conjecture on number symbolism in ancient texts, also see McClain, *The Myth of Invariance: The Origin of the Gods, Mathematics and Music from the Ṛg Veda to Plato* (Boulder: Shambhala, 1978), as well as the critical reviews of this work by Jodi Vogel in *Notes*, 2nd series, 34.3 (March 1978): 614–616, and by David Konstan in *Isis* 70.4 (December 1979): 599–600.

> and minor parts was based on content. For example, in many passages the major or minor part is a speech and the other part (minor or major respectively) is a narrative or description. From this analysis, Duckworth concluded that the *Aeneid* contains "hundreds of Golden Mean ratios."[106]

Naturally, there are several potential problems here. Why choose the number of lines in the passages in the text? Why not choose letters? What might a "numerical analysis" of those reveal? Livio provides his own critique, mostly mathematical, of Duckworth's methods. But perhaps most important is the question of hidden content.[107] Even if one could assume that Vergil really did hide something in his text, why hide the Golden Ratio? On a lark? To prove that he could do so? Why not hide something more interesting or more complex? This book as argued that cultures might have encoded knowledge in the form of stories, poetry, and so on, but that was for a specific purpose — to retain the knowledge in cultural memory and as an efficient mode of transmission of that knowledge from one generation to the next. That is quite different from hiding something in a text simply as a demonstration of intellectual dexterity.

But the discussion above about Saussure and Duckworth concerns readings of texts where the language itself — indeed, the alphabetic system — has been utilized as a structure to hide something. In the studies presented in this book, the concern is with not quite so deep a level, in a sense; the Chinese texts, and even Plato's text, are to be read straight for content, and what is "hidden" is the writer intended meaning. This takes us back, in a way, to the question of whether these early texts are dealing with the recording and recounting of history in some form. The sinologist Derk Bodde notes:

106 Ibid., 199.

107 For more on Duckworth's studies, see George E. Duckworth, "Mathematical Symmetry in Vergil's Aeneid", *Transactions and Proceedings of the American Philological Association* 91 (1960): 184–220, as well as three reviews: R.D. Williams, "[review of] George E. Duckworth, *Structural Patterns and Proportions in Vergil's 'Aeneid': A Study in Mathematical Composition*", *Classical Philology* 58.4 (October 1963): 248–251; Robert B. Lloyd, "Reviewed Work: *Structural Patterns and Proportions in Vergil's Aeneid*", *The American Journal of Philology* 85.1 (January 1964): 71–77; and M.L. Clarke, "Virgil and the Golden Section — Reviewed Work: *Structural Patterns and Proportions in Vergil's Aeneid* by George E. Duckworth", *The Classical Review*, new series, 14.1 (March 1964): 43–45. Also see Janet P. Bews, "'Aeneid' I and .618?", *Phoenix* 24.2 (Summer, 1970): 130–143. On a similar concept of mathematical patterns in classical texts, see Robert J. Getty, "Neopythagoreanism and Mathematical Symmetry in Lucan, *De bello civili* 1", *Transactions and Proceedings of the American Philological Association* 91 (1960): 310–323.

> The theory to which Euhemerus has given his name maintains that the origin of myth is to be found in actual history, and that the gods and demigods of mythology were, to start with, actual human beings. As commonly used by writers on Chinese mythology, however, "euhemerization" denotes precisely the opposite process: the transformation of what were once myths and gods into seemingly authentic history and human beings. Unquestionably, a fair amount of what purports to be early Chinese history has been subjected to this kind of euhemerization, the literal acceptance of which by most people until recent years has led to gross misunderstandings concerning the beginnings of Chinese civilization. Not infrequently, to be sure, the literalists might encounter certain mythological elements not wholly concealed beneath their euhemerist dress, but when this happened, these could always be explained as mere later accretions to what in essence was genuine history.[108]

There is much more insightful analysis of this kind in this same section of Bodde's book on various topics concerning Chinese civilization. Even in ancient times, Bodde says, the Chinese recognized that this process of "reverse euhemerization" was going on.[109] He notes that Confucian philosophy in particular sought to eliminate all supernatural elements from the ancient stories they had inherited:

> [I]t is precisely the Confucianists who, more than any other school of thought, were historically minded and assumed prime responsibility for conserving and editing the ancient texts which eventually became the Chinese classics. In so doing they were, on the one hand, always intensely interested in the search for historical precedents which would confirm their own social and political doctrines; on the other hand, their own strong humanism tended to make them either indifferent toward supernatural matters, or to seek to explain them in purely rationalistic terms.[110]

Bodde's primary point is that there was a kind of reductionism towards myth, with the Confucianists stripping myths down, and the Daoists, for example, employing myth largely as metaphor to illustrate philosophical arguments.

---

108 Derk Bodde, *Essays on Chinese Civilization* (Princeton: Princeton University Press, 1981), 48.

109 Ibid., 49.

110 Ibid., 51.

This means that the stories presented here, stories of flying vehicles and strange mirrors, survived through an intellectual and cultural environment that was actually hostile to fanciful and mythical stories. This implies that the writers were not treating these accounts as mythical or invented, but rather as some kind of actual historical record, and one worth repeating. Bodde comments that the Chinese sometimes took actual myths — in the sense of invented stories — and "euhemerized" them into "what came to be accepted as authentic history".[111] But this does not mean that in this interplay of myth and history, Chinese writers failed to preserve accounts of actual events.

9 Towards the Future

In a discussion of archaeological sites and our aesthetic understanding of them, the scholar Francis Sparshott points out that even in what we consider ancient times, the cultures of the period "looked back, just as we do, to the days of Sophocles and Plato".[112] These cultures had their own understanding of their place in a very long — and sometimes interrupted — historical continuum. Sparshott also makes another interesting point, in the context of ancient buildings and the ruins we see today:

> There are three ways in which monuments of the past may be preserved. They may survive as cores or shells within and around later modifications that have converted the whole into something fit for contemporary use. Or they may be preserved with careful piety, repaired and restored on the traditional pattern so that they are as if new. Or, finally, they may survive as ruins through their own toughness or the indolence of posterity — life simply moves away from them, they are irrelevant, they are too much trouble to pull down, they are left to stand or collapse in their own time. They are, essentially, places in which everyone has lost interest.[113]

Reflecting upon Sparshott's description, one could readily replace "monuments" with "narratives". Narratives about the past undergo the same kind of preservation — or abuse and neglect — as ancient buildings. If we restore

111 Ibid., 79; concerning this "reverse euhemerization", see the discussion in the first chapter of Sarah Allan, *The Shape of the Turtle: Myth, Art, and Cosmos in Early China* (Albany: State University of New York Press, 1991).

112 See p. 87 of Francis Sparshott, "The Antiquity of Antiquity", *Journal of Aesthetic Education* 19.1 (Spring 1985): 87–98.

113 Ibid., 91.

them incorrectly, moreover, misreading them, we fail to gain insight into what they really were like. If we let them collapse, they are lost forever. So, for these old narratives, preservation with "careful piety" may be in order, with a respect for the "traditional pattern" that was used in their original construction.

In fact, Sparshott himself, while noting that the past often appears to us in the form of ruins, argues that it is not inaccessible. He hints at a possible way to understand those remote times known as "antiquity" — and the possibility that they are not so remote:

> Works ... casually acknowledged as "ancient" may present themselves as bearers of a sensibility that seems to us to be fully revealed even in the opacity of its reference. We do not know what the "lovers" in the "Tomb of the Diver"[114] are likely to be saying to each other, any more than we know whether a gesticulating couple glimpsed across a crowded restaurant are quarrelling, or plotting a murder, or merely conversing. But they are comfortable and vividly present as part of a painted world, as our fellow diners are present in our social world. We are not more estranged from their imagined lives than we are from the vividly realized and elastically articulated forms that bring them so acceptably before us ... We belong immediately to the same world of directly shared meaningfulness.[115]

So, the past is the past but it is not so remote from us as we might fear. The perspectives concerning place and time held by earlier cultures indeed are understandable to us. The only need is for a method: as the historian Moses Finley puts it,

> All art is dialogue. So is all interest in the past ... The more precisely we listen and the more we become aware of its pastness, even of its near-inaccessibility, the more meaningful the dialogue becomes.[116]

---

114 The "Tomb of the Diver" ("Tomba del Tuffatore") is a tomb in Paestum, Italy, dating from the fifth century BCE, with a number of wall paintings, including one of a man diving into the water and one depicting two lovers reclining; see the brief note on p. 206 of George Emmanuel Mylonas, "The Frescoes from the Tomb of the Diver at Paestum", *American Journal of Archaeology* 75.2 (April 1971): 194–217, and the more detailed analysis in R. Ross Holloway, "High Flying at Paestum: A Reply", *American Journal of Archaeology* 81.4 (Autumn 1977): 554–555.

115 Sparshott, 93.

116 M.I. Finley, *Aspects of Antiquity: Discoveries and Controversies* (New York: Viking, 1968), 6.

The art historian George Kubler, cited earlier, argued for the following approach, which may strike the reader as not unlike that of Finley:

> When we scan things for traces of the shape of the past, everything about them deserves our attention. Yet this conclusion, which is self-evident once we recognize that things alone allow us to know the past, is generally ignored under the demands of specialized study. Archaeological studies and the history of science are concerned with things only as technical products, while art history has been reduced to a discussion of the meanings of things without much attention to their technical and formal organization. The task of the present generation is to construct a history of things that will do justice both to meaning and being, both to the plan and to the fullness of existence, both to the scheme and to the thing.[117]

Kubler wrote those words in 1962, so the term "present generation" must be shifted forward yet again; the project of constructing a proper "history of things" — or a "history of texts", one could say — that "will do justice both to meaning and being" is still in progress. In the present book, there has been attempt both to present "things ... as technical products", such as the description of strange flying machines, and to start an investigation into the meaning of the fact that these types of technologies are discussed in such a broad range of cultures and over such a vast span of time.

To borrow a phrase from the philosopher Michel Foucault, we must consider the use of an *archéologie du savoir* — an "archaeology of knowledge" — a system for carrying out investigations of the past in the way that Kubler advocates.[118] This book has aimed to introduce an initial set of source materials for that process, and to suggest a new methodology for their analysis. There remain many more stories, narratives, chronicles, and other sources out there, awaiting a careful exploration of the knowledge — lost, hidden, or simply yet to be examined — therein.

117 Kubler, 126.
118 Michel Foucault, *L'Archéologie du savoir* (Paris: Gallimard, 1969).

Bibliography

Aarne, Antti. *Vergleichende Märchenforschungen*. Helsingfors: Druckerei der Finnischen Literaturgesellschaft, 1908.

Adam, James. *The Republic of Plato: Volume 1, Books I–V*. Cambridge: Cambridge University Press, 1905.

Adams, Mark. *Meet Me in Atlantis: My Obsessive Quest to Find the Sunken City*. New York: Dutton, 2015.

Adler, William. "Berossus, Manetho, and '1 Enoch' in the World Chronicle of Panodorus." *The Harvard Theological Review* 76.4 (October 1983): 419–442.

Adler, William. *Time Immemorial: Archaic History and Its Sources in Christian Chronography from Julius Africanus to George Syncellus*. Washington, D.C.: Dumbarton Oaks Research Library and Collection, 1989.

Adrados, Francisco Rodriquez. "Coherencia e Incoherencia en la Forma y Contenido del *Timeo*." In *Interpreting the Timaeus–Critias: Proceedings of the IV Symposium Platonicum — Selected Papers*, edited by Tomás Calvo and Luc Brisson, 37–47. Sankt Augustin, Germany: Academia Verlag, 1997.

Aelian. *Historical Miscellany*. Translated by Nigel G. Wilson. Cambridge, MA: Harvard University Press, 1997.

Agrippa von Nettesheim, Heinrich Cornelius. *Three Books of Occult Philosophy*. Edited by Donald Tyson. Translated by James Freake. St. Paul, MN: Llewellyn, 1993.

Aitken, Ellen Bradshaw, and Jennifer K. Berenson Maclean. *Philostratus's Heroikos: Religion and Cultural Identity in the Third Century CE*. Leiden: Brill, 2004.

Alcock, Susan E., John F. Cherry, and Jaś Elsner, eds. *Pausanias: Travel and Memory in Roman Greece: Travel and Memory in Roman Greece*. Oxford: Oxford University Press, 2001.

Allan, Sarah. *The Shape of the Turtle: Myth, Art, and Cosmos in Early China*. Albany: State University of New York Press, 1991.

Allen, John M. *Atlantis: The Andes Solution*. New York: St. Martin's Press, 1999.

Allen, M., W. Ambrisco, M. Anastasiou, D. Bate, Y. Bitsakis, A. Crawley, M.G. Edmunds, D. Gelb, R. Hadland, P. Hockley, A. Jones, T. Malzbender, H. Mangou, X. Moussas, A. Ramsey, J.H. Seiradakis, J.M. Steele, A. Tselikas, and M. Zafeiropoulou. *Almagest: Special Issue — The Inscriptions of the Antikythera Mechanism* 7.1 (May 2016).

Allen, Mark, and John H. Fisher, eds. *The Complete Poetry and Prose of Geoffrey Chaucer*. Boston: Wadsworth, 2012.

Ames, Roger T., ed. *Wandering at Ease in the Zhuangzi*. Albany: State University of New York Press, 1998.

Andrews, P.B.S. "Larger than Africa and Asia?" *Greece & Rome*, 2nd series, 14.1 (April 1967): 76–79.

Apollonius of Rhodes. *Argonautica*. Edited and translated by William H. Race. Cambridge, MA: Harvard University Press, 2009.

Apuleius. *The Apologia and Florida of Apuleius of Madaura*. Translated by H.E. Butler. Oxford: Clarendon Press, 1909.

Aristotle. *Art of Rhetoric*. Translated by J.H. Freese. Cambridge, MA: Harvard University Press, 1926.

Aristotle. *Metaphysics*. Translated by C.D.C. Reeve. Cambridge: Hackett, 2016.

Aristotle. *Metaphysics: Book Iota*. Translated by Laura M. Castelli. Oxford: Clarendon Press, 2018.

Aristotle. *Meteorologica*. Translated by H.D.P. Lee. Cambridge, MA: Harvard University Press, 1952.

Aristotle. *Minor Works: On Colours. On Things Heard. Physiognomics. On Plants. On Marvellous Things Heard. Mechanical Problems. On Indivisible Lines. The Situations and Names of Winds. On Melissus, Xenophanes, Gorgias*. Translated by W.S. Hett. Cambridge, MA: Harvard University Press, 1936.

Árnason, Arnar, Nicolas Ellison, Jo Vergunst, and Andrew Whitehouse, eds. *Landscapes Beyond Land: Routes, Aesthetics, Narratives*. New York: Berghahn, 2012.

Ashe, Geoffrey. Dawn Behind the Dawn: A Search for the Earthly Paradise. New York: Henry Holt, 1992.

Ashkenazi, Michael. *Handbook of Japanese Mythology*. Santa Barbara, CA: ABC-CLIO, 2003.

Ashurst, David. "Journey to the Antipodes. Cosmological and Mythological Themes in *Alexanders Saga*." In *Proceedings of the Eleventh International Saga Conference*, edited by Geraldine Barnes and Margaret Clunies Ross, 1–13. Sydney: University of Sydney, Centre for Medieval Studies, 2000.

Aslan, Reza. *No god but God: The Origins, Evolution, and Future of Islam*. New York: Random House, 2005.

Augustine. *The City of God*. Translated by Marcus Dods. Peabody, MA: Hendrickson, 2009.

Ayrton, William E. "The Mirror of Japan and Its Magic Quality." *Nature* 19.493 (10 April 1879): 539–542.

Ayrton, William E., and John Perry. "The Magic Mirror of Japan." *Proceedings of the Royal Society of London* 28 (1878–1879): 127–148.

Bacon, Francis. *New Atlantis and The Great Instauration*. Edited by Jerry Weinberger. Arlington Heights, IL: Harlan Davidson, 1989.

Bacon, Roger. *Opera hactenus inedita Rogeri Bacon*. 16 vols. Edited by Robert Steele. Oxford: Clarendon Press, 1909–1941.

Bacon, Roger. *The Opus Majus of Roger Bacon*. 2 vols. Translated by Robert Belle Burke. New York: Russell & Russell, 1962.

Bacon, Roger. *Roger Bacon's Letter Concerning the Marvelous Power of Art and of Nature and Concerning the Nullity of Magic*. Translated by Tenney L. Davis. Easton, PA: The Chemical Publishing Company, 1923.

Bacon, Roger. *Rogeri Bacon, Opera quaedam hactenus inedita*. Edited by J.S. Brewer. London: Longman, Green, Longman and Roberts, 1859.

Baert, Barbara. "Caliban as a Wild-Man: An Iconographical Approach." In *Constellation Caliban: Figurations of a Character*, edited by Nadia Lie and Theo D'Haen, 43–60. Amsterdam: Rodopi, 1997.

Baker-Brian, Nicholas J. *Manichaeism: An Ancient Faith Rediscovered*. London: T. & T. Clark, 2011.

Baker, Donald C., ed. *The Variorum Edition of the Works of Geoffrey Chaucer*. Vol. 2, *The Canterbury Tales*. Part 12, *The Squire's Tale*. (Norman, OK: University of Oklahoma Press, 1990).

Ball, Sydney H. "Luminous Gems, Mythical and Real." *The Scientific Monthly* 47.6 (December 1938): 496–505.

Baltrušaitis, Jurgis. *Le miroir: essai sur une légende scientifique: révélations, science-fiction et fallacies*. Paris: Elmayan, 1978.

Baltzly, Dirk, trans. *Proclus: Commentary on Plato's Timaeus. Volume 4, Book 3, Part II: Proclus on the World Soul*. Cambridge: Cambridge University Press, 2009.

Barber, Elizabeth Wayland, and Paul T. Barber. *When They Severed Earth from Sky: How the Human Mind Shapes Myth*. Princeton: Princeton University Press, 2004.

Barker, Ernest. *Greek Political Theory: Plato and His Predecessors*. London: Methuen & Co., 1951.

Barnes, Geraldine, and Margaret Clunies Ross, eds. *Proceedings of the Eleventh International Saga Conference*. Sydney: University of Sydney, Centre for Medieval Studies, 2000.

Barrow, John. *Some Account of the Public Life and a Selection from the Unpublished Writings of the Earl of Macartney*. 2 vols. London: T. Cadell and W. Davies, 1807.

Barry, John Conor David. "The Seal of the Author: Paradigm, *Logos* and Myth in Plato's *Sophist* and *Statesman*." PhD diss., University of Ottawa, 2014.

Bartlett, Robert. *The Natural and the Supernatural in the Middle Ages*. Cambridge: Cambridge University Press, 2008.

Bascom, William R. *African Dilemma Tales*. The Hague: Mouton Publishers, 1975.

Baudrillard, Jean. *Simulacra and Simulation*. Translated by Sheila Faria Glaser. Ann Arbor: The University of Michigan Press, 1994.

Bautch, Kelley Coblentz. *A Study of the Geography of 1 Enoch 17–19: "No One Has Seen What I Have Seen"*. Leiden: Brill, 2003.

Bayanov, Dmitri, and Igor Bourtsev. "On Neanderthal vs. Paranthropus." *Current Anthropology* 17.2 (June 1976): 312–318.

Beall, Stephen M. "Homo Fandi Dulcissimus: The Role of Favorinus in the Attic Nights of Aulus Gellius." *American Journal of Philology* 122.1 (Spring 2001): 87–106.

Becker, Carl, ed. *Vermächtnis der Antike: Gesammelte Essays zur Philosophie und Geschichtsschreibung*. Göttingen: Vandenhoeck and Ruprecht, 1960.

Bender, Cora, Thomas Hensel, and Erhard Schüttpelz, eds. *Schlangenritual: Der Transfer der Wissensformen vom Tsu'ti'kive der Hopi bis zu Aby Warburgs Kreuzlinger Vortrag*. Berlin: Akademie Verlag, 2007.

Benford, Gregory. *Deep Time: How Humanity Communicates Across Millennia*. New York: Bard, 1999.

Bennett, Bradley C. "Doctrine of Signatures: An Explanation of Medicinal Plant Discovery or Dissemination of Knowledge?" *Economic Botany* 61.3 (Autumn 2007): 246–255.

Bennett, Matthew. *The Hutchinson Dictionary of Ancient & Medieval Warfare*. Chicago: Fitzroy Dearborn, 1998.

Benson, Elizabeth P., ed. *The Olmec and Their Neighbors: Essays in Memory of Matthew W. Stirling*. Washington, D.C.: Dumbarton Oaks Research Library and Collections, 1981.

Berg, Jacob Albert, Annemaré Kotzé, Tobias Nicklas, and Madeleine Scopello, eds. "In Search of Truth": Augustine, Manichaeism and other Gnosticism — Studies for Johannes van Oort at Sixty. Leiden: Brill, 2011.

Bergemann, H. "Jivaka — Schädelchirurg und 'Röntgenarzt' im alten Indien." *Blätter für Zahnheilkunde / Bulletin Dentaire* 29.7 (July 1968): 112–115.

Berlo, Janet C., ed. *Art, Ideology, and the City of Teotihuacan*. Washington, D.C.: Dumbarton Oaks Research Library and Collections, 1993.

Bernheimer, Richard. *Wild Men in the Middle Ages: A Study in Art, Sentiment, and Demonology*. Cambridge, MA: Harvard University Press, 1952.

Berry, Craig A. "Flying Sources: Classical Authority in Chaucer's 'Squire's Tale'." *ELH* 68.2 (Summer 2001): 287–313.

Berry, Michael V. "Oriental Magic Mirrors and the Laplacian Image." *European Journal of Physics* 27.1 (January 2006): 109–118.

Besant, Annie, and Charles W. Leadbeater. *Occult Chemistry: Clairvoyant Observations On the Chemical Elements*. London: The Theosophical Publishing Company, 1919.

Bews, Janet P. "'Aeneid' I and .618?" *Phoenix* 24.2 (Summer, 1970): 130–143.

Bharadvaja, Rishi. *Vymaanika-Shaastra Aeronautics*. Mysore: Coronation Press, 1973.

Bichakjian, Bernard H., Tatiana Chernigovskaya, Adam Kendon, and Anke Möller, eds. *Becoming Loquens: More Studies in Language Origins*. Frankfurt am Main: Peter Lang Europäischer Verlag der Wissenschaften, 2000.

Bickerman, Elias J. "Origines Gentium." *Classical Philology* 47.2. (April 1952): 65–81.

Birch, Cyril, ed. *Anthology of Chinese Literature: From Early Times to the 14th Century*. New York: Grove, 1965.

Birrell, Anne. "The Four Flood Myth Traditions of Classical China." *T'oung Pao* 83.4–5 (1997): 213–259.

Bittlestone, Robert. *Odysseus Unbound: The Search for Homer's Ithaca*. Cambridge: Cambridge University Press, 2005.

Black, Matthew. *The Book of Enoch Or I Enoch: A New English Edition*. Leiden: Brill, 1985.

Blair, Lawrence. *Ring of Fire: An Indonesian Odyssey*. Singapore: Didier Millet, 2010.

Blavatksy, Helena Petrovna. *Isis Unveiled: A Master-Key to the Mysteries of Ancient and Modern Science and Theology*. Vol. 1, *Science*. 6th ed. New York: J.W. Bouton, 1892.

Blavatksy, Helena Petrovna. *The Secret Doctrine: The Synthesis of Science, Religion, and Philosophy*. Vol. 1, *Cosmogenesis*. 2nd ed. London: The Theosophical Publishing Company, 1888.

Blavatksy, Helena Petrovna. *The Secret Doctrine: The Synthesis of Science, Religion, and Philosophy*. Vol. 2, *Anthropogenesis*. London: The Theosophical Publishing Company, 1888.

Blavatksy, Helena Petrovna. *The Voice of the Silence and Other Chosen Fragments from the Book of Golden Precepts*. Sioux Falls, SD: NuVision, 2007.

Blumenthal, David R., ed. and trans. *The Commentary of R. Ḥōṭer ben Shelōmō to the Thirteen Principles of Maimonides*. Leiden: Brill, 1974.

Bodde, Derk. *Essays on Chinese Civilization*. Princeton: Princeton University Press, 1981.

Bøe, Sverre. *Gog and Magog: Ezekiel 38–39 as Pre-text for Revelation 19, 17–21 and 20, 7–10*. Tuebingen: Mohr Siebeck, 2001.

Bolton, H. Carrington. "A Modern Oracle and Its Prototypes: A Study in Catoptromancy." *The Journal of American Folklore* 6.20 (January–March 1893): 25–37.

Bonner, Campbell. "The Technique of Exorcism." *The Harvard Theological Review* 36.1 (January 1943): 39–49.

Borelli, Giovanni Alfonso. *The Flight of Birds*. Edited by Thomas O'Brien Hubbard and John H. Ledeboer. London: Printed and Published for the Aëronautical Society of Great Britain by King, Sell & Olding, 1911.

Borelli, Giovanni Alfonso. *On the Movement of Animals*. Translated by Paul Maquet. Berlin: Springer-Verlag, 1989.

Bosc de Vèze, Ernest. *Petite encyclopédie synthétique des sciences occultes*. Nice: Bureau de la Curiosité, 1904.

Boyer, Alain-Michel Boyer, ed. *Modernités: Mondes Perdus*. Talence: Presses universitaires de Bordeaux, 1991.

Bray, Francesca, Vera Dorofeeva-Lichtmann, and Georges Métailié, eds. *Graphics and Text in the Production of Technical Knowledge in China: the Warp and the Weft*. Leiden: Brill, 2007.

Brecher, Kenneth, and Michael Feirtag, eds. *Astronomy of the Ancients*. Cambridge, MA: MIT Press, 1979.

Brewer, Keagan. *Prester John: The Legend and its Sources*. Farnham: Ashgate, 2015.

Brigham, David R. *Public Culture in the Early Republic: Peale's Museum and Its Audience.* Washington, D.C.: Smithsonian Institution Press, 1995.

Brisson, Luc. *How Philosophers Saved Myths: Allegorical Interpretation and Classical Mythology*. Translated by Catherine Tihanyi. Chicago: University of Chicago Press, 2004.

Brisson, Luc. *Plato the Myth Maker*. Translated and edited by Gerard Naddaf. Chicago: University of Chicago Press, 1998.

Brisson, Luc. *Platon: Les mots et les mythes*. Paris: F. Maspero, 1982.

Broadie, Sarah. "Theodicy and Pseudo-History in the *Timaeus*." *Oxford Studies in Ancient Philosophy* 21 (2001): 1–28.

Brooks-Davies, Douglas. *The Mercurian Monarch: Magical Politics from Spenser to Pope.* Manchester: Manchester University Press, 1983.

Brown, P., T. Sutikna, M.J. Morwood, R.P. Soejono, Jatmiko, E. Wayhu Saptomo, and R.A. Due. "A New Small-bodied Hominin from the Late Pleistocene of Flores, Indonesia." *Nature* 431 (2004): 1055–1061.

Brumbaugh, Robert S. "Note on the Numbers in Plato's *Critias*." *Classical Philology* 43.1 (January 1948): 40–42.

Brumbaugh, Robert S. *Plato's Mathematical Imagination: The Mathematical Passages in the Dialogues and their Interpretation*. Bloomington: Indiana University Press, 1954.

Brumm, Adam, Gerrit D. van den Bergh, Michael Storey, Iwan Kurniawan, Brent V. Alloway, Ruly Setiawan, Erick Setiyabudi, Rainer Grün, Mark W. Moore, Dida Yurnaldi, Mika R. Puspaningrum, Unggul P. Wibowo, Halmi Insani, Indra Sutisna, John A. Westgate, Nick J.G. Pearce, Mathieu Duval, Hanneke J.M. Meijer, Fachroel Aziz, Thomas Sutikna, Sander van der Kaars, Stephanie Flude, and M.J. Morwood. "Age and Context of the Oldest Known Hominin Fossils from Flores." *Nature* 534 (2016): 249–253.

Buckley, Jorunn Jacobsen. *The Mandaeans: Ancient Texts and Modern People*. Oxford: Oxford University Press, 2002.

Buckley, Tim. "A Historical Cycle of Hermeneutics in Proclus' *Platonic Theology*." In *Reading Plato in Antiquity*, edited by Harold Tarrant and Dirk Baltzly, 125–134. London: Duckworth, 2006.

Bulatkin, Eleanor Webster. *Structural Arithmetic Metaphor in the Oxford "Roland"*. Columbus, OH: Ohio State University Press, 1972.

Bulling, Anneliese. *The Decoration of Mirrors of the Han Period: A Chronology*. Ascona, Switzerland: Artibus Asiae Publishers, 1960.

Bulwer-Lytton, Edward. *The Coming Race*. Edinburgh and London: W. Blackwood and Sons, 1871.

Burkert, Walter. *Structure and History in Greek Mythology and Ritual.* Berkeley: University of California Press, 1979.

Burnyeat, M.F. "Plato on Why Mathematics is Good for the Soul." *Proceedings of the British Academy* 103 (2000): 1–81.

Burton, Richard Francis. *A Plain and Literal Translation of the Arabian Nights' Entertainments, Now Entituled* [sic] *The Book of The Thousand Nights and a Night.* 10 vols. Benares: Kamashastra Society, 1885–1888.

Bury, R.G. "Plato and History." *The Classical Quarterly*, new series, 1.1/2 (January–April 1951): 86–93.

Calame, Claude. "The Dithyramb, a Dionysiac Poetic Form: Genre Rules and Cultic Contexts." In *Dithyramb in Context*, edited by Barbara Kowalzig and Peter Wilson, 332–352. Oxford: Oxford University Press, 2013.

Calvo, Tomás, and Luc Brisson, eds. *Interpreting the Timaeus–Critias: Proceedings of the IV Symposium Platonicum — Selected Papers.* Sankt Augustin, Germany: Academia Verlag, 1997.

Cameron, Alan. "Crantor and Posidonius on Atlantis." *The Classical Quarterly*, new series, 33.1 (1983): 81–91.

Cameron, Alan. *Greek Mythology in the Roman World.* Oxford: Oxford University Press, 2004.

Cammann, Schuyler. "Chinese Mirrors and Chinese Civilization." *Archaeology* 2.3 (September 1949): 114–120.

Campany, Robert Ford. *Strange Writing: Anomaly Accounts in Early Medieval China.* Albany: State University of New York Press, 1996.

Campbell, Killis, ed., *The Seven Sages of Rome: Edited from the Manuscripts, with Introduction, Notes, and Glossary.* Genève: Slatkin Reprints, 1975.

Canning, Joseph, Edmund J. King, and Martial Staub, eds. *Knowledge, Discipline and Power in the Middle Ages: Essays in Honour of David Luscombe.* Leiden: Brill, 2011.

Caracostea, Daniel. "*Louis-Francois Jacolliot* (1837–1890): A Biographical Essay." *Theosophical History* 9.1 (January 2003): 12–39.

Carlson, John B. "Olmec Concave Iron-Ore Mirrors: The Aesthetics of a Lithic Technology and the Lord of the Mirror." In *The Olmec and Their Neighbors: Essays in Memory of Matthew W. Stirling*, edited by Elizabeth P. Benson, 117–148. Washington, D.C.: Dumbarton Oaks Research Library and Collections, 1981.

Carra de Vaux, Bernard. *L'abrege des merveilles, traduit de l'arabe d'apres les manuscrits de la Bibliotheque Nationale de Paris par le Baron Carra de Vaux.* Paris: Klincksieck, 1898.

Castriota, David. *Myth, Ethos, and Actuality: Official Art in Fifth-century BC Athens.* Madison, WI: University of Wisconsin Press, 1992.

Ceasar, Michael. "Atlantis of the Altiplano." *Mercator's World* 4.2 (March–April 1999): 24–31.

Censorinus. *The Birthday Book.* Translated by Holt N. Parker. Chicago: University of Chicago Press, 2007.

Censorinus. *De die natale* (*"The Natal Day"*). Translated by William Maude. New York: Cambridge Encyclopedia Co., 1900.

Ch'oe, Nam-sŏn. *Kosat'ong* (故事通 "Compilation of Old Events"). Kyŏngsong: Samjungdang, 1943.

Chabot, J.B. "Notice sur une mappemonde syrienne du XIIIe siècle." *Bulletin de géographie historique et descriptive* (1897): 98–112.

Chakravarty, Kalyan Kumar, and Robert G. Bednarik. *Indian Rock Art and Its Global Context*. Delhi: Motilal Banarsidass, 1997.

Chamberlain, Basil Hall. "Aino Folk-Lore." *The Folk-Lore Journal* 6.1 (January–December 1888): 1–51.

Chamberlain, Basil Hall. *Aino Folk-Tales*. London: Folk-Lore Society, 1888.

Chan, Leo Tak-hung. "Text and Talk: Classical Literary Tales in Traditional China and the Context of Casual Oral Storytelling." *Asian Folklore Studies* 56.1 (1997): 33–63.

Chang, Shu-Hong, and Cheng-Xian Li. *The Flying Devis of Dunhuang* (Beijing: China Travel and Tourism Press, 1980).

Charbit, Yves. "The Platonic City: History and Utopia." Translated by Arundhati Virmani. *Population* (*English Edition*) 57.2 (March–April 2002): 207–235.

Chauvin, Victor. *Bibliographie des ouvrages arabes ou relatifs aux Arabes, publiés dan l'Europe chrétienne de 1810 à 1885*. 12 vols. Liège: H. Vaillant-Carmanne, 1892–1922.

Chavalas, Mark. "Genealogical History as 'Charter': A Study of Old Babylonian Period Historiography and the Old Testament." In *Faith, Tradition, and History: Old Testament Historiography in its Near Eastern Context*, edited by A.R. Millard, J.K. Hoffmeier, and D.W. Baker, 103–128. Winona Lake, IN: Eisenbrauns, 1994.

Chen, Fan-Pen Li. *Chinese Shadow Theatre: History, Popular Religion, and Women Warriors*. Montréal: McGill-Queen's University Press, 2007.

Chen, Fan-Pen Li. "Shadow Theaters of the World." *Asian Folklore Studies* 62.1 (2003): 25–64.

Chen, Jinguo, and Jun Lin. 明教的新發現—福建霞浦縣摩尼教史跡辨析 ["New Manichaean Discoveries — An Analysis of the Relics of Manichaeism in Xiapu County, Fujian"]. In *Bu Zhi Yu Yi: Zhongyang Mei Yuan "Yiwen Ketang" Mingjia Jiangyan Lu* (不止於藝—中央美院 "藝文課堂" 名家講演錄 "More than Art — Famous 'Arts and Literature' Lectures, Central Academy of Fine Arts"), edited by Shaowen Li, 343–389. Beijing: Beijing daxue chubanshe, 2010.

Chen, Jue. "History and Fiction in the *Gujing Ji* (Record of an Ancient Mirror)." *Monumenta Serica* 52 (2004): 161–197.

Chen, Jue. "The Mystery of an 'Ancient Mirror': An Interpretation of *Gujing Ji* in the Context of Medieval Chinese Cultural History." *East Asian History* 27 (June 2004): 33–50.

Chen, Jue. *Record of an Ancient Mirror: An Interdisciplinary Reading*. Wiesbaden: Harrassowitz, 2010.

Chen, Jue. "Wang Ji's Travel Route in the South — An Itinerary Historical or Fictional?" In *Reisen in chinesischer Geschichte und Gegenwart: Erfahrungen, Berichte, Zeugnisse*, edited by Mechtild Leutner and Klaus Mühlhahn, 53–69. Wiesbaden: Harrassowitz, 2008.

Chen, Pingyuan. "From Popular Science to Science Fiction: An Investigation of Flying Machines." In *Translation and Creation: Readings of Western Literature in Early Modern China, 1840–1918*, edited by David Pollard, 209–240. Philadelphia: John Benjamin, 1998.

Chiri, Mashiho. *Ainu bungaku* (アイヌ文学 "Ainu Literature"). Tokyo: Gengensha, 1955.

Chiri, Mashiho. *Bunrui Ainugo jiten* (分類アイヌ語辭典 "Classified Dictionaries of the Ainu Language"), 3 vols. Tokyo: Nihon Jōmin Bunka Kenkyūjo: 1953–1954; 1962.

Chiri, Mashiho. *Chiri Mashiho chosakushū* (知里真志保著作集 "Collected Works of Chiri Mashiho"), 6 vols. Tokyo: Heibonsha, 1973–1976.

Chiri, Mashiho. 樺太アイヌの説話 ["Folktales of the Sakhalin Ainu"]. *Karafuto-chō hakubutsukan ihō* (樺太庁博物館彙報 "Karafuto Prefecture Museum Bulletin") 3.1 (1944): 1–146.

Chiri, Mashiho. 樺太アイヌの説話 ["Folktales of the Sakhalin Ainu"]. *Minzoku-gaku kenkyū* (民族学研究 "Ethnology Research") 12.4 (1948): 46–56.

Choe, Sang-su. *The Survey of Korean Kites*. Seoul: Korea Books Publishing Co., 1958.

Choksey, K.M. *Dentistry in Ancient India*. Bombay: Popular Book Depot, 1953.

Chŏn, Kyu-t'ae. *Korean Heritage*. Seoul: Jeong Eum Sa, 1975.

Christenson, Allen J. *Popol Vuh: The Sacred Book of the Maya: The Great Classic of Central American Spirituality, Translated from the Original Maya Text*. Norman, OK: University of Oklahoma Press, 2007.

Christmas, Henry. *The Cradle of the Twin Giants, Science and History*. 2 vols. London: Richard Bentley, 1849.

Christopher, Kevin. "Atlantis Behind the Myth: No Way, No How, No Where." *Skeptical Inquirer* 26.1 (January–February 2002): 44–45.

Clarke, Arthur C. *Profiles of the Future*. New York: Harper and Row, 1973.

Clarke, M.L. "Virgil and the Golden Section — Reviewed Work: *Structural Patterns and Proportions in Vergil's Aeneid* by George E. Duckworth." *The Classical Review*, new series, 14.1 (March 1964): 43–45.

Clay, Diskin. "The Plan of Plato's *Critias*." In *Interpreting the Timaeus–Critias: Proceedings of the IV Symposium Platonicum — Selected Papers*, edited by Tomás Calvo and Luc Brisson, 49–54. Sankt Augustin, Germany: Academia Verlag, 1997.

Clay, Diskin. "Plato's Atlantis: The Anatomy of a Fiction." In *Proceedings of the Boston Area Colloquium in Ancient Philosophy, 1999*, edited by John J. Cleary and Gary M. Gurtler, 1–22. Leiden: Brill, 2000.

Clay, Diskin, and Andrea Purvis. *Four Island Utopias: Being Plato's Atlantis, Euhemeros of Messene's Panchaia; Iamboulos' Island of the Sun; Sir Francis Bacon's New Atlantis*. Newburyport, MA: Focus Publishing, 1999.

Cleary, John J. "Proclus as a Reader of Plato's *Timaeus*." In *Reading Plato in Antiquity*, edited by Harold Tarrant and Dirk Baltzly, 135–150. London: Duckworth, 2006.

Cleary, John J., and Gary M. Gurtler, eds. *Proceedings of the Boston Area Colloquium in Ancient Philosophy, 1999*. Leiden: Brill, 2000.

Clouston, W.A. "On the Magical Elements in Chaucer's 'Squire's Tale,' with Analogues." In *John Lane's Continuation of Chaucer's "Squire's Tale"*, edited by Frederick J. Furnivall, 263–471. London: Published for the Chaucer Society by K. Paul, Trench, Trübner & Co., 1888, 1890.

Cohen-Mor, Dalya. *A Matter of Fate: The Concept of Fate in the Arab World as Reflected in Modern Arabic Literature*. Oxford: Oxford University Press, 2001.

Cohen, Adam Max. *Shakespeare and Technology: Dramatizing Early Modern Technological Revolutions*. New York: Palgrave Macmillan, 2006.

Cohen, Alvin P. "Documentation Relating to the Origins of the Chinese Shadow-Puppet Theater." *Asia Major*, 3rd series, 13.1 (2000): 83–108.

Cohen, Ivan M. "Herodotus and the Story of Gyges: Traditional Motifs in Historical Narrative." *Fabula* 45.1–2 (March 2004): 55–68.

Cole, John R. "Cult Archaeology and Unscientific Method and Theory." In *Advances in Archaeological Method and Theory*, 11 vols., edited by Michael B. Schiffer, 3.1–33. New York: Academic Press, 1978–1987.

Collina-Girard, Jacques. *L'Atlantide retrouvée? Enquête scientifique autour d'un mythe*. Paris: Belin, 2009.

Collina-Girard, Jacques. "La crise finiglaciaire à Gibraltar et l'Atlantide: tradition et géologie." *Préhistoire anthropologie méditerranéennes* 10–11 (2001–2002): 53–60.

Collins, Andrew. *Gateway to Atlantis: The Search for the Source of a Lost Civilization*. New York: Carroll & Graf Publishers, 2000.

Colston, Stephen A. "'No Longer Will There Be a Mexico': Omens, Prophecies, and the Conquest of the Aztec Empire." *American Indian Quarterly* 9.3 (Summer 1985): 239–258.

Columbus, Ferdinand. *The Life of the Admiral Christopher Columbus by his Son Ferdinand*. Edited and translated by Benjamin Keen. New Brunswick, NJ: Rutgers University Press, 1959.

Commager, Steele. *The Odes of Horace: A Critical Study*. New Haven: Yale University Press, 1962.

Compagni, V. Perrone, ed. *De Occulta Philosophia*. Leiden: Brill, 1992.

Comparetti, Domenico. *Virgilio nel medio evo*. 2 vols. Firenze: Bernardo Seeber, 1896.

Connerton, Paul. *How Societies Remember*. Cambridge: Cambridge University Press, 1989.

Cook, Michael. "Pharaonic History in Medieval Egypt." *Studia Islamica* 57 (1983): 67–103.

Coomaraswamy, Ananda K. and Rama P. Coomaraswamy. *The Door in the Sky: Coomaraswamy on Myth and Meaning*. Princeton: Princeton University Press, 1997.

Cooper, John M., and D.S. Hutchinson, eds. *Plato: Complete Works*. Indianapolis: Hackett, 1997.

Coppens, Philip. *The Lost Civilization Enigma: A New Inquiry Into the Existence of Ancient Cities, Cultures, and Peoples Who Pre-Date Recorded History*. Pompton Plains, NJ: New Page Books, 2013.

Cornell, Christine. "Remembering the Ancients: Observations on Technoscience in Čapek's *R.U.R.*" *Bulletin of Science, Technology & Society* 31.2 (April 2011): 103–112.

Correale, Robert M., and Mary Hamel, eds. *Sources and Analogues of the Canterbury Tales*. 2 vols. Cambridge: D.S. Brewer, 2002–2005.

Cory, Preston. *Cory's Ancient Fragments of the Phoenician, Carthaginian, Babylonian. Egyptian and Other Writers*. London: Reeves & Turner, 1876.

Courlander, Harold. *Hopi Voices: Recollections, Traditions, and Narratives of the Hopi Indians*. Albuquerque, NM: University of New Mexico Press, 1982.

Coyle, John Kevin. *Manichaeism and Its Legacy*. Leiden: Brill, 2009.

Crane, Susan. *Animal Encounters: Contacts and Concepts in Medieval Britain*. Philadelphia: University of Pennsylvania Press, 2013.

Crane, Susan. *Gender and Romance in Chaucer's "Canterbury Tales"*. Princeton: Princeton University Press, 1994.

Critchlow, Keith. *Time Stands Still: New Light on Megalithic Science*. New York: St. Martin's Press, 1982.

Crook, Nigel, ed. *The Transmission of Knowledge in South Asia: Essays on Education, Religion, History, and Politics*. New York: Oxford University Press, 1996.

Cruttenden, Walter. *Lost Star of Myth and Time*. Pittsburgh: St. Lynn's Press, 2006.

Crystal, Eric. "Myth, Symbol and Function of the Toraja House." *Traditional Dwellings and Settlements Review* 1.1 (Fall 1989): 7–17.

Cuomo, Serafina. *Technology and Culture in Greek and Roman Antiquity*. Cambridge: Cambridge University Press, 2007.

Curnoe, Darren, Xueping Ji, Andy I.R. Herries, Kanning Bai, Paul S.C. Taçon, Hende Bao, David Fink, Yunsheng Zhu, John Hellstrom, Yun Luo, Gerasimos Cassis, Bing Su, Stephen Wroe, Shi Hong, William C.H. Parr, Shengmin Huang, and Natalie Rogers. "Human Remains from the Pleistocene-Holocene Transition of Southwest China Suggest a Complex Evolutionary History for East Asians." *PLoS ONE* 7.3 (March 2012): e31918. doi:10.1371/journal.pone.0031918.

Currell, David. *Shadow Puppets & Shadow Play*. Ramsbury: Crowood Press, 2007.

Curry, Walter Clyde. *Chaucer and the Mediaeval Sciences*. New York: Barnes & Noble, 1960.

d'Amiens, Girart. *Meliacin ou le Cheval de fust*, edited by Antoinette Saly. Aix-en-Provence: CUERMA, 1990.

d'Huy, Julien. "A Cosmic Hunt in the Berber Sky: A Phylogenetic Reconstruction of a Palaeolithic Mythology." *Les Cahiers de l'AARS* [*Association des Amis de l'Art Rupestre Saharien*] 16 (2013): 93–106.

d'Huy, Julien. "Peut-on tester scientifiquement la phylogénétique des mythes? A propos de la chasse cosmique." *Mythologie française* 263 (June 2016): 20–24.

d'Huy, Julien. "A Phylogenetic Approach of Mythology and its Archaeological Consequences." *Rock Art Research* 30.1 (May 2013): 115–118.

d'Huy, Julien. "Polyphemus (*Aa. Th. 1137*): A Phylogenetic Reconstruction of a Prehistoric Tale." *Nouvelle Mythologie Comparée* 1 (2013): 3–18.

Daly, Herman E. "The Economic Thought of Frederick Soddy." *History of Political Economy* 12.4 (1980): 469–488.

Danzig, Gabriel. "Rhetoric and the Ring: Herodotus and Plato on the Story of Gyges as a Politically Expedient Tale." *Greece & Rome*, 2nd series, 55.2 (October 2008): 169–192.

Dathorne, O.R. *Imagining the World: Mythical Belief Versus Reality in Global Encounters*. Westport, CT: Bergin & Garvey, 1994.

Davenport, David W., and Ettore Vincenti. *2000 a.C., Distruzione Atomica*. Milan: Sugarco, 1979.

Davies, Mansel. "Frederick Soddy: The Scientist as Prophet." *Annals of Science* 49.4 (July 1992): 351–367.

Davis, Gerald J. *The Canterbury Tales: The New Translation*. Bridgeport, CT: Insignia, 2016.

Davis, John Francis. *The Chinese: a General Description of the Empire of China and Its Inhabitants*. 2 vols. London: Charles Knight & Co., 1836.

De Camp, L. Sprague. *Lost Continents: The Atlantis Theme in History, Science, and Literature*. New York: Dover, 1970.

de Roo, Peter. *History of America Before Columbus According to Documents and Approved Authors*. Vol. 1, *American Aborigines*. Philadelphia: J.B. Lippincott, 1900.

de Santillana, Giorgio, and Hertha von Dechend. *Hamlet's Mill: An Essay on Myth and the Frame of Time*. Boston: Gambit, 1969.

De Vorsey, Louis, Jr. "The Importance of Native American Maps in the Discovery and Exploration of North America." *Terrae Incognitae* 42.1 (September 2010): 5–18.

Delatte, Louis. *Textes Latins et Vieux Francais Relatifs aux Cyranides*. Paris: Droz, 1942.

Delcourt, Marie, and Robert L. Rankin. "The Last Giants." *History of Religions* 4.2 (Winter 1965): 209–242.

Demiéville, Paul. "Le miroir spirituel." *Sinologica: Zeitschrift für chinesische Kultur und Wissenschaft* 1.2 (1947): 112–137.

Despeux, Catherine. "The Body Revealed: The Contribution of Forensic Medicine to Knowledge and Representation of the Skeleton in China." In *Graphics and Text in the Production of Technical Knowledge in China: the Warp and the Weft*, edited by Francesca Bray, Vera Dorofeeva-Lichtmann, and Georges Métailié, 635–684. Leiden: Brill, 2007.

Detienne, Marcel. *The Writing of Orpheus: Greek Myth in Cultural Context*. Translated by Janet Lloyd. Baltimore: The Johns Hopkins University Press, 2002.

Di Pasquale, Giovanni. "Vitruvius's Image of the Universe: Architecture and Mechanics." In *Vitruvianism: Origins and Transformations*, edited by Paolo Sanvito, 49–64. Berlin: Walter de Gruyter, 2016.

Diény, Jean-Pierre, ed. *Hommage à Kwong Hing Foon, Études d'histoire culturelle de la Chine*. Paris: Collège de France, Institut des hautes études chinoises, 1995.

DiMarco, Vincent. "The Dialogue of Science and Magic in Chaucer's *Squire's Tale*." In *Dialogische Strukturen/Dialogic Structures: Festschrift für Willi Erzgräber zum 70. Geburtstag*, edited by Thomas Kuhn and Ursula Schaefer, 50–68. Tübingen: Gunter Narr, 1996.

DiMarco, Vincent. "The Squire's Tale." In *Sources and Analogues of the Canterbury Tales*, 2 vols, edited by Robert M. Correale and Mary Hamel, 1.169–1.210. Cambridge: D.S. Brewer, 2002–2005.

Dohrenwend, Doris. "The Early Chinese Mirror." *Artibus Asiae* 27.1–2 (1964): 79–98.

Dönitz, Saskia. "Alexander the Great in Medieval Hebrew Traditions." In *A Companion to Alexander Literature in the Middle Ages*, edited by Z. David Zuwiyya, 21–40. Leiden: Brill, 2011.

Donnelly, Ignatius. *Atlantis: The Antediluvian World*. New York: Harper, 1882.

Doshi, P.K. "History of Stereotactic Surgery India." In *Textbook of Stereotactic and Functional Neurosurgery*, edited by A.M. Lozano, P.L. Gildenberg, and R.R. Tasker, 155–169. Berlin: Springer, 2009.

Drummond, William. "Memoir on the Antiquity of the Zodiacs of Esneh and Dendera — Part II." *The Classical Journal* 24 (September and December 1821), 251–282.

Duckworth, George E. "Mathematical Symmetry in Vergil's Aeneid." *Transactions and Proceedings of the American Philological Association* 91 (1960): 184–220.

Duling, Dennis C. "The Eleazar Miracle and Solomon's Magical Wisdom in Flavius Josephus's 'Antiquitates Judaicae' 8.42–49." *The Harvard Theological Review* 78.1–2 (January–April 1985): 1–25.

Dull, Jack L. "Anti-Qin Rebels: No Peasant Leaders Here." *Modern China* 9.3 (July 1983): 285–318.

Dundes, Alan. "The Motif-Index and the Tale Type Index: A Critique." *Journal of Folklore Research* 34.3 (September–December 1997): 195–202.

Dunn, W.A. "The Ancient and Modern Sciences of the Radioactive Transformations of Nature: A Vindication of the Teachings of H.P. Blavatsky — Part Three." *The Theosophical Path* 14.1 (January 1918): 53–61.

Dušanić, Slobodan. "Plato's Atlantis." *L'Antiquité Classique* 51 (1982): 25–52.

Dwyer, Eugene. "Excess." In *Encyclopedia of Comparative Iconography: Themes Depicted in Works of Art*, 2 vols., edited by Helene E. Roberts, 1.293–1.297. Chicago: Fitzroy Dearborn, 1998.

Eck, Diana L. *India: A Sacred Geography*. New York: Three Rivers Press, 2012.

Edelstein, Ludwig. "The Function of the Myth in Plato's Philosophy." *Journal of the History of Ideas* 10.4 (October 1949): 463–481.

Edwards, E.D. *Chinese Prose Literature of the T'ang Period, AD 618–906*. 2 vols. London: Arthur Probsthain, 1937–1938.

Eisenstein, Elizabeth L. "Clio and Chronos: An Essay on the Making and Breaking of History-Book Time." *History and Theory* 5, suppl. 6 (1966): 36–64.

El-Shamy, Hasan M. *A Motif Index of The Thousand and One Nights*. Bloomington: Indiana University Press, 2006.

El-Shamy, Hasan M. "Siblings in *Alf layla wa-layla*." In *The Arabian Nights in Transnational Perspective*, edited by Ulrich Marzolph, 83–102. Detroit: Wayne State Press, 2007.

Elad, Amikam. *The Rebellion of Muḥammad al-Nafs al-Zakiyya in 145/762*. Leiden: Brill, 2016.

Eliade, Mircea. "Symbolisme du 'Vol magique'." *Numen* 3.1 (January 1956): 1–13.

Ellegård, Alvar. "Stone Age Science in Britain?" *Current Anthropology* 22.2 (April 1981): 99–125.

Ellenberger, C. Leroy. "A Lesson from Velikovsky." *Skeptical Inquirer* 10.4 (Summer 1986): 380–381.

Ellingson-Waugh, Ter. "Musical Flight in Tibet." *Asian Music* 5.2 (1974): 3–44.

Ellis, Richard. *Imagining Atlantis*. New York: Alfred A. Knopf, 1998.

Ellis, Robert P. *Francis Bacon: The Double-Edged Life of the Philosopher and Statesman*. Jefferson, NC: McFarland & Company, 2015.

Elsner, Jaś. "Structuring 'Greece': Pausanias' *Periegesis* as a Literary Construct." In *Pausanias: Travel and Memory in Roman Greece: Travel and Memory in Roman Greece*, edited by Susan E. Alcock, John F. Cherry, and Jaś Elsner, 3–20. Oxford: Oxford University Press, 2001.

Elton, Oliver, trans., *The Nine Books of the Danish History of Saxo Grammaticus*. 2 vols. London: Norrœna Society, 1905.

Endō, Shiho, and Hiroshi Nakagawa, eds. *Motozo Nabesawa's Ainu-language Notes (1928–1959) at the National Museum of Ethnology* [= *Senri Ethnological Reports* 134]. Osaka: National Museum of Ethnology, 2016.

Enoch, Jay M. "History of Mirrors Dating Back 8000 Years." *Optometry and Vision Science* 83.10 (October 2006): 775–781.

Erikson, Kai T. "Out of Sight, Out of Our Minds." *The New York Times Magazine* (6 March 1994): 36–41; 50; 61.

Esmailpour, Abolqasen. *Manichaean Gnosis and Creation Myth*. Philadelphia: University of Pennsylvania, Department of East Asian Languages and Civilizations, 2005.

Etter, Carl. *Ainu Folklore: Traditions and Culture of the Vanishing Aborigines of Japan 1949*. Chicago: Wilcox & Follett, 1949.

Evans, Harry B. *Exploring the Kingdom of Saturn: Kircher's* Latium *and its Legacy*. Ann Arbor: University of Michigan Press, 2012.

Evans, James. *The History and Practice of Ancient Astronomy*. New York: Oxford University Press, 1998.

Fagan, Garrett G. *Archaeological Fantasies*. New York: Routledge, 2006.

Fagan, Garrett G. "Diagnosing Pseudoarchaeology." In *Archaeological Fantasies*, edited by Garrett G. Fagan, 23–46. New York: Routledge, 2006.

Fagan, Garrett G., and Chris Hale. "The New Atlantis and the Dangers of Pseudohistory." *Skeptic* 9.1 (Spring 2001): 78–87.

Faraone, Christopher A. "From Magic Ritual to Semiotic Game: The Transformation of Neo-Assyrian Love Spells in Classical and Hellenistic Greece." In *Ideologies as Intercultural Phenomena: Proceedings of the Third Annual Symposium of the Assyrian and Babylonian Intellectual Heritage Project*, edited by Antonio Panaino and Giovanni Pettinato, 61–74. Milan: Mimesis Edizioni, 2002.

Faridany-Akhavan, Zahra. "All the King's Toys." *Muqarnas* 10 (1993): 292–298.

Feder, Kenneth L. "Diagnosing the Atlantis Syndrome." *Skeptical Inquirer* 26.3 (May–June 2002): 53–54.

Feder, Kenneth L. *Frauds, Myths, and Mysteries: Science and Pseudoscience in Archaeology*, 2nd ed. Mountain View, CA: Mayfield Publishing Company, 1995.

Federal Aviation Administration. *Glider Flying Handbook*. Washington, D.C.: Federal Aviation Administration, 2003.

Ferdowsi, Abolqasem. *Shahnameh: The Persian Book of Kings*. Translated by Dick Davis. New York: Penguin, 2016.

Ferguson, Arthur B. *Utter Antiquity: Perceptions of Prehistory in Renaissance England*. Durham, NC: Duke University Press, 1993.

Ferrão, J.B. Tolentino. "Barcos e trens a vapor, balões aerostaticos, e outras invenções entre os antigos Hindus." *Vasco da Gama* 1 (1872): n.p.

Fessenden, Helen M. *Fessenden: Builder of Tomorrows*. New York: Coward-McCann, 1940.

Fessenden, Reginald A. *The Deluged Civilization of the Caucasus Isthmus*. Boston: T.J. Russell, 1923.

Finch, Michael. *Min Yông-hwan: A Political Biography*. Honolulu: University of Hawai'i Press, 2002.

Finley, M.I. *Aspects of Antiquity: Discoveries and Controversies*. New York: Viking, 1968.

Fischer, J.L. "The Sociopsychological Analysis of Folktales." *Current Anthropology* 4.3 (June 1963): 235–295.

Fitzgerald, John. *Awakening China: Politics, Culture, and Class in the Nationalist Revolution*. Stanford, CA: Stanford University Press, 1996.

Flem-Ath, Rand, and Rose Flem-Ath. *When the Sky Fell: In Search of Atlantis*. New York: St. Martin's Press, 1995.

Fodor, Sándor. "The Origins of the Arabic Legends of the Pyramids." *Acta Orientalia Academiae Scientiarum Hungaricae* 23.3 (September 1970): 335–363.

Fodor, Sándor. "The Origins of the Arabic Sūrīd Legend." *Zeitschrift für Ägyptische Sprache und Altertumskunde* 96 (December 1970): 103–109.

Forrester, John, and John Henry, eds. *Jean Fernel's* On the Hidden Causes of Things: *Forms, Souls, and Occult Diseases in Renaissance Medicine*. Leiden: Brill, 2005.

Forth, Gregory. "Images of the Wildman Inside and Outside Europe." *Folklore* 118.3 (December 2007): 261–281.

Foucault, Michel. *L'Archéologie du savoir*. Paris: Gallimard, 1969.

Fowler, Robert L. *Early Greek Mythology*. 2 vols. Oxford: Oxford University Press, 2013.

Franck, Adolphe. *The Kabbalah: The Religious Philosophy of the Hebrews*. Secacus, NJ: Citadel Press, 1967.

Frank, Adam, and Gavin A. Schmidt. "The Silurian Hypothesis: Would It Be Possible to Detect an Industrial Civilization in the Geological Record?" *International Journal of Astrobiology* (2018): 1–9. https://doi.org/10.1017/S1473550418000095.

Frankfurter, David. "Apocalypses Real and Alleged in the Mani Codex." *Numen* 44.1 (January 1997): 60–73.

Fraser, J.T., N. Lawrence, and F.C. Haber, eds. *Time. Science, and Society in China and the West: The Study of Time V*. Amherst: University of Massachusetts Press, 1986.

Frazer, James G. "Some Folk-Lore of Ferns." *American Fern Journal* 24.1 (January–March, 1934): 10–15.

Frazer, James G., ed. and trans. *Fastorum libri sex: The Fasti of Ovid*. 2 vols. Cambridge: Cambridge University Press, 2015.

Fredericks, S. Casey. "Plato's Atlantis: A Mythologist Looks at Myth." In *Atlantis: Fact or Fiction?*, edited by Edwin S. Ramage, 81–99. Bloomington, IN: Indiana University Press, 1978.

Freeth, T. "The Antikythera Mechanism 2. Is it Posidonius' Orrery?" *Mediterranean Archeology and Archeaometry* 2.2 (2002): 45–58.

Freeth, T., Y. Bitsakis, X. Moussas, J.H. Seiradakis, A. Tselikas, H. Mangou, M. Zafeiropoulou, R. Hadland, D. Bate, A. Ramsey, M. Allen, A. Crawley, P. Hockley, T. Malzbender, D. Gelb, W. Ambrisco, and M.G. Edmund. "Decoding the Ancient Greek Astronomical Calculator Known as the Antikythera Mechanism." *Nature* 444 (30 November 2006): 587–591.

Fridman, Eva Jane Neumann, and Mariko Namba Walter, eds. *Shamanism: An Encyclopedia of World Beliefs, Practices, and Culture*. 2 vols. Santa Barbara, CA: ABC-CLIO, 2004.

Friedman, John Block. *The Monstrous Races in Medieval Art and Thought*. Syracuse: Syracuse University Press, 2000.

Friedrich, Walter L. *Fire in the Sea: The Santorini Volcano: Natural History and the Legend of Atlantis*. Cambridge: Cambridge University Press, 2000.

Fritze, Ronald H. *Invented Knowledge: False History, Fake Science and Pseudo-religions*. London: Reaktion Books, 2009.

Fudge, Bruce. "Signs of Scripture in the 'City of Brass'." *Journal of Qur'anic Studies* 8.1 (2006): 88–118.

Furnivall, Frederick J., ed. *John Lane's Continuation of Chaucer's "Squire's Tale"*. London: Published for the Chaucer Society by K. Paul, Trench, Trübner & Co., 1888, 1890.

Gabba, Emilio. "True History and False History in Classical Antiquity." *The Journal of Roman Studies* 71 (1981): 50–62.

Gallaga, Emiliano, and Marc G. Blainey, eds. *Manufactured Light: Mirrors in the Mesoamerican Realm*. Boulder: University Press of Colorado, 2016.

Gantenbein, Urs Leo. "Converging Magical Legends: Faustus, Paracelsus, and Trithemius." In *The Faustian Century: German Literature and Culture in the Age of Luther and Faustus*, edited by James M. Van der Laan and Andrew Weeks, 93–124. Rochester: Camden House, 2013.

Gantevoort, Michelle, Duane W. Hamacher, and Savannah Lishick, "Reconstructing the Star Knowledge of Aboriginal Tasmanians." *Journal of Astronomical History and Heritage* 19.3 (2016): 327–347.

Gardner, Iain. *The Kephalaia of the Teacher: The Edited Coptic Manichaean Texts in Translation with Commentary*. Leiden: Brill, 1995.

Garfield, Susan. "Oslo Conference Suggests that World Religions Carry Nuclear Waste Danger Warnings into the Far Future." *The Nuclear Guardianship Library* (1992). http://www.nonukes.org/w28relig.htm.

Garfield, Susan. "Proposed 'Atomic Priesthood' is NOT Nuclear Guardianship! A Critique of Thomas Sebeok's Vision of the Future." *Nuclear Guardianship Forum* 3 (Spring 1994): n.p.

Garnier, John. *The Worship of the Dead or the Origin and Nature of Pagan Idolatry and its Bearing upon the Early History of Egypt and Babylonia*. London: Chapman & Hall, 1904.

Gaukroger, Stephen, ed., *The Uses of Antiquity: The Scientific Revolution and the Classical Tradition*. Dordrecht: Kluwer, 1991.

Ge, Hong. *Baopuzi Neipian* (抱朴子内篇 "Master Embracing Simplicity — Inner Chapters"). Taipei: Taiwan shang wu yin shu guan, 1983.

Ge, Hong, and James R. Ware. *Alchemy, Medicine, and Religion in the China of AD 320, the Nei P'ien of Ko Hung (Pao-p'u-tzu)*. Cambridge, MA: M.I.T. Press, 1966.

Ge, Zhaoguang. *An Intellectual History of China, Volume One: Knowledge, Thought, and Belief before the Seventh Century CE*. Translated by Michael S. Duke and Josephine Chiu-Duke. Leiden: Brill, 2014.

Gee, Henry. "Flores, God and Cryptozoology." *Nature News* (27 October 2004). doi:10.1038/news041025-2.

Geertz, Armin W. "A Reed Pierced the Sky: Hopi Indian Cosmography on Third Mesa, Arizona." *Numen* 31.2 (December 1984): 216–241.

Geertz, Armin W. "Tsu'ngyam Tradition: Men, Women and Snakes in Hopi Theology." In *Schlangenritual: Der Transfer der Wissensformen vom Tsu'ti'kive der Hopi bis zu Aby Warburgs Kreuzlinger Vortrag*, edited by Cora Bender, Thomas Hensel, and Erhard Schüttpelz, 103–118. Berlin: Akademie Verlag, 2007.

Geiger, Joseph K. *Transition: An Overview of History from a Christian Perspective*. Victoria, BC: Trafford Publishing, 2006.

Gellius, Aulus. *The Attic Nights of Aulus Gellius*. 3 vols. Translated by John C. Rolfe. Cambridge, MA: Harvard University Press, 1927.

Geraghty, Paul. "Pulotu, Polynesian Homeland." *The Journal of the Polynesian Society* 102.4 (December 1993): 343–384.

Gerhardt, Mia I. *The Art of Story-telling: A Literary Study of the Thousand and One Nights*. Leiden: Brill, 1963.

Getty, Robert J. "Neopythagoreanism and Mathematical Symmetry in Lucan, *De bello civili* 1." *Transactions and Proceedings of the American Philological Association* 91 (1960): 310–323.

Gibbon, Edward. *The History of the Decline and Fall of the Roman Empire*. 6 vols. London: Strahan & Cadell, 1776–1789.

Gibson, Edgar C.S., trans. *A Select Library of Nicene and Post-Nicene Fathers of the Christian Church*. 2nd series. Vol. 11. New York: The Christian Literature Company, 1894.

Gibson, Roy, and Ruth Morello, eds. *Pliny the Elder: Themes and Contexts*. Leiden: Brill, 2011.

Gil, Moshe. "The Babylonian Encounter and the Exilarchic House in the Light of Cairo Geniza Documents and Parallel Arab Sources." In *Judaeo-Arabic Studies: Proceedings of the Founding Conference of the Society for Judaeo-Arabic Studies*, edited by Norman Golb, 135–174. Amsterdam: Harwood Academic Publishers, 1997.

Giles, H.A. *A Chinese Biographical Dictionary*. London: Bernard Quaritch, 1898.

Gill, Christopher. "The Genre of the Atlantis Story." *Classical Philology* 72.4 (October 1977): 287–304.

Gill, Christopher. *Plato: The Atlantis Story* (*Timaeus 17–27 / Critias*). Bristol, England: Bristol Classical Press, 1980.

Gill, Christopher. "Plato's Atlantis Story and the Birth of Fiction." *Philosophy and Literature* 3.1 (1979): 64–78.

Gill, Robin D. *Mad in Translation: A Thousand Years of* Kyōka, *Comic Japanese Poetry in the Classic* Waka *Mode*. Key Biscayne, FL: Paraverse Press, 2009.

Ginzberg, Louis. *Legends of the Jews*. 2 vols. Philadelphia: Jewish Publication Society, 2003.

Ginzburg, Carlo. *Myths, Emblems, Clues*. London: Hutchinson Radius, 1990.

Giovanni Reale. "Plato's Doctrine of the Origin of the World, with Special Reference to the *Timaeus*." In *Interpreting the Timaeus–Critias: Proceedings of the IV Symposium*

Platonicum — Selected Papers, edited by Tomás Calvo and Luc Brisson, 149–164. Sankt Augustin, Germany: Academia Verlag, 1997.

Gittes, Tobias Foster. *Boccaccio's Naked Muse: Eros, Culture, and the Mythopoeic Imagination*. Toronto: University of Toronto Press, 2008.

Glenn, Justin. "The Polyphemus Folktale and Homer's Kyklôpeia." *Transactions and Proceedings of the American Philological Association* 102 (1971): 133–181.

Godwin, Joscelyn. *Athanasius Kircher: A Renaissance Man and the Quest for Lost Knowledge*. London: Thames and Hudson, 1979.

Golas, Peter J. *Picturing Technology in China: From Earliest Times to the Nineteenth Century*. Hong Kong: Hong Kong University Press, 2015.

Golb, Norman, ed. *Judaeo-Arabic Studies: Proceedings of the Founding Conference of the Society for Judaeo-Arabic Studies*. Amsterdam: Harwood Academic Publishers, 1997.

Goldberg, Benjamin. *The Mirror and Man*. Charlottesville: University of Virginia Press, 1985.

Goldberg, Christine. "Dilemma Tales in the Tale Type Index: The Theme of Joint Efforts." *Journal of Folklore Research* 34.3 (September–December 1997): 179–193.

Goldin, Paul R. "The Myth that China Has No Creation Myth." *Monumenta Serica* 56 (2008): 1–22.

Goldman, Robert P., and Sally Sutherland Goldman, trans. *The Rāmāyaṇa of Vālmīki: An Epic of Ancient India*. Vol. 7, *Uttarakāṇḍa*. Princeton: Princeton University Press, 2017.

Goldman, Robert P., Sally Sutherland Goldman, and Barend A. van Nooten, trans. *The Rāmāyaṇa of Vālmīki: An Epic of Ancient India*. Vol. 6, *Yuddhakāṇḍa*. Princeton: Princeton University Press, 2009.

Gooday, Graeme J.N., and Morris F. Low. "Technology Transfer and Cultural Exchange: Western Scientists and Engineers Encounter Late Tokugawa and Meiji Japan." *Osiris*, 2nd series, 13 (1998): 99–128.

Goodheart, Benjamin J. "Tracing the History of the Ornithopter: Past, Present, and Future." *Journal of Aviation/Aerospace Education & Research* 21.1 (Fall 2011): 31–44.

Goodrick-Clark, Nicholas, ed. *Helena Blavatsky*. Berkeley: North Atlantic Books, 2004.

Gordin, Michael D. *The Pseudoscience Wars: Immanuel Velikovsky and the Birth of the Modern Fringe*. Chicago: University of Chicago Press, 2012.

Gottheil, Richard J.H. *The Literature of Persia*. 2 vols. New York: Colonial Press, 1900.

Gould, Stephen Jay. *Ever Since Darwin: Reflections in Natural History*. New York: Norton, 1977.

Goulding, Robert. *Defending Hypatia: Ramus, Savile, and the Renaissance Rediscovery of Mathematical History*. Dordrecht: Springer, 2010.

Grabes, Herbert. *The Mutable Glass: Mirror-imagery in Titles and Texts of the Middle Ages and English Renaissance*. Translated by Gordon Collier. Cambridge: Cambridge University Press, 2009.

Graham, A.C., trans. *Chuang-tzu: The Seven Inner Chapters and Other Writings from the Book Chuang-tzu*. London: George Allen and Unwin, 1981.

Graham, A.C., and Nathan Sivin. "A Systematic Approach to Mohist Optics." In *Chinese Science: Explorations of an Ancient Tradition*, edited by Shigeru Nakayama and Nathan Sivin, 105–152. Cambridge, MA, 1973.

Graves, Carl, Gabrielle Heffernan, Luke McGarrity, Emily Millward, and Marsia Sfakianou Bealby, eds. *Current Research in Egyptology: Proceedings of the Thirteenth Annual Symposium University of Birmingham, 2012*. Oxford: Oxbow Books, 2013.

Greene, Mott T. *Natural Knowledge in Preclassical Antiquity*. Baltimore and London: The Johns Hopkins University Press, 1992.

Greenhill, Alfred George. "The Legend of Alexander and His Flying Machine." *Nature: A Weekly Illustrated Journal of Science* 101.2524 (14 March 1918): 25–26.

Gregory, Richard. *Mirrors in Mind*. New York: W.H. Freeman, 1996.

Guénon, René. *The King of the World*. Edited by Samuel D. Fohr and translated by Henry D. Fohr. Hillsdale, NY: Sophia Perennis, 2004.

Guilhem, Olivier. *Mockeries and Metamorphoses of an Aztec God: Tezcatlipoca, "Lord of the Smoking Mirror"*. Boulder: University Press of Colorado, 2003.

Guilhem, Olivier, and Leonardo López Lujan. "Images of Moctezuma and His Symbols of Power." In *Moctezuma: Aztec Ruler*, edited by Colin McEwan and Leonardo López Lujan, 78–91. London: British Museum Press, 2009.

Gulácsi, Zsuzsanna. "The Central Asian Roots of a Chinese Manichaean Silk Painting in the Collection of the Yamato Bunkakan, Nara, Japan." In *"In Search of Truth": Augustine, Manichaeism and other Gnosticism — Studies for Johannes van Oort at Sixty*, edited by Jacob Albert Berg, Annemaré Kotzé, Tobias Nicklas, and Madeleine Scopello, 315–338. Leiden: Brill, 2011.

Gulácsi, Zsuzsanna. "Searching for Mani's Picture Book in Textual and Pictorial Sources." *Transcultural Studies* 1 (2011): 233–262.

Gulácsi, Zsuzsanna, and Jason BeDuhn. "Picturing Mani's Cosmology: An Analysis of Doctrinal Iconography on a Manichaean Hanging Scroll from 13th/14th-Century Southern China." *Bulletin of the Asia Institute* 25 (October 2015): 55–105.

Haase, Wolfgang, and Meyer Reinhold, eds. *The Classical Tradition and the Americas*. Vol. 1, *European Images of the Americas and the Classical Tradition*. Part 1. Berlin: Walter de Gruyter, 1994.

Hall, A.R. "The Early Significance of Chinese Mirrors." *Journal of the American Oriental Society* 55.2 (June 1935): 182–189.

Hall, David L., and Roger T. Ames. *Thinking from the Han: Self, Truth, And Transcendence in Chinese and Western Culture*. Albany: State University of New York Press, 2003.

Halliday, William Reginald. *Greek Divination: A Study of Its Methods and Principles*. London: Macmillan, 1913.

Hallion, Richard P. *Taking Flight: Inventing the Aerial Age, from Antiquity Through the First World War*. Oxford: Oxford University Press, 2003.

Halpin, Marjorie, and Michael M. Ames, eds. *Manlike Monsters on Trial: Early Records and Modern Evidence*. Vancouver: University of British Columbia Press, 1980.

Halsberghe, Nicole. "Introduction and Development of the Screw in Seventeenth-Century China: Theoretical Explanations and Practical Applications by Ferdinand Verbiest." *East Asian Science, Technology, and Medicine* 34 (2011): 163–193.

Hamacher, Duane W., and Ray P. Norris. "Meteors in Australian Aboriginal Dreamings." *WGN — Journal of the International Meteor Organization* 38.3 (June 2010): 87–98.

Hamori, Andras. "An Allegory from the Arabian Nights: The City of Brass." *Bulletin of the School of Oriental and African Studies, University of London* 34.1 (1971): 9–19.

Han, Zhenhuan. 我國古代海航用的量天尺 ["The Use of the Celestial Measuring Ruler in Maritime Activities of Pre-Modern China"]. *Wenwu Jikan* (文物集刊 "Collection of Cultural Relics") 2 (1980): 217–221.

Hancock, Graham. *Heaven's Mirror: Quest for the Lost Civilization*. New York: Crown Publishers, 1998.

Hancock, Graham. *Underworld: The Mysterious Origins of Civilization*. New York: Crown Publishers, 2002.

Hancock, Graham, and Robert Bauval. *The Message of the Sphinx: A Quest for the Hidden Legacy of Mankind*. New York: Crown Publishers, 1996.

Hanegraaff, Wouter J. *New Age Religion and Western Culture: Esotericism in the Mirror of Secular Thought*. Albany: State University of New York Press, 1998.

Hanfmann, George M.A. "Lydiaka." *Harvard Studies in Classical Philology* 63 (1958): 65–88.

Hankey, Rosalie. "Ghosts and Shamanism in Kwangtung." *California Folklore Quarterly* 2.4 (October 1943): 303–308.

Hansen, William F. *Ariadne's Thread: A Guide to International Tales Found in Classical Literature*. Ithaca: Cornell University Press, 2002.

Hansen, William F. *Phlegon of Tralles' Book of Marvels*. Exeter: University of Exeter Press, 1996.

Hansen, William. "Strategies of Authentication in Ancient Popular Fiction." In *The Ancient Novel and Beyond*, edited by Stelios Panayotakis, Maaike Zimmerman, and Wytse Keulen, 301–314. Leiden: Brill, 2003.

Hapgood, Charles H. *Maps of the Ancient Sea Kings: Evidence of Advanced Civilization in the Ice Age*. Philadelphia: Chilton Books, 1966.

Harley, J.B., and David Woodward, eds., *The History of Cartography*. Vol. 1, *Cartography in Prehistoric, Ancient, and Medieval Europe and the Mediterranean*. Chicago: University of Chicago Press, 1987.

Harper, Donald. "Iatromancy, Diagnosis, and Prognosis in Early Chinese Medicine." In *Innovation in Chinese Medicine*, edited by Elisabeth Hsu, 99–120. Cambridge: Cambridge University Press, 2001.

Harper, Donald. "Physicians and Diviners: The Relation of Divination to the Medicine of the *Huangdi neijing*." *Extrême-Orient, Extrême-Occident* 21 (1999): 91–110.

Hart, Clive. *Kites: An Historical Survey*. Mount Vernon, NY: P.P. Appel, 1982.

Hart, Clive. *The Prehistory of Flight*. Berkeley: University of California Press, 1985.

Hartlaub, G.F. *Zauber des Spiegels: Geschichte und Bedeutung des Spiegels in der Kunst*. Munich: R. Piper, 1951.

Hartmann, Andreas. "*Cui vetustas fidem faciat*: Inscriptions and Other Material Relics of the Past in Graeco-Roman Antiquity." In *Inscriptions and Their Uses in Greek and Latin Literature*, edited by Peter Liddel and Polly Low, 33–64. Oxford: Oxford University Press, 2013.

Hawes, Greta. *Rationalizing Myth in Antiquity*. Oxford: Oxford University Press, 2014.

Hawkes, David, John Minford, and Siu-Kit Wong. *Classical, Modern, and Humane: Essays in Chinese Literature*. Shatin, N.T., Hong Kong: Chinese University Press, 1989.

Hawley, Samuel. *The Imjin War: Japan's Sixteenth-Century Invasion of Korea and Attempt to Conquer China*. Seoul: Royal Asiatic Society, Korea Branch, 2005.

Heaney, Michael. "The Mongolian Almas: A Historical Re-evaluation of the Sighting by Baradin." *Cryptozoology* 2 (1983): 40–52.

Heffernan, Carol F. *The Orient in Chaucer and Medieval Romance*. Cambridge: D.S. Brewer, 2003.

Heilbroner, Robert L. *The Future as History*. New York: Harper & Row, 1960.

Heinberg, Richard. *Memories and Visions of Paradise: Exploring the Universal Myth of a Lost Golden Age*. Wheaton, IL: Quest Books, 1995.

Hennessy, William M., ed., *Annala Uladh: Annals of Ulster, otherwise, Annala Senait, Annals of Senat; a chronicle of Irish affairs from AD 431, to AD 1540* … 4 vols. Dublin: Printed for H.M. Stationery Office by A. Thom & Co., 1887–1901.

Henning, W.B. "A Sogdian Fragment of the Manichaean Cosmogony." *Bulletin of the School of Oriental and African Studies, University of London* 12.2 (1948): 306–318.

Henry, Albert. *Les Œuvres d'Adenet le Roi, Tome V: Cléomades*. Genève: Slatkine Reprints, 1996.

Herodotus. *The History*. Translated by David Grene. Chicago: University of Chicago Press, 1987.

Hewitt, J.N.B., ed. *Thirty-Second Annual Report of the Bureau of American Ethnology to the Secretary of the Smithsonian Institution 1910–1911*. Washington, D.C.: Smithsonian Institution, 1918.

Heyerdahl, Thor. *Early Man and the Ocean: A Search for the Beginnings of Navigation and Seaborne Civilizations*. Garden City, NY: Doubleday, 1979.

Heyerdahl, Thor. *Kon-Tiki: Across the Pacific by Raft*. Translated by F.H. Lyon. Chicago: Rand McNally, 1950.

Heyerdahl, Thor. *The Ra Expeditions*. Translated by Patricia Crampton. Garden City, NY: Doubleday, 1971.

Higbie, Carolyn. "Hellenistic Mythographers." In *The Cambridge Companion to Greek Mythology*, edited by Roger D. Woodward, 237–254. Cambridge: Cambridge University Press, 2007.

Hilger, M. Inez. *Together with the Ainu, a Vanishing People*. Norman, OK: University of Oklahoma Press, 1971.

Hinckley, Henry Barrett. *Notes on Chaucer: A Commentary on the Prolog and Six Canterbury Tales*. Northampton, MA: Nonotuck Press, 1907.

Hirai, Hiro. *Medical Humanism and Natural Philosophy: Renaissance Debates on Matter, Life and the Soul*. Leiden: Brill, 2011.

Hirth, Friedrich. "Chinese Metallic Mirrors, With Notes on Some Ancient Specimens of the Musée Guimet." In Boas Anniversary Volume: Anthropological Papers Written in Honor of Franz Boas, Professor of Anthropology in Columbia University, Presented to Him on the Twenty-Fifth Anniversary of his Doctorate, edited by Berthold Laufer and H.A. Andrews, 208–256. New York: G.E. Stechert & Co., 1906.

Hockney, David. *Secret Knowledge: Rediscovering the Lost Techniques of the Old Masters*. New York: Viking Studio, 2006.

Holay, Prashant Prabhaka. *Machines in Samskrita Literature*. New Delhi: Samskrita Bharati, 2000.

Holford-Strevens, Leofranc. *Aulus Gellius: An Antonine Scholar and his Achievement*. Oxford: Oxford University Press, 2003.

Holloway, R. Ross. "High Flying at Paestum: A Reply." *American Journal of Archaeology* 81.4 (Autumn 1977): 554–555.

Homer. *The Iliad*. 2 vols. Translated by A.T. Murray. London: Heineman, 1965.

Honda, Katsuichi. *Harukor: An Ainu Woman's Tale*. Translated by Kyoko Selden. Berkeley: University of California Press, 2000.

Horky, Philip Sidney. *Plato and Pythagoreanism*. Oxford: Oxford University Press, 2013.

Houdebert, Aurélie. "L'histoire du cheval d'ébène, de Tolède à Paris: propositions sur les modalités d'une transmission." *Actes du colloque des 26–29 novembre 2012 "Byzance et l'Occident: Rencontre de l'Est et de l'Ouest"* (Budapest: Collège Eötvös József ELTE, 2013): 143–156.

Houston, Drusilla Dunjee. *Wonderful Ethiopians of the Ancient Cushite Empire*. Oklahoma City: The Universal Publishing Co., 1926.

Houston, Stephen D., David Stuart, and Karl A. Taube. *The Memory of Bones: Body, Being, and Experience among the Classic Maya*. Austin: University of Texas Press, 2006.

Howland, Jacob. "Plato's Reply to Lysias: *Republic* 1 and 2 and *Against Erastosthenes*." *American Journal of Philology* 125.2 (Summer 2004): 179–208.

Howley, Joseph A. *Aulus Gellius and Roman Reading Culture: Text, Presence, and Imperial Knowledge in the* Nocetes Atticae. Cambridge: Cambridge University Press, 2018.

Hsu, Elisabeth, ed. *Innovation in Chinese Medicine*. Cambridge: Cambridge University Press, 2001.

Hsu, Elisabeth, ed. "Pulse Diagnostics in the Western Han: How *mai* and *qi* Determine *bing*'." In *Innovation in Chinese Medicine*, edited by Elisabeth Hsu, 51–91. Cambridge: Cambridge University Press, 2001.

Hsu, Elisabeth, ed. *The Transmission of Chinese Medicine*. Cambridge: Cambridge University Press, 1999.

Huang, Chun-chieh and Erik Zürcher, eds. *Time & Space in Chinese Culture*. Leiden: Brill, 1995.

Huffman, Carl A. *Archytas of Tarentum: Pythagorean, Philosopher and Mathematician King*. Cambridge: Cambridge University Press, 2005.

Hughes, Jeff. "Occultism and the Atom: The Curious Story of Isotopes." *Physics World* 16.9 (September 2003): 31–35.

Hugill, Andrew. "Imaginary Music Technologies: A Study." Paper presented at the *Technologie imaginaire: l'approche pataphysique de la musique* conference, Université de Paris-IV Sorbonne, 2005.

Hungerford, T.A.G., ed. *Australian Signpost: An Anthology*. Melbourne: F.W. Cheshire, 1956.

Husband, Timothy, and Gloria Gilmore-House. *The Wild Man: Medieval Myth and Symbolism*. New York: The Metropolitan Museum of Art, 1980.

Hussey, Maurice, A.C. Spearing, and James Winny. *An Introduction to Chaucer*. Cambridge: Cambridge University Press, 1965.

Hutchinson, H.N. "Earthquakes." *Folklore* 19.4 (30 December 1908): 472–473.

Ibn al-Zubayr, Aḥmad ibn al-Rashīd. *Book of Gifts and Rarities* (*Kitab al-Hadāyā wa al-Tuḥaf*): *Selections Compiled in the Fifteenth Century from an Eleventh-Century Manuscript on Gifts and Treasures*. Translated by Ghāda al Hijjāwī al-Qaddūmī. Cambridge, MA: Harvard University Press, 1996.

Ibn Khaldūn. *The Muqaddimah: An Introduction to History*. Translated by Franz Rosenthal and edited by N.J. Dawood. Princeton: Princeton University Press, 1967.

Irwin, Robert. *The Arabian Nights: A Companion*. London: Allen Lane 1994.

Isaac, Ephraim. "New Light upon the Book of Enoch from Newly-Found Ethiopic Mss." *Journal of the American Oriental Society* 103.2 (April–June 1983): 399–411.

Isser, Stanley. *The Sword of Goliath: David in Heroic Literature*. Leiden: Brill, 2003.

Jaang, Li. "Long-Distance Interactions as Reflected in the Earliest Chinese Bronze Mirrors." In *The Lloyd Cotsen Study Collection of Chinese Bronze Mirrors.* Vol. 2,

Studies, edited by Lothar von Falkenhausen, 34–49. Los Angeles: Cotsen Occasional Press, UCLA Cotsen Institute of Archaeology Press, 2011.

James, Peter. *The Sunken Kingdom: The Atlantis Mystery Solved*. London: Pimlico, 1996.

Janhunen, Juha. "Tracing the Bear Myth in Northeast Asia." *Acta Slavica Iaponica* 20 (2003): 1–24.

Jay, Jennifer W. "Li, Princess Wencheng." In *Biographical Dictionary of Chinese Women: Tang Through Ming, 618–1644*, edited by Lily Xiao Hong Lee and Sue Wiles, 204–205. Armonk, NY: M.E. Sharpe, 2014.

Jensen, Minna Skafte. *The Homeric Question and the Oral-formulaic Theory*. Copenhagen: Museum Tusculanum Press, 1980.

Jochim, Chris. "Just Say No to 'No Self' in *Zhuangzi*." In *Wandering at Ease in the Zhuangzi*, edited by Roger T. Ames, 35–74. Albany: State University of New York Press, 1998.

Johansen, Thomas K. *Plato's Natural Philosophy: A Study of the* Timaeus-Critias. Cambridge: Cambridge University Press, 2004.

Johansen, Thomas K. "Truth, Lies and History in Plato's *Timaeus-Critias*." *Histos* 2 (1998): 192–215.

Johnston, Charles. "Theosophy." *The Theosophical Quarterly* 21.1 (July 1923): 21–24.

Jones, Alexander. *A Portable Cosmos: Revealing the Antikythera Mechanism, Scientific Wonder of the Ancient World*. Oxford: Oxford University Press, 2017.

Jones, Alexander. "Pseudo-Ptolemy, De Speculis." *SCIAMVS*: Sources and Commentaries in Exact Sciences 2 (2001): 145–186.

Jones, Alexander. "On Some Borrowed and Misunderstood Problems in Greek Catoptrics." *Centaurus* 30 (1987): 1–17.

Jones, Christopher P. "The Emperor and the Giant." *Classical Philology* 95.4 (October 2000): 476–481.

Jones, Christopher P. "Philostratus' *Heroikos* and Its Setting in Reality." *The Journal of Hellenic Studies* 121 (2001): 141–149.

Jones, G. Hartwell. *Celtic Britain and the Pilgrim Movement*. London: Honourable Society of Cymmrodorion, 1912.

Jones, H.S.V. "The Cléomadès, the Méliacin, and the Arabian Tale of the 'Enchanted Horse'." *The Journal of English and Germanic Philology* 6.2 (January 1907): 221–243.

Jones, H.S.V. "The *Cléomadès* and Related Folk-Tales." *PMLA* (*Publications of the Modern Language Association of America*) 23.4 (1908): 557–598.

Jones, H.S.V. "Some Observations upon the Squire's Tale." *PMLA* (*Publications of the Modern Language Association of America*) 20.2 (1905): 346–359.

Jones, Steven S. "'The Rarest Thing in the World': Indo-European or African?" *Research in African Literatures* 7.2 (Autumn 1976): 200–210.

Jordan, Paul. "Esoteric Egypt." In *Archaeological Fantasies*, edited by Garrett G. Fagan, 109–128. New York: Routledge, 2006.

Joseph, John K. and V.T. Anthony. "Ethnobotanical investigations in the genus *Momordica* L. in the Southern Western Ghats of India." *Genetic Resources and Crop Evolution* 55 (2008): 713–721.

Josephus, Flavius. *Against Apion*. Translated by H. St. J. Thackeray. Cambridge, MA: Harvard University Press, 1926.

Josephus, Flavius. *Jewish Antiquities*. 9 vols. Translated by Louis H. Feldman, Ralph Marcus, H. St. J. Thackeray, and Allen Paul Wikgren. Cambridge, MA: Harvard University Press, 1930–1965.

Joyce, Richard. *The Myth of Morality*. London: Cambridge University Press, 2003.

Kabitoglou, E. Douka. *Plato and the English Romantics: Dialogoi*. London: Routledge, 1990.

Kalinowski, Marc. "Mythe, cosmogénèse et théogonie dans la Chine ancienne." *L'Homme* 36.137 (January–March 1996): 41–60.

Kaltenmark, Max. "Miroirs magiques." In *Mélanges de sinologie offerts à Monsieur Paul Demiéville*. 2 vols., 2.151–2.166. Paris: Presses universitaires de France, 1966–1974.

Kane, Panduranga Vamana, ed. *The Harshacarita of Bāṇabhaṭṭa: Text of Uchchhvāsas I–VIII*. 2nd ed. Delhi: Motilal Banarsidass, 1965.

Kang, Minsoo. *Sublime Dreams of Living Machines: The Automaton in the European Imagination*. Cambridge, MA: Harvard University Press, 2011.

Kanga, D.D. *Where Theosophy and Science Meet: A Stimulus to Modern Thought — Part 1*. Adyar, Madras: Adyar Library Association, 1938.

Kanjilal, Dileep Kumar. *Vimana in Ancient India*. Calcutta: Sanskrit Pustak Bhandar, 1985.

Kao, Hsin-Sheng C. *Li Ju-Chen*. New York: Twayne Publishers, 1981.

Karlgren, Bernhard. "Early Chinese Mirror Inscriptions." *Bulletin of the Museum of Far Eastern Antiquities* 6 (1934): 9–74.

Kauffman, George B., ed. *Frederick Soddy (1877–1956): Early Pioneer in Radiochemistry*. Dordrecht: D. Reidel. 1986.

Kayano, Shigeru. *Honoo no uma: Ainu minwashū* (炎の馬：アイヌ民話集 "Horse of Fire: A Collection of Ainu Folklore"). Tokyo: Suzusawa Shoten, 1977.

Kehoe, Alice Beck. *Controversies in Archaeology*. Walnut Creek, CA: Left Coast Press, 2008.

Kelly, Douglas. "Imitation, Metamorphosis, and Froissart's Use of the Exemplary *Modus tractandi*." In *Froissart Across the Genres*, edited by Donald Maddox and Sara Sturm-Maddox, 101–118. Gainesville: University of Florida Press, 1998.

Kenna, Ralph, Máirín MacCarron, and Pádraig MacCarron, eds. *Maths Meets Myths: Quantitative Approaches to Ancient Narratives*. Switzerland: Springer, 2016.

Kim, Seok-hwan. 해설: 실학과 (實學) 비차변증설에 (飛車辨證說) 대한 고찰 (1) ["Review: Korean Classic Pragmatism and Flying Vehicles (1)"]. *Journal of the Korean Society for Aeronautical and Space Sciences* 16.4 (1988): 4070.

Kim, Seok-hwan. 해설: 실학과 (實學) 비차변증설에 (飛車辨證說) 대한 고찰 (2) ["Review: Korean Classic Pragmatism and Flying Vehicles (2)"]. *Journal of the Korean Society for Aeronautical and Space Sciences* 17.1 (1989): 1080.

Kindaichi, Kyōsuke, ed. *Ainu bunka — minzokugaku* (アイヌ文化・民俗学 "Ainu Culture — Folklore"). Tokyo: Sanseidō, 1993.

Kindaichi, Kyōsuke, ed. *Ainu seiten* (アイヌ聖典 "Ainu Scriptures"). Tokyo: Kaizōsha, 1930.

Kindaichi, Kyōsuke, ed. *Kitaezo koyō ihen* (北蝦夷古謡遺篇 "Ancient Folklore of Northern Ezo"). Tokyo: Kōin Sōsho Kankōjo, 1914.

King, Charles William. *Antique Gems: Their Origin, Uses, And Value As Interpreters of Ancient History; And As Illustrative Of Ancient Art: With Hints To Gem Collectors*. London: John Murray, 1860.

Kinser, Samuel. "Saussure's Anagrams: Ideological Work." *Modern Language Notes* 94.5 (December 1979): 1105–1138.

Kirkland, Russell. "Tales Of Thaumaturgy: T'ang Accounts of the Wonder-Worker Yeh Fa-shan." *Monumenta Serica* 40 (1992): 47–86.

Kish, George, ed. *A Source Book in Geography*. Cambridge, MA: Harvard University Press, 1978.

Kitagawa, Joseph M. *On Understanding Japanese Religion*. Princeton: Princeton University Press, 1987.

Knibb, Michael A. *The Ethiopic Book of Enoch: A New Edition in the Light of the Aramaic Dead Sea Fragments*. 2 vols. Oxford: Clarendon, 1978.

Knight, Christopher, and Robert Lomas. *Uriel's Machine: The Ancient Origins of Science*. London: Arrow, 2000.

Koffler, Pauline Bentley. "The Story of the Magic Mirror (*Gujingji*) by Wang Du." In *Hommage à Kwong Hing Foon, Études d'histoire culturelle de la Chine*, edited by Jean-Pierre Diény, 165–214. Paris: Collège de France, Institut des hautes études chinoises, 1995.

Kohl, Philip L., Mara Kozelsky, and Nachman Ben-Yehuda, eds. *Selective Remembrances: Archaeology in the Construction, Commemoration, and Consecration of National Pasts*. Chicago: University of Chicago Press, 2007.

Kohn, Livia. *Daoism Handbook*. Leiden: Brill, 2000.

Kole, Gwin J. "Johnson's 'Dissertation on Flying' and John Wilkins' *Mathematical Magick*." *Modern Philology* 47.1 (August 1949): 24–31.

Kolker, Robert, ed. *The Oxford Handbook of Film and Media Studies*. Oxford: Oxford University Press, 2008.

Konstan, David. "Review of The Myth of Invariance: The Origin of the Gods, Mathematics and Music from the Ṛg Veda to Plato by Ernest G. McClain". *Isis* 70.4 (December 1979): 599–600.

Kósa, Gábor. "The Sea of Fire as a Chinese Manichaean Metaphor: Source Materials for Mapping an Unnoticed Image." *Asia Major*, 3rd series, 24.2 (2011): 1–52.

Kósa, Gábor. "The Sun, the Moon and Paradise — An Interpretation of the Upper Section of the Chinese Manichaean Cosmology Painting." *Journal of Inner Asian Art and Archaeology* 6 (2011): 179–203.

Kósa, Gábor. "Translating the *Eikōn*. Some Considerations on the Relation of the Chinese Cosmology Painting to the *Eikōn*." In *Vom Aramäischen zum Alttürkischen. Fragen zur Übersetzung von manichäischen Texten*, edited by Jens Peter Laut and Klaus Röhrborn, 49–84. Berlin: Walter de Gruyter, 2014.

Kovacs, David, trans. *Euripides: Children of Heracles, Hippolytus, Andromache, Hecuba*. Cambridge, MA: Harvard University Press, 1995.

Kovacs, Jürgen, and Paul U. Unschuld. *Essential Subtleties on the Silver Sea: The* Yin-Hai Jing-Wei: *A Chinese Classic on Ophthalmology*. Berkeley: University of California Press, 1998.

Kowalzig, Barbara, and Peter Wilson, eds. *Dithyramb in Context*. Oxford: Oxford University Press, 2013.

Kramer, Joost. "Wrestling with the Past: The Origins of Pharaonic History in the Islamic Middle Ages." In *Current Research in Egyptology: Proceedings of the Thirteenth Annual Symposium University of Birmingham, 2012*, edited by Carl Graves, Gabrielle Heffernan, Luke McGarrity, Emily Millward, and Marsia Sfakianou Bealby, 94–102. Oxford: Oxbow Books, 2013.

Krause, Kenneth W. "Pathology or Paradigm Shift? Human Evolution, *Ad Hominem* Science, and the Anomalous Hobbits of Flores." *Skeptical Inquirer* 33.4 (July–August 2009): 31–39.

Kubler, George. *The Shape of Time: Remarks on the History of Things*. New Haven: Yale University Press, 1962.

Kunz, George F. *Rings for the Finger, from the Earliest Times to the Present*. Philadelphia and London: J.B. Lippincott, 1917.

Kuriyama, Shigehisa. *The Expressiveness of the Body and the Divergence of Greek and Chinese Medicine*. New York: Zone Books, 1999.

Kurke, Leslie. *Coins, Bodies, Games, and Gold: The Politics of Meaning in Archaic Greece*. Princeton: Princeton University Press, 1999.

Kurtz, Joachim. "Silken Strands: Making Technology Work in China." In *Cultures of Knowledge: Technology in Chinese History*, edited by Dagmar Schäfer, 45–74. Leiden: Brill, 2012.

Kutash, Emilie. *Ten Gifts of the Demiurge: Proclus on Plato's Timaeus*. London: Bloomsbury Academic, 2011.

Kwŏn, Tŏk-kyu. *Chosŏn ŏmun kyŏngwi* (朝鮮語文經緯 "Choson Language Reading Methods"). Kyŏngsong: Kwangmunsa, 1923.

LaFargue, Michael. *Tao and Method: A Reasoned Approach to the* Tao Te Ching. Albany: SUNY Press, 1994.

L'Ancre, Pierre de Rostéguy. *L'incredulité et mescreance du sortilege plainement convaincue, où il est amplement et curieusement traicté de la vérité ou illusion du sortilège et d'une infinité d'autres rares et nouveaux subjects*. Paris: N. Buon, 1622.

Laird, Andrew. "Ringing the Changes on Gyges: Philosophy and the Formation of Fiction in Plato's *Republic*." *The Journal of Hellenic Studies* 121 (2001): 12–29.

Lana Terzi, Francesco. *The Aerial Ship*. Edited by Thomas O'Brien Hubbard and John Henry Ledeboer. London: Printed and Published for the Aëronautical Society of Great Britain [by] King, Sell & Olding, 1910.

Lana Terzi, Francesco. *Prodromo, ouero, Saggio di alcune inuentione nuoue, premesso all'arte maestra / opera che prepara il p. Francesco Lana Bresciano della Compagnia di Giesu., per mostrare li piu reconditi principij della naturale filosofia, riconosciuti con accurata teorica nelle piu segnalate inuentioni, ed isperienze sin'hora ritrouate da gli scrittori di questa materia & altra nuoue dell'autore medesimo*. Brescia: Per li Rizzardi, 1670.

Lane, Edward William, trans., *The Arabian Nights Entertainments; or, The Thousand and One Nights*. New York: George Routledge & Sons, 1886.

Lane, Edward William, trans., *The Thousand and One Nights, Commonly Called, in England, The Arabian Nights Entertainments*. 3 vols. London: Charles Knight and Co., 1841.

Lasch, Richard. "Die Ursache und Bedeutung der Erdbeben im Volksglauben und Volksbrauch." *Archiv für Religionswissenschaft* (1902): 236–257 and 369–383.

Laufer, Berthold. *The Prehistory of Aviation*. Chicago: Field Museum of Natural History, 1928.

Laufer, Berthold. "The Prehistory of Television." *The Scientific Monthly* 27.5 (November 1928): 455–459.

Laufer, Berthold, and H.A. Andrews, eds. Boas Anniversary Volume: Anthropological Papers Written in Honor of Franz Boas, Professor of Anthropology in Columbia University, Presented to Him on the Twenty-Fifth Anniversary of his Doctorate. New York: G.E. Stechert & Co., 1906.

Laut, Jens Peter, and Klaus Röhrborn, eds. *Vom Aramäischen zum Alttürkischen. Fragen zur Übersetzung von manichäischen Texten*. Berlin: Walter de Gruyter, 2014.

Lavelle, B.M. "The Apollodoran Date for Archilochus." *Classical Philology* 97.4 (October 2002): 344–351.

Lavoie, Amy. "Human History Precedes Writing." *Harvard University Gazette Online* (13 December 2007). http://www.news.harvard.edu/gazette/2007/12.13/35-deephistory.html.

Law, Bimala Churn. *A History of Pali Literature*. 2 vols. London: Kegan Paul, Trench, Trubner & Co., 1933.

Layard, J.W. "Shamanism: An Analysis Based on Comparison with the Flying Tricksters of Malekula." *The Journal of the Royal Anthropological Institute of Great Britain and Ireland* 60 (July–December 1930): 525–550.

le Roi, Adenet, and André Henri Constant van Hasselt. *Li roumans de Cléomadès*. 2 vols. Bruxelles: Victor Devaux et cie., 1865–1866.

Leach, Jerry W., and Edmund Leach, eds. *The Kula: New Perspectives on Massim Exchange*. Cambridge: Cambridge University Press, 1983.

Lee, A.J. "Islamic Star Patterns." *Muqarnas: An Annual on Islamic Art and Architecture* 4 (1987): 182–197.

Lee, Ki-baik. *A New History of Korea*. Translated by Edward W. Wagner. Cambridge, MA: Harvard University Press, 1984.

Lee, Lily Xiao Hong, and Sue Wiles, eds. *Biographical Dictionary of Chinese Women: Tang Through Ming, 618–1644*. Armonk, NY: M.E. Sharpe, 2014.

Leonard, Jane Kate. *Wei Yuan and China's Rediscovery of the Maritime World*. Cambridge, MA: Council on East Asian Studies, Harvard University Press, 1984.

Leskosky, Richard J. "Phenakiscope: 19th Century Science Turned to Animation." *Film History* 5.2 (June 1993): 176–189.

Lesourd, Philip S. "The Passamaquoddy 'Witchcraft Tales' of Newell S. Francis." *Anthropological Linguistics* 42.4 (Winter 2000): 441–498.

Lessa, William A. *More Tales from Ulithi Atoll: A Content Analysis*. Berkeley: University of California Press, 1980.

Lester, Katherine Morris, Bess Viola, and Helen Westermann Oerke. *Accessories of Dress: An Illustrated Encyclopedia*. Mineola, NY: Dover, 2004.

Leutner, Mechtild, and Klaus Mühlhahn, eds. *Reisen in chinesischer Geschichte und Gegenwart: Erfahrungen, Berichte, Zeugnisse*. Wiesbaden: Harrassowitz, 2008.

Lévêque, Pierre, and Pierre Vidal-Naquet. *Clisthène l'Athénien: essai sur la représentation de l'espace et du temps dans la pensée politique grecque, de la fin du VIe siècle à la mort de Platon*. Paris: Macula, 1964.

Levitan, William. "Dancing at the End of the Rope: Optation Porfyry and the Field of Roman Verse." *Transactions of the American Philological Association* 115 (1985): 245–269.

Lewis, Charlton T., and Charles Short. *Harper's Latin Dictionary: A New Latin Dictionary Founded on the Translation of Freund's Latin-German Lexicon*. New York: Harper & Brothers, 1907.

Li, Shaowen, ed. *Bu Zhi Yu Yi: Zhongyang Mei Yuan "Yiwen Ketang" Mingjia Jiangyan Lu* (不止於藝—中央美院 "藝文課堂" 名家講演錄 "More than Art — Famous 'Arts and Literature' Lectures, Central Academy of Fine Arts"). Beijing: Beijing daxue chubanshe, 2010.

Li, Shizhen. *Bencao gangmu* (本草綱目 "Compendium of Materia Medica"). 6 vols. Shanghai: Shangwu yinshuguan, 1959.

Liddel, Peter, and Polly Low, eds. *Inscriptions and Their Uses in Greek and Latin Literature*. Oxford: Oxford University Press, 2013.

Lie, Nadia, and Theo D'Haen, eds. *Constellation Caliban: Figurations of a Character*. Amsterdam: Rodopi, 1997.

Lin, Jian-Liang, and Hong-Sen Yan. *Decoding the Mechanisms of Antikythera Astronomical Device*. Heidelberg: Springer, 2016.

Lionarons, Joyce Tally. "Magic, Machines, and Deception: Technology in the 'Canterbury Tales'." *The Chaucer Review* 27.4 (1993): 377–386.

Little, A.G. *Franciscan Papers, Lists, and Documents*. Manchester: Manchester University Press, 1943.

Little, Stephen, and Shawn Eichman. *Taoism and the Arts of China*. Chicago: The Art Institute of Chicago, 2000.

Liu, Jianqiang. *Tibetan Environmentalists in China: The King of Dzi*. Translated by Ian Rowen, Cyrus K. Hui, and Emily T. Yeh. Lanham, MD: Lexington Book, 2015.

Liu, Xianzhou. *Zhongguo Jixie Gongcheng Faming Shi* (中國機械工程發明史 "The History of Inventions in Chinese Mechanical Engineering"). Beijing: Kexue chubanshe, 1962.

Liu, Xin. 西京雜記 [*Miscellaneous Records of the Western Capital*]. Taipei: Shangwu, 1979.

Liu, Zhaomin. 我國古代有關飛行及航空器的構想 ["The Concept of Flight and Aircraft in Ancient China"]. *Zhonghua Keji Shixue Huixuekan* (中華科技史學會學刊 "Journal of the Chinese Society of Science and Technology") 16 (2011): 53–59.

Livio, Mario. *The Golden Ratio: The Story of Pi, the World's Most Astonishing Number*. New York: Broadway Books, 2002.

Lloyd, Robert B. Review of *Structural Patterns and Proportions in Vergil's Aeneid* by George E. Duckworth. *The American Journal of Philology* 85.1 (January 1964): 71–77.

Lo, Vivienne. "The Influence of Nurturing Life Culture on the Development of Western Han Acumoxa Therapy." In *Innovation in Chinese Medicine*, edited by Elisabeth Hsu, 19–50. Cambridge: Cambridge University Press, 2001.

Lockhart, James, trans. and ed. *We People Here: Nahuatl Accounts of the Conquest of Mexico*. Berkeley: University of California Press, 1993.

Loewe, Michael. "The Cycle of Cathay: Concepts of Time in Han China and Their Problems." In *Time & Space in Chinese Culture*, edited by Chun-chieh Huang and Erik Zürcher, 305–328. Leiden: Brill, 1995.

Loewe, Michael, ed. *Early Chinese Texts: A Bibliographical Guide*. Berkeley: Society for the Study of Early China and Institute of East Asian Studies, University of California, Berkeley, 1993.

Long, Pamela O. *Openness, Secrecy, Authorship: Technical Arts and the Culture of Knowledge from Antiquity to the Renaissance*. Baltimore: Johns Hopkins University Press, 2001.

Lowes, John Livingston. "The Squire's Tale and the Land of Prester John." *Washington University Studies* 1 (October 1913): 3–18.

Lozano, A.M., P.L. Gildenberg, and R.R. Tasker, eds. *Textbook of Stereotactic and Functional Neurosurgery*. Berlin: Springer, 2009.

Lu, Peter J., and Paul J. Steinhardt. "Decagonal and Quasi-Crystalline Tilings in Medieval Islamic Architecture." *Science* 315.5815 (23 February 2007): 1106–1110.

Lu, Yilu. 《山海經》中的一足形象與圖像探析－兼論奇肱國與奇股國問題 ["Investigating the Images and Graphics of 'One Foot' in the *Shan Hai Jing* and an Analysis of the 'One Arm' Country and the 'One Foot' Country"]. *Danjiang Zhongwen Xuebao* (淡江中文學報 "Tamkang Journal of Chinese Literature") 29 (2013): 29–50.

Luce, Gordon H. *Old Burma — Early Pagán*. 3 vols. Locust Valley, NY: J.J. Augustin, 1969–1970.

Lucian. *Lucian, with an English Translation*. 8 vols. Translated by A.M. Harmon, K. Kilburn, and M.D. Macleod. London: William Heinemann, 1913–1967.

Lucian. *The Works of Lucian of Samosata*. 4 vols. Translated by H.W. Fowler and F.G. Fowler. Oxford: The Clarendon Press, 1905.

Lucretius. *On the Nature of Things*. Translated by Martin Ferguson Smith. Indianapolis, IN: Hackett, 2001.

Lukacher, Brian. "Joseph Gandy and the Mythography of Architecture." *The Journal of the Society of Architectural Historians* 53.3 (September 1994): 280–299.

Lunazzi, José J. "On How Mirrors Would Have Been Employed in the Ancient Americas." In *Manufactured Light: Mirrors in the Mesoamerican Realm*, edited by Emiliano Gallaga and Marc G. Blainey, 125–142. Boulder: University Press of Colorado, 2016.

Lynch, Kathryn L. "East Meets West in Chaucer's Squire's and Franklin's Tales." *Speculum* 70.3 (July 1995): 530–551.

Lyon, Jean. *Just Half a World Away: My Search for the New India*. New York: Thomas Y. Crowell Company, 1954.

Lyons, Malcolm C. *The Arabian Epic: Heroic and Oral Story-Telling*. 3 vols. Cambridge: Cambridge University Press, 1995.

Lyons, Malcolm C. *The Arabian Nights: Tales of 1001 Nights*. 3 vols. London: Penguin, 2008.

Macaulay, David. *Motel of the Mysteries*. Boston: Houghton Mifflin, 1979.

MacDonell, Arthur A. *A Sanskrit-English Dictionary*. London: Longmans, Green, and Co., 1893.

Maciocia, Giovanni. *The Foundations of Chinese Medicine: A Comprehensive Text for Acupuncturists and Herbalists*. Edinburgh: Churchill Livingstone, 1989.

Mack, Robert L., ed. *Arabian Nights' Entertainments*. Oxford: Oxford University Press, 1995.

Macrobius. *Commentary on the Dream of Scipio*. Translated by William Harris Stahl. New York: Columbia University Press, 1990.

Maddox, Donald, and Sara Sturm-Maddox, eds. *Froissart Across the Genres*. Gainesville: University of Florida Press, 1998.

Magraner, Jordi. "Oral Statements Concerning Living Unknown Hominids: Analysis, Criticism, and Implication for Language Origins." In *Becoming Loquens: More Studies in Language Origins*, edited by Bernard H. Bichakjian, Tatiana Chernigovskaya, Adam Kendon, and Anke Möller, 339–385. Frankfurt am Main: Peter Lang Europäischer Verlag der Wissenschaften, 2000.

Mahony, William K. *The Artful Universe: An Introduction to the Vedic Religious Imagination*. Albany: SUNY Press, 1998.

Majewicz, Alfred F., ed. *The Collected Works of Bronisław Piłsudski*. Vol. 1, *The Aborigines of Sakhalin*. Berlin: Mouton de Gruyter, 1998.

Mak, Se-yuen, and Din-yan Yip. "Secrets of the Chinese Magic Mirror Replica." *Physics Education* 36.2 (March 2001): 102–107.

Malalasekera, G.P. *Dictionary of Pali Proper Names*. 2 vols. London: J. Murray, 1937–1938.

Malinowski, Bronislaw. *Argonauts of the Western Pacific: An Account of Native Enterprise and Adventure in the Archipelagoes of Melanesian New Guinea*. London: Routledge, 2002.

Malotki, Ekkehart, and Michael Lomatuway'ma. *Earth Fire: A Hopi Legend of the Sunset Crater Eruption*. Walnut, CA: Kiva, 2005.

Malotki, Ekkehart, Ken Gary, and Karen Knorowski. *Hopi Stories of Witchcraft, Shamanism, and Magic*. Lincoln, NE: University of Nebraska Press, 2001.

Malotki, Ekkehart, Michael Lomatuway'ma, Lorena Lomatuway'ma, and Sidney Namingha. *Hopi Ruin Legends: Kiqötutuwutsi*. Lincoln, NE: University of Nebraska Press, 1993.

Malotki, Ekkehart, Michael Lomatuway'ma, Lorena Lomatuway'ma, and Sidney Namingha. *Hopi Tales of Destruction*. Lincoln, NE: University of Nebraska Press, 2002.

Marcellinus, Ammianus. *History*. 3 vols. Translated by J.C. Rolfe. Cambridge, MA: Harvard University Press, 1950–1952.

Markale, Jean. *Celtic Civilization*. New York: Gordon & Cremonesi, 1978.

Marlin, Demelza. "Saussure and the Elusive Question of the Origin." *Semiotica* 2008.172 (January 2008): 185–200.

Maroney, Tim. *The Book of Dzyan*. Oakland, CA: Chaosium Books, 2000.

Martijin, Marije. "The *eikôs mythos* in Proclus' Commentary on the *Timaeus*." In *Reading Plato in Antiquity*, edited by Harold Tarrant and Dirk Baltzly, 151–167. London: Duckworth, 2006.

Martijin, Marije. *Proclus on Nature: Philosophy of Nature and Its Methods in Proclus' Commentary on Plato's* Timaeus. Leiden: Brill, 2010.

Marzolph, Ulrich, ed. *The Arabian Nights in Transnational Perspective*. Detroit: Wayne State Press, 2007.

Marzolph, Ulrich, and Richard van Leeuwen. *The Arabian Nights Encyclopedia*. 2 vols. Santa Barbara: ABC-CLIO, 2004.

Maten, Eric P. *Budhasvāmin's Bṛhatkathāślokasaṅgraha: A Literary Study of an Ancient Indian Narrative*. Leiden: Brill, 1973.

Maurice, Thomas. *The History of Hindostan and its Arts, and its Sciences, as Connected with the History of the Other Great Empires of Asia, During the Most Ancient Periods of the World. With Numerous Illustrative Engravings*. 2 vols. London: W. Bulmer and W. Nicol, 1820.

Mavor, James W. Jr., "Atlantis and Catastrophe Theory." *Oceanus* 28.1 (Spring 1985): 44–51.

Mayor, Adrienne. *The First Fossil Hunters: Paleontology in Greek and Roman Times*. Princeton: Princeton University Press, 2000.

Mayor, Adrienne, and Michael Heaney. "Griffins and Arimaspeans." *Folklore* 104.1/2 (1993): 40–66.

McBride, Richard D. II "A Koreanist's Musings on the Chinese *Yishi* Genre." *Sungkyun Journal of East Asian Studies* 6.1 (April 2006): 31–59.

McClain, Ernest G. *The Myth of Invariance: The Origin of the Gods, Mathematics and Music from the Ṛg Veda to Plato*. Boulder: Shambhala, 1978.

McClain, Ernest G. "A New Look at Plato's Timaeus." *Journal of Musicological Research* 1.4 (October–December 1975): 341–360.

McCrindle, J.W. *The Christian Topography of Cosmas, An Egyptian Monk*. London: Hakluyt Society, 1897.

McCune, Shannon. "World Maps by Korean Cartographers." *Journal of Social Sciences and Humanities* 45 (June 1977): 1–8.

McDougall, Russell. "Blind Man's Bluff in a Sandstorm? Literary Education in the Late Condominium Sudan." In *Experiences of Freedom in Postcolonial Literatures and Cultures*, edited by Annalisa Oboe and Shaul Bassi, 59–70. London: Routledge, 2011.

McEwan, Colin, and Leonardo López Lujan, eds. *Moctezuma: Aztec Ruler*. London: British Museum Press, 2009.

McIntosh, Gregory C. *The Piri Reis Map of 1513*. Athens, GA: University of Georgia Press, 2000.

McKusick, Marshall, and Eugene A. Shinn. "Bahamian Atlantis Reconsidered." *Nature* 287.5777 (September 1980): 11–12.

McLaughlin, Thomas J. "Aristotelian Mover-Causality and the Principle of Inertia." *International Philosophical Quarterly* 38.2 (1998): 137–151.

McPherson, Robert S. *Viewing the Ancestors: Perceptions of the Anaasází, Mokwič, and Hisatsinom*. Norman, OK: University of Oklahoma Press, 2014.

Mead, G.R.S. *Concerning H.P. Blavatsky*. Kila, MT: Kessinger, 1997.

Meena, Ravi. *Saffronising Education: "Acche Din"!* Lucknow: Onlinegatha, 2015.

Meletinski, Eleazar M. *The Poetics of Myth*. Translated by Guy Lanoue and Alexandre Sadetsky. New York: Garland, 1998.

Merricks, Linda. *The World Made New: Frederick Soddy, Science, Politics, and Environment*. Oxford: Oxford University Press, 1996.

Millard, A.R., J.K. Hoffmeier, and D.W. Baker, eds. *Faith, Tradition, and History: Old Testament Historiography in its Near Eastern Context*. Winona Lake, IN: Eisenbrauns, 1994.

Miller, Barbara Stoler, trans. *The Bhagavad-Gita: Krishna's Counsel in Time of War*. New York: Bantam Dell, 2004.

Miller, Irving Elgar. *The Significance of the Mathematical Element in the Philosophy of Plato*. Chicago: University of Chicago Press, 1904.

Miller, Walter M. Jr., *A Canticle for Leibowitz*. Philadelphia: Lippincott, 1959.

Minford, John, and Joseph S.M. Lau, eds. *Classical Chinese Literature: An Anthology of Translations*. 2 vols. Hong Kong: The Chinese University Press, 2000.

Mizelle, Brett. "'Man Cannot Behold It Without Contemplating Himself': Monkeys, Apes and Human Identity in the Early American Republic." *Pennsylvania History: A Journal of Mid-Atlantic Studies* 66 (1999): 144–173.

Mizobata, Kiyokazu. "Caxton's Revisions: the *Game of Chess*, the *Mirror of the World*, and *Reynard the Fox*." In *Arthurian and Other Studies: Presented to Shunichi Noguchi*, edited by Takashi Suzuki and Tsuyoshi Mukai, 257–262. Cambridge: D.S. Brewer, 1993.

Mobley, Gregory. "The Wild Man in the Bible and the Ancient Near East." *Journal of Biblical Literature* 116.2 (Summer 1997): 217–233.

Moes, Robert, and Anne Pike Tay. *Mingei: Japanese Folk Art from the Brooklyn Museum Collection*. New York: Universe Books, 1985.

Molland, A.G. "Roger Bacon as Magician." *Traditio* 30 (1974): 445–460.

Monod, Jacques. *Le Hasard et la Nécessité: Essai sur la philosophie naturelle de la biologie moderne*. Paris: Éditions du Seuil, 1970.

Moore, Steven. *The Novel: An Alternative History: Beginnings to 1600*. New York: Continuum, 2010.

Moreh, Shmuel. "The Shadow Play ('Khayāl al-Zill') in the Light of Arabic Literature." *Journal of Arabic Literature* 18 (1987): 46–61.

Morgan, Kathryn A. "Designer History: Plato's Atlantis Story and Fourth-Century Ideology." *The Journal of Hellenic Studies* 118 (1998): 101–118.

Morgan, Kathryn A. *Myth and Philosophy from the Presocratics to Plato*. Cambridge: Cambridge University Press, 2000.

Morley, Henry, ed. *Early Prose Romances*. London: George Routledge and Son, 1889.

Morris, Bertram. "Review of *The Economy of Literature* by Marc Shell." *Technology and Culture* 21.1 (January 1980): 140–142.

Morris, Ian, and Barry B. Powell. *A New Companion to Homer*. Leiden: Brill, 1997.

Morrisson, Mark S. *Modern Alchemy: Occultism and the Emergence of Modern Atomic Theory*. Oxford: Oxford University Press, 2007.

Morwood, M.J. and Penny Van Oosterzee. *A New Human: The Startling Discovery and Strange Story of the "Hobbits" of Flores, Indonesia*. New York: Smithsonian Books, 2007.

Morwood, M.J., R.P. Soejono, R.G. Roberts, T. Sutikna, C.S. Turney, K.E. Westaway, W.J. Rink, J.X. Zhao, G.D. van den Bergh, R.A. Due, D.R. Hobbs, M.W. Moore, M.I. Bird, and L.K. Fifield. "Archaeology and Age of a New Hominin from Flores in Eastern Indonesia." *Nature* 431 (2004): 1087–1091.

Motse. *The Ethical and Political Works of Motse. Translated from the Original Chinese Text by Yi-Pao Mei*. London: Probsthain, 1929.

Mukunda, H.S., S.M. Deshpande, H.R. Nagendra, A. Prabhu, and S.P. Govindaraju. "A Critical Study of the Work *Vymanika Shastra*." *Scientific Opinion* (May 1974): 5–12.

Munday, Anthony. *John a Kent and John a Cumber: A Comedy*. London: Shakespeare Society, 1851.

Mundy, Talbot. *The Nine Unknown*. Indianapolis: Bobbs-Merrill, 1924.

Mungello, D.E. *The Great Encounter of China and the West, 1500–1800*. Lanham, MD: Rowman & Littlefield, 2009.

Muroga, Nobuo, and Kazutaka Unno. "The Buddhist World Map in Japan and its Contact with European Maps." *Imago Mundi* 16 (1962): 49–69.

Murray, Julia K., and Suzanne E. Cahill. "Recent Advances in Understanding the Mystery of Ancient Chinese 'Magic Mirrors': A Brief Summary of Chinese Analytical and Experimental Studies." *Chinese Science* 8 (1987): 1–8.

Mylonas, George Emmanuel. "The Frescoes from the Tomb of the Diver at Paestum." *American Journal of Archaeology* 75.2 (April 1971): 194–217.

Naddaf, Gerard. "The Atlantis Myth: An Introduction to Plato's Later Philosophy of History." *Phoenix* 48.3 (Autumn 1994): 189–209.

Nakamura, Hiroshi. "Old Chinese World Maps Preserved by the Koreans." *Imago Mundi* 4 (1947): 3–22.

Nakano, Toru, Tseng Yuho Ecke, and Suzanne Cahill. *Bronze Mirrors from Ancient China: Donald H. Graham, Jr. Collection*. Honolulu: Donald H. Graham, Jr., 1994.

Nakayama, Shigeru and Nathan Sivin, eds. *Chinese Science: Explorations of an Ancient Tradition*. Cambridge, MA: MIT Press, 1973.

Nallino, Carlo Alfonso. "Al-Khuwarizmi e il suo rifacimento della Geografia di Tolomeo." In *Raccolta di Scritti Editi Inediti*. Vol. 5, *Astrologia — Astronomia — Geografia*, edited by Maria Nallino, 458–532. Rome: Istituto per l'Oriente, 1944.

Nasr, Seyyed Hossein. *Sufi Essays*. Albany: SUNY Press, 1972.

Needham, Joseph. *Science and Civilisation in China*. Vol. 2, *History of Scientific Thought*. Cambridge: Cambridge University Press, 1956.

Needham, Joseph. *Science and Civilisation in China.* Vol. 4, *Physics and Physical Technology, Part I: Physics.* Cambridge: Cambridge University Press, 1962.

Needham, Joseph. *Science and Civilisation in China.* Vol. 4, *Physics and Physical Technology, Part II: Mechanical Engineering.* Cambridge, England: Cambridge University Press, 1965.

Németh, György, ed. *Kalendae: Studia sollemnia in memoriam Johannis Sarkady.* Budapest-Debrecen: University of Debrecen, 2008.

Nesselrath, Heinz-Günther. *Platon und die Erfindung von Atlantis.* Munich: K.G. Saur, 2002.

Nesselrath, Heinz-Günther. "'Where the Lord of the Sea Grants Passage to Sailors through the Deep-Blue Mere No More': The Greeks and the Western Seas." *Greece & Rome*, 2nd series, 52.2 (October 2005): 153–171.

Newbold, R.F. "Pliny HN 2.199." *Classical Philology* 68.3 (July 1973): 211–213.

Newcomb, W.W. Jr., *The Rock Art of Texas Indians.* Austin: University of Texas Press, 1967.

Nichols, Mary P. "Glaucon's Adaptation of the Story of Gyges & Its Implications for Plato's Political Teaching." *Polity* 17.1 (Autumn 1984): 30–39.

Nienhauser, W.H. Jr., *The Indiana Companion to Traditional Chinese Literature.* 2 vols. Bloomington: Indiana University Press, 1986–1998.

Nienkamp, Jean, ed. *Plato on Rhetoric and Language: Four Key Dialogues.* Mahwah, NJ: Lawrence Erlbaum, 1999.

North, John. *Stonehenge: A New Interpretation of Prehistoric Man and the Cosmos.* New York: The Free Press, 1996.

North, Richard. *Pagan Words and Christian Meanings.* Amsterdam: Rodopi, 1991.

Northup, Clark Sutherland, ed. *The Essays of Francis Bacon.* Cambridge, MA: Houghton, Mifflin and Company, 1908.

Nunn, Patrick D. "On the Convergence of Myth and Reality: Examples from the Pacific Islands." *The Geographical Journal* 167.2 (June 2001): 125–138.

Nur, Amos. *Apocalypse: Earthquakes, Archaeology, and the Wrath of God.* Princeton: Princeton University Press, 2008.

O'Brien, James F. "Some Medieval Anticipations of Inertia." *New Scholasticism* 44 (Summer 1970): 345–371.

Oboe, Annalisa, and Shaul Bassi, eds. *Experiences of Freedom in Postcolonial Literatures and Cultures.* London: Routledge, 2011.

Ogden, Daniel. *Magic, Witchcraft, and Ghosts in the Greek and Roman Worlds.* Oxford: Oxford University Press, 2002.

Ogden, Daniel. *In Search of the Sorcerer's Apprentice: The Traditional Tales of Lucian's Lover of Lies.* Swansea: The Classical Press of Wales, 2007.

Ohnuki-Tierney, Emiko. "Concepts of Time among the Ainu of the Northwest Coast of Sakhalin." *American Anthropologist*, new series, 71.3 (June 1969): 488–492.

Ohnuki-Tierney, Emiko. "A Northwest Coast Sakhalin Ainu World View." 2 vols. PhD diss., University of Wisconsin, 1968.

Ohnuki-Tierney, Emiko. *Sakhalin Ainu Folklore*. Washington, D.C.: American Anthropological Association, 1969.

Olshin, Benjamin B. "A Revealing Reflection: The Case of the Chinese Emperor's Mirror." *ICON: Journal of the International Committee for the History of Technology* 18 (2012): 122–141.

Olshin, Benjamin B. "Sophistical Devices: Leonardo da Vinci's Investigations of Perpetual Motion." *ICON: Journal of the International Committee for the History of Technology* 15 (2009): 1–39.

Ophir, Adi. *Plato's Invisible Cities: Discourse and Power in the Republic*. London: Routledge, 1991.

Orchard, Andy. *Pride and Prodigies: Studies in the Monsters of the* Beowulf-*Manuscript*. Toronto: University of Toronto Press, 2003.

Orlov, Andrei A. "Overshadowed by Enoch's Greatness: 'Two Tablets' Traditions from the *Book of Giants* to *Palea Historica*." *Journal for the Study of Judaism* 32.2 (2001): 137–158.

Orwin, Clifford. "Review Essay on Thucydides." *Political Theory* 28.6 (December 2000): 861–869.

Oser, Wendy. "So That Posterity May Know." Review of *Warnungen an die ferne Zukunft: Atommüll als Kommunikationsproblem* ["Warnings to the Far Future: Nuclear Waste As Communication Problem"], edited by Roland Posner. *Nuclear Guardianship Forum* 3 (Spring 1994): 3.

Osle, Margaret J., ed. *Rethinking the Scientific Revolution*. Cambridge: Cambridge University Press, 2000.

Osmun, George F. "Palaephatus — Pragmatic Mythographer." *Classical Journal* 52.3 (December 1956) 131–137.

Ovidius Naso, Publius. *Fasti*. Translated by James G. Frazer and revised by G.P. Goold. Cambridge, MA: Harvard University Press, 1931.

Ovidius Naso, Publius. *The Metamorphoses of Ovid*. Translated by Michael Simpson. Amherst: University of Massachusetts Press, 2001.

Palaephatus. *Περὶ ἀπίστων / On Unbelievable Tales*. Translated by Jacob Stern. Wauconda, IL: Bolchazy-Carducci, 1996.

Panaino, Antonio, and Giovanni Pettinato, eds. *Ideologies as Intercultural Phenomena: Proceedings of the Third Annual Symposium of the Assyrian and Babylonian Intellectual Heritage Project*. Milan: Mimesis Edizioni, 2002.

Panayotakis, Stelios, Maaike Zimmerman, and Wytse Keulen, eds. *The Ancient Novel and Beyond*. Leiden: Brill, 2003.

Papamarinopoulos, Stavros P. "Atlantis in Homer and Other Authors Prior to Plato." In *Science and Technology in Homeric Epics*, edited by S.A. Paipetis, 469–508. Dordrecht: Springer, 2008.

Papamarinopoulos, Stavros P. "Atlantis in Spain V." *Bulletin of the Geological Society of Greece* 43.1 (2010): 138–146.

Paparazzo, Ernesto. "Philosophy and Science in the Elder Pliny's *Naturalis Historia*." In *Pliny the Elder: Themes and Contexts*, edited by Roy Gibson and Ruth Morello, 89–112. Leiden: Brill, 2011.

Paracelsus, Theophrastus Bombastus von Hohenheim. "Astronomia Magna: oder die ganze Philosophia Sagax der grossen und kleinen Welt." In *Sämtliche Werke*, edited by Karl Sudhoff, 12.1–1444. Munich: R. Oldenbourg, 1929.

Park, David. *The Grand Contraption: The World as Myth, Number and Chance*. Princeton: Princeton University Press, 2005.

Park, Seong-Rae. 尹達圭의 飛車: 韓國 飛行機의 前史 ["Yun Dalgyu's Flying Machine: Prehistory of Korean Air Flight"]. 역사문화연구 ("Journal of History and Culture") 18 (2003): 1–20.

Parker, Henry. *Village Folk-tales of Ceylon*, 3 vols. London: Luzac & Co., 1910–1914.

Pataricza, Dóra. "Phlegon and the Giants." In *Kalendae: Studia sollemnia in memoriam Johannis Sarkady*, edited by György Németh, 271.281. Budapest-Debrecen: University of Debrecen, 2008.

Pausanias. *Description of Greece*. 4 vols. Translated by W.H.S. Jones and H.A. Ormerod. Cambridge, MA: Harvard University Press, 1918–1935.

Pauwels, Louis, and Jacques Bergier. *Le matin des magiciens: Introduction au réalisme fantastique*. Paris: Gallimard 1960.

Pavia, Mario N. *Drama of the Siglo de Oro: A Study of Magic, Witchcraft and Other Occult Beliefs*. New York: Hispanic Institute in the United States, 1959.

Pease, Arthur Stanley. "Some Aspects of Invisibility." *Harvard Studies in Classical Philology* 53 (1942): 1–36.

Peirano, Irene. *The Rhetoric of the Roman Fake: Latin Pseudepigrapha in Context*. Cambridge: Cambridge University Press, 2012.

Pellegrino, Charles R. *Unearthing Atlantis: An Archaeological Odyssey*. New York: Random House, 1991.

Peltonen, Markku, ed. *The Cambridge Companion to Bacon*. Cambridge: Cambridge University Press, 1996.

Pendergast, Mark. *Mirror Mirror: A History of the Human Love Affair with Reflection*. New York: Basic Books, 2003.

Pentikäinen, Juha, ed. *Shamanism and Northern Ecology*. Berlin: Walter de Gruyter & Co., 1996.

Penzer, Norman Mosley, ed. *The Ocean of Story Being C.H. Tawney's Translation of Somadeva's Katha Sarit Sagara (or Ocean of Streams of Story)*. 10 vols. London: Sawyer, 1924.

Pérez, Patrick. "Elements of an Amerindian Landscape: The Arizona Hopi." In *Landscapes Beyond Land: Routes, Aesthetics, Narratives*, edited by Arnar Árnason, Nicolas Ellison, Jo Vergunst, and Andrew Whitehouse, 83–97. New York: Berghahn, 2012.

Perry, Ben Edwin. *The Ancient Romances: A Literary-historical Account of Their Origins*. Berkeley: University of California Press, 1967.

Philippi, Donald L. *Songs of Gods, Songs of Humans: The Epic Tradition of the Ainu*. Tokyo: University of Tokyo Press, 1979.

Pieprzak, Katarzyna. "Ruins, Rumors and Traces of the City of Brass: Moroccan Modernity and Memories of the Arab Global City." *Research in African Literatures* 38.4 (Winter 2007): 187–203.

Piłsudski, Bronisław. "On the Bear Festival of the Ainu on the Land of Sakhalin." In *The Collected Works of Bronisław Piłsudski*. Vol. 1, *The Aborigines of Sakhalin*, edited by Alfred F. Majewicz, 438–561. Berlin: Mouton de Gruyter, 1998.

Pinault, David. *Story-Telling Techniques in the Arabian Nights*. Leiden: Brill, 1992.

Pitcher, John, ed., *Francis Bacon: The Essays*. London: Penguin, 1985.

Plato. *Critias*. Translated by Diskin Clay. In *Plato: Complete Works*, edited by John M. Cooper and D.S. Hutchinson, 1293–1306. Indianapolis: Hackett, 1997.

Plato. *Epinomis*. Translated by Richard D. McKirahan. In *Plato: Complete Works*, edited by John M. Cooper and D.S. Hutchinson, 1617–1633. Indianapolis: Hackett, 1997.

Plato. *Laws*. Translated by Trevor J. Saunders. In *Plato: Complete Works*, edited by John M. Cooper and D.S. Hutchinson, 1318–1616. Indianapolis: Hackett, 1997.

Plato. *Lesser Hippias*. Translated by Nicholas D. Smith. In *Plato: Complete Works*, edited by John M. Cooper and D.S. Hutchinson, 922–936. Indianapolis: Hackett, 1997.

Plato. *Phaedrus*. Translated by Alexander Nehamas and Paul Woodruff. In *Plato: Complete Works*, edited by John M. Cooper and D.S. Hutchinson, 506–556. Indianapolis: Hackett, 1997.

Plato. *Philebus*. Translated by Dorothea Frede. In *Plato: Complete Works*, edited by John M. Cooper and D.S. Hutchinson, 398–456. Indianapolis: Hackett, 1997.

Plato. *Republic*. Translated by G.M.A. Grube and revised by C.D.C. Reeve. In *Plato: Complete Works*, edited by John M. Cooper and D.S. Hutchinson, 971–1221. Indianapolis, IN: Hackett, 1997.

Plato. *Statesman*. Translated by C.J. Rowe. In *Plato: Complete Works*, edited by John M. Cooper and D.S. Hutchinson, 294–358. Indianapolis: Hackett, 1997.

Plato. *Timaeus*. Translated by Donald J. Zeyl. In *Plato: Complete Works*, edited by John M. Cooper and D.S. Hutchinson, 1224–1291. Indianapolis: Hackett, 1997.

Platt, Verity. *Facing the Gods: Epiphany and Representation in Graeco-Roman Art, Literature and Religion*. Cambridge: Cambridge University Press, 2011.

Pleyte, C.M. "An Unpublished Batak Creation Legend." *The Journal of the Anthropological Institute of Great Britain and Ireland* 26 (1897): 103–109.

Pliny the Elder. *Natural History*. 10 vols. Translated by D.E. Eichholz, W.H.S. Jones, and H. Rackham. Cambridge, MA: Harvard University Press, 1938–1963.

Plutarch. *Lives*, 11 vols. Translated by Bernadotte Perrin. London: W. Heinemann, 1914–1926.

Plutarch. *Moralia*. 15 vols. Translated by Harold F. Cherniss, Frank Cole Babbitt, W.C. Hembold, and F.H. Sandbach. Cambridge, MA: Harvard University Press, 1927–1969.

Podlecki, A.J. "Archilochus and Apollo." *Phoenix* 28.1, *Studies Presented to Mary E. White on the Occasion of Her Sixty-Fifth Birthday* (Spring 1974): 1–17.

Pollard, David, ed. *Translation and Creation: Readings of Western Literature in Early Modern China, 1840–1918*. Philadelphia: John Benjamin, 1998.

Pollock, Sheldon I., trans. and Robert P. Goldman, ed. *The Rāmāyaṇa of Vālmīki: An Epic of Ancient India*, vol. 3, *Araṇyakāṇḍa*. Vol. 3, *Aranyakanda*. Princeton: Princeton University Press, 1991.

Popkin, Richard H. "Newton and Spinoza and the Bible Scholarship of the Day." In *Rethinking the Scientific Revolution*, edited by Margaret J. Osle, 297–314. Cambridge: Cambridge University Press, 2000.

Porshnev, B.F., Dmitri Bayanov, and Igor Bourtsev. "The Troglodytidae and the Hominidae in the Taxonomy and Evolution of Higher Primates." *Current Anthropology* 15.4 (December 1974): 449–456.

Porter, Deborah Lynn. *From Deluge to Discourse: Myth, History, and the Generation of Chinese Fiction*. Albany: State University of New York Press, 1996.

Porter, James I. "Ideals and Ruins: Pausanias, Longinus, and the Second Sophistic." In *Pausanias: Travel and Memory in Roman Greece*, edited by Susan E. Alcock, John F. Cherry, and Jas Elsner, 63–92. Oxford: Oxford University Press, 2001.

Power, Amanda. *Roger Bacon and the Defence of Christiandom*. Cambridge: Cambridge University Press, 2013.

Power, Amanda. "Seeking Remedies for Great Danger: Contemporary Appraisals of Roger Bacon's Expertise." In *Knowledge, Discipline and Power in the Middle Ages: Essays in Honour of David Luscombe*, edited by Joseph Canning, Edmund J. King, and Martial Staub, 63–78. Leiden: Brill, 2011.

Preston, Elizabeth Winter. *The Earth and Its Cycles: A Comparison Between the Results of Modern Geological and Archaeological Research and the Statements of the Secret Doctrine*. London: Theosophical Publishing House, 1931.

Price, A. Lindsay. *Swans of the World: In Nature, History, Myth & Art*. Tulsa, OK: Council Oaks Books, 1994.

Price, Bronwen. ed. *Francis Bacon's* New Atlantis: *New Interdisciplinary Essays*. Manchester: Manchester University Press, 2002.

Prioreschi, Plinio. *A History of Medicine*. Vol. 3, *Roman Medicine*. Omaha, NE: Horatius Press, 1998.

Pritchard, Paul W. *Plato's Philosophy of Mathematics*. Sankt Augustin: Academia Verlag, 1995.

Pseudo-Aristotle. *On the Cosmos*. Translated by D.J. Furley. Cambridge, MA: Harvard University Press, 1978.

Pujari, R.M., Khle Pradeep, and N.R. Kumar. *Pride of India: A Glimpse into India's Scientific Heritage*. New Delhi: Samskrita Bharati, 2006.

Quintilian. *The Institutio Oratoria of Quintialian, with an English Translation*. 4 vols. Translated by H.E. Butler. London: William Heinemann, 1920–1922.

Radet, Georges Albert. *La Lydie et le monde grec au temps des Mermnades (687–546)*. Paris: Thorin et fils, 1893.

Raghavan, Venkatarama. *Yantras or Mechanical Contrivances in Ancient India* (*Transactions of the Indian Institute of Culture*, no.10). Basavangudi: Indian Institute of Culture, 1952.

Ramage, Edwin S., ed. *Atlantis: Fact or Fiction?* Bloomington, IN: Indiana University Press, 1978.

Raman, Varadaraja V. *Indic Visions in an Age of Science*. New York: Metanexus, 2011.

Ramaswamy, Sumathi. *The Lost Land of Lemuria: Fabulous Geographies, Catastrophic Histories*. Berkeley: University of California Press, 2004.

Rao, S.K. Ramachandra, ed. *Encyclopaedia of Indian Medicine*. 6 vols. Mumbai: Popular Prakashan, 1985.

Raphael, Simcha Paull. *Jewish Views of the Afterlife*. Lanham, MD: Rowman & Littlefield, 2009.

Rasanayagam, C. *Ancient Jaffna. Being a Research into the History of Jaffna from Very Early Times to the Portuguese Period*. New Delhi: Asian Educational Services, 1984.

Reader, Ian, and George J. Tanabe, Jr. *Practically Religious: Worldly Benefits and the Common Religion of Japan*. Honolulu: University of Hawai'i Press, 1998.

Reardon, B.P., ed. *Collected Ancient Greek Novels*. Berkeley: University of California Press, 1989.

Reed, Carrie E. "Motivation and Meaning of a "Hodge Podge': Duan Chengshi's *Youyang zazu*." *Journal of the American Oriental Society* 123.1 (January–March 2003): 121–145.

Reed, Carrie E. *A Tang Miscellany: An Introduction to the Youyang Zazu*. New York, 2003.

Reeves, Eileen. *Galileo's Glassworks: The Telescope and the Mirror*. Cambridge, MA: Harvard University Press, 2008.

Reiche, Harald A.T. "The Language of Archaic Astronomy: A Clue to the Atlantis Myth?" In *Astronomy of the Ancients*, edited by Kenneth Brecher and Michael Feirtag, 155–189. Cambridge, MA: MIT Press, 1979.

Reiche, Harald A.T. Review of *Hamlet's Mill. An Essay on Myth and the Frame of Time*, by Giorgio de Santillana and Hertha von Dechend. *The Classical Journal* 69.1 (October–November 1973): 81–83.

Reid, Constance. *The Search for E.T. Bell: Also Known as John Taine*. Cambridge: Cambridge University Press, 1993.

Reid, Daniel P. *Chinese Herbal Medicine*. Boston: Shambhala, 1987.

Reinhardt, Karl. "Gyges und sein Ring." In *Vermächtnis der Antike: Gesammelte Essays zur Philosophie und Geschichtsschreibung*, edited by Carl Becker, 175–183. Göttingen: Vandenhoeck and Ruprecht, 1960.

Ren, Fang. *Shuyiji* (述異記 "Records of Strange Things"). Taipei: Taiwan shang wu yin shu guan, 1983.

Ren, Ke. "Fin-de-Siècle Diplomat: Chen Jitong (1852–1907) and Cosmopolitan Possibilities in the Late Qing World." PhD diss., Johns Hopkins University, 2014.

Rice, Michael. *The Archaeology of the Arabian Gulf, c. 5000–323 BC*. London: Routledge, 1994.

Richardson, Ernest C. "Antediluvian Libraries." *Papers and Proceedings of the Eleventh General Meeting of the American Library Association, Held at the Fabyan House, White Mountains, September 8–11, 1890* (Boston: Library Bureau, 1890): 40–44.

Richardson, Ernest C. *The Beginnings of Libraries*. Princeton: Princeton University Press, 1914.

Richardson, Hugh. "Mun Sheng Kong Co and Kim Sheng Kong Co: Two Chinese Princesses in Tibet." *The Tibet Journal* 22.1 (Spring 1997): 3–11.

Richardson, William A.R. "South America on Maps before Columbus? Martellus's 'Dragon's Tail' Peninsula." *Imago Mundi* 55 (2003): 25–37.

Richter, Siegfried G., Charles Horton, and Klaus Ohlhafer, eds. Mani in Dublin: Selected Papers from the Seventh International Conference of Manichaean Studies in the Chester Beatty Library, Dublin, 8–12 September 2009. Leiden: Brill, 2015.

Roberts, Alexander, and James Donaldson, eds. *The Ante-Nicene Fathers: Translations of the Writings of the Fathers down to AD 325*. Vol. 5, *Hippolytus, Cyprian, Caius, Novatian, Appendix*. Buffalo: The Christian Literature Company, 1886.

Roberts, Helene E., ed. *Encyclopedia of Comparative Iconography: Themes Depicted in Works of Art*. 2 vols. Chicago: Fitzroy Dearborn, 1998.

Robinson, Christopher. *Lucian and his Influence in Europe*. Chapel Hill, North Carolina: University of North Carolina Press, 1979.

Rocher, Ludo. "Concepts of Time in Classical India." In *Time and Temporality in the Ancient World*, edited by Ralph M. Rosen, 91–110. Philadelphia: University of Pennsylvania Museum of Archaeology and Anthropology, 2004.

Rockhill, W. Woodville, trans. *The Life of the Buddha and the Early History of His Order: Derived from Tibetan Works in the Bkah-Hgyur and Bstan-Hgyur*. London: Kegan Paul, Trench, Trübner & Co., 1907.

Roller, Duane W. *Through the Pillars of Herakles: Greco-Roman Exploration of the Atlantic*. London: Routledge, 2006.

Romm, James S. *The Edges of the Earth in Ancient Thought: Geography, Exploration, and Fiction*. Princeton: Princeton University Press, 1992.

Romm, James S. "New World and '*novos orbes*': Seneca in the Renaissance Debate over Ancient Knowledge of the Americas." In *The Classical Tradition and the Americas*. Vol. 1, *European Images of the Americas and the Classical Tradition*. Part 1, edited by Wolfgang Haase and Meyer Reinhold, 77–116. Berlin: Walter de Gruyter, 1994.

Roochnik, David. *Of Art and Wisdom: Plato's Understanding of Techne*. University Park, PA: Pennsylvania State University Press, 1996.

Rosen, Ralph M., ed. *Time and Temporality in the Ancient World*. Philadelphia: University of Pennsylvania Museum of Archaeology and Anthropology, 2004.

Rosenbaum, Michael. *Kata and the Transmission of Knowledge in Traditional Martial Arts*. Boston: YMAA Publications, 2004.

Rosenberg, Donna. *Folklore, Myths, and Legends: A World Perspective*. Lincolnwood, IL: LTC Publishing Group, 1997.

Rosenmeyer, Thomas G. "The Numbers in Plato's *Critias*: A Reply." *Classical Philology* 44.2 (April 1949): 117–120.

Rosenmeyer, Thomas G. "Plato's Atlantis Myth: 'Timaeus' or 'Critias'?" *Phoenix* 10.4 (Winter 1956): 163–172.

Ross, Margaret Clunies. *A History of Old Norse Poetry and Poetics*. Cambridge: D.S. Brewer, 2005.

Rossi, Paolo. "Bacon's Idea of Science." In *The Cambridge Companion to Bacon*, edited by Markku Peltonen, 25–46. Cambridge: Cambridge University Press, 1996.

Rothaus, Richard M. "The Single Burial of Polyneices." *The Classical Journal* 85.3 (February–March 1990): 209–217.

Rothenberg, Jeff. "Ensuring the Longevity of Digital Documents." *Scientific American* 272.1 (January 1995): 42–47.

Rotstein, Andrea. *The Idea of Iambos*. Oxford: Oxford University Press, 2010.

Rowe, Christopher. "On Plato, Homer, and Archaeology." *Arion*, 3rd series, 5.3 (Winter 1998): 134–144.

Roy, Mira. "The Concept of *Yantra* in the *Samarāṅgaṇa-Sūtradhāra* of Bhoja." *Indian Journal of History of Science* 19.2 (1984): 118–124.

Ruitenbeek, Klaas. *Carpentry and Building in Late Imperial China: A Study of the Fifteenth-Century Carpenter's Manual* Lu Ban Jing (Leiden: Brill, 1993).

Rupert, Milan, and Oliver Julian Todd. *Chinese Bronze Mirrors: A Study based on the Todd Collection of 1,000 Bronze Mirrors Found in the Five Northern Provinces of Suiyuan, Shensi, Shansi, Honan, and Hopei, China*. Peiping: San Yu Press, 1935.

Russo, Lucio. *The Forgotten Revolution: How Science was Born in 300 BC and Why It Had to Be Reborn*. Berlin: Springer, 2004.

Rusten, Jeffrey. "Living in the Past: Allusive Narratives and Elusive Authorities in the World of the *Heroikes*." In *Philostratus's Heroikos: Religion and Cultural Identity in the Third Century CE*, edited by Ellen Bradshaw Aitken and Jennifer K. Berenson Maclean, 143–158. Leiden: Brill, 2004.

Ryder, Arthur W., trans. *The Panchatantra*. Chicago: University of Chicago Press, 1925.

Sailey, Jay. *The Master Who Embraces Simplicity: A Study of the Philosopher Ko Hung, AD 283–343*. San Francisco: Chinese Materials Center, 1978.

Saines, G., and M.G. Tomilin. "Magic Mirrors of the Orient." *Journal of Optical Technology* 66.8 (August 1999): 758–765.

Saldaña, José, ed. *Science and Cultural Diversity: Filling a Gap in the History of Science*. Mexico City: Sociedad Latinoamericana de Historia de las Ciencias y la Tecnología, 2001.

Salguero, C. Pierce. "The Buddhist Medicine King in Literary Context: Reconsidering an Early Medieval Example of Indian Influence on Chinese Medicine and Surgery." *History of Religions* 48.3 (February 2009): 183–210.

Salmony, Alfred. "Chinese Metal Mirrors: Origin, Usage, and Decoration." *Magazine for the Buffalo Museum of Science* 25 (1945): 96–104.

Salvini, Mattia. "The *Samarāṅganasūtradhāra*: Themes and Context for the Science of Vāstu." *Journal of the Royal Asiatic Society*, series 3, 22.1 (January 2012): 35–55.

Sanders, Carol, ed. *The Cambridge Companion to Saussure*. Cambridge: Cambridge University Press, 2004.

Sandy, Gerald N., trans. "Wonders Beyond Thule." In *Collected Ancient Greek Novels*, edited by B.P. Reardon, 775–782. Berkeley: University of California Press, 1989.

Sanvito, Paolo, ed. *Vitruvianism: Origins and Transformations*. Berlin: Walter de Gruyter, 2016.

Sastri, Kali Kumar Dutta. "A Mediaeval Account of Mechanical Devices in India." *Proceedings of the Indian History Congress* 24 (1961): 74–78.

Sato-Rossberg, Nana. "Chiri Mashiho's Performative Translations of Ainu Oral Narratives." *Japanese Studies* 28.2 (2008): 135–148.

Saunders, Nicholas J. "A Dark Light: Reflections on Obsidian in Mesoamerica." *World Archaeology* 33.2 (October 2001): 220–236.

Saunders, Nicholas J. "Stealers of Light, Traders in Brilliance: Amerindian Metaphysics in the Mirror of Conquest." *RES: Anthropology and Aesthetics* (33: *Pre-Columbian States of Being*) (Spring 1998): 225–252.

Saunders, Virginia. "Magic in the Sanskrit Drama." *Journal of the American Oriental Society* 45 (1925): 110–114.

Schäfer, Dagmar, ed. *Cultures of Knowledge: Technology in Chinese History*. Leiden: Brill, 2012.

Schafer, Edward H. *The Golden Peaches of Samarkand: A Study of T'ang Exotics*. Berkeley: University of California Press, 1963.

Schafer, Edward H. "A T'ang Taoist Mirror." *Early China* 4 (1978–1979): 56–59.

Schiffer, Michael B., ed. *Advances in Archaeological Method and Theory*. 11 vols. New York: Academic Press, 1978–1987.

Schmidt, Nathaniel. "Traces of Early Acquaintance in Europe with the Book of Enoch." *Journal of the American Oriental Society* 42 (1922): 44–52.

Schmidt, Nathaniel. "The Two Recensions of Slavonic Enoch." *Journal of the American Oriental Society* 41 (1921): 307–312.

Sclove, Richard E. "Alchemy to Atomic War: Frederick Soddy's 'Technology Assessment' of Atomic Energy, 1900–1915." *Science, Technology, & Human Values* 14.2 (Spring 1989): 163–194.

Scott-Eliot, W. *The Story of Atlantis and the Lost Continent of Lemuria*. London: Theosophical Publishing House, 1925.

Screech, Timon. *The Lens Within the Heart: The Western Scientific Gaze and Popular Imagery in Later Edo Japan*. Honolulu: University of Hawai'i Press, 2002.

Screech, Timon. *Secret Memoirs of the Shoguns: Isaac Titsingh and Japan, 1779–1822*. London: Routledge, 2006.

Seaford, Richard. *Reciprocity and Ritual: Homer and Tragedy in the Developing City-State*. Oxford: Oxford University Press, 1994.

Seery, John Evan. "Politics as Ironic Community: On the Themes of Descent and Return in Plato's Republic." *Political Theory* 16.2 (May 1988): 229–256.

Seneca the Younger. *Natural Questions*. 2 vols. Translated by Thomas H. Corcoran. Cambridge, MA: Harvard University Press, 1971–1972.

Seregin, D.A., A.G. Seregin, and M.G. Tomilin. "Method of Shaping the Front Profile of Metallic Mirrors with a Given Relief of its Back Surface." *Journal of Optical Technology* 71.2 (February 2004): 121–122.

Seth, Michael J. *A Concise History of Korea: From the Neolithic Period Through the Nineteenth Century*. Lanham, Maryland: Rowman & Littlefield, 2016.

Settegast, Mary. *Plato Prehistorian: 10,000 to 5000 BC in Myth, Religion, Archaeology*. Cambridge, MA: Rotenberg Press, 1987.

Settegast, Mary. *When Zarathustra Spoke: The Reformation of Neolithic Culture and Religion*. Costa Mesa, CA: Mazda Publishers, 2005.

Sharma, Sudarshan Kumar, trans. *Samarāṅgana Sūtradhāra of Bhojadeva (Paramāra Ruler of Dhārā): An Ancient Treatise on Architecture* ... 2 vols. Delhi: Parimal Publications, 2007.

Shattuck, Roger. *Forbidden Knowledge: From Prometheus to Pornography*. New York: St. Martin's Press, 1996.

Shcheglov, Dimitry A. "Eratosthenes' Parallel of Rhodes and the History of the System of *Climata*." *Klio* 88.2 (February 2006): 351–359.

Shell, Marc. *The Economy of Literature*. Baltimore: Johns Hopkins University Press, 1978.

Shermer, Michael. "Time Capsules." *American Scientist* 93.3 (May–June 2005): 276–277.

Shibatani, Masayoshi. *The Languages of Japan*. Cambridge: Cambridge University Press, 1990.

Shinn, Eugene A. "A Geologist's Adventure with Bimini Beachrock and Atlantis True Believers." *Skeptical Inquirer* 28.1 (January–February 2004): 38–44.

Shnirelman, Victor A. "Archaeology and Ethnic Politics: The Discovery of Arkaim." *Museum International* 50.2 (April–June 1998): 33–39.

Shnirelman, Victor A. "Russian Response: Archaeology, Russian Nationalism, and the 'Arctic Homeland'." In *Selective Remembrances: Archaeology in the Construction, Commemoration, and Consecration of National Pasts*, edited by Philip L. Kohl, Mara Kozelsky, and Nachman Ben-Yehuda, 31–70. Chicago: University of Chicago Press, 2007.

Shorey, Paul. "Hippias Paidagogos." *The School Review* 17.1 (January 1909): 1–9.

Shruthi, K.R., and Rajani Jairam. "Mechanical Contrivances and Dharu Vimanas Described in 'Samarangana Sutradhara' of Bhojadeva." *IOSR Journal of Humanities and Social Science*, 20.12 (December 2015): 16–20.

Shukla, Dvijendra Nath. *Samarangana-Sutradhara, Pt. II: Royal Palace and Royal Arts*. Lucknow: Vastuvanmaya-Prakasna-Sala, 1967.

Shushan, Gregory. *Conceptions of the Afterlife in Early Civilizations: Universalism, Constructivism, and Near-Death Experience*. London: Continuum, 2009.

Siculus, Diodorus. *Library of History*. 12 vols. Translated by R.M. Geer, C.H. Oldfather, C.L. Sherman, Francis R. Walton, and C. Bradford Welles. Cambridge, MA: Harvard University Press, 1933–1967.

Siegel, Rudolph E. *Galen's System of Physiology and Medicine: An Analysis of his Doctrines and Observations on Bloodflow, Respiration, Tumors and Internal Diseases*. Basel: S. Karger, 1968.

Singer, Bayla. *Like Sex with Gods: An Unorthodox History of Flying*. College Station, TX: Texas A&M University Press, 2003.

Sinnett, A.P. *The Occult World*. 4th ed. London: Trübner, 1884.

Siouris, George M. *Missile Guidance and Control Systems*. New York: Springer, 2004.

Sitchin, Zecharia. *The 12th Planet*. New York: Avon, 1976.

Sitchin, Zecharia. *The Wars of Gods and Men*. New York: Avon, 1985.

Sivin, Nathan. "On the Limits of Empirical Knowledge in Chinese and Western Science." In *Time. Science, and Society in China and the West: The Study of Time V*, edited by J.T. Fraser, N. Lawrence, and F.C. Haber, 151–169. Amherst: University of Massachusetts Press, 1986.

Sivin, Nathan. "On the Pao p'u tzu nei p'ien and the Life of Ko Hung (283–343)." *Isis* 60.3 (Autumn 1969): 388–391.

Sivin, Nathan. "Review of Science and Civilisation in China. Volume 7, The Social Background. Part 2, General Conclusions and Reflections, by Joseph Needham." *China Review International* 12.2 (2005): 297–307.

Sivin, Nathan. "Shen Kua: A Preliminary Assessment of His Scientific Thought and Achievements." *Sung Studies Newsletter* 13 (1977): 31–56.

Sivin, Nathan. "Taoism and Science." In *Medicine, Philosophy and Religion in Ancient China: Researches and Reflections*, edited by Nathan Sivin, 1–72. Aldershot: Variorum, 1995.

Sivin, Nathan. "On the Word 'Taoist' as a Source of Perplexity. With Special Reference to the Relations of Science and Religion in Traditional China." *History of Religions* 17.3–4 (1978): 303–330.

Slessarev, Vsevolod. *Prester John: The Letter and the Legend*. Minneapolis: The University of Minnesota Press, 1959.

Smith, Kirby Flower. "The Literary Tradition of Gyges and Candaules." *The American Journal of Philology* 41.1 (1920): 1–37.

Smith, Kirby Flower. "The Tale of Gyges and the King of Lydia [part 1]." *American Journal of Philology* 23.3 (1902): 261–282.

Smith, Kirby Flower. "The Tale of Gyges and the King of Lydia [part 2]." *American Journal of Philology* 23.4 (1902): 361–387.

Smith, Martyn. "Pyramids in the Medieval Islamic Landscape: Perceptions and Narratives." *Journal of the American Research Center in Egypt* 43 (2007): 1–14.

Smith, Oliver D. "The Atlantis Story: An Authentic Oral Tradition?" *Shima: The International Journal of Research into Island Cultures* 10.2 (2016), 10–17.

Smith, Richard J. *Fortune-Tellers & Philosophers: Divination in Traditional Chinese Society*. Boulder: Westview Press, 1991.

Smith, William, ed. *A Dictionary of Greek and Roman Geography*. 2 vols. London: John Murray, 1878.

Smyser, Hamilton M. "A View of Chaucer's Astronomy." *Speculum* 45.3 (July 1970): 359–373.

Snodgrass, Anthony M. *The Dark Age of Greece: An Archaeological Survey of the Eleventh to the Eighth Centuries BC*. Edinburgh: Edinburgh University Press, 1971).

Soddy, Frederick. *The Interpretation of Radium and the Structure of the Atom*. 4th ed. New York: G.P. Putnam's Sons, 1920.

Soddy, Frederick. *Science and Life*. New York: E.P. Dutton, 1920.

Soddy, Frederick. *The Story of Atomic Energy*. London: Nova Atlantis, 1949.

Spalinger, Anthony J. "The Date of the Death of Gyges and Its Historical Implications." *Journal of the American Oriental Society* 98.4 (October–December 1978): 400–409.

Sparshott, Francis. "The Antiquity of Antiquity." *Journal of Aesthetic Education* 19.1 (Spring 1985): 87–98.

Špelda, Daniel. "The Search for Antediluvian Astronomy: Sixteenth- and Seventeenth-Century Astronomers' Conceptions of the Origins of the Science." *Journal for the History of Astronomy* 44.3 (August 2013): 337–362.

Špelda, Daniel. "Veritas filia temporis: The Origins of the Idea of Scientific Progress." *Annals of Science* 73.4 (October 2016): 375–391.

Stafford, Emma, and Judith Herrin, eds. *Personification in the Greek World: from Antiquity to Byzantium*. Aldershot: Ashgate, 2005.

Stanner, W.E.H. "The Dreaming." In *Australian Signpost: An Anthology*, edited by T.A.G. Hungerford, 51–65. Melbourne: F.W. Cheshire, 1956.

Starobinski, Jean. *Words Upon Words: The Anagrams of Ferdinand de Saussure*. New Haven: Yale University Press, 1979.

Steindorf, Sally Anne. "Walking Against the Wind: Negotiating Television and Modernity in Rural Rajasthan." PhD diss., Syracuse University, 2008.

Steiner, George. *Grammars of Creation*. New Haven: Yale University Press, 2001.

Stephens, John Lloyd. *Incidents of Travel in Central America, Chiapas, and Yucatan ... Revised from the Latest American Edition, with Additions by Frederick Catherwood*. London: Arthur Hall, Virtue, & Co., 1854.

Stephens, Susan A., and John J. Winkler, eds. *Ancient Greek Novels: The Fragments*. Princeton: Princeton University Press, 1995.

Stern, Jacob. "Rationalizing Myth: Methods and Motives in Palaephatus." In *From Myth to Reason? Studies in the Development of Greek Thought*, edited by Richard Buxton, 215–222. Oxford: Oxford University Press, 1999.

Sterne, Jonathan. *The Audible Past: Cultural Origins of Sound Reproduction*. Durham: Duke University Press, 2003.

Stevens, Wesley M. "Science at the Edge of the Earth." In *Science and Cultural Diversity: Filling a Gap in the History of Science*, edited by José Saldaña, 51–58. Mexico City: Sociedad Latinoamericana de Historia de las Ciencias y la Tecnología, 2001.

Steward, Eric C. *Gathered Around Jesus: An Alternative Spatial Practice in the Gospel of Mark*. Cambridge: James Clarke, 2009.

Stiebing, William. *Ancient Astronauts, Cosmic Collisions and Other Popular Theories About Man's Past*. Buffalo, NY: Prometheus Press, 1984.

Stoczkowski, Wiktor. *Explaining Human Origins: Myth, Imagination, and Conjecture*. Cambridge: Cambridge University Press, 2002.

Stone, Michael E. *Selected Studies in Pseudepigrapha and Apocrypha: With Special Reference to the Armenian Tradition*. Leiden: Brill, 1991.

"Storing Information: Come Up and See my Etchings." *The Economist* (30 May 1998): 77.

Strabo. *The Geography of Strabo*. Translated by Horace Leonardo Jones. London: William Heinemann, 1917.

Strasenburgh, Gordon. "On Paranthropus and 'Relic Hominoids'." *Current Anthropology* 16.3 (September 1975): 486–487.

Strutynski, Udo. "The Survival of Indo-European Mythology in Germanic Legendry: Toward an Interdisciplinary Nexus." *The Journal of American Folklore* 97.383 (January–March 1984): 43–56.

Stubelius, Svante. *Airship, Aeroplane, Aircraft: Studies in the History of Terms for Aircraft in English*. Göteborg: Almqvist & Wiksell, 1958.

Stubelius, Svante. *Balloon, Flying-Machine, Helicopter: Further Studies in the History of Terms for Aircraft in English.* Göteborg: Almqvist & Wiksell, 1960.

Su, E. *Duyang Zabian* (杜陽雜編 "Compilation of Miscellanea from Duyang"). Beijing shi: Zhonghua shuju, 1985.

Suzuki, Takashi, and Tsuyoshi Mukai, eds. *Arthurian and Other Studies: Presented to Shunichi Noguchi.* Cambridge: D.S. Brewer, 1993.

Swallow, R.W. *Ancient Chinese Bronze Mirrors.* Peiping: Henri Vetch, 1937.

Swan, Charles, trans. *Gesta Romanorum.* London: George Bell & Sons, 1906.

Sweeney, Amin. *Malay Shadow Puppets: The Wayang Siam of Kelantan.* London: British Museum, 1972.

Takeuchi, Kimpei. "Ancient Chinese Bronze Mirrors." *The Burlington Magazine for Connoisseurs* 19.102 (September 1911): 311–313 and 316–319.

Tambiah, S.J. "On Flying Witches and Flying Canoes: The Coding of Male and Female Values." In *The Kula: New Perspectives on Massim Exchange*, edited by Jerry W. Leach and Edmund Leach, 171–200. Cambridge: Cambridge University Press, 1983.

Tanaka, Sakurako (Sherry). "Ainu Shamanism." In *Shamanism: An Encyclopedia of World Beliefs, Practices, and Culture*, 2 vols., edited by Eva Jane Neumann Fridman and Mariko Namba Walter, 1.657–665. Santa Barbara, CA: ABC-CLIO, 2004.

Tanaka, Sakurako (Sherry). "The Ainu of Tsugaru: The Indigenous History and Shamanism of Northern Japan." PhD diss., University of British Columbia, 2000.

Tarán, Leonardo. "Proclus on the Old Academy." *Collected Papers: (1962–1999)*, edited by Leonardo Tarán, 564–622. Leiden: Brill, 2001.

Tarán, Leonardo. "The Creation Myth in Plato's *Timaeus*." In *Collected Papers: (1962–1999)*, edited by Leonardo Tarán, 303–340. Leiden: Brill, 2001.

Tarrant, Harold. "Atlantis: Myths Ancient and Modern." *The European Legacy* 12.2 (April 2007): 159–172.

Tarrant, Harold, ed. and trans. *Proclus: Commentary on Plato's Timaeus. Volume 1, Book I: Proclus on the Socratic State and Atlantis* (Cambridge: Cambridge University Press, 2007).

Tarrant, Harold, and Dirk Baltzly, eds. *Reading Plato in Antiquity.* London: Duckworth, 2006.

Tatia, Nathmal, and Angraj Chaudhary, eds. *The Suttanipata-Atthakatha.* Bihar: Nava Nalanda Mahavihara, 1975.

Taube, Karl A. "The Iconography of Mirrors at Teotihuacan." In *Art, Ideology, and the City of Teotihuacan*, edited by Janet C. Berlo, 169–204. Washington, D.C.: Dumbarton Oaks Research Library and Collections, 1993.

Taves, Brian. *Talbot Mundy, Philosopher of Adventure: A Critical Biography.* Jefferson, NC: McFarland & Company, 2005.

Taylor, Christopher Eric. "Unknowing the Middle Ages: How Middle English Poetics Rewrote Literary History." PhD diss., University of Texas at Austin, 2014.

Taylor, Paul Beekman. *Chaucer Translator*. Lanham, MD: University Press of America, 1998.

Taylor, Rabun. *The Moral Mirror of Roman Art*. Cambridge: Cambridge University Press, 2008.

Tehrani, Jamshid J., and Felix Riede. "Toward an Archaeology of Pedagogy: Learning, Teaching and the Generation of Material Culture Traditions." *World Archaeology* 40.3 (2008): 316–331.

Teiser, Stephen F. *The Scripture on the Ten Kings: And the Making of Purgatory in Medieval Chinese Buddhism*. Honolulu: University of Hawai'i, 1994.

ter Haar, B.J. *The White Lotus Teachings in Chinese Religious History*. Honolulu: University of Hawai'i, 1999.

Teresi, Dick. *Lost Discoveries: The Ancient Roots of Modern Science — From the Babylonians to the Maya*. New York: Simon and Schuster, 2002.

The Five Strange Wonders, in the North and West of England: as they were communicated to divers honourable members of Parliament … London: Printed for W. Thomas, 1659.

Thiersch, Hermann. *Pharos, Antike, Islam und Occident; ein Beitrag zur Architekturgeschichte*. Leipzig: B.G. Teubner, 1909.

Thomas, Jr., Robert McGill. "Thomas Banyacya, 89, Teller of Hopi Prophecy to World." *The New York Times*, 15 February 1999, sec. B, p. 8, col. 5.

Thompson, Stith. *Motif-Index of Folk Literature*. 6 vols. Bloomington: Indiana University Press, 1955–1958.

Thompson, William Irwin. *At the Edge of History*. New York: Harper & Row, 1971.

Thompson, William Irwin. *Imaginary Landscape: Making Worlds of Myth and Science*. New York: St. Martin's Press, 1989.

Thorstenberg, Edward. "'Duke Frederick of Normandy,' an Arthurian Romance." *Modern Philology* 7.3 (January 1910): 395–409.

Todd, James Henthorn, ed. and trans. *The Irish Version of the Historia Britonum of Nennius*. Dublin: Irish Archaeological Society, 1848.

Trauth, Kathleen M., Stephen C. Hora, and Robert C. Guzowski. *Expert Judgement on Markers to Deter Inadvertent Human Intrusion into the Waste Isolation Pilot Plant — Sandia National Laboratories Report SAND92-1382 / UC-721*. Albuquerque, NM: Sandia National Laboratories, 1993.

Trenn, Thaddeus J. *The Self-Splitting Atom: The History of the Rutherford-Soddy Collaboration*. London: Taylor and Francis, 1977.

Trigger, Bruce G. *A History of Archaeological Thought*. Cambridge: Cambridge University Press, 1989.

Truitt, E.R. *Medieval Robots: Mechanism, Magic, Nature, and Art*. Philadelphia: University of Pennsylvania Press, 2015.

Turkle, Sherry. *Alone Together: Why We Expect More from Technology and Less from Each Other*. New York: Basic Books, 2011.

Turkle, Sherry. *Reclaiming Conversation: The Power of Talk in Digital Age*. New York: Penguin Press, 2015.

Turnbull, David. *Masons, Tricksters and Cartographers: Comparative Studies in the Sociology of Scientific and Indigenous Knowledge*. London: Routledge, 2003.

Tyrrell, William Blake, and Larry J. Bennett. *Recapturing Sophocles' Antigone*. Lanham, Maryland: Rowman and Littlefield, 1998.

Uberoi, J.P. Singh. *Politics of the Kula Ring: An Analysis of the Findings of Bronislaw Malinowski*. Manchester: Manchester University Press, 1971.

Ueda, Toshi. *Ishikari o mamoru kumo-shin no yumewomlta paroatte* (石狩を守るクモ神の夢を見たパロアッテ "Paroatte Dreamed and Saw the Spider Goddess Who Protected Ishikari"). Hokkaido: Ainu Museum, 15 June 1999.

Umehara, Sueji and Jiro Harada. "The Late Mr. Moriya's Collection of Ancient Chinese Mirrors." *Artibus Asiae*, 18.3–4 (1955): 238–256.

Ure, Percy Neville. *The Origin of Tyranny*. Cambridge: Cambridge University Press, 1922.

Uricchio, William. "Television's First Seventy-Five Years: The Interpretive Flexibility of a Medium in Transition." In *The Oxford Handbook of Film and Media Studies*, edited by Robert Kolker, 286–305. Oxford: Oxford University Press, 2008.

van den Bergh, Gerrit D., Yousuke Kaifu, Iwan Kurniawan, Reiko T. Kono, Adam Brumm, Erick Setiyabudi, Fachroel Aziz, and M.J. Morwood. "*Homo floresiensis*-like Fossils from the Early Middle Pleistocene of Flores." *Nature* 534 (2016): 245–248.

van der Horst, Pieter W. *Japheth in the Tents of Shem: Studies on Jewish Hellenism in Antiquity*. Leuven: Peeters, 2002.

Van der Laan, James M., and Andrew Weeks, eds. *The Faustian Century: German Literature and Culture in the Age of Luther and Faustus*. Rochester: Camden House, 2013.

Varty, Kenneth, ed. *Reynard the Fox: Cultural Metamorphoses and Social Engagement in the Beast Epic from the Middle Ages to the Present*. New York: Berghahn Books, 2000.

Veith, Ilza, trans. *The Yellow Emperor's Classic of Internal Medicine*. Berkeley: University of California Press, 2002.

Velikovsky, Immanuel. *Worlds in Collision*. Garden City, NY: Doubleday, 1950.

Verbrugghe, Gerald P., and John M. Wickersham. *Berossos and Manetho, Introduced and Translated: Native Traditions in Ancient Mesopotamia and Egypt*. Ann Arbor: University of Michigan Press, 1996.

Verdicchio, Massimo. "A Reader Like Phaedrus." *Diacritics* 14.1 (Spring 1984): 24–35.

Vermeir, Koen. "The Magic of the Magic Lantern (1660–1700): On Analogical Demonstration and the Visualization of the Invisible." *The British Journal for the History of Science* 38.2 (June 2005): 127–159.

Veyne, Paul. *Did the Greeks Believe in Their Myths? An Essay on the Constitutive Imagination*. Translated by Paula Wissing. Chicago: University of Chicago Press, 1988.

Vickers, Brian, ed. *Francis Bacon: The Major Works*. Oxford: Oxford University Press, 2002.

Vidal-Naquet, Pierre. "Athènes et l'Atlantide. Structure et signification d'un mythe platonicien." *Revue des Études Grecques* 77.366–368 (July–December 1964): 420–444.

Vidal-Naquet, Pierre. *L'Atlantide: Petite histoire d'un mythe platonicien*. Paris: Les Belles Lettres, 2005.

Vijayalakshmy, R. *A Study of the Peruṅkatai: An Authentic Version of the Story of Udayana*. Madras: International Institute of Tamil Studies, 1981.

Vitruvius. *The Ten Books on Architecture*. Translated by Morris Hicky Morgan. Cambridge, MA: Harvard University Press, 1914.

Vogel, Jean Philippe. *The Goose in Indian Literature and Art*. Leiden: Brill, 1962.

Vogel, Jodi. "Review of The Myth of Invariance: The Origin of the Gods, Mathematics and Music from the Ṛg Veda to Plato by Ernest G. McClain." Notes, 2nd series, 34.3 (March 1978): 614–616.

von Falkenhausen, Lothar, ed. *The Lloyd Cotsen Study Collection of Chinese Bronze Mirrors*. Vol. 2, *Studies*. Los Angeles: Cotsen Occasional Press, UCLA Cotsen Institute of Archaeology Press, 2011.

von Liebenau, Theodor. "Der Ring des Gyges in der Schweiz." *Schweizer Archiv für Volkskunde* 4 (1900): 220–221.

von Meyer, Ernst. *A History of Chemistry from Earliest Times to the Present Day*. Translated by George McGowan. London: Macmillan, 1906.

von Schiefner, F. Anton. *Tibetan Tales, Derived from Indian Sources: Translated from the Tibetan of the Kahgyur. Done into English from the German by W.R.S. Ralston*. London: Kegan Paul, Trench, Trübner & Co., Ltd., 1906.

Voth, H.R. *The Traditions of the Hopi*. Chicago: Field Columbian Museum, 1905.

Wackers, Paul. "Medieval French and Dutch Renardian Epics: Between Literature and Society." In *Reynard the Fox: Cultural Metamorphoses and Social Engagement in the Beast Epic from the Middle Ages to the Present*, edited by Kenneth Varty, 55–72. New York: Berghahn Books, 2000.

Wada, Kan. "Some Shamanistic Features of Ainu Religion." In *Shamanism and Northern Ecology*, edited by Juha Pentikäinen, 305–312. Berlin: Walter de Gruyter & Co., 1996.

Wade, Nicholas J. *A Natural History of Vision*. Cambridge, MA: MIT Press, 1998.

Walker, Alicia. *The Emperor and the World: Exotic Elements and the Imaging of Middle Byzantine Imperial Power, Ninth to Thirteenth Centuries CE*. Cambridge: Cambridge University Press, 2012.

Walker, Benjamin. *The Hindu World: An Encyclopedic Survey of Hinduism*. 2 vols. London: Allen & Unwin, 1968.

Walshe, W. Gilbert. "Some Chinese Funeral Customs." *Journal of the North China Branch of the Royal Asiatic Society* 35 (1903–1904): 26–64.

Wang, An-Chi. *Gulliver's Travels and Ching-Hua Yuan Revisited: A Menippean Approach*. New York: Peter Lang Publishing, 1995.

Wang, Chong. *Lunheng*. 2 vols. Translated by Alfred Forke. Leipzig: O. Harrassowitz, 1907–1911.

Wang, Eugene Yuejin. "Mirror, Death, and Rhetoric: Reading Later Han Chinese Bronze Artifacts." *The Art Bulletin* 76.3 (September 1994): 511–534.

Wang, Eugene Yuejin. "Mirror, Moon, and Memory in Eighth-Century China: From Dragon Pond to Lunar Palace." *Cleveland Studies in the History of Art* 9 (2005): 42–67.

Wang, Renyu. *Kaiyuan Tianbao Yishi* (開元天寶遺事 "Anecdotes of the Kaiyuan and Tianbao Periods"). Taipei: Yiwen Yinshuguan Yinxing, 1965–1970.

Wang, Zhenduo. 葛洪《抱朴子》中飛車的複原 ["The Reconstruction of Flying Vehicles from Ge Hong's *Baopuzi*"]. *Zhongguo Lishi Bowuguan Guankan* (中國歷史博物館館刊 "Journal of the Museum of Chinese History") 6 (1984): 48–51.

Warner, Marina. *Stranger Magic: Charmed States and the Arabian Nights*. Cambridge, MA: Harvard University Press, 2012.

Wasilewska, Ewa. *Creation Stories of the Middle East*. London: Jessica Kingsley Publishers, 2000.

Wasserstrom, Steven M. *Between Muslim and Jew: The Problem of Symbiosis Under Early Islam*. Princeton: Princeton University Press, 1995.

Wei, Yuan. *Haiguo Tuzhi* (海國圖志 "Illustrated Gazetteer of the Maritime Nations"). 60 *chuan*. Yangzhou: Guweitang, 1847.

Weinberger, Jerry. "On the Miracles in Bacon's *New Atlantis*." In *Francis Bacon's* New Atlantis: *New Interdisciplinary Essays*, edited by Bronwen Price, 106–128. Manchester: Manchester University Press, 2002.

Weiss, Herold. *A Day of Gladness: The Sabbath among Jews and Christians in Antiquity*. Columbia, SC: University of South Carolina Press, 2003.

Wenley, A.G. "A Chinese Sui Dynasty Mirror." *Artibus Asiae* 25.2–3 (1962): 141–148.

Werner, Daniel S. *Myth and Philosophy in Plato's Phaedrus*. Cambridge: Cambridge University Press, 2012.

Wessels, Tom. *The Myth of Progress: Toward a Sustainable Future*. Hanover, NH: University Press of New England, 2013.

West, Martha Ullman. "Dancers as Living Archives." *The Chronicle Review* 52.31 (7 April 2006), B14.

Wheeler, Stephen M. *Narrative Dynamics in Ovid's Metamorphoses*. Tübingen: Gunter Narr Verlag, 2000).

Whipple, Chris G. "Can Nuclear Waste Be Stored Safely at Yucca Mountain?" *Scientific American* 274.6 (June 1996): 72–79.

White, Jr., Lynn. Review of *Hamlet's Mill: An Essay on Myth and the Frame of Time*, by Giorgio de Santillana and Hertha von Dechend. *Isis* 61.4 (Winter 1970): 540–541.

White, K.D. *Greek and Roman Technology*. Ithaca: Cornell University Press, 1984.

Whitley, James. *The Archaeology of Ancient Greece*. Cambridge: Cambridge University Press, 2001.

Whitney, W.N. “Notes on the History of Medical Progress in Japan.” *Transactions of the Asiatic Society of Japan* 12.4 (July 1885): 245–470.

Wilford, John Noble. “In Medieval Architecture, Signs of Advanced Math.” *The New York Times*, 27 February 2007, F2.

Wilkins, John. *Mathematical Magick, or, The Wonders that May be Performed by Mechanical Geometry*. London: Printed for Edw. Gellibrand, 1680.

Williams, R.D. Review of *Structural Patterns and Proportions in Vergil's 'Aeneid': A Study in Mathematical Composition* by George E. Duckworth. *Classical Philology* 58.4 (October 1963): 248–251.

Wilson, Colin. *The Atlantis Blueprint: Unlocking the Ancient Mysteries of a Long-Lost Civilization*. New York: Delacorte Press, 2001.

Wilson, Colin. *Atlantis and the Kingdom of the Neanderthals: 100,000 Years of Lost History*. Rochester, VT: Bear & Co., 2006.

Winiarczyk, Marek. *Euhemeros von Messene*. Munich: K.G. Saur, 2002.

Winny, James. “Chaucer's Science.” In *An Introduction to Chaucer*, edited by Maurice Hussey, A.C. Spearing, and James Winny, 153–184. Cambridge: Cambridge University Press, 1965.

Winks, Robin W., ed. *The Historian as Detective: Essays on Evidence*. New York: Harper & Row, 1970.

Wise, John. *A System of Aeronautics, Comprehending Its Earliest Investigations and Modern Practice and Art*. Philadelphia: Joseph A. Speel, 1850.

Wojcik, Jan W. “Pursuing Knowledge: Robert Boyle and Isaac Newton.” In *Rethinking the Scientific Revolution*, edited by Margaret J. Osle, 183–200. Cambridge: Cambridge University Press, 2000.

Wolters, Xavier François Marie Gérard. *Notes on Antique Folklore on the Basis of Pliny's Natural History Bk. XXVIII 22–29*. Amsterdam: H.J. Paris, 1935.

Wood, Florence, and Kenneth Wood. *Homer's Secret Iliad: The Epic of the Night Skies Decoded*. London: John Murray, 1999.

Wood, Juliette. “Virgil and Taliesin: The Concept of the Magician in Medieval Folklore.” *Folklore* 94.1 (1983): 91–104.

Wormald, B.H.G. *Francis Bacon: History, Politics & Science, 1561–1626*. Cambridge: Cambridge University Press, 1993.

Wu, Fusheng. “*Sao* Poetry: *The Lyrics of Chu* (*Chuci*).” In *How to Read Chinese Poetry: A Guided Anthology*, edited by Zong-qi Cai, 36–56. New York: Columbia University Press, 2008.

Wunderli, Peter. “Saussure's Anagrams and the Analysis of Literary Texts.” In *The Cambridge Companion to Saussure*, edited by Carol Sanders, 174–185. Cambridge: Cambridge University Press, 2004.

Xenophon. *Xenophon: Scripta Minora*. Translated by E.C. Marchant and G.W. Bowersock. Cambridge, MA: Harvard University Press, 1925.

Xi, Zezong. "Chinese Studies in the History of Astronomy, 1949–1979." *Isis* 72.3 (September 1981): 456–470.

Xiang, Xinyang, and Keren Liu, eds. *Xijing Zaji Jiaozhu* (西京雜記校注 "Annotated Edition of Miscellaneous Records of the Western Capital"). Shanghai: Guji chubanshe, 1991.

Yamada, Takako. *The World View of the Ainu: Nature and Cosmos Reading from Language*. London: Kegan Paul, 2001.

Yan, Hongsen. *Reconstruction Designs of Lost Ancient Chinese Machinery*. Dordrecht: Springer, 2007.

Yetts, W.P. "Two Chinese Mirrors." *The Burlington Magazine for Connoisseurs* 74.430 (January 1939): 22–24 and 26–28.

Yi, Kyu-gyŏng. *Oju Yŏnmun Changjŏn San'go* (五洲衍文長箋散稿 "Scattered Manuscripts of Glosses and Comments of Oju"). 2 vols. Sŏul t'ŭkpyŏl-si: Tong'guk munhwasa, 1959.

Yi, Sang-baek, ed. *Han'guksa* (韓國史 "The History of Korea"). 7 vols. Seoul: Ŭryu munhwasa, 1961–1971.

Yi, Shitong. 量天尺考 ["On the 'Foot' Used in Celestial Measuring"]. *Wenwu* (文物 "Cultural Relics") 2 (1978): 10–16.

Yoshida, Yutaka. "Southern Chinese Version of Mani's Picture Book Discovered?" In Mani in Dublin: Selected Papers from the Seventh International Conference of Manichaean Studies in the Chester Beatty Library, Dublin, 8–12 September 2009, edited by Siegfried G. Richter, Charles Horton, and Klaus Ohlhafer, 389–398. Leiden: Brill, 2015.

Yu, Ning. "Metaphor, Body, and Culture: The Chinese Understanding of Gallbladder and Courage." *Metaphor and Symbol* 18.1 (2003): 13–31.

Yuan, Haiwang. *The Magic Lotus Lantern and Other Tales from the Han Chinese*. Westport, CT: Libraries Unlimited, 2006.

Yücesoy, Hayrettin. "Translation as Self-Consciousness: Ancient Sciences, Antediluvian Wisdom, and the 'Abbāsid Translation Movement." *Journal of World History* 20.4 (December 2009): 523–557.

Zapp, Ivar, and George Erikson. *Atlantis in America: Navigators of the Ancient World*. Kempton, IL: Adventures Unlimited Press, 1998.

Zhang, Baichun, Tian Miao, and Liu Qiang. "The Evolution of Editions of *Yuanxi Qiqi Tushuo Luzui* (A Record of the Best Illustrations and Descriptions of Extraordinary Devices of the Far West) and *Xinzhi Zhuqi Tushuo* (Illustration and Descriptions of Several Newly-built Devices)." *The Chinese Journal for the History of Science and Technology* 27.2 (2006): 115–136.

Zhang, Jing. "In His Thievish Eyes: The Voyeur/Reader in Li Yu's "The Summer Pavilion." *Southeast Review of Asian Studies* 34 (2012): 25–42.

Zhou, Zhongfu, Shuyun Sun, Rubin Han, and T. Ko. "A Study of the Lacquerish Patina (*Qi Gu*) on Ancient Bronze Mirrors." In *Historical Perspectives on East Asian Science, Technology and Medicine*, edited by Alan K.L. Chan, George K. Clancey, and Hui-Chieh Loy, 465–469. Singapore: Singapore University Press, 2002.

Zink, David. *The Stones of Atlantis*. New York: Prentice Hall, 1990.

Zŏng, In-sŏb, ed. and trans. *Folk Tales from Korea*. 3rd ed. Elizabeth, NJ: Hollym International, 1982.

Zuwiyya, Z. David, ed. *A Companion to Alexander Literature in the Middle Ages*. Leiden: Brill, 2011.

Zysk, Kenneth G. *Asceticism and Healing in Ancient India: Medicine in the Buddhist Monastery*. Oxford: Oxford University Press, 1991.

Index

www.ingramcontent.com/pod-product-compliance
Lightning Source LLC
LaVergne TN
LVHW010638300926
849161LV00005B/976